W9-AXW-001

Fiber Optic Test and Measurement

ISBN 0-13-534330-5

90000

9 780135 343302

Fiber Optic Test and Measurement

Dennis Derickson, Editor

Christian Hentschel
Joachim Vobis
Loren Stokes
Paul Hernday
Val McOmber

Douglas M. Baney
Wayne V. Sorin
Josef Beller
Christopher M. Miller
Stephen W. Hinch

To join a Prentice Hall PTR Internet mailing list, point to:
http://www.prenhall.com/mail_lists/

Prentice Hall PTR
Upper Saddle River, New Jersey 07458

Library of Congress Cataloging-in-Publication Data

Derickson, Dennis.
    Fiber optic test and measurement / Dennis Derickson.
      p.   cm.
    Includes index.
    ISBN 0–13–534330–5
    1. Optical communications—Testing.   2. Optical fibers—Testing.
I. Title.
    TK5103.59.D47   1998
    621.382'75'0287—dc21   97–20323
                                             CIP

Acquisitions editor: Bernard M. Goodwin
Cover designer: Wanda España
Cover design director: Jerry Votta
Manufacturing manager: Alexis R. Heydt
Marketing manager: Miles Williams
Compositor/Production services: Pine Tree Composition, Inc.

© 1998 by Prentice-Hall, Inc.
Upper Saddle River, New Jersey 07458

Prentice Hall books are widely used by corporations and government agencies for training, marketing, and resale.

The publisher offers discounts on this book when ordered in bulk quantities. For more information contact:

> Corporate Sales Department
> Phone: 800–382–3419
> Fax: 201–236–7141
> E-mail: corpsales@prenhall.com
>
> Or write:
>
> Prentice Hall PTR
> Corp. Sales Dept.
> One Lake Street
> Upper Saddle River, New Jersey 07458

All rights reserved. No part of this book may be reproduced, in any form or by any means, without permission in writing from the publisher.

Printed in the United States of America
10 9 8 7 6 5

ISBN 0-13-534330-5

PRENTICE-HALL INTERNATIONAL (UK) LIMITED, *LONDON*
PRENTICE-HALL OF AUSTRALIA PTY. LIMITED, *SYDNEY*
PRENTICE-HALL CANADA INC., *TORONTO*
PRENTICE-HALL HISPANOAMERICANA, S.A., *MEXICO*
PRENTICE-HALL OF INDIA PRIVATE LIMITED, *NEW DELHI*
PRENTICE-HALL OF JAPAN, INC., *TOKYO*
PEARSON EDUCATION ASIA PTE. LTD., *SINGAPORE*
EDITORA PRENTICE-HALL DO BRASIL, LTDA., *RIO DE JANEIRO*

This book is dedicated to the families of all the authors:
Sara, Dagmar, Sigrid, Basia, Oliwia, Isabelle, Douglas David, Eva,
Peter, Sue, Brian, Brett, Ann, Natasha, Aaron, Nicki, Greg, Juliana,
Kenna, David, Tim, Heather, Laura, Anneliese, Kay-Mario,
Venessa, and Yasmin.

Contents

Preface   **xix**

1 Introduction to Fiber Optic Systems and Measurements   **1**

   1.1 Introduction   1
   1.2 Fiber Optic Links: The Basics   3
      1.2.1   Digital Communication Links   3
   1.3 Digital Communication Links   6
      1.3.1   Optic Fiber   6
      1.3.2   Optical Amplifiers and/or Optical Repeaters   8
      1.3.3   O/E Converters   8
   1.4 Wavelength Division Multiplexed Systems   9
      1.4.1   Wavelength Division Multiplexed Systems   9
   1.5 Analog Links   12
      1.5.1   Analog Links   12
   1.6 Characterization of Digital Fiber-Optic Links   13
      1.6.1   Bit Error Ratio   13
      1.6.2   Waveform Analysis   13
      1.6.3   Link Jitter   14
      1.6.4   Summary   14
   1.7 Optical Fibers and Two-Part Optical Components   15
      1.7.1   Step-Index Multimode Fiber   16
      1.7.2   Graded-Index Multimode Fiber   17
      1.7.3   Singlemode Fiber   18
      1.7.4   Optical Fiber Amplifiers and Two-Part Optical Components   20
   1.8 Measurement of Optical Fiber and Two-Part Optical Components   21
      1.8.1   Insertion Loss   21
      1.8.2   Amplifier Gain and Noise Figure Measurement   23
      1.8.3   Chromatic Dispersion   23
      1.8.4   Polarization-Related Measurements   25
      1.8.5   Reflection Measurements   26

1.9   Optical Transmitters 27
    1.9.1   Fabry-Perot Lasers 27
    1.9.2   Distributed Feedback Lasers (DFBs) 28
    1.9.3   Vertical Cavity Surface-Emitting Laser (VCSEL) 29
    1.9.4   DFB With Electrooptic Modulator 31
    1.9.5   DFB With Integrated Electroobsorption Modulator 32
1.10   LEDS 32
    1.10.1   Surface-Emitting LEDs 33
    1.10.2   Edge-Emitting LED 33
    1.10.3   Comparison of Optical Sources 34
1.11   Optical Receivers 34
    1.11.1   p-i-n Photodetectors 34
    1.11.2   APD Detectors 36
1.12   Optical Transmitter and Receiver Measurements 37
    1.12.1   Power 37
    1.12.2   Polarization 39
    1.12.3   Optical Spectrum Analysis 41
    1.12.4   Accurate Wavelength Measurement 42
    1.12.5   Linewidth and Chirp Measurement 43
    1.12.6   Modulation Analysis: Frequency Domain 44
    1.12.7   Modulation Analysis: Stimulus-Response Measurement 46
    1.12.8   Modulation Analysis: Time Domain 47
    1.12.9   Optical Reflection Measurements 48
1.13   Organization of the Book 50
Appendix: Relationships Between Wavelength and Frequency 52
References 53

2   Optical Power Measurement   55

    2.1   Introduction 55
    2.2   Power Meters with Thermal Detectors 56
    2.3   Power Meters with Photodetectors 58
        2.3.1   p-i-n diode Operation 59
        2.3.2   Spectral Responsivity 61
        2.3.3   Temperature Stabilization 62
        2.3.4   Spatial Homogeneity 62
        2.3.5   Power Range and Nonlinearity 63
        2.3.6   Polarization Dependence 66
        2.3.7   Optical Reflectivity and Interference Effects 66
        2.3.8   Compatibility with Different Fibers 67
    2.4   Absolute Power Measurement 71
        2.4.1   LED-Power Measurement 72
        2.4.2   High-Power Measurement 73
        2.4.3   Uncertainties in Absolute-Power Measurement 74

2.5   Responsivity Calibration   76
    2.5.1   Traceability and Uncertainty in Responsivity Calibrations   78
2.6   Linearity Calibration   80
    2.6.1   Linearity Calibration Based on Comparison   80
    2.6.2   Linearity Calibration Based on Superposition   83
2.7   Summary   85
Acknowledgments   86
References   86

3 Optical Spectrum Analysis   87

3.1   Introduction to Optical Spectrum Analysis   87
3.2   Types of Optical Spectrum Analyzers   88
    3.2.1   Basic Block Diagram   88
    3.2.2   Fabry-Perot Interferometers   88
    3.2.3   Interferometer-Based Optical Spectrum Analyzers   90
    3.2.4   Diffraction-Grating-Based Optical Spectrum Analyzers   90
3.3   Anatomy of a Diffraction-Grating-Based Optical Spectrum Analyzer   91
    3.3.1   Basic OSA Block Diagram   91
    3.3.2   The Entrance or Input Slit   92
    3.3.3   The Collimating Optics   93
    3.3.4   The Diffraction Grating   95
    3.3.5   The Focusing Optics   98
    3.3.6   The Exit or Output Slit   99
    3.3.7   The Detector   100
    3.3.8   Single Monochromator Summary   100
    3.3.9   Single Monochromator Versus Double Monochromator   101
    3.3.10   Double Monochromator   101
    3.3.11   Double-Pass Monochromator   101
    3.3.12   Littman Double-Pass Monochromator   102
3.4   Operation and Key Specifications of Diffraction-Grating-Based Optical
       Spectrum Analyzers   104
    3.4.1   Wavelength Accuracy   104
    3.4.2   Wavelength-Calibration Techniques   104
    3.4.3   Wavelength Resolution and Dynamic Range   109
    3.4.4   Sensitivity and Sweep Time   112
    3.4.5   Input Polarization Sensitivity   114
3.5   Spectral Measurements on Modulated Signals   115
    3.5.1   Signal Processing in an OSA   115
    3.5.2   Zero-Span Mode   117
    3.5.3   Trigger Sweep Mode   117
    3.5.4   ADC-Trigger Mode   118
    3.5.5   ADC-AC Mode   118
    3.5.6   Gated-Sweep Mode   119

  3.6   **OSA Application Examples 120**
         3.6.1   Light-Emitting Diodes (LEDs) 120
         3.6.2   Fabry-Perot Lasers 122
         3.6.3   Distributed Feedback (DFB) Lasers 123
         3.6.4   Optical Amplifier Measurements 125
         3.6.5   Recirculating Loop 126
  3.7   **Summary 128**
         Acknowledgments 129
         References 129

4   **Wavelength Meters**   **131**

  4.1   **Introduction 131**
         4.1.1   Wavelength Definition 132
         4.1.2   Methods of Accurate Wavelength Measurement 132
  4.2   **The Michelson Interferometer Wavelength Meter 133**
         4.2.1   Fringe-Counting Description of Wavelength Meter
                 Operation 134
         4.2.2   Doppler-Shift Approach to Understanding Wavelength Meter
                 Operation 137
         4.2.3   Accurate Measurement of Distance, Velocity, and Time 138
         4.2.4   Wavelength Measurement with Respect to a Wavelength
                 Standard 139
         4.2.5   Summary of Michelson-Interferometer Wavelength Meter
                 Operation 90
  4.3   **Wavelength Meters in Multiple Signal Environments 140**
  4.4   **Absolute Wavelength Accuracy Considerations for Michelson-**
         Interferometer Wavelength Meters 143
         4.4.1   The Ability to Count Many Fringes and to Count Them Accurately 144
         4.4.2   Index of Refraction of Air and Dispersion of Air 145
         4.4.3   Accuracy of the Reference-Laser Wavelength 150
         4.4.4   Dependence on the Signal Spectral Width 151
         4.4.5   Optical Alignment Issues 152
         4.4.6   Diffraction Effects 153
         4.4.7   Summary of Wavelength Accuracy Factors 153
  4.5   **Michelson Wavelength-Meter Measurement Considerations 153**
         4.5.1   Relative Wavelength Resolution 154
         4.5.2   Wavelength Coverage 155
         4.5.3   Sensitivity and Measurement Speed, Measurement Range,
                 and Dynamic Range 156
         4.5.4   Amplitude Accuracy 157
         4.5.5   Single-Mode and Multimode Fiber Input Considerations for Wavelength
                 Meters 158
         4.5.6   Measurement of Pulsed Signals 158

    4.6   Alternate Wavelength Meter Techniques 159
        4.6.1   Fabry-Perot Filters 159
        4.6.2   Static Fabry-Perot Interferometer Wavelength Meter 162
        4.6.3   Static Fizeau Interferometer Wavelength Meter 163
        4.6.4   Wavelength Discriminators 164
    4.7   Summary 165
    References 166

5   High Resolution Optical Frequency Analysis   169
    5.1   Introduction 169
    5.2   Basic Concepts 170
        5.2.1   Linewidth and Chirp 173
        5.2.2   Interference between Two Optical Fields 175
    5.3   Laser Linewidth Characterization 179
        5.3.1   Heterodyne Using a Local Oscillator 179
        5.3.2   Delayed Self-Heterodyne 185
        5.3.3   Delayed Self-Homodyne 188
        5.3.4   Photocurrent Spectrum: Coherence Effects 189
        5.3.5   Coherent Discriminator Method 194
        5.3.6   Comparison of Techniques 201
    5.4   Optical Spectral Measurement of a Modulated Laser 202
        5.4.1   Heterodyne Method 204
        5.4.2   Gated Delayed Self-Homodyne 205
    5.5   Laser Chirp Measurement 208
    5.6   Frequency Modulation Measurement 213
    5.7   Summary 217
    References 218

6   Polarization Measurements   220
    6.1   Introduction 220
    6.2   Polarization Concepts 221
        6.2.1   A General Description of Polarized Light 221
        6.2.2   A Polarization Coordinate System 223
        6.2.3   The Polarization Ellipse 223
        6.2.4   The Jones Calculus 224
        6.2.5   The Stokes Parameters 226
        6.2.6   Degree of Polarization 228
        6.2.7   The Poincaré Sphere 229
        6.2.8   The Polarimeter and Polarization Analyzer 231
        6.2.9   The Mueller Matrix 232
    6.3   Retardance Measurement 234
        6.3.1   Introduction 234

6.3.2   Measurement of Retardance   235
6.3.3   The Poincaré Sphere Method   236
6.3.4   The Jones Matrix Method   237

6.4   Measurement of Cross-Talk in Polarization-Maintaining Fiber   237
6.4.1   Introduction   237
6.4.2   The Crossed-Polarizer Cross-Talk Measurement   239
6.4.3   The Polarimetric Cross-Talk Measurement   240
6.4.4   Measurement of Cross-Talk Along a PM Fiber   242
6.4.5   Cross-Talk Measurement of PM Fiber Interfaces   243
6.4.6   Measurement of the Polarization Stability of Cascaded
          PM Fibers   243

6.5   Summary   243
6.6   References   245

**7   Intensity Modulation and Noise Characterization
    of Optical Signals   246**

7.1   Modulation Domain Analysis   246
7.1.1   Simplified Transmission Systems   247
7.1.2   Lightwave Transmission Components   247
7.1.3   Intensity-Modulated Waveform and Spectrum   250
7.1.4   Modulation-Frequency-Domain Measurements   251

7.2   Modulation Transfer Function   252
7.2.1   Lightwave Component Analyzer   253
7.2.2   E/O Transfer Function Measurements   254
7.2.3   O/E Transfer Function Measurements   260
7.2.4   O/O Transfer Function Measurements   263

7.3   Modulation Signal Analysis   263
7.3.1   Lightwave Signal Analyzer   264
7.3.2   Intensity Modulation and Modulation Depth   265
7.3.3   Distortion   266

7.4   Intensity Noise Characterization   269
7.4.1   Intensity Noise Measurement Techniques   269
7.4.2   Relative Intensity Noise   272

7.5   Modulation Domain Calibration Techniques   275
7.5.1   Optical Impulse Response   276
7.5.2   Optical Heterodyning   279
7.5.3   Two-Tone Technique   280
7.5.4   Optical Intensity Noise   280
7.5.5   Comparison of Calibration Techniques   281

References   282

8   Analysis of Digital Modulation on Optical Carriers   284

   8.1   Digital Fiber-Optic Communications Systems   284
          8.1.1   SONET/SDH Standards   285
          8.1.2   Performance Analysis of Fiber Optic Systems   287
   8.2   Bit-Error Ratio   288
          8.2.1   BER Measurement   289
          8.2.2   BERT Design   290
          8.2.3   Test Patterns for Out-of-Service Testing   291
          8.2.4   Clock Recovery   293
          8.2.5   Example Measurements   293
   8.3   Eye-Diagram Analysis   298
          8.3.1   Eye-Diagram Generation   298
          8.3.2   Digital Sampling Oscilloscope Architectures   300
          8.3.3   Real-Time Sampling   301
          8.3.4   Equivalent-Time Sampling   302
          8.3.5   Oscilloscopes for Eye-Diagram Analysis   306
          8.3.6   Eye-Parameter Analysis   307
          8.3.7   Eyeline Diagrams   311
          8.3.8   Extinction Ratio   316
   8.4   Mask Measurements   324
          8.4.1   Mask Definition   324
          8.4.2   Mask Margins   325
          8.4.3   Mask Alignment   325
   8.5   Jitter Testing   326
          8.5.1   Introduction   326
          8.5.2   Jitter Issues   329
          8.5.3   Jitter Mathematical Representation   330
          8.5.4   Jitter Measurement Categories   330
          8.5.5   Jitter Measurement Techniques   335
   References   337

9   Insertion Loss Measurements   339

   9.1   Introduction   339
   9.2   How the Component Influences the Measurement Technique   339
          9.2.1   Measurement of Pigtailed and Connectorized Components   340
          9.2.2   Measurement of Flange-Mount Components   340
          9.2.3   Measurement of Components with Bare-Fiber Pigtails   341
          9.2.4   Insertion-Loss Measurement of Integrated Optics Components   342
          9.2.5   Imaging Techniques   343
   9.3   Single-Wavelength Loss Measurements   343
   9.4   Uncertainties of Single-Wavelength Loss Measurements   345
          9.4.1   Power-Meter-Related Uncertainties   345

9.4.2   Uncertainty Caused by Polarization-Dependent
        Loss (PDL)   346
9.4.3   Uncertainty Caused by Optical Interference   347
9.4.4   Uncertainty Caused by the Wavelength Characteristic
        of the Source   350
9.4.5   Uncertainty Caused by Incompatible Fibers   354
9.5   PDL Measurement   354
9.5.1   Polarization-Scanning Method   354
9.5.2   Mueller Method   356
9.6   Introduction to Wavelength-Dependent Loss Measurements   358
9.7   Wavelength-Dependent Loss Measurements Using
      a Tunable Laser   359
9.7.1   Loss Measurements with a Tunable Laser and a Power
        Meter   359
9.7.2   Loss Measurements with a Tunable Laser and an Optical Spectrum
        Analyzer   366
9.8   Wavelength-Dependent Loss Measurements Using
      a Broadband Source   368
9.8.1   Broadband Light Sources   370
9.8.2   Receiver Characteristics Relevant to Loss Measurement
        with Broadband Sources   377
9.8.3   Examples of Filter Measurements Using Broadband Sources   379
9.9   Summary   381
Literature   381

10   Optical Reflectometry for Component Characterization   383

10.1   Introduction   383
10.1.1   Motivation for High Resolution Measurements   385
10.2   Total Return Loss Technique   387
10.2.1   Reflection Sensitivity   387
10.2.2   Multiple Reflections   389
10.3   Basic Concepts for Spatially Resolved Reflectometry   391
10.3.1   Spatial Resolution   391
10.3.2   Dispersion Limit   393
10.3.3   Rayleigh Backscatter and Spatial Resolution   394
10.3.4   Rayleigh Backscatter and Coherent Speckle   396
10.3.5   Coherent vs. Direct Detection   399
10.4   Optical Low Coherence Reflectometry   401
10.4.1   Introduction   401
10.4.2   Description of Operation   402
10.4.3   Special Considerations   406
10.5   Survey of Different Techniques   420
10.5.1   Direct Detection OTDR   420

        10.5.2   Photon Counting OTDR   422
        10.5.3   Incoherent Frequency Domain Techniques   423
        10.5.4   Coherent Frequency Domain Techniques   425
   10.6 **Comparison of Techniques**   430
   References   431

11 **OTDRs and Backscatter Measurements**   **434**

   Overview   434
   11.1 Introduction   435
   11.2 Principle of OTDR Operation   435
        11.2.1   OTDR Fiber Signature   438
        11.2.2   Level Diagram   439
        11.2.3   Performance Parameters   441
        11.2.4   Tradeoff between Dynamic Range and Resolution   444
        11.2.5   Ghost Features Caused by Multiple Reflections   446
   11.3 Fiber Loss, Scatter, and Backscatter   447
        11.3.1   Loss in Fiber   447
        11.3.2   Backscatter Signal Analysis   449
   11.4 Measuring Splice- and Connector Loss   454
        11.4.1   Fusion Splice Loss   454
        11.4.2   Different Fibers   455
        11.4.3   Insertion Loss of Reflective Events   457
        11.4.4   Bending Loss   457
        11.4.5   Uncertainty of Loss Measurements   458
        11.4.6   A Variable Splice-Loss Test Setup   459
   11.5 Return Loss and Reflectance   461
        11.5.1   Return-Loss Measurements   461
        11.5.2   Reflectance Measurements   461
        11.5.3   Accumulative Return Loss   466
   11.6 Automated Remote Fiber Testing   467
        11.6.1   Link Loss Comparison   467
        11.6.2   Dark Fiber Testing   468
        11.6.3   Active Fiber Testing   468
   11.7 Outlook   472
   References   472

12 **Dispersion Measurements**   **475**

   12.1 Introduction   475
   12.2 Measurement of Intermodal Dispersion   476
        12.2.1   Introduction   476

12.2.2   The Pulse Distortion Method   477
12.2.3   The Frequency Domain Method   478
12.3   Measurement of Chromatic Dispersion   479
12.3.1   Introduction   479
12.3.2   Causes of Chromatic Dispersion   479
12.3.3   Definitions and Relationships   479
12.3.4   Control of Chromatic Dispersion   480
12.3.5   The Modulation Phase-Shift Method   482
12.3.6   The Differential Phase-Shift Method   485
12.3.7   The Baseband AM Response Method   486
12.4   Polarization-Mode Dispersion   487
12.4.1   Introduction   487
12.4.2   Causes of PMD   490
12.4.3   Mode Coupling and the Principal States
          of Polarization   490
12.4.4   Definitions and Relationships   492
12.4.5   Statistical Characterization of PMD in Mode-
          Coupled Fiber   494
12.4.6   A Brief Summary of PMD Measurement Methods   495
12.4.7   The Fixed-Analyzer Method   495
12.4.8   The Jones-Matrix Eigenanalysis Method   502
12.4.9   The Interferometric Method   507
12.4.10 The Poincaré Arc Method   511
12.4.11 The Modulation Phase-Shift Method   512
12.4.12 The Pulse-Delay Method   513
12.4.13 Agreement between PMD Measurement Method   514
12.5   Summary   515
References   516

13   Characterization of Erbium-Doped Fiber Amplifiers   519

Introduction   519
13.1   Fiber Amplifiers   520
13.1.1   Basic Concept   520
13.2   Gain   525
13.2.1   Small-Signal Gain   529
13.2.2   Saturated Gain   530
13.2.3   Polarization Hole-Burning   530
13.2.4   Spectral Hole-Burning   531
13.2.5   Gain Tilt, Gain Slope   532
13.3   Noise   533
13.3.1   Optical Noise   533
13.3.2   Intensity/Photocurrent Noise   534

13.4   Noise Figure   542
13.4.1 Noise-Figure Definition   543
13.5   Characterization of Gain and Noise Figure   546
13.5.1 Amplifier Gain   547
13.5.2 Measurement of Noise Figure   550
13.6   Other Types of Optical Amplifiers   583
13.6.1 Rare-Earth Doped Fiber Amplifiers   583
13.6.2 Gain from Fiber Nonlinearities   587
13.6.3 Semiconductor Amplifiers   589
13.6.4 Measurements of Other Types of Optical Amplifiers   590
13.7   Sources of Measurement Errors   590
13.8   Useful Constants for EDFA Measurements   591
13.9   Summary   591
References   592

Appendix A   Noise Sources in Optical Measurements   597

A.1   Electrical Thermal Noise   598
A.2   Optical Intensity Noise   601
A.3   Photocurrent Shot Noise   604
A.4   Optical-Phase-Noise to Intensity-Noise Conversion   608
A.5   Summary   612
References   613

Appendix B   Nonlinear Limits for Optical Measurements   614

B.1   Raman Limit   614
B.2   Self-Phase Modulation   616
B.3   Brillouin Limit   618
B.4   Summary   619
References   620

Appendix C   Fiber Optic Connectors and Their Care   621

C.1   Background   621
C.2   Connector Styles   622
C.3   Connector Design   624
C.4   Connector Care   630
C.5   Cleaning Procedures   633
Reference   638

Index   639

Preface

There are many excellent books in the area of fiber-optic communication systems. This book tackles one important subset of this broad field: fiber optic test and measurement techniques. It focuses specifically on the measurement and testing of fiber-optic communication links and the components that make up the link. This book also demonstrates methods to characterize the interactions between these components that have dramatic effects on system performance. The area of lightwave measurement technology is rapidly evolving as is the entire fiber optics industry. It is difficult to keep up with the myriad of measurement demands generated in both the telecommunications and data communications area. The contents of this book provide a detailed coverage of measurement principles that are needed to design and maintain fiber optic systems now and in the future.

It became clear to the authors of this book that no single source of information is available in the broad subject area of lightwave test and measurement for fiber optic systems. The authors are with the lightwave test and measurement divisions at Hewlett-Packard. This book combines the collective experience of the lightwave staff at Hewlett-Packard together in a single source. The material in this book has been developed from application notes, seminars, conference presentations, journal publications, Ph.D. theses, and unpublished works from the last ten years. Much of this material has not had wide circulation to date.

The book will be useful for technicians, engineers, and scientists involved in the fiber optics industry or who want to become familiar with it. The book is designed to address the needs of people new to the field and to those intimately familiar with it. Chapter

1 describes the operation of a fiber optic link and its components. It then briefly describes the most common measurement needs of the fiber optic link and components with measurement block diagrams and example results. The chapter will be particularly valuable for the reader who wants a basic introduction but is not ready to dive into the more detailed coverage given in the following chapters. The basic concepts of the measurements will be presented with a minimum of mathematical detail. The main chapters of the book are designed so that the first section of each chapter provides an overview. Graphical aids are used whenever possible to help in understanding. Later sections of each chapter are geared to cover the material in greater depth for more advanced readers.

Chapters 2 to 6 cover the fundamental areas of fiber optic measurements:

a. Optical power measurements (Chapter 2);

b. Spectral measurements (Chapter 3 to 5);

c. Polarization measurements (Chapter 6).

The measurement of power is fundamental to most every lightwave measurement. Chapter 2 covers methods of power measurement and associated accuracy concerns. The area of optical spectral measurements is quite broad and is divided into three chapters. Chapter 3 covers the most common method using a diffraction-grating based optical spectrum analyzer. Chapter 4 covers the area of wavelength meters. Wavelength meters are the electrical analogue of frequency counters because they allow very accurate measurements of laser wavelength. Chapter 5 covers the area of very high wavelength-resolution spectral measurements.

The coverage of polarization in fiber optic systems (Chapter 6) is an area that historically has been underemphasized. This situation has changed since data rates have increased and optical amplifiers have been installed into fiber optic systems.

Chapters 7 and 8 focus on the measurement of the modulation on lightwave signals. In Chapter 7, the emphasis is on frequency-domain analysis of intensity modulation. The measurement of laser modulation bandwidth, distortion, and intensity noise are covered. Chapter 8 discusses time-domain measurements of the modulation. Here subjects such as eye-diagram analysis, temporal signal jitter, and bit-error-ratio measurements are covered.

Chapters 9 to 13 cover measurement topics that are common to two-port optical devices. Two-port optical devices include optical fiber, optical amplifiers, filters, couplers, isolators, and any other device where light enters and leaves. Chapter 9 covers the techniques used to measure insertion loss. Chapter 10 and 11 cover methods of measuring the reflectivity of components. Chapter 12 covers the measurement of chromatic and polarization mode dispersion. Finally, Chapter 13 covers erbium-doped fiber amplifier (EDFA) testing.

Fiber Optic Test
and Measurement

C H A P T E R

1

Introduction to Fiber Optic Systems and Measurements

Dennis Derickson

1.1 INTRODUCTION

This book explores test and measurement techniques for fiber optic systems. Its primary area of concern is the fiber optic link and its related components: the optical transmitter, the fiber medium, two-port optical components such as filters and isolators, and the optical receiver.

The field of fiber optic test has become more complex as fiber optic systems have evolved. The pioneers in fiber optic research must be credited with hard work and perseverance. First they developed the basic technology of low-loss fiber, produced reliable lasers, fabricated photodetectors, and demonstrated modest bandwidth link capabilities. Because the technology was not yet commercialized, they also had to develop the test and measurement technology required to sustain the research. With deployment of fiber optic systems in the late-1970s, commercial fiber optic measurement equipment became available. The test needs for these first fiber optic systems were modest. A loss test set consisting of a source and a power meter was used to check for the presence of signals or the loss in a fiber component. An optical time-domain reflectometer could be used to find the location of a break in an optical fiber.

What are the changes in the industry that caused fiber optic communications and the associated measurement technology to become more sophisticated?

Wavelength now matters. Early fiber optic systems did not have to be especially concerned with the spectral content of the laser sources. The center wavelength range for a laser to be qualified was often over a 1260 to 1340 nm window. Today's telecommunication systems can incorporate multiple wavelength channels spaced at 100 GHz (0.8 nm) increments that are co-propagating on a single fiber. These multiple wavelength systems

1

are referred to as wavelength division multiplexed (WDM) systems. Each of the channels must be stable in wavelength and power. Simple, accurate, and inexpensive instruments to test the power, signal to noise ratio, and wavelength of each channel are now needed.

The data rates used in fiber have dramatically increased. The commercial state of the art for high-speed telecommunication systems has pushed beyond the previous 2.5 Gb/s rate to 10 Gb/s. Researchers are demonstrating 40 Gb/s technology and beyond. The push toward 10 Gb/s was not an easy one for fiber optic systems. The width of the optical spectrum from laser sources must be carefully analyzed and controlled. The chromatic dispersion and polarization mode dispersion of installed fiber must be checked for compatibility with high data rates. The bandwidth response of packaged optical transmitters, receivers, and associated components had to be optimized.

Optical amplifiers are now basic building blocks of fiber optic networks. The development of erbium-doped fiber amplifiers (EDFAs) has caused a significant change in the design of fiber optic links. An optical amplifier is used to directly boost the optical signal to compensate for loss in the optical fiber. Previous systems used repeaters to regenerate the digital signals when the signal-to-noise ratio was degraded by fiber loss. Repeaters convert the optical signal to an electrical signal, reproduce the digital pattern and retransmit the signal back onto the fiber with an optical transmitter. The economics of optical amplifiers are so compelling that working repeaters are often removed and replaced with amplifiers. Optical amplifiers have motivated the switch from 1300 nm to the lower loss 1550 nm wavelength in optical fiber. Optical amplifiers are an enabling technology that allows WDM systems to be practical over long spans. Each wavelength channel is simultaneously amplified within a single amplifier. The use of optical amplifiers also allows the use of optical components that have significant loss such as wavelength multiplexers and demultiplexers.

Short link-length data communications systems are displacing copper cables. Data communication is moving towards fiber as the dominant transmission medium. Early fiber-optic data communications systems used surface-emitting light-emitting diodes (LEDs) with data rates in the 10 to 622 Mb/s range. Vertical cavity surface-emitting lasers now allow data communication rates over 1.5 Gb/s at very low cost. The range of applications for data communications is expanding. Fiber optic links are now cost competitive with even very short copper-cable links. Fiber optics are used to interconnect arrays of computer workstations and hard disk drives. The local area network backbone for many businesses uses fiber optic links with very high bandwidths.

The complexity of fiber optic systems continues to grow. Researchers are now working to give fiber optics a more pervasive role in telecommunications. Previously, the role of a fiber optic cable was to provide a low-loss link between point A and point B. Future systems will utilize optical methods to route the path of a signal through a telecommunication network. Passive optical network concepts would allow a signal to stay in the optical domain over the majority of the network path. Optical routers allow the signals to be directed to the correct destinations. Multiple wavelength systems are a key enabler allowing these advanced functions to be implemented.

It is necessary to understand some of the basic features of a fiber-optic communication link before launching into a chapter-by-chapter analysis of fiber optic test and measurement requirements. It is the goal of this chapter to provide some basic background in fiber optic systems and measurement techniques that will introduce readers to the measurement-specific chapters that will follow. There are many textbooks[1-3] that provide a much more detailed presentation that are highly recommended reading. Palais[1] recommends that only 7 out of his 12 chapters be covered in a one-semester college course. This Chapter 1 gives a condensed summary of the operational principles of fiber optic links that should take only a few hours to read. It also introduces some of the measurement issues associated with fiber-optic link design, manufacture, installation, and maintenance. The level of discussion in this introduction is intentionally very qualitative. Hopefully those who are not familiar with some of the measurement issues will gain an introduction and appreciation before studying the measurement-specific chapters that follow.

What is coming next in this chapter? A general fiber optic link is described. Important measurements methods for fiber-optic link characterization are then described. The link is then broken up into its individual components for description and measurement analysis: the optical transmitter, the fiber medium and optical components, and the optical receiver. Finally, a chapter-by-chapter description of the book's contents is given.

1.2 FIBER OPTIC LINKS: THE BASICS

The fiber optic link exists to provide the lowest-loss connection between two points. Fiber optic links provide the lowest cost and highest performance solution for everything but the shortest link lengths. Fiber optic links are capable of transmitting digital- and analog-formatted signals. Section 1.2.1 will discuss the basic features of digital communication links. Section 1.2.2 will describe WDM systems where more than one signal is present on the fiber optic cable. Section 1.2.3 will describe analog communication systems and associated testing requirements. After this short introduction to fiber optic links, Section 1.3 will discuss some of the basic measurements for end-to-end characterization of the link.

1.2.1 Digital Communication Links

Figure 1.1 shows a diagram of a digital fiber-optic link. Signal waveforms are shown at several key points along the signal path. These waveforms illustrate some of the fundamental characteristics of a fiber optic link. After reading this chapter, it will be useful to return to this figure to see how well the basic measurement concepts of a fiber optic system are illustrated in a single figure. Let us now discuss the link operation as the signal flow proceeds from the electrical input to the electrical output.

In this example, the input electrical data stream to the optical transmitter is a non-return to zero (NRZ) digital pattern of 0, 1, 0, 0, 1, 1. The electrical signal is converted to an optical signal with the electrical-to-optical (E/O) converter, the optical transmitter. The optical transmitter is essentially a current-to-power converter. In a perfect E/O converter every electron injected into the source should produce a single output photon. Semiconductor laser (light amplified by stimulated emission of radiation) diodes are the predomi-

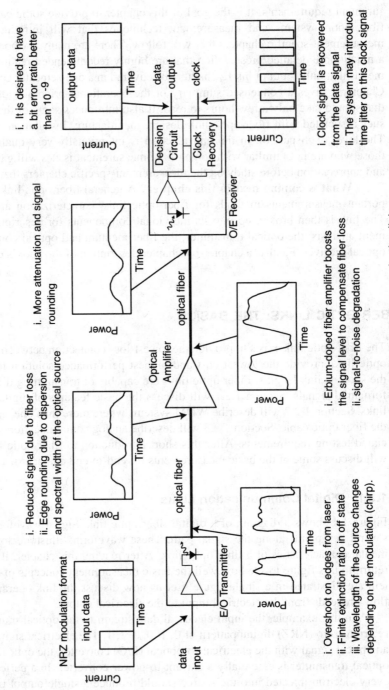

Figure 1.1 A digital fiber-optic communication link.

NRZ modulation format

input data

Current

Time

data input

E/O Transmitter

Power

Time

i. Overshoot on edges from laser
ii. Finite extinction ratio in off state
iii. Wavelength of the source changes depending on the modulation (chirp).

optical fiber

Power

Time

i. Reduced signal due to fiber loss
ii. Edge rounding due to dispersion and spectral width of the optical source

Optical Amplifier

Power

Time

i. Erbium-doped fiber amplifier boosts the signal level to compensate fiber loss
ii. signal-to-noise degradation

optical fiber

Power

Time

i. More attenuation and signal rounding

O/E Receiver

Decision Circuit

Clock Recovery

data output

clock

Current

Time

i. Clock signal is recovered from the data signal
ii. The system may introduce time jitter on this clock signal

output data

Current

Time

i. It is desired to have a bit error ratio better than 10 -9

4

nant sources for long-distance, high-speed digital systems. Lasers provide high optical power, narrow spectral width, and modulation rates beyond 10 Gb/s. LEDs are common lightwave sources for shorter distances and lower data rates. Section 1.6 compares the available fiber-optic transmitter types.

International standards such as Synchronous Optical Network (SONET) and Synchronous Digital Hierarchy (SDH)[4-6] specify data rates and formats used for digital modulation in telecommunication systems. Data rates from 51.84 Mb/s to 9.95 Gb/s are specified in the SONET/SDH standards. Each of these high-speed channels contains lower data rate signals that have been time-interleaved using a method called time division multiplexing (TDM). A different set of standards exist for data communication fiber-optic links. Gigabit ethernet and fibre channel[7,8] are examples of high speed data communication interface specifications. Fibre channel systems are becoming very popular for computer peripheral applications with up to 1.063 Gb/s data rates. Fiber distributed data interface (FDDI) is another standard used for local area networks.[9]

Figure 1.1 shows that the output of the optical transmitter is not a perfect replica of the input electrical signal. A significant test and measurement challenge is to characterize the waveform and signal-to-noise degradation present in E/O converters. The laser transmitter in this example has significant overshoot during the transition from the zero to one state. This ringing of the laser output power is caused by interactions between the round-trip time of photons in the laser and the speed at which the optical gain can be changed by a step change in current.[10] The characteristic frequency of the ringing is referred to as the relaxation oscillation frequency. It has a typical value between 2 to 30 GHz depending on laser design parameters.

Figure 1.1 shows how the edges of the digital waveform become more rounded after propagation through the fiber optic cable. This waveform spreading leads to distance limitations in an optical fiber link. The spectral width of an optical source is a major consideration for waveform spreading. Excess laser bandwidth is important due to chromatic dispersion in optical fibers. The different wavelengths contained in the optical signal travel at different velocities due to chromatic dispersion. After propagating through long fiber lengths, adjacent data bits start to overlap leading to intersymbol interference and errors. Fabry-Perot lasers and LEDs are optical source examples that have wide spectral width. The spectral width of these sources may limit their usefulness for long fiber-optic links.

Even narrow spectral-width sources have concerns. Distributed feedback (DFB) lasers are the dominant source for high-performance telecommunication applications. During the modulation transition between the one and zero power levels, the wavelength of DFBs is pulled away from its nominal operating value. This wavelength shift is commonly referred to as "chirp." Chirp causes the spectral width of the laser to be much wider than is necessary to transmit the information. A 2.5 Gb/s digitally modulated signal should occupy an optical bandwidth of less than 10 GHz to accommodate the information. A chirped laser could have an optical bandwidth of 30 GHz or more.

The optical transmitter establishes the wavelength of the fiber optic link. The choice of laser wavelength is driven by the loss characteristics of optical fiber cables. This leads us into a discussion of the characteristics of optical fiber.

1.3 DIGITAL COMMUNICATION LINKS

1.3.1 Optical Fiber

The premier feature of optical fiber is its extremely low loss. This has made it the dominant transmission medium for long link lengths. Section 1.4 describes fiber optic cable operation in more detail. An overview is given here.

Loss characteristics. The loss versus wavelength of optical fiber is shown in Figure 1.2a for many different types of fibers. The loss characteristics of fiber determines where optical communication is practical. At 1550 nm, singlemode optical fiber has an attenuation of 0.2 dB/km. This allows fiber optic signals to be propagated through very long lengths of fiber without regeneration. A fiber optic link at a 2.5 Gb/s rate can tolerate a 25 dB signal loss before system noise becomes a problem. This means that a distance of 125 km could be spanned with such a link. This is roughly 1000 times lower loss than a coaxial cable at 2.5 Gb/s. This low insertion loss is the driving force behind the fiber optic revolution.

Figure 1.2 also illustrates the wavelengths used in fiber optic systems. Axes are given in wavelength frequency and wavenumber. These variables are related by the following equations.

$$\text{frequency} = \frac{c}{\lambda_{vac}} \tag{1.1}$$

$$\text{wavenumber} = \frac{1}{\lambda_{vac}} \tag{1.2}$$

Here λ_{vac} is the wavelength of light in a vacuum and c is the speed of light in a vacuum. Fiber optic transmitters emit in the wavelength range between 630 nm (476 THz) and 1650 nm (182 THz). As a reference, the visible portion of the optical spectrum lies in the range of 400 to 700 nm.

Digital fiber-optic links are categorized by their end applications. Telecommunication systems use the 1300 and 1550 nm windows for lowest loss in the fiber. Since a telecommunication system must cover a very large distance, loss in the fiber is of high importance. Attenuation values as low as 0.2 dB/km are available at 1550 nm using singlemode fiber. Erbium-doped fiber amplifiers (discussed in Chapter 13 later) are also available at 1550 nm, making this the highest performance wavelength available. Erbium-doped amplifiers require optical pump lasers at either 980 nm or 1480 nm. 1650 nm is an important wavelength that is used to monitor the loss characteristics of a fiber while live traffic is being carried.

Data communication systems use multimode glass-fiber at either 780 to 850 nm, 980 nm or 1300 nm bands. Since shorter distances are being covered, higher loss can be tolerated. The 780 to 850 nm wavelength range has developed because of the availability of inexpensive optical fiber components and for historical reasons. These systems generally use data rates up to 1 Gb/s over links that are less than a few km. Plastic fiber-optic links are used with 650 nm sources. The data rate is typically less than 100 Mb/s with a

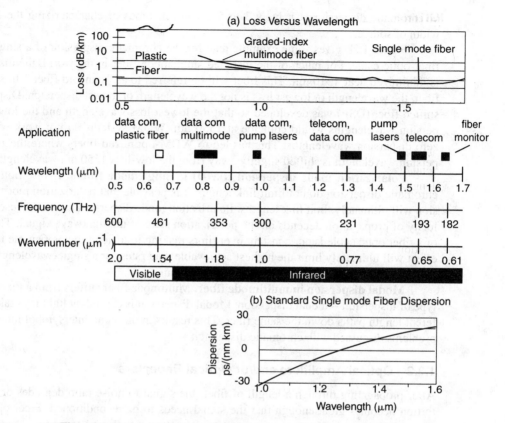

Figure 1.2 (a) Loss in optical fiber waveguides and important wavelengths for fiber optic communications. (b) Chromatic dispersion of standard singlemode fiber.

range of tens of meters limited by the higher loss of plastic fiber and dispersion, which leads us to the next section.

Chromatic dispersion in singlemode fibers. The loss characteristics of optical fiber often limit the distance that a signal can propagate. This is not always the case. In singlemode fibers, chromatic dispersion can limit the distance over which fiber optic signals can propagate. Chromatic dispersion describes the fact that the speed of signal propagation in the fiber depends on the wavelength of the light. Figure 1.1 illustrates the detrimental effects of fiber dispersion. Notice that as the signal propagates through a long length of fiber, the edges of the waveform start to become more rounded. Eventually, the adjacent bits start to overlap in time causing the digital waveform to have poor readability. Pulse spreading causes adjacent digital bits to interfere with each other. The amount of signal rounding depends on the amount of chromatic dispersion in the cable and the spectral width of the laser transmitter. Chapter 12 describes how to make measurements of chromatic dispersion. Chapters 3 to 5 discuss methods of characterizing the spectral width of sources.

Figure 1.2b gives some example data for the chromatic dispersion of a singlemode fiber-optic cable. For much of the installed singlemode fiber in the world, the dispersion is minimized near 1310 nm. This fiber will be referred to as "standard fiber." In standard fiber, the wavelength of lowest loss is not the wavelength of lowest dispersion. Dispersion-shifted fiber (DSF) was developed so that the lowest loss wavelength and the lowest dispersion wavelengths would coincide at 1550 nm. WDM systems do not work well near zero dispersion wavelengths. This has led to WDM-optimized fibers where the zero dispersion wavelength is shifted slightly away from the low-loss 1550 nm wavelength.

Polarization mode dispersion (PMD) in singlemode fibers. If insertion loss or chromatic dispersion do not limit link length, it is possible that polarization mode dispersion will ultimately limit link lengths. Polarization mode dispersion arises because the velocity of propagation depends on the polarization state of the lightwave signal. The PMD of a fiber optic cable is not constant over time, making it difficult to compensate for. This effect will ultimately limit the highest achievable data rates for a single-wavelength fiber-optic system. Chapter 12 covers PMD measurements.

Modal dispersion in multimode fiber. Multimode fiber suffers from a more severe type of dispersion—modal dispersion. Modal dispersion is caused by light rays taking different length paths down the same fiber. This results in the same intersymbol interference problems discussed with chromatic dispersion.

1.3.2 Optical Amplifiers and/or Optical Repeaters

After propagating through a length of fiber, the signal-to-noise ratio degrades or the distortion becomes great enough that the signal needs to be reconditioned. Fiber optic systems often use optoelectronic repeaters to detect the signal, and regenerate, retime, and then retransmit the signal onto the fiber with another optical transmitter.

Fiber optic systems also use optical amplifiers to boost the signal midspan if long-link distances is required. Optical amplifiers can compensate for fiber loss at the expense of adding a small amount of noise and distortion to the signal. Erbium-doped fiber amplifiers (EDFAs) are the predominant optical amplifier technology.[11] These amplifiers have revolutionized the way fiber optic systems are designed. EDFAs offer low noise contribution (3 to 4 dB noise figure), high gain (to 40 dB), and high output-power capability (approximately 50 mW of output power) near 1550 nm. Undersea telecommunication systems with hundreds of cascaded optical amplifiers have proven that long link-length optically amplified systems are practical.[12] The TPC-5 link across the Pacific Ocean successfully uses 260 cascaded optical amplifiers to cover an 8600 km span.

1.3.3 O/E Converters

The optical signal is then converted back into the electrical domain with an O/E converter. The O/E converter is an optical power to electrical current converter. Ideally, each received photon would be converted to an electron. p-i-n diode detectors are commonly used. They can have extremely high bandwidth, but the sensitivity of high-speed photore-ceivers is often limited by the noise contribution of the electronics following the detector. Avalanche photodetectors (APDs) are used for high-sensitivity systems. APDs can pro-duce several electrons for each received photon. This electronic gain in the APD allows the sensitivity of the system to be set by the APD noise characteristics and not by the noise of the following electronics. This results in higher sensitivity, especially at high data rates.

In addition to O/E conversion, the optical receiver must recreate the input data stream and clock signal. Since the received signal power can vary, the receiver must have an automatic gain control stage or limiting amplifier. It is important to accurately extract the clock from the data with very good timing accuracy. The clock signal is used to trig-ger the decision circuit that decides if a 1 or a 0 has been received.

1.4 WAVELENGTH DIVISION MULTIPLEXED SYSTEMS

1.4.1 Wavelength Division Multiplexed Systems

Figure 1.1 shows how a single wavelength digital waveform is sent through a fiber optic link. Increasing the data rate of a single wavelength channel is one strategy to increase the throughput on optical fibers. The data rate for a single optical channel will eventually reach its limits due to chromatic dispersion and/or polarization mode dispersion. An im-portant strategy to further increase the available bandwidth is to add multiple wavelength channels. Multiple wavelength systems are referred to as being wavelength division multiplexed (WDM). Early WDM systems used a wide wavelength spacing. It was common to increase the bandwidth of a 1310 nm link by adding a 1550 nm channel. The installation of WDM systems is often driven by economic reasons. It is less expensive to update the terminal equipment to WDM capability than to install new fiber optic ca-bles. The introduction of the Erbium-doped fiber amplifier has moved nearly all WDM activity to the 1530 to 1565 nm wavelength window. More recent WDM installations are referred to as dense WDM (DWDM) systems due to the narrow spacing between optical channels.

Figure 1.3 shows the diagram of a DWDM fiber optic link. Lasers are combined in a wavelength multiplexer to a single fiber. International standards organizations are defin-ing standard wavelength channels for DWDM systems.[13] A common standard is a DWDM laser wavelength spacing of 100 GHz between channels. Systems use 4, 8, 16, or 32 channels located on the 100 GHz channel grid shown in Table 1.1.

Wavelength multiplexers are used to combine and separate the wavelength channels onto a single fiber with low loss. EDFA booster amplifiers are used to counteract the in-sertion loss of the wavelength multiplexers at the transmitter and the receiver. A single EDFA can amplify all of the wavelength channels simultaneously.

With the introduction of WDM systems, new roles emerged for the use of fiber op-tics in telecommunication systems. Previously, when a set of signals needed routing to a

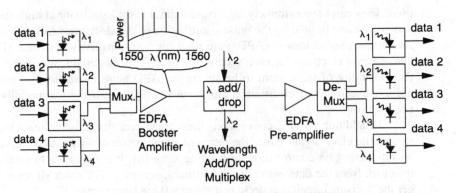

Figure 1.3   A wavelength division multiplexed (WDM) fiber optic link.

particular location, the optical signal was detected and then routed electronically. The coming trend is to allow the optics to do some of the routing functions. The ability to add and drop certain wavelength channels from a fiber while sending on other channels undisturbed is an example of optical routing capability.

At the receiver, an EDFA is used to compensate for the loss of the demultiplexer filters and to increase receiver sensitivity. The demultiplexer sorts the wavelengths out to the individual receivers for detection.

The switch from single wavelength to multiple wavelength systems has introduced measurement challenges:

1. In single wavelength systems, the absolute wavelength of a laser and associated components was not critical. Laser wavelengths now need to be measured with less than 0.01 nm accuracy. Studies on aging of semiconductor lasers require wavelength resolutions of 0.001 nm.

2. The loss versus wavelength of optical components has taken on new importance. The wavelength demultiplexor at the receiver must reject the signal from adjacent channels. This requires accurate, high-dynamic range measurement capability. Cross talk between channels can also be introduced through fiber nonlinearities that can cause power from one channel to be converted to another channel. Fiber nonlinearities are discussed in Appendix B.

3. The monitoring of wavelength, channel power, and signal-to-noise ratio is required for network management. Single wavelength systems require power measurements at various points in the network to find fault conditions. Simple power measurement is not sufficient in WDM systems. Spectral monitors are now necessary to sort out performance of each channel.

Table 1.1   ITU-T Defined WDM Wavelength Channels

Frequency (THz)	Wavelength (nm)
195.6	1532.7
195.5	1533.5
195.4	1534.9
195.3	1535.0
195.2	1535.8
195.1	1536.6
195.0	1537.4
194.9	1538.2
194.8	1539.0
194.7	1539.8
194.6	1540.6
194.5	1541.3
194.4	1542.1
194.3	1542.9
194.2	1543.7
194.1	1544.5
194.0	1545.3
193.9	1546.1
193.8	1546.9
193.7	1547.7
193.6	1548.5
193.5	1549.3
193.4	1550.1
193.3	1550.9
193.2	1551.7
193.1	1552.5
193.0	1553.3
192.9	1554.1
192.8	1554.9
192.7	1555.7
192.6	1556.6
192.5	1557.4
192.4	1558.2
192.3	1559.0
192.2	1559.8
192.1	1560.6

1.5 ANALOG LINKS

1.5.1 Analog Links

There are many systems that transmit analog signals over the fiber optic link. Figure 1.4 shows a typical diagram of an analog fiber-optic link and a measurement system. The goal of an analog link is to reproduce an identical version of the input signal at the output. A small amount of noise and signal distortion can appear at the output due to fiber optic contributions. It is a significant design challenge to keep the degradation levels low.

One very important application of analog fiber-optic links is in cable-TV video distribution. Cable TV systems use analog signals that directly connect to the television receivers. Older cable TV systems use long cascades of electronic amplifiers that result in noise build-up and poor reception near the fringes of the system coverage. Analog fiber-optic links are used to decrease the number of electronic repeaters between the system head end and the subscriber. A star configuration of fiber optic cables feeds the signal to local neighborhoods. The signal is then converted back to analog electronics and distributed to the subscribers over coaxial cable. The analog fiber-optic links in this application must have very low intermodulation and harmonic distortion levels to insure the quality of reception at the consumer. Laser transmitters and receivers developed for cable TV systems must be linear. Important fiber optic transmitter and receiver parameters such as the composite second-order and composite triple-beat distortions are described in Darcie et al.[14] The laser transmitter must be tested with a complex input signal that can contain dozens of video and data channels. Optical reflections in analog systems can cause laser frequency variations to be coupled into an increase in system noise (see Appendix A). Chapter 7 in this book addresses many of the measurement categories important for these applications.

Analog links are also used for low-loss distribution of microwave signals. An example application is antenna remoting and distribution of signals for phased array radars.[15]

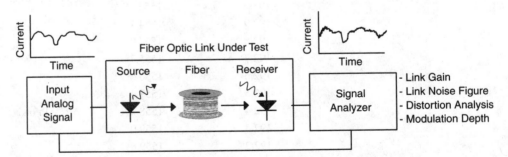

Figure 1.4 An analog fiber-optic link.

1.6 CHARACTERIZATION OF DIGITAL FIBER-OPTIC LINKS

Section 1.2 introduced some of the basic features of fiber optic links. It must be emphasized again that the primary purpose of a fiber optic link is to provide a low-loss data connection between two points. This section will outline techniques used to verify that acceptable end-to-end performance is achieved.

1.6.1 Bit Error Ratio

The most important parameter of a digital system is the rate at which errors occur in the system. A common evaluation method is the bit error ratio test as shown in Figure 1.5. A custom digital pattern is injected into the system. It is important to use a data pattern that simulates data sequences most likely to cause system errors. A pseudo-random binary sequence (PRBS) is often used to simulate a wide range of bit patterns. The PRBS sequence is a "random" sequence of bits that repeats itself after a set number of bits. A common pattern is $2^{23}-1$ bits in length. The output of the link under test is compared to the known input with an error detector. The error detector records the number of errors and then ratios this to the number of bits transmitted. A bit error ratio of 10^{-9} is often considered the minimum acceptable bit error ratio for telecommunication applications. Data communications have more stringent requirements where 10^{-13} is often considered the minimum. A common measurement is to test the bit error ratio as a function of loss in the optical fiber. Chapter 8 describes bit error ratio measurements in detail.

1.6.2 Waveform Analysis

Bit error ratio measurements provide a pass/fail criteria for the system and can often identify particular bits that are in error. It is then necessary to troubleshoot a digital link to find the cause of the error or to find the margin in performance that the system provides. Digital waveforms at the input and output of the system can be viewed with high-speed oscilloscopes to identify and troubleshoot problem bit patterns as is shown in Figure 1.6. The pattern generator used in bit error ratio analysis can be coordinated with the oscilloscope to view the particular pattern that caused an error. Chapter 8 discusses methods of digital waveform analysis.

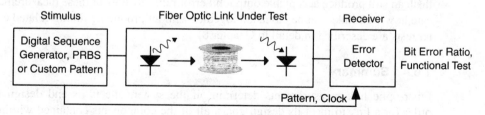

Figure 1.5 Bit error ratio measurements and functional test.

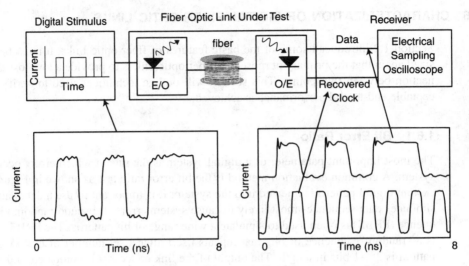

Figure 1.6   Waveform analysis of a fiber optic link.

1.6.3   Link Jitter

Another link evaluation is clock jitter measurement as shown in Figure 1.7. Control of jitter is important to insure that many fiber optic links can be successfully cascaded together.[16,17] A perfect clock waveform would have a uniform bit period (unit interval) over all time. The fiber optic system can add variability to the unit interval period which is referred to as jitter. Three types of jitter measurements are performed: absolute jitter, jitter transfer, and jitter tolerance. Absolute jitter measurements look at the jitter produced at the link output with a low-jitter clock signal at the input. Jitter transfer characterizes the amount of jitter produced at the output of the device under test compared to the amount of jitter at its input as a function of frequency. An example of a jitter transfer measurement is given in Figure 1.7b. The plot shows jitter modulation frequencies where the jitter at the output is increased compared to the jitter at the input (unacceptable). Standards such as SONET and SDH have specified levels of jitter transfer that can be tolerated as a function of the jitter modulation frequency.[4]

Jitter tolerance measurements find the highest level of jitter at the input of a system that can still produce acceptable output bit error ratio. Results of these measurements indicate how well links can be cascaded together without producing timing-related errors. Jitter tests are described in detail in Chapter 8.

1.6.4   Summary

Fiber-optic link measurements determine if the system meets its end design goals. In order for a link to meet its design goals, all of the components contained within the link must be characterized and specified to guarantee system performance. The following sections of Chapter 1 describe the characteristics of each component along with an analysis

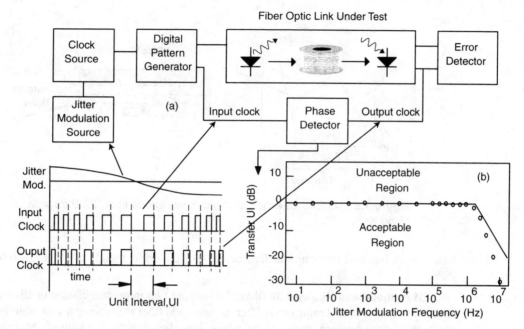

Figure 1.7 (a) System to characterize link jitter. (b) A jitter transfer function measurement example.

of the testing needs for these components. The discussion starts with optical fibers and amplifiers. The discussion then continues with optical transmitters and receivers. In each section, a brief description of each component type is made. This is then followed by a description of the testing requirements on each component. References are then made to sections of the book where more detail can be found on the measurement methods.

1.7 OPTICAL FIBERS AND TWO-PORT OPTICAL COMPONENTS

Figure 1.8 illustrates the basic features of an optical fiber. Figure 1.8a shows the end view of the fiber and the corresponding index of refraction profile. The central part of the fiber is called the core and the outer portion is called the cladding. The fiber is most often made from silica (SiO_2) glass. The center core of the fiber is doped to produce a slight increase in the index of refraction and a corresponding decrease in signal velocity. A common core dopant is Ge. The index of refraction difference between the core and cladding is very small, typically around 0.01.

The optical signal is confined to the core due to the principle of total internal reflection. In total internal reflection, a ray of light incident on the core-to-cladding boundary will be 100% reflected back toward the core as long as a critical angle is not exceeded. This confined reflection of light rays is illustrated in the fiber side-view drawing in Figure

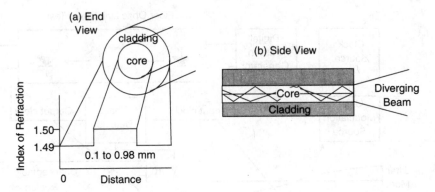

Figure 1.8   (a) End view of a step-index multimode fiber. (b) Side view of the fiber.

1.8b. For an index of refraction difference of 0.01, the critical angle is a shallow 6.6 degrees.

Multimode and singlemode fiber.   There are two important classes of fiber used in links: singlemode and multimode fiber. Singlemode fiber is associated with high bandwidth telecommunication applications where long link length is required. Multimode fiber is associated with data communication systems, lower data rates, and short link lengths.

Singlemode and multimode terminologies refer to the profile of light found in the end view of the fiber and the paths that light takes along the fiber length. Figure 1.8b shows a side view of a multimode fiber and three example rays that propagate through the fiber. One ray could take a path following the center of the fiber. Another ray could take a path that bounces off of the cladding. These two example rays have a very different path length through the fiber. In a singlemode fiber, the size of the core is reduced until only one path (mode) is allowed to propagate through the fiber.

1.7.1   Step-Index Multimode Fiber

Figure 1.8 illustrates a step-index multimode fiber. When specifying a fiber type, the dimensions and the modal character of the fiber are specified. A common step-index fiber has a core diameter of 100 µm and a cladding diameter of 140 µm. The index of refraction is abruptly stepped from the core to the cladding. This multimode fiber would be referred to as 100/140 SI MM fiber. The first two dimensions are the core and cladding diameters, the SI refers to step index and MM refers to multimode. Multimode fibers have the advantage of ease of use and cost of associated components when compared to singlemode fiber. The larger fiber-optic core dimensions make fiber coupling to transmitters and receivers less stringent. It also becomes easier for operators to prepare optical connectors.

980/1000 SI MM plastic fibers are used in very low-cost and short link-length applications. The primary advantage of these fibers is that they are easy to install and maintain. Very simple tools can be used to add connectors. Plastic fibers use a low-loss window at 650 nm. Here fiber loss is 0.2 dB/m, almost 1,000 times greater than the loss minimum for glass optical fibers.

1.7.2 Graded-Index Multimode Fiber

Figure 1.9 illustrates the operation of a graded-index multimode fiber. In graded-index fiber, the refractive index changes gradually between the core and the cladding. Multimode fibers for data communication at 850 nm and 1300 nm use either a 50 μm or 62.5 μm core diameter. Common graded index fibers are 50/125 GI MM and 62.5/125 GI MM types with 62.5/125 being more common. The outer diameter of the glass fiber is 125 μm. Additional coatings are added over the glass cladding to protect the fiber resulting in an increase in the diameter to 250 μm.

Multimode Fiber Dispersion.   A consequence of multimode fiber designs is that an optical pulse will quickly spread out in time due to the multiple paths available for the signal. The signal spreading limits the maximum link length or maximum data rate that the fiber can accommodate. Modal dispersion is a measure of this multipath signal distortion. Figure 1.9c shows how an input data sequence can be spread out in time causing adjacent bits to overlap. Errors will occur if the data is allowed to spread too widely. Multimode dispersion is characterized by a pulse-broadening parameter. It is stated in terms of the ps of pulse broadening per unit of fiber length. It can also be related to a bandwidth-distance product. For a 500 MHz-km product (a typical value for multimode fiber), you could make a 622 MHz link over a 1 km length or a 155.5 MHz link over 4 km length with equal pulse broadening penalty.

Multimode Fiber Speckle Pattern.   If a multimode fiber is illuminated with a narrow spectral-width optical source, a dark and light "speckle pattern" is formed in the end view as shown in Figure 1.9a. The dark and light patterns form because light from different optical paths add constructively or destructively. If a multimode fiber cable is

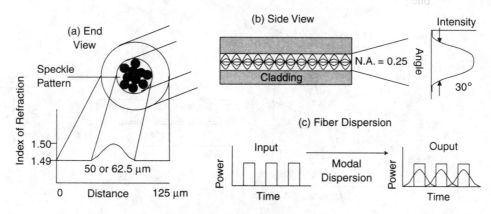

Figure 1.9   (a) End view of a graded-index multimode fiber. (b) Side view of the fiber. (c) The effects of modal dispersion.

Table 1.2 Comparison of Multimode Fiber Characteristics

Multimode Fiber	Core (μm)	Cladding (μm)	Numerical Aperture	Attenuation dB/km	Pulse Broadening
62.5/125 GI	62.5	125	0.27	0.7 @ 1300 nm	500–1200 MHz-km
50/125 GI	50	125	0.21	0.5 @ 1300 nm	500–1500 MHz-km
100/140 SI	100	140	0.2–0.3	5 @ 850 nm	20 MHz-km (22 nm/km)
980/1000 SI plastic	980	1000	0.5	0.2 dB/m @ 660 nm	(110 ns/km)

wiggled, the speckle-pattern distribution will change. This change in speckle pattern can lead to mode selective loss if optical components in the system have loss dependent on the exact cross-section modal structure. If a wide spectral width source excites a multimode fiber, the speckle pattern becomes less defined. This mode selective loss property of multimode fiber data communication systems encourages the use of lightwave sources with moderately broad spectral widths.

Divergence Angle.   Light diverges from fibers in a cone-shaped radiation pattern. The steepness of the cone depends on the difference in refractive index between the core and the cladding and the diameter of the waveguide. In general, multimode fibers have a fairly steep divergence angle. Figure 1.9b shows the far-field power versus angle that is found in a 50/125 GI MM fiber.

Table 1.2 compares the characteristics of multimode fibers.

1.7.3 Singlemode Fiber

Singlemode fiber provides the lowest loss and lowest dispersion optical waveguide. Figure 1.10 illustrates a singlemode fiber. The diameter of the core in a standard singlemode fiber used in telecommunication applications is approximately 9 μm. The outer diameter

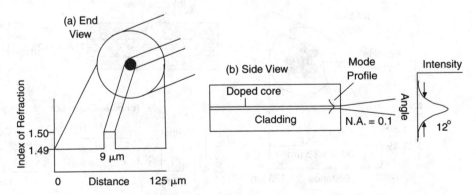

Figure 1.10   (a) End view of a singlemode fiber. (b) Side view of the fiber.

of singlemode fiber is 125 μm. This fiber would be referred to as 9/125 SI SM where SM refers to singlemode.

A singlemode fiber has a single lobed cross-section. This single lobe is associated with a single path that the light can take along the core of the fiber. When light exits the core, it begins to diverge with a cone-shaped radiation pattern. For singlemode fiber, the angle of the cone is about 12 degrees. The divergence angle is often quoted in terms of the numerical aperture (NA). The NA is defined to be the sin of half of the divergence angle. The NA of singlemode fiber is sin (12/2) = 0.1. Singlemode fiber can also be produced at shorter wavelengths (into the visible spectrum) by reducing the diameter of the core region and changing the doping profile. As an example, singlemode fiber for a red wavelength (632 nm) has a 3 μm core diameter.

Chromatic Dispersion.   Since only a single path for light exists in singlemode fiber, the modal dispersion of multimode fiber is not present. Chromatic dispersion dominates in singlemode fibers. The glass fiber has a propagation velocity that is wavelength dependent. Standard singlemode fiber has its minimum total dispersion at 1310 nm. Dispersion is often quoted in terms of the dispersion parameter, D, which has units of ps/(nm km). Multiplying the dispersion parameter by the spectral width of the optical transmitter and by the link distance gives an indication of how much pulse broadening will occur. As an example, standard singlemode fiber has a dispersion parameter of 20 ps/(nm km) at 1550 nm. A high performance DFB laser has a spectral width of 0.2 nm when modulated. If the signal were propagated over a 50 km length, each bit would be spread by 200 ps. At the SONET standard rate of OC-48 (2.5 Gb/s), the bit period is 400 ps. The dispersion in this link example would result in a large degradation in link performance.

There are several types of singlemode optical fiber used for the telecommunications wavelengths between 1300 nm and 1550 nm. Five important fiber types are listed below. Table 1.3 lists some of the important characteristics of these singlemode fibers.

- **Standard singlemode fiber:** This fiber has its dispersion minimum at 1310 nm. This is the most common fiber in use today.

- **Dispersion-shifted singlemode fiber:** This fiber has its dispersion minimum shifted to 1550 nm. This allows fiber to have low loss and low dispersion at the

Table 1.3   Singlemode Fiber Parameters

Fiber Type	Attenuation		Chromatic Dispersion	
	1310 nm	*1550 nm*	*1310 nm*	*1550 nm*
9/125 Conventional	0.35	0.25	0	17
9/125 Dispersion shifted	0.35	0.25	−15	0
9/125 WDM optimized	0.35	0.25	−12	3

same wavelength. Much larger data rates and link lengths are available in this fiber. This fiber was deployed in the late-1980s and early-1990s.

- **WDM-optimized fiber:** This fiber offers low, but nonzero chromatic dispersion at 1550 nm. It is being used in newly installed systems that will use WDM communication techniques. The introduction of a finite amount of dispersion minimizes channel cross-talk due to nonlinearities in fibers. Nonlinearities in optical fibers are discussed in Appendix B. This fiber started deployment in the mid-1990s.

- **Polarization maintaining (PM) fiber:** This fiber is optimized so that the polarization of a signal does not change with distance. PM fiber is often used for short distance interconnections between optical components that have polarization dependencies. An example is a laser connected to an external modulator that is polarization dependent. Chapter 6 discusses measurements associated with polarization-maintaining fiber.

- **Erbium-doped singlemode fiber.** Erbium added to the core of singlemode fiber allows for the formation of an optical amplifier near 1550 nm. The erbium atoms must be excited with an optical pumping signal in order to provide amplification. Section 1.4.3 and Chapter 13 discuss EDFAs in more detail.

1.7.4 Optical Fiber Amplifiers and Two-Port Optical Components

EDFAs are such important optical components that they deserve special mention. A description of EDFAs will also illustrate some of the passive optical components used in fiber optic systems. Figure 1.11 shows a diagram of a typical EDFA and supporting optical components. A pump laser at either 980 nm or 1480 is coupled into the erbium-doped fiber through a WDM coupler. This coupler allows the pump to be coupled with minimum loss while at the same time providing a low-loss path for signals at the amplification wavelength. The 980 nm and 1480 nm wavelengths are chosen because erbium atoms absorb energy well at these wavelengths. If enough optical pump energy is absorbed, the erbium doping provides optical gain in the region between 1530 and 1570 nm. The gain in the amplifier is equal in either direction through the doped fiber. An isolator is used to suppress gain in one direction.

The pumping configuration of the amplifier determines many of the optical amplifier parameters. The optical pump is often inserted both at the input and the output. The highest output power is achieved with a pump at the output. The lowest noise amplifier

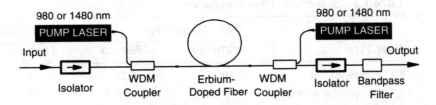

Figure 1.11   Erbium-doped fiber amplifier (EDFA) block diagram.

occurs with a pump at the input if low-loss components are used. 980 nm pumps provide the lowest noise-contribution amplifiers.

A high-gain EDFA produces a large amount of noise power at its output. Randomly generated light in the amplifier is boosted to high levels at the amplifier output. This noise is referred to as amplified spontaneous emission (ASE). The noise is produced over a broad range of wavelengths corresponding to the gain bandwidth of the amplifier. An optical bandpass filter is often used in conjunction with the amplifier so that the amount of amplified noise is reduced to a narrow window around the desired signal wavelength. The ASE from optical amplifiers can be a useful measurement tool as is described in Chapter 9.

The EDFA is used to allow longer fiber-optic link lengths without resorting to regeneration. The utility is actually more far-reaching. WDM technology is greatly aided by the addition of amplifiers. The optical amplifier can be used to compensate for losses in filters and other optical components. More complex amplifier configurations have two-gain stages with functional elements such as add-drop multiplexers placed between the stages. This allows high-loss optical components to be inserted into the system without an overall system performance degradation. Chapter 13 gives a good introduction to the basics of EDFAs and methods for their characterization.

1.8 MEASUREMENT OF OPTICAL FIBER AND TWO-PORT OPTICAL COMPONENTS

In 1.4, a description of two-port optical components was given. This section outlines the types of measurements performed on these components. Figures 1.12 to 1.16 illustrate measurement categories. This book will not discuss measurement of some of the physical properties of fiber such as pull-strength, fiber concentricity, or core parameters. For this coverage, refer to References 2, 18, and 19.

1.8.1 Insertion Loss

Figure 1.12a shows a typical measurement set-up and result for an insertion loss measurement. To make an insertion loss measurement, both a source and receiver are necessary.

For the source, either a wavelength tunable laser or a broadband source is used. The tunable laser provides a high power, narrow spectral width signal. Broadband sources produce a signal over a wider wavelength range than the tunable laser source but at a much lower power per unit wavelength.

The receiver is either an optical power meter or an optical spectrum analyzer (OSA). The power meter is a calibrated optical to electrical converter. It does not provide wavelength information. The OSA has a tunable bandpass filter ahead of the power meter. The OSA is capable of measuring power and wavelength.

One insertion loss measurement method uses a tunable laser as the source and an optical power meter (OPM) as the receiver. This method features large measurement range and fine wavelength resolution in most measurements. Tunable lasers have wave-

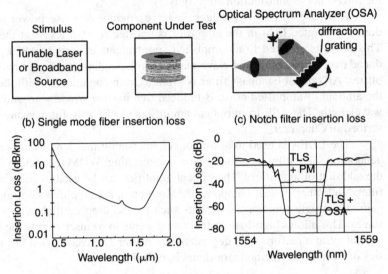

Figure 1.12 (a) Insertion loss measurement of two-port optical devices. (b) Measurement of singlemode fiber loss with a white light source and optical spectrum analyzer (OSA). (c) Measurement of a notch filter with tunable laser source (TLS) and a power meter (PM) or an OSA.

length tuning ranges of less than 200 nm. Detection of broadband noise from the tunable source is a major limitation of this measurement case.

Broadband emission sources can be used with an OSA to cover a broader wavelength range. Tungsten lamp emitters can cover the entire fiber-optic communication wavelength range in a single source. Optical amplifiers can provide broadband emission over narrower wavelength ranges, but with much higher power. The broadband source/OSA solution offers wide wavelength range coverage, moderate measurement range, and fast measurement speed. Figure 1.12b shows an example measurement of the loss of a singlemode fiber. A tungsten lamp source is used to cover this wide wavelength range. The tungsten lamp source has the disadvantage of having a very low power level per unit of optical bandwidth.

The highest performance solution is to use a tunable laser as the source and an OSA as the receiver. The use of a tunable laser provides the highest possible resolution due to the narrow spectral width of the source. The OSA provides additional filtering to reject some of the broadband noise emission that unintentionally emits from tunable lasers. Figure 1.12c shows the measurement of a notch filter with a tunable laser source and either a OPM or an OSA as a receiver. Measurements on deep-notch filters is one of the most difficult insertion loss measurements. Chapter 9 covers insertion loss measurement methods in detail.

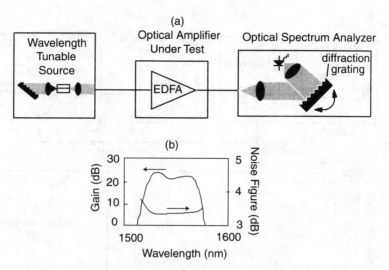

Figure 1.13   (a) Optical amplifier gain and noise figure measurements. (b) Gain and noise figure versus wavelength for an EDFA.

1.8.2 Amplifier Gain and Noise Figure Measurement

For active optical devices, both amplifier gain and amplifier noise contributions must be characterized. Figure 1.13 shows a test configuration used to measure gain and noise figure of optical amplifiers. Figure 1.13b shows a typical result from such a measurement.

Gain measurements are often done in large signal conditions when the amplifier is experiencing gain saturation. This requires the use of a high-power excitation source.

Characterization of the noise contribution can be done with optical domain or electrical domain measurements. Optical techniques measure the level of the ASE coming from the amplifier. Electrical techniques use a photodetector and an electrical spectrum analyzer to characterize the total amount of detected noise produced by the system. Care must be taken in the measurement of the amplifier noise. The noise generated by the optical amplifier must be distinguished from the amplified noise generated from the source at the input to the amplifier.

For WDM systems, the amplifier may need to be characterized using the same signal-loading conditions as will be seen in the actual application. Several new WDM amplifier characterization techniques have been demonstrated that reduce the number of sources necessary for amplifier characterization. Chapter 13 covers the characterization of optical amplifiers.

1.8.3 Chromatic Dispersion

Chromatic dispersion measurements characterize how the velocity of propagation in fiber or components changes with wavelength. This measurement of chromatic dispersion is accomplished by analyzing the group delay through the fiber as a function of wavelength

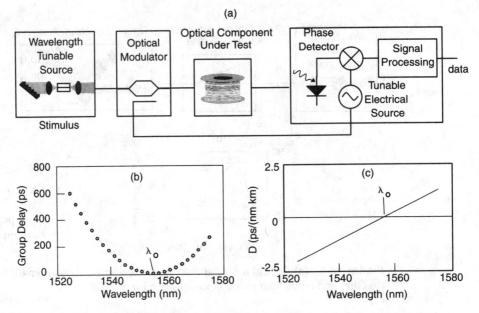

Figure 1.14 (a) Chromatic dispersion measurement of two-port optical devices. (b) Relative group delay versus wavelength. (c) Dispersion parameter versus wavelength.

as is shown in Figure 1.14a. A wavelength tunable optical source is intensity modulated as the stimulus for the device under test. The phase of the detected modulation signal is compared to that of the transmitted modulation. The wavelength of the tunable source is then incremented and the phase comparison is made again. By calculating how the phase delay changes with wavelength, the group delay of the fiber is measured. Figure 1.14b shows data for relative group delay versus wavelength for 1550 nm dispersion-shifted singlemode fiber. The group delay data is used to calculate the dispersion parameter results shown in Figure 1.14c.

The accurate characterization of the minimum fiber dispersion wavelength, λ_o, is important in the design of high-speed TDM and WDM communication systems. WDM systems do not operate well if there is extremely low dispersion. Fiber nonlinearities such as four-wave mixing cause the WDM channels to cross-couple leading to degradation in bit error performance. Undersea fiber installations use dispersion management techniques to reduce nonlinear effects and still achieve low average dispersion. A dispersion-managed fiber system has low dispersion on average with segments of positive and negative dispersion connected together.

High-speed TDM systems often suffer from chromatic dispersion limitations, especially at data rates of 10 Gb/s and above. Dispersion compensation components such as chirped Bragg grating filters and dispersion compensation fibers require accurate measurement of dispersion. Chapter 12 covers measurement of chromatic dispersion in detail.

1.8.4 Polarization-Related Measurements

The polarization state of a lightwave signal can have a significant effect on the performance of a lightwave system. Polarization of the lightwave signal refers to the orientation of the electric field in space. It is important to understand how the insertion loss and group delay of a two-port optical component vary as a function of the input signal polarization. The concept of polarization and how it affects lightwave systems is discussed in detail in Chapter 6.

The polarization state of the lightwave signal can vary significantly along the length of the fiber. A linearly polarized signal that is introduced into a fiber will evolve through elliptical and linearly polarized states as it propagates down the fiber. This is due to the fact that the velocity of propagation in a fiber can be dependent on the polarization of the input light source.

Figure 1.15 illustrates a measurement technique to characterize the polarization transfer function of optical fiber and optical fiber components. The key equipment piece is a polarization analyzer. The polarization analyzer is able to measure the polarization state that is incident on it. The polarization state of a signal can be represented by a Jones polarization-state vector. The Jones state vector contains two complex numbers that quantify the amplitude and phase of the vertical and horizontal components of the optical field. The goal of this measurement is to characterize the polarization-transforming properties of the optical fiber. The polarization transfer function is represented by the Jones transfer matrix illustrated in Figure 1.15. This two-by-two complex matrix will predict the output polarization state for any input polarization state. The Jones matrix is experimentally measured by applying three well-known polarization states at the input to the component under test and then characterizing the resulting output polarization state in the polarization analyzer. The Jones matrix can be used to find the worst-case

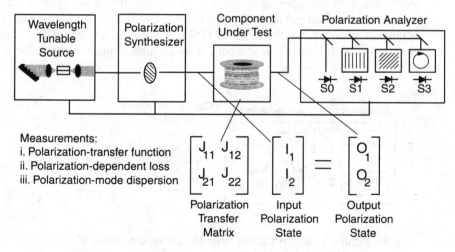

Figure 1.15 Polarization transfer function measurements.

polarization-dependent loss (PDL) and polarization-mode dispersion (PMD) of the device under test.

PMD is a term used to describe the fact that the group delay is polarization dependent. PMD, in addition to chromatic dispersion, can cause adjacent digital bits to interfere. PMD can limit the highest bit rate that is achievable in a fiber optic system. Chapter 12 covers characterization methods for PMD.

PDL is another polarization-related effect in two-port optical devices. PDL occurs when the loss through a device depends on the input state of polarization. Small flexures in an optical fiber can dramatically change the polarization transfer function of a fiber. A component with PDL will convert these polarization fluctuations into system loss changes. It is therefore desirable to reduce PDL in system components. There are several methods of PDL measurement covered in this book. One method shown in Chapter 6 uses the Jones polarization transfer function to predict the highest- and lowest-loss case. Chapter 9 discusses methods that use a variant of the measurement set-up of Figure 1.12. By adding a polarization synthesizer to the insertion-loss measurement, the lowest- and highest-loss polarization input states can be found experimentally.

1.8.5 Reflection Measurements

When fiber optic cables are deployed, it is important to monitor insertion-loss performance as well as to look for major faults such as a fiber breaks. Optical time-domain reflectometry (OTDR) is the primary tool for these applications. Figure 1.16 shows an OTDR measurement block diagram. OTDRs inject a pulsed signal onto the fiber optic cable. A small amount of the pulsed signal is continuously reflected back in the opposite direction by irregularities in the optical fiber structure. This reflected signal is referred to as Raleigh backscatter. The magnitude of the Raleigh backscattered signal is surprisingly large. A pulse that occupies 1 m of optical fiber will return a signal 73 dB lower than the incident signal. By measuring the amount of backscatter signal versus time, the loss versus distance of the fiber optic cable is measured. If a fiber optic break occurs, the backscatter will stop and the break location is detected. Chapter 11 describes OTDR measurements in detail. For traditional OTDRs, accuracy of several meters is sufficient for finding faults. In optical components, a much higher distance resolution is necessary.

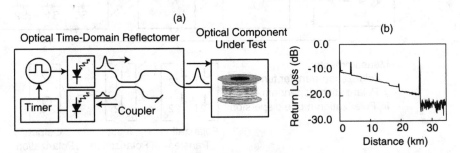

Figure 1.16   (a) Optical time-domain reflectometer (OTDR). (b) Example OTDR display.

Chapter 10 describes some of the techniques that are used for high resolution OTDRs. High resolution OTDRs are capable of detecting reflections with 10 μm resolution.

1.9   OPTICAL TRANSMITTERS

Optical sources are one of the most researched areas in fiber optic communications. The characteristics of the source often determine the maximum length of a fiber link and the data rate that is achievable. This section will describe the types of optical transmitters used in telecommunications and data communications applications. Section 1.12 will describe the measurements that are made on optical transmitters.

1.9.1   Fabry-Perot Lasers

The Fabry-Perot (FP) laser diode is the most widely used source for lightwave telecommunication systems. This popularity is due to the simplicity of fabrication and low cost. FP lasers perform best when used with low chromatic-dispersion fibers due to their substantial spectral width.

Figure 1.17a illustrates a sectional view of an FP laser parallel to the direction of light emission. The laser has two parts: a semiconductor optical amplifier to provide gain and mirrors to form a resonator around the amplifier. The semiconductor optical amplifier is formed by applying current to a lower-bandgap active layer surrounded by higher-bandgap materials. The surrounding higher-bandgap materials confine electrons and holes into a small volume. The applied current excites electrons from the low energy valence band state to the higher energy conduction band state. As the high-energy electrons lose

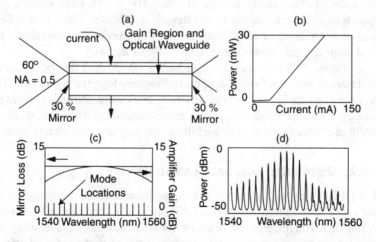

Figure 1.17 (a) Cross-section of a Fabry-Perot (FP) laser diode. (b) Light versus current characteristic. (c) Amplifier gain, mirror loss, and longitudinal mode location. (d) Power versus wavelength for the FP laser.

energy, they give off light. The low-bandgap and high-bandgap layers form an optical waveguide to direct the light that is produced. The active region composition is designed to emit at the desired wavelength. For lasers emitting in the 1100 to 1700 nm region, compounds of In, Ga, As, and P are grown on top of InP substrates. For the 780 to 980 nm range, compounds of In, Al, Ga, and As are grown on top of GaAs substrates. The semiconductor crystal is cleaved to form a mirror. The air-to-semiconductor interface at the cleavage plane provides 30% reflectivity and 70% transmission. Light emits from the laser with a steep-angle cone-shaped radiation pattern.

Figure 1.17b shows the optical output power versus current that is measured from an FP laser. As the current increases, the optical amplifier gain increases. When the amplifier gain equals the mirror loss, the lasing threshold is reached. The laser breaks into oscillation above threshold current. The slope of the power versus current curve above threshold indicates how efficient the laser converts injected electrons into emitted photons. FP lasers are capable of producing many mW of output power. FP lasers for EDFA-pump applications are designed to produce 100 mW into a singlemode fiber. FP lasers can be modulated to very high rates by varying the current to the laser diode.

Fig. 1.17c plots some of the FP-laser parameters that define its operation. The gain versus wavelength, the mirror reflectivity versus wavelength, and the location in wavelength of the longitudinal modes of the laser are shown when pumped above threshold. The optical amplifier needs a net gain of 11 ($11 = 1/0.3 \times 1/0.3$) at threshold to overcome the loss in the two 30% reflectivity mirrors. The distance between the two mirrors determines the spacing between the possible lasing wavelengths as is illustrated by the tick marks in Figure 1.17c. These wavelengths are called the longitudinal modes of the laser. The longitudinal mode spacing (in Hz) is the velocity divided by twice the length of the cavity. A typical FP laser is 300 μm long, has an index of refraction of 3.3, and therefore has a longitudinal mode spacing of 150 GHz. A 150 GHz spacing in frequency corresponds to a 1.25 nm spacing in wavelength at 1550 nm. The mirror reflectivity is independent of wavelength in an FP-laser design. The gain versus wavelength function of the optical amplifier determines the spectral shape of the laser.

Figure 1.17d shows the power versus wavelength that occurs in a typical FP laser. FP lasers operate with significant power in many longitudinal modes. The power distribution between these longitudinal modes is constantly shifting, resulting in extra low frequency intensity variations called mode-partition noise. The relatively wide spectral width of FP lasers limits their usefulness for long-distance communication.

1.9.2 Distributed Feedback Lasers (DFBs)

The DFB laser was designed to overcome the spectral shortcomings of the FP laser.[20] Figure 1.18a shows the structure of a DFB laser. It is very similar to an FP laser with the addition of a Bragg reflector structure located near the light-emitting active region. The Bragg reflector grating provides a periodic change in the index of refraction in the waveguide. Each period of the grating reflects a small amount of light back in the opposite direction. The Bragg grating forms an efficient mirror at the wavelength where the grating period is one-half of the wavelength of light in the semiconductor material. At 1550 nm,

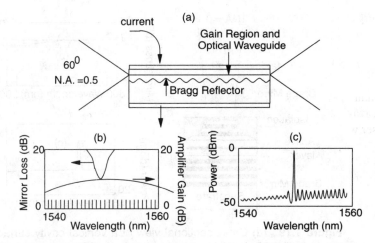

Figure 1.18   (a) Cross-section of a distributed feedback (DFB) laser. (b) Bragg grating reflectivity, amplifier gain, and longitudinal mode location. (c) Power versus wavelength for the DFB laser.

the Bragg grating period is 116 nm. The small dimensions of the grating make the device fabrication technology more critical than an FP laser. Extra wafer processing steps are necessary to etch the Bragg grating into the semiconductor material and to grow new material on top. This results in a more expensive laser diode.

The effect of the Bragg grating on the mirror reflectivity shape is shown in Figure 1.18b. The Bragg grating reflection passband is only a few nanometers wide. As the current is increased in this laser, only a narrow wavelength range can reach the threshold condition. This frequency-dependent mirror reflectivity forces all of the laser power to emit in a single longitudinal mode. Figure 1.18c shows the single longitudinal mode spectral shape of a DFB laser. DFB lasers are the industry standard for use in long-distance fiber-optic links.

1.9.3   Vertical Cavity Surface-Emitting Laser (VCSEL)

The vertical cavity surface-emitting laser (VCSEL) was originally developed as a low-cost alternative to FP and DFB lasers.[21] The first commercial application of these lasers is in the area of high-speed data communication links replacing LEDs. Vertical cavity lasers emit perpendicular to the top plane of a semiconductor wafer. The VCSEL uses a multilayer dielectric mirror that is grown directly on the semiconductor surface as shown in Figure 1.19a. This mirror consists of alternations of high and low index of refraction layers to form a Bragg reflector. The distinguishing feature of this structure is its extremely short optical amplifier length (on the order of 100 nm). This length is compared to the 300 μm length typical of an FP or DFB laser. This short amplifier length limits the available gain from the amplifier to a very small value. Figure 1.19b shows the gain, mirror reflectivity, and mode locations of the VCSEL design. The mirror reflectivity is very high so

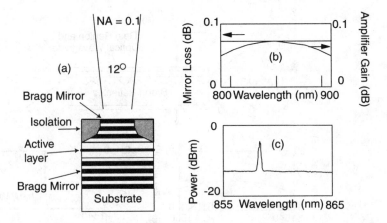

Figure 1.19 (a) Cross-sectional view of a vertical cavity surface emitting laser (VCSEL). (b) Mirror reflectivity, amplifier gain, and longitudinal mode location. (c) Optical spectrum of a VCSEL.

that the low-gain optical amplifier can achieve threshold. Only a single longitudinal mode is available for lasing because of the small distance between mirrors. For a 1 μm cavity length, the longitudinal mode spacing would be 45 THz. As can be seen in Figure 1.19c, there is no sign of closely spaced longitudinal modes as is seen in the DFB laser of Figure 1.18c.

In data communication applications, the diameter of the laser is chosen to be large enough to support multiple modes in the plane perpendicular to the lasing direction. This causes the central lasing peak to spectrally broaden. The wider spectral width is often designed into VCSELs to avoid mode-selective loss in multimode fiber applications. The major advantage of the VCSEL design is the ease of fabrication, simplified on-wafer testing, and simple packaging resulting in lowered costs. The VCSEL has higher output power and higher modulation rates than surface-emitting LEDs for data communications. To date, this type of laser is commercially available in the 780 to 980 nm wavelength range. The 1300 nm and 1550 nm versions are still in the laboratory stage.

Sources with External Modulators: Narrow Spectral Width Under Modulation. It is important to have a narrow spectral width source while modulation is applied to a laser. Wide spectral width sources in conjunction with chromatic dispersion leads to intersymbol interference in optical fibers. Although DFB lasers have a single lasing wavelength, they are not a perfect source for fiber optic systems. If a high-speed current modulation is applied to a DFB laser, the laser's center wavelength is pulled up and down in wavelength during the modulation. This laser wavelength pulling is referred to as chirp. Chirp causes substantial broadening of the modulated laser linewidth. Chapter 5 covers measurement techniques for modulated laser linewidth.

The solution to reducing chirp and modulated spectral width to small values is to separate the functions of lasing and of modulation. This is accomplished with DFB lasers followed by external modulators. The next two sections describe two important types of external modulator lasers.

1.9.4 DFB With Electrooptic Modulator

Figure 1.20a shows a Mach-Zehnder (MZ) modulator[22] that follows a DFB laser. The MZ modulator consists of an integrated optical waveguide on a material that can exhibit the electrooptic effect. Electrooptic materials have an index of refraction that can be changed with the application of voltage.

MZ modulators operate using interferometry techniques. The optical signal is branched into two separate paths and is then recombined at the output. The two paths of the interferometer are nearly, but not exactly, the same length. When light from the two paths is combined at the output, the two signals will have slightly different phases. If the two signals are exactly in phase, the light will combine in the output waveguide with low loss. If the two combining signals are 180 degrees out of phase, the light will not propagate in the output waveguide resulting in light radiation from the waveguide to the surrounding substrate. The electrooptic effect makes the velocity of propagation in each arm of the interferometer dependent on a voltage applied to an electrode. Figure 1.20b shows the transmission versus voltage for a MZ interferometer. Depending on the modulation voltage, the light will propagate with high or low loss at the output combining waveguide. The modulator is normally biased in the half-on, half-off state which is called quadrature.

Well designed MZ modulators do not contribute to spectral broadening of the DFB laser signal. An important design consideration for MZ modulators is the V_π of the modulator as shown in Figure 1.20b. It is desirable to keep V_π as low as possible to reduce the modulator drive requirements. MZ modulators have demonstrated modulation bandwidths above 50 GHz. Both $LiNbO_3$ and GaAs material systems have been used for modulator fabrication. The main disadvantage of external modulators is cost and ease of use. MZ modulators must have a specific input polarization state to function properly. The polar-

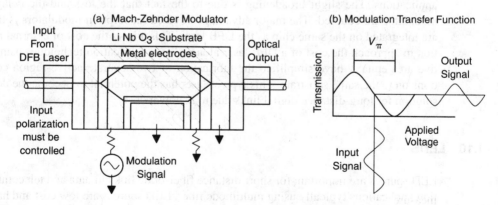

Figure 1.20 (a) Mach-Zehnder (MZ) interferometer modulator. (b) Modulation transfer function for a Mach-Zehnder interferometer.

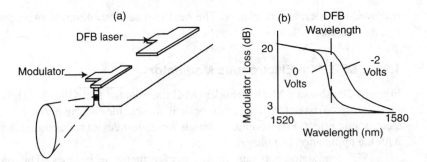

Figure 1.21 (a) DFB laser followed by an integrated electroabsorption external modulator. (b) Transmission characteristics of the electroabsorption modulator as a function of reverse-bias voltage.

ization state must be controlled between the DFB and the modulator. DFB lasers with MZ modulators are the highest performance solution for long link lengths. MZ modulators are also used in high-performance analog communication links.

1.9.5 DFB With Integrated Electroabsorption Modulator

A DFB laser with an integrated electroabsorption modulator (DFB/EA) is shown in Figure 1.21a. The electroabsorption principle of operation is illustrated in Figure 1.21b. This modulator uses the effect that the band gap of a semiconductor can be adjusted in wavelength as a function of reverse-bias voltage. The DFB's lasing wavelength is placed at the absorption edge. The modulator's absorption edge is adjusted to a high-loss or low-loss condition depending on the modulation voltage. The electroabsorption modulator causes a slight broadening of the input DFB spectral width but is adequate for most long-distance applications. The slight broadening is due to the fact that the loss and the delay in the modulator are coupled. The major advantage of electroabsorption modulators is that they are integrated on the same chip as the DFB laser. This allows the cost of external modulation to approach the cost of a DFB laser alone. Clever fabrication techniques using selective area epitaxy have simplified the fabrication of two separate active-region compositions on to the same substrate. This type of laser has the potential to become the dominant solution for long-distance digital links due to its lower cost.

1.10 LEDS

LED sources are important for short-distance fiber-optic links in data and telecommunication applications typically using multimode fiber. LED sources are low cost and have adequate performance for many applications. The two most common wavelengths for multimode data communication systems are 850 nm and 1300 nm.

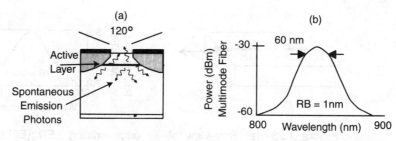

Figure 1.22 (a) Cross-section of a surface-emitting LED (SLED). (b) Power versus wavelength for the SLED as coupled into a 50/125 graded-index multimode fiber. The optical spectrum analyzer bandwidth is 1 nm.

1.10.1 Surface-Emitting LEDs

The most common multimode source is the surface-emitting LED (SLED) shown in Figure 1.22a. The SLED has low-bandgap semiconductor materials sandwiched between high-bandgap materials as is found in semiconductor lasers. The major difference in an LED is that there are no mirrors to provide feedback. Current is passed through the active region to create hole and electron pairs in the low-bandgap active region. The electrons from the conduction band lose energy spontaneously and emit photons in all directions. A fraction of the generated light is coupled into the multimode fiber. Figure 1.22b shows the spectral characteristics of the SLED as coupled into a 50/125 GI multimode fiber. The spectral width is over 60 nm. Compare this with the 0.1 nm spectral width of a DFB laser. The wide SLED spectral width limits its usefulness for very long link lengths. The broad spectral content of an SLED is an advantage in reducing speckle-related mode-selective loss in multimode fiber systems. The advantage of the SLED is its reliability and low cost. The power from surface emitting SLEDs is usually much less than 1 mW coupled into a multimode fiber. The bandwidth of surface emitting LEDs is limited by the carrier lifetime in the active region. Data rates up to 622 Mb/s have been achieved with these devices.[2,23]

1.10.2 Edge-Emitting LED

Figure 1.23a shows a diagram of an edge-emitting LED (EELED). The EELED is very similar to an FP laser without mirrors. This EELED configuration shows two segments. One segment is forward biased to produce gain in a semiconductor optical amplifier. The other segment is reverse biased to produce an optical absorber. The absorber prevents the optical amplifier from becoming an FP laser. The output of the semiconductor optical amplifier is also antireflection-coated to further prevent a mirror from forming. The EELED optical amplifier produces ASE. Spontaneously emitted light at the input of the amplifier produces ASE at the amplifier output. The spectral shape of the EELED is shown in Figure 1.23b when coupled into singlemode fiber. The spectral width of the EELED is 60 nm in this example centered at 1550 nm. The spectral shape of the EELED is similar to the

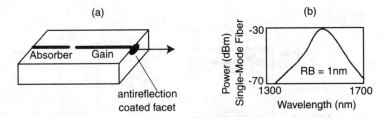

Figure 1.23 (a) Side-view of an edge-emitting LED (EELED). (b) Power versus wavelength for the EELED as coupled into a 9/125 μm singlemode fiber. The OSA bandwidth is 1 nm.

gain versus wavelength function of the optical amplifier. EELEDs are capable of producing several mW of power. EELEDs are limited in their modulation speed by carrier lifetime in the active region to a few hundred MHz. Chapter 9 describes the use of EELEDs in detail along with measurement applications.

1.10.3 Comparison of Optical Sources

Table 1.4 shows a comparison of the characteristics of the lightwave sources that were introduced in this section. The large range of source characteristics reflect the diversity of applications for fiber optic links.

1.11 OPTICAL RECEIVERS

The optical receiver detects the lightwave signal and then conditions the resulting electrical signal to the appropriate levels with receiving electronics. Figure 1.1 showed a diagram of a digital optical receiver. Photodetectors are used for optical to electrical conversion. A variable gain amplifier or limiting amplifier is used to get a specified signal level to a decision circuit. The clock must also be extracted from the incoming signal. The simplest clock extraction circuit is a bandpass filter and a frequency doubling circuit. More advanced phase-locked loop circuits are often used for clock extraction. The clock signal is then used to trigger a sampling circuit that decides the digital value of the incoming data. A good coverage of optical receiver design is given in Miller and Kaminov.[2] Sections 1.11.1 and 1.11.2 describe the operation of p-i-n and avalanche photodetectors in more detail.

1.11.1 p-i-n Photodetectors

Figure 1.24a shows a diagram of a p-i-n photodetector. The p-i-n designations refer to the doping of the top three layers of the detector structure. The i, or intrinsic layer, is chosen to be a low-bandgap material that absorbs incoming photons. The bandgap of this layer determines the longest wavelength light that can be absorbed. The p-i-n doping structure

Table 1.4 Comparison of Lightwave Sources

Source Type	Application	Associated Fiber Type	Data Rate	CW Spectral Width	Modulated Linewidth, Chirp	Power into Fiber mW
Fabry-Perot laser	datacom, telecom	singlemode/multimode	to 10 Gb/s	5 nm multiple line	medium	2–100 S.M.
DFB laser	telecom, analog	singlemode	to 10 Gb/s	10 MHz single line	good	2–20 S.M.
DFB with electro-absorbtion	telecom	singlemode	to 10 Gb/s	10 MHz single line	excellent	0.2 S.M.
DFB with Mach-Zehnder	telecom, analog	singlemode	> 10 Gb/s	10 MHz single line	highest performance	0.2 S.M.
Vertical cavity laser	datacom	multimode	to 5 Gb/s	1 GHz for multiple mode	medium for multiple mode	0.1 M.M.
Surface-emitting LED	datacom	multimode	622 Mb/s	10 THz broadband	broadband source	0.1 M.M.
Edge-emitting LED	datacom	singlemode/multimode	155 Mb/s	5 THz broadband	broadband source	0.1 S.M.

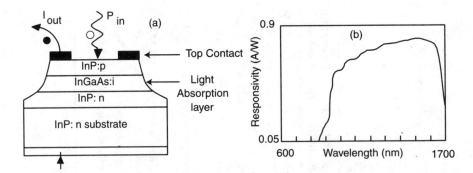

Figure 1.24 (a) p-i-n photodetector diagram. (b) Responsivity versus wavelength for a InGaAs/InP detector.

causes a very high electric field to be developed across the intrinsic layer. Once a photon is absorbed, a hole and electron are created in the layer. The electric field sweeps these two charged particles out to be collected in an external electrical circuit. Thus a p-i-n detector is a photon to electron converter. The amount of current produced per unit of input power is called the detector responsivity.

A common material for the p-i-n photodetector is an InGaAs undoped layer surrounded by doped p and n InP material. An example of absorption versus wavelength function for an InGaAs/InP p-i-n detector is shown in Figure 1.24b. Well designed p-i-n detectors can convert over 90% of the incoming photons to electrons. These detectors are capable of responding to signals that are modulated to very high rates. The capacitance and the transit time to sweep out electrons and holes determine the ultimate bandwidth capability. Detectors have been designed with over 100 GHz of modulation bandwidth.[24] Chapter 2 describes the operation of this type of detector in more detail.

1.11.2 APD Detectors

A diagram of an avalanche photodetector (APD) detector is shown in Figure 1.25.[25,26] The APD was designed to improve the sensitivity compared to a p-i-n-based receiver. The sensitivity issue surrounding p-i-n receivers is that it is often difficult to design a wideband electronic amplifier after the photodetector with a low enough noise contribution. The APD addresses this problem by providing low-noise, high-bandwidth electronic amplification that is designed into the detection process. The APD has a low-bandgap semiconductor absorption region very similar to that of a p-i-n detector. The electrons are then accelerated to very high velocities toward a separate multiplication region. The high-energy electrons collide with the lattice to form new free electrons in an avalanche multiplication process. The ideal low-noise APD multiplies electrons but does not multiply holes in the avalanche process. This condition results in minimum noise contribution from the multiplication process. APD detectors result in receivers that have better sensitivity than p-i-n-based receivers when used in high-speed systems. APDs require a high-bias voltage to produce avalanche conditions. The multiplication process is also temperature-

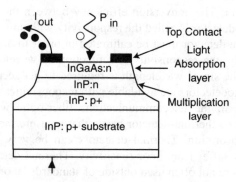

Figure 1.25   Avalanche photodetector diagram with separate absorption and multiplication regions.

dependent making the supporting circuitry more difficult to design. APDs are commonly used in systems up to a rate of 2.5 Gb/s. Higher bandwidth devices are being developed in research laboratories.

1.12   OPTICAL TRANSMITTER AND RECEIVER MEASUREMENTS

Sections 1.9 and 1.11 described the operational principles of sources and receivers for fiber optic systems. Knowledge of component operation is important for understanding the testing requirements for these devices. This section examines some of the measurement areas important for characterizing optical transmitters and receivers. The discussion is organized by measurement areas. Sections 1.12.1 to 1.12.5 describe three fundamental source characterization areas of optical power, polarization, and spectral measurements. Sections 1.12.6 to 1.12.8 cover techniques used to measure the modulation characteristics of lightwave components. Section 1.12.9 describes reflectometry techniques for optical transmitters and receivers.

1.12.1   Power

Power measurement is fundamental to all areas of optical characterization. Figure 1.26 illustrates a basic power-meter instrument diagram. Light from the optical fiber is imaged onto a diode photodetector. The photodetector is used to convert the optical power into a

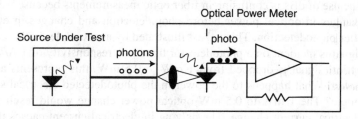

Figure 1.26   Optical power measurement.

proportional electrical current. The conversion efficiency between the input power and the output current of a photodetector is called the responsivity with units of Amps/Watt.

The responsivity of the detector must be calibrated in order to make optical power measurements. Unfortunately, the responsivity is a function of wavelength for all photodetectors. Knowledge of the signal wavelength is required to get accurate power measurements. Fortunately, photodetectors are available with responsivities that are relatively independent of wavelength in the telecommunications fiber-optic wavelength bands. Standards labs have developed thermal-detector heads that measure the temperature rise caused by optical signal absorption. Thermal detectors can be very accurate and are wavelength-independent but suffer from poor sensitivity. Thermal detectors are used to calibrate photodetectors but are not often used outside of standards laboratories.

Photodiode detectors are capable of measuring optical power levels of approximately -110 dBm to $+10$ dBm. The upper power limit is determined by saturation effects in the photodetector that cause the responsivity to decrease. Higher power levels can be measured using calibrated optical attenuators. The low power level is limited by the averaging time of the measurement and the dark current of the photodetector. The dark current of a photodetector is the amount of current measured in the absence of an optical input signal. Ultimately, the lower power level will be limited by the particle nature of light. For the most sensitive measurements, individual photons of light can be counted. Photon-counting detectors that can count individual photons are difficult to achieve in the infrared range of 1200 to 1700 nm. Sensitivity of optical receivers is covered in detail in Appendix A.

Care must be taken in the design of power meters so that the power measurement is independent of the input polarization state of the signal. Any movement of an optical fiber can alter the polarization state that is incident on the detector.

Another consideration is the reflectivity of the optical head. Large reflections can feedback signal to the optical source potentially altering the power and spectral characteristics of the incoming light. Chapter 2 covers the area of optical power measurements and calibration.

dB Optical and dB Electrical.   The decibel (dB) is a commonly used unit for comparing power levels in systems.

$$dB = 10 \log\left(\frac{\text{Power}_1}{\text{Power}_2}\right) \qquad (1.3)$$

The use of dBs is confusing in fiber optic measurements because we often talk about both changes in optical power before photodetection and changes in electrical signal power after photodetection. This is best illustrated by an example. Let's say that a 1 mW optical signal is incident on a photodetector that has a responsivity of 1 A/W. If the power of the optical signal is dropped from 1 mW to 0.5 mW, this represents a 3 dB drop in optical power. What happens to the power in the photodetected electrical signal in this same example? The 1 mW to 0.5 mW optical power change would result in a 1 mA to 0.5 mA electrical current change. The halving in electrical current causes the electrical power to drop by 6 dB since the electrical power dissipation is proportional to the current squared. Any dB change in optical power will result in a 2 dB change in electrical power. To eliminate confusion in dB, it is customary to refer to the optical power drop in terms of dB optical and the electrical power drop after detection in terms of dB electrical. This distinction of electrical and optical dBs is especially important for the modulation analysis description given in Sections 1.12.6 to 1.12.8.

1.12.2 Polarization

The polarization state of a source is a second fundamental property of the lightwave signal. Polarization refers to the electric field orientation of a lightwave signal. Laser sources are predominantly linear polarized sources. LEDs have no preferred direction of polarization and are predominantly unpolarized. A polarized lightwave signal will go through dramatic changes in its polarization state as it propagates through optical fiber. This change in polarization state in fiber is due to a weak change in the velocity of propagation as a function of polarization state. Stress in the fiber or ovality of the waveguide shape causes this polarization dependence of fiber. Components in fiber-optic communication systems may have polarization-dependent loss, gain, or velocity. A system with a polarized input and a large polarization-dependent loss will perform unpredictably depending on temperature or fiber stress. It is therefore important to understand the polarization characteristics of optical sources.

The goals of a polarization measurement are to determine the fraction of the total light power that is polarized and to determine the orientation of the polarized component. Figure 1.27 illustrates a polarization analyzer instrument that characterizes the polarization state of a lightwave signal. This instrument consists of four power meters with polarization characterizing optical components in front of them. The polarization analyzer measures the Stokes parameters; S_0, S_1, S_2, and S_3. The total power of the signal is the parameter S_0. All of the other Stokes parameters are normalized to S_0. S_1 indicates the power difference between vertical and horizontal polarization components. S_2 indicates the power difference between +45 and −45 degrees linear polarization. S_1 and S_2 are measured with polarizers in front of detectors. S_3 indicates the power difference between right-hand and left-hand circular polarization. S_3 is measured with a waveplate in front of a detector.

Polarization State.   The polarization state of a source is conveniently visualized using a Poincaré sphere representation. Figure 1.27 shows a diagram of the Poincaré sphere. The axes of the Poincaré sphere are the Stokes parameters normalized to the total power S_1, S_2, and S_3. The polarization state of a source can then be represented by the three-dimensional coordinates (S_1, S_2, S_3). The value of any of the coordinates is between zero and one. The outer surface of the sphere represents signals that are highly polarized, such as most lasers. The equator of the sphere contains all of the linear polarization states. The poles show circular polarization states. Points between the equator and the poles are elliptically polarized.

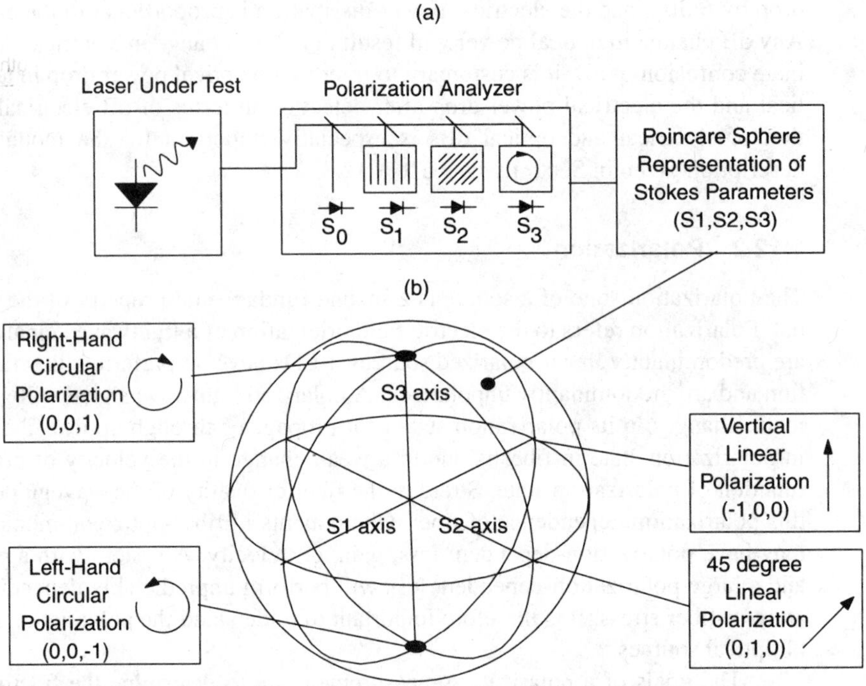

(a)

(b)

Figure 1.27 (a) Equipment used to measure polarization state. (b) Poincaré sphere representation of polarization state.

Degree of Polarization. The interior of the sphere is used to represent partially polarized light. The center of the sphere is for unpolarized lightwave signals such as those from a light bulb or the sun. The degree of polarization (DOP) is used to indicate the extent of polarization in a source. 100% DOP is found on the outer surface and 0% DOP is found in the center. Most lasers will have a DOP of greater than 95%. Most LEDs have very low degrees of polarization.

The Poincaré sphere provides a convenient tool for visualizing how the polarization of a source changes as it passes through different parts of a lightwave system. Optical fiber is slightly birefringent. Birefringent materials have a velocity that is polarization dependent. A laser that is coupled into an optical fiber may start out with linear polarization. After traveling through a length of optical fiber, the polarization state could be found anywhere on the surface of the Poincaré sphere. Since the polarization of an optical signal is constantly changing along the length of a long optical fiber, it is very important that all optical components have performance that is polarization independent. Parameters such as polarization-dependent loss (PDL) and polarization mode dispersion (PMD) are very important in today's lightwave systems. Chapter 6 is devoted to polarization measurements for fiber optic systems. Chapter 12 covers the PMD measurement area.

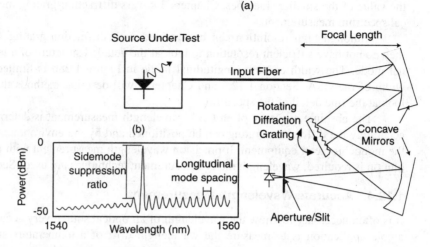

Figure 1.28   (a) Diagram of an OSA. (b) Measurement parameters of a DFB laser.

1.12.3 Optical Spectrum Analysis

The third fundamental property of a signal is its spectral content. An OSA is used to measure the power versus wavelength coming from an optical source. Figure 1.28a shows an OSA that uses a diffraction grating to accomplish wavelength filtering. An OSA consists of a tunable bandpass filter and an optical power meter. Diffraction grating filters are used for most general purpose fiber-optic input OSAs. The light from the input fiber is collimated (light rays made parallel) and applied to the diffraction grating. The diffraction grating has a series of finely spaced reflectors (for example, 1000 reflectors/mm). The diffraction grating separates the input light into different angles depending on wavelength. The light from the grating is then focused onto an output slit. The grating is rotated to select the wavelength that reaches the optical detector. The filter bandwidth for such an instrument is determined by the diameter of the optical beam that is incident on the diffraction grating and on the aperture sizes at the input and output of the optical system. The filter bandwidth of an OSA is limited to the 0.01 to 0.1 nm range due to physical size constraints for a portable instrument.

OSAs that use FP interference filters can also be used. FP filters offer the possibility of very narrow wavelength resolution. The disadvantage is that these filters have a passband that regularly repeats versus wavelength. Chapter 4 covers OSAs using these types of filters.

Figure 1.28b shows a spectral plot for a DFB laser that is being modulated with 2.5 Gb/s digital data. This plot shows that the laser emits in one predominant longitudinal mode. Other modes are suppressed by over 30 dB. The spacing between modes in this diagram is 1.4 nm. The OSA for this measurement must have a very narrow passband and steep skirts. A filter stopband that is at least 50 dB down is needed to accurately measure the value of the smaller sidelobes. Chapters 3 covers diffraction-grating methods of optical spectrum measurement.

Although high resolution can be obtained with diffraction-grating spectrometers, they do not have sufficient resolution to look at the detailed structure of a laser longitudinal mode. The width of each longitudinal mode in Figure 1.28b is limited by the filter width of the OSA. Section 1.12.5 and Chapter 5 will describe methods that are used to look at the fine detail of the spectrum.

The absolute accuracy of an OSA wavelength measurement is determined by how accurately the diffraction grating can be positioned and by the environmental stability of the monochromer equipment. If precision wavelength measurement with picometer resolution is required, wavelength meter measurement techniques are used (Section 1.12.4).

1.12.4 Accurate Wavelength Measurement

It is often necessary to know the wavelength of an optical source very accurately. An example application is to measure the wavelength drift of a temperature-stabilized DFB laser versus ambient temperature. Resolution of 0.001 nm would be desired in such an application. The OSA techniques discussed in Chapter 3 are not sufficiently accurate to measure the wavelength of a laser with picometer resolution. Figure 1.29a illustrates a method by which very accurate wavelength measurements can be made. The most common wavelength meter implementation uses a Michelson interferometer configuration. The light from the unknown source is split into two paths that are then recombined at a photodetector. One of the path lengths is variable and the other is fixed in length. As the variable arm is moved, the photodetector current varies due to constructive and destructive interference as is shown in Figure 1.29b. The period of this interference is one half of the wavelength of light in the medium of the interferometer. To accurately measure the wavelength of the unknown signal, a reference laser with a known wavelength is introduced into the interferometer. The wavelength meter compares the interference pattern of

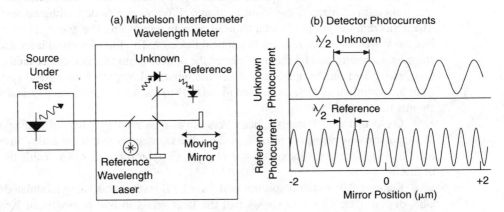

Figure 1.29 (a) Michelson interferometer wavelength measurement. (b) Photocurrent waveforms from the unknown and reference detectors as a function of mirror position.

the known laser to that of the unknown laser to determine the wavelength of the unknown signal. Since the unknown signal and the wavelength reference take the same path through the interferometer, the measurement method is less sensitive to environmental changes. Fourier transforms can be done on the interference photocurrent to determine the full spectral characteristics of the optical source. Wavelength meters have limited dynamic range compared to grating-based OSAs.

Helium-neon (HeNe) lasers emitting at 632.9907 nm are often used as wavelength references. HeNe lasers have a well-known wavelength that is relatively insensitive to temperature. Using HeNe references, wavelength accuracies of less then 1 part per million are possible. A 1550 nm laser can be measured to an accuracy of better than ± 0.0015 nm. Chapter 4 covers accurate wavelength measurement methods.

1.12.5   Linewidth and Chirp Measurement

Neither OSAs nor wavelength meters offer sufficient wavelength resolution to display the details of each longitudinal mode of a laser. The linewidth of an unmodulated DFB laser is most often less than 10 MHz. The width of the laser lines in Figure 1.28b is limited by the finite filter width of the OSA which in this plot is 12 GHz. Resolution improvements of 1000 times are required to study the fine spectral detail of laser sources. Heterodyne and homodyne analysis tools are used to examine the fine structure of optical signals. These analysis methods allow the measurement of modulated and unmodulated spectral shapes of the longitudinal modes in laser transmitters.

Heterodyne.   Figure 1.30a illustrates a heterodyne measurement set-up. In heterodyne measurements, the unknown signal is combined with a stable, narrow-linewidth local oscillator (LO) laser. The LO signal is adjusted to be within 50 GHz of the unknown signal to be detected by conventional electronic instrumentation. The LO must have the same polarization as the unknown signal for best conversion efficiency. The unknown signal and the LO mix in the photodetector to produce a difference frequency (IF signal) in the 0 to 50 GHz region. The intermediate frequency (IF) signal is analyzed with an electronic signal analyzer such as a spectrum analyzer.

Figure 1.30b shows the result of a heterodyne measurement of a laser under sinusoidal modulation at 500 MHz. The carrier and the two modulation sidebands are easily resolved with this technique. The major limitation of heterodyne techniques is the availability of very stable LO signals.

Homodyne.   Homodyne analysis techniques give more limited information on the optical spectrum, but are much easier to perform. Homodyne techniques are similar to heterodyne analysis except that the LO is a time-delayed version of itself. If an optical signal is delayed in time by more than the inverse of the source spectral width (measured in Hz), the signal becomes phase independent of the original signal, allowing it to be an effective LO. The signal and the delayed version of the signal are combined in a manner similar to that of Figure 1.30a. The intermediate frequency is centered around 0 Hz because both signals have the same center wavelength. Figure 1.30c shows an example of a homodyne measurement of an unmodulated DFB laser. The measured linewidth is 20

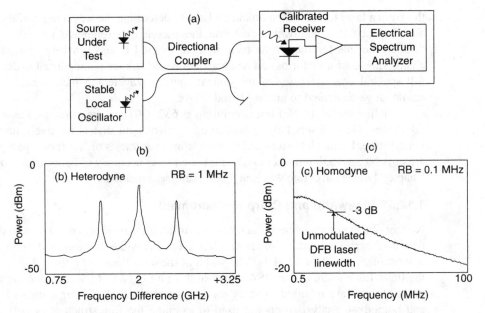

Figure 1.30 (a) Measurement configuration for Heterodyne spectrum analysis. (b) Intensity modulation sidebands for a DFB laser modulated at 1 GHz using Heterodyne measurement techniques. (c) Unmodulated linewidth measurement of a DFB laser using Homodyne techniques.

MHz. A limitation of this technique is that the asymmetries of the optical spectrum can not be seen. Homodyne analysis does not offer any information about the center wavelength of a laser.

Homodyne and Heterodyne techniques are also useful in characterizing laser chirp. Chapter 5 discusses homodyne and heterodyne spectrum analysis techniques.

1.12.6 Modulation Analysis: Frequency Domain

Most lightwave systems use intensity modulation to convey information on to the optical signals. Techniques are necessary to analyze parameters associated with this modulation. This section describes characterization methods that display information as a function of the modulation frequency.

Figure 1.31a shows a diagram of a lightwave signal analyzer. It consists of a photodetector followed by a preamplifier and an electrical spectrum analyzer to display the spectral content of the modulation on the lightwave signal. The modulation frequency response of the optical receiver, amplifier, and the electrical spectrum analyzer must be accurately calibrated as a unit. The display of Figure 1.31b shows the power of the modulation signal as a function of the modulation frequency. Do not confuse this information with the display of an OSA as shown in Figure 1.28b.

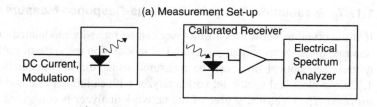

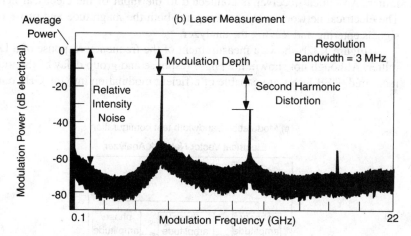

Figure 1.31   (a) Frequency domain analysis of the optical modulation. (b) DFB laser measurement example showing modulation depth, distortion, and intensity noise parameters.

This modulation domain signal analyzer allows the measurement of several characteristics of the modulation:

1. Depth of optical modulation.
2. Intensity noise.
3. Distortion.

Figure 1.31b shows a measurement example of a DFB laser modulated at 6 GHz. The display is presented in dB electrical units. A comparison of the average power and the 6 GHz modulation line indicates that this source has a small depth of modulation. The magnitude of the lines at 12 and 18 GHz compared to the 6 GHz modulation show the second and third harmonic distortion levels of the optical source. Distortion measurements are critical for lasers designed to work in analog cable-TV signal distribution applications. Lasers also add intensity noise to the optical signal. The relative intensity noise (RIN) of a source is characterized by ratioing the noise level at a particular modulation frequency to the average power of the signal. RIN measurements are normalized to a 1 Hz bandwidth. A DFB laser without modulation may have a RIN level of −145 dB/Hz. Chapter 7 covers frequency domain modulation measurements and calibration techniques in detail.

1.12.7 Modulation Analysis: Stimulus-Response Measurement

It is important to know how fast a laser can be intensity modulated or the upper modulation rate limits of an optical receiver. The modulation response of optical receivers, transmitters, and optical links can be measured using the instrumentation shown in Figure 1.32a. An electrical vector network analyzer with calibrated O/E and E/O converters is illustrated. The electrical source of the network analyzer is connected to the optical transmitter. An optical receiver is connected to the input of the electrical network analyzer. The electrical network analyzer compares both the magnitude and phase of the electrical signals entering and leaving the analyzer.

Figure 1.32b shows a measurement of the frequency response of a DFB laser transmitter. Although not shown, the phase response and group delay of the transmitter is also measured. DFB lasers are capable of efficient modulation into the GHz range. The modu-

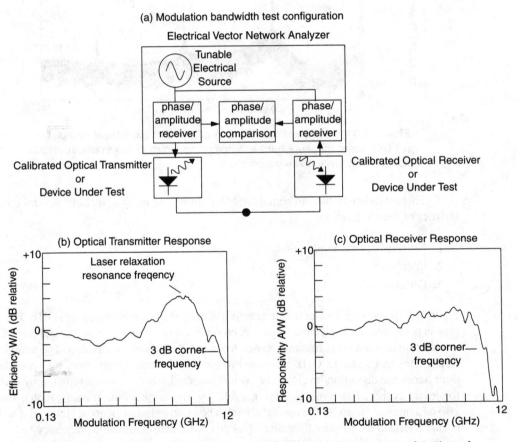

(a) Modulation bandwidth test configuration

Electrical Vector Network Analyzer

Tunable Electrical Source

phase/amplitude receiver — phase/amplitude comparison ← phase/amplitude receiver

Calibrated Optical Transmitter or Device Under Test

Calibrated Optical Receiver or Device Under Test

(b) Optical Transmitter Response

Efficiency W/A (dB relative)

Laser relaxation resonance freqency

3 dB corner frequency

+10

0

-10

0.13   Modulation Frequency (GHz)   12

(c) Optical Receiver Response

Responsivity A/W (dB relative)

3 dB corner frequency

+10

0

-10

0.13   Modulation Frequency (GHz)   12

Figure 1.32 (a) Equipment to measure modulation efficiency as a function of modulation frequency. (b) DFB laser modulation response. (c) Optical receiver modulation response.

lation response often has peaking near the upper modulation frequency limits of the laser. This peaking occurs at the laser relaxation oscillation frequency.[10] Figure 1.32c shows a measurement of the modulation response of an optical receiver.

Calibration of the O/E and E/O converters in both magnitude and phase response is one of the major challenges in making such a measurement. Calibration of the magnitude and phase response of the electrical vector network analyzer is well established. Calibration of the O/E response is much more difficult. Let us assume that the device under test is an optical receiver. In order to measure the optical receiver magnitude and phase response, it is necessary to know the magnitude and phase response of the optical transmitter. The calibration of an optical transmitter magnitude and phase response with high accuracy is a considerable challenge. Chapter 7 covers methods of stimulus-response calibrations and related measurement issues.

1.12.8   Modulation Analysis: Time Domain

The majority of fiber optic links use digital modulation techniques. The shape of the modulation waveform as it progresses through a link is of great interest. The optical power versus time can be characterized by detecting the lightwave signal and applying the electrical signal to an oscilloscope as is shown in Figure 1.33a. High speed sampling oscillo-

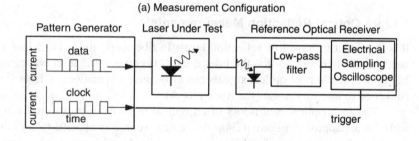

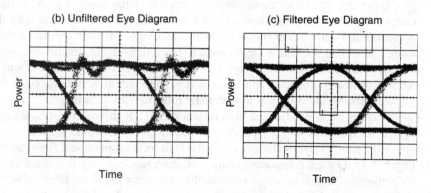

Figure 1.33   (a) Time-domain analysis of the optical modulation. (b) Eye diagram of digitally modulated DFB laser with high bandwidth optical receiver. (c) Eye diagram with SONET reference optical receiver.

scopes are often used due to the gigabit per second data rates involved in both telecommunication and data communication systems.

Figures 1.33b and 1.33c illustrate eye diagram measurements which are often used to characterize the quality of a lightwave signal. The clock waveform is applied to the trigger of the oscilloscope. The laser output is applied to the input of the oscilloscope through a calibrated optical receiver. The display shows all of the digital transitions overlaid in time. This eye-diagram display can be used to troubleshoot links that have poor bit-error ratio performance. Several laser characteristics can lead to poor error-rate performance. Optical sources have significant overshoot and ringing on the waveform transitions. Random- and pattern-dependent transition times cause jitter between the input electrical waveform and the output optical waveform. International standards such as SONET (Synchronous Optical NETwork), SDH (Synchronous Digital Hierarchy), and Fibre Channel specify limits on the amount of waveform distortion and time jitter that are acceptable. Figure 1.34c is an example of an eye-diagram measurement using a standardized receiver as specified by SONET and SDH. SONET/SDH reference receivers specify an optical receiver with a tightly controlled modulation response that is filtered at $3/4$ of the bit rate. A mask constructed around the eye diagram is shown in Figure 1.33c. The measurement system can count the number of samples that fall within the mask window region. Chapter 8 covers time-domain measurements of lightwave signals.

1.12.9 Optical Reflection Measurements

It is important to have low reflection levels in fiber optic links. DFB lasers must be isolated from reflections so that the laser doesn't become unstable in wavelength or power. Multiple reflections of optical signals can also cause fluctuations in laser frequency to be converted to intensity noise (see Appendix A).

The total optical return-loss of an optical receiver or transmitter can be measured with the apparatus of Figure 1.34a. An optical source is applied to a device under test through a directional coupler. Light is then reflected back toward the source from the device under test. The reflected signal is separated from the incident signal in the directional coupler. By comparing the forward and reverse signal levels, the total optical return-loss is measured. Figure 1.34b shows the return-loss versus wavelength for a packaged laser using a tunable laser source for excitation. The return-loss is frequency dependent in this optical device because of the many reflective surfaces found in the component. If a total return-loss level is found to be too large, it is important to be able to locate the reflecting surfaces. This requires optical time-domain reflectometry techniques.

Figure 1.35a illustrates a high resolution OTDR measurement based on broadband source interferometry. The spacing between reflections inside optical components requires different OTDR techniques than for fault location in optical fibers as was shown in Figure 1.16. Optical component characterization requires very fine distance resolution in the millimeter to micron range compared to the meter range for fiber reflectometry. This technique uses a Michelson interferometer and a broadband light source to locate reflections with 20 µm accuracy. When a broadband source is used to excite a Michelson interferometer, interference occurs only when the movable mirror to directional coupler dis-

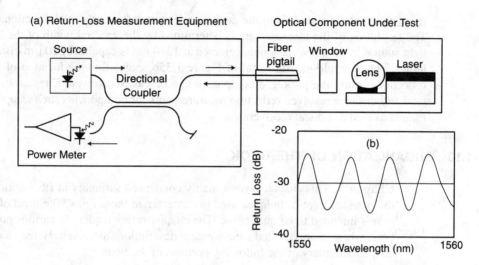

Figure 1.34   (a) Total return-loss measurement setup. (b) Example measurement of reflectivity as a function of wavelength for a packaged laser.

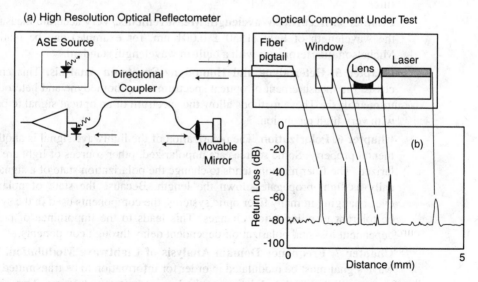

Figure 1.35   (a) High resolution optical reflectometer for measuring return-loss versus distance. (b) Example measurement for a packaged laser showing locations of various reflecting surfaces.

tance equals the distance from the device under test reflection to the directional coupler. The resolution of the measurement is determined by the spectral width of the broadband light source. A 50 nm wide source centered at 1310 nm is capable of 20 μm distance resolution. The example measurement of Figure 1.35b shows that the location of individual reflections inside the packaged component can be easily resolved. Chapter 10 describes total and distance resolved reflection measurements for components such as optical transmitters as well as optical receivers.

1.13 ORGANIZATION OF THE BOOK

Chapter 1. This chapter gives a highly condensed summary of fiber optic links and the measurement techniques used to characterize these links. The level of presentation is intended to be qualitative. The chapter refers readers to various points in the book where more detailed measurement descriptions are given. Here is a chapter by chapter summary of the following portions of the book.

Chapter 2. Power Measurement. Power is one of the fundamental properties of a lightwave signal. The photodetector is the most common device for measuring power. Several different types of detectors are available to cover the fiber optic wavelength bands.

Chapter 3. Optical Spectrum Analysis. The measurement of the wavelength of light is another fundamental lightwave property. This chapter concentrates on the most common measurement technique of using a diffraction grating as an optical filter.

Chapter 4. Accurate Wavelength Measurements. Very accurate measurement of the wavelength of light (1550 ± 0.001 nm, for example) can be made using a Michelson interferometer with a built-in wavelength standard.

Chapter 5. Heterodyne and Homodyne Spectrum Analysis. This chapter discusses the measurement of optical spectrum using homodyne and heterodyne spectrum analysis. These methods allow the spectrum of an optical signal to be resolved with very high resolution.

Chapter 6. Polarization. The polarization of the lightwave signal is another fundamental property. Some sources are unpolarized, other sources of light are 100% polarized. The fiber medium tends to change the polarization state of a signal dramatically as light propagates down the length. Because the state of polarization is ever-changing in many fiber optic systems, the components used in the system must be tolerant to polarization changes. This leads to the importance of polarization-dependent loss and polarization-dependent delay through components.

Chapter 7. Frequency Domain Analysis of Lightwave Modulation. The lightwave signal must be modulated in order for information to be transmitted. The most common lightwave-modulation method is intensity modulation. This chapter will analyze the modulation of the lightwave signal in the frequency domain. Noise and distortion are commonly measured parameters in the frequency domain. This chapter will include the modulation stimulus-response behavior of optical receivers and transmitters.

Chapter 8. Time Domain Analysis of Lightwave Modulation. The most common intensity modulation is on–off digital modulation. This type of modulation is often best studied in the time domain. Oscilloscopes with optical front ends are used to look at the eye diagram in order to set up parameters such as extinction ratio and eye-mask conformance. Jitter measurement in lightwave systems is also covered.

Chapter 9. Insertion Loss Measurements. The measurement of the loss of multi-port optical devices is covered. This measurement area has become very important due to the development of WDM systems. There are several methods to resolve loss versus wavelength. One can use a wavelength tunable laser and a power meter, a white light source in conjunction with an OSA or a tunable laser in conjunction with an OSA. Each of these measurement combinations has its strengths and weaknesses.

Chapter 10. Return Loss Measurements. The reflectivity of optical components is an important measurement. Systems are often intolerant of reflections. Total return-loss measurements are covered first. Instruments that allow high-resolution spatially resolved measurement of reflectivity are the primary focus of this chapter.

Chapter 11. Optical Time-Domain Reflectometry. This chapter covers one of the most important measurement areas in fiber optic communications. When a fiber is broken in a system, optical time-domain reflectometers are used to locate the break. Recent developments in the area now have OTDRs used as preventive maintenance tools for the installed fiber base.

Chapter 12. Chromatic Dispersion and Polarization-Mode Dispersion Measurements. Chromatic dispersion measurements of fiber and dispersion compensation devices is important for WDM and high-speed TDM systems. New fibers and components are being developed that allow dispersion management of new and upgraded fiber optic links. Polarization mode dispersion is becoming a primary effect that may limit the ultimate speed from TDM systems.

Chapter 13. Erbium-Doped Fiber Amplifier Measurements. Erbium-doped fiber amplifiers have changed the way new telecommunication systems are designed. Because of this special role, an entire chapter is devoted to EDFA description and characterization. The development of WDM systems has caused the measurements of EDFAs to be more complex.

Appendix A. Limits on Low-Level Optical Measurements: Noise. This section discusses the types of noise and their origins for optical and electronic systems.

Appendix B. Limits on High-level Optical Measurements: Nonlinearities. This section discusses the nonlinear behavior of optical fiber and how it limits the range of measurements.

Appendix C. Connector Care. One of the most common failure modes of fiber-optic test equipment is damage to optical connectors. This chapter will give examples of connector damage and discuss methods to prevent it.

APPENDIX: RELATIONSHIPS BETWEEN WAVELENGTH AND FREQUENCY

Authors in this book will be using both wavelength and frequency to describe optical measurement parameters. It is useful to be familiar with the relationships between these variables. This is especially useful when authors are quoting bandwidths in terms of small wavelength or small frequency deviations.

Wavelength and Frequency are related by the following relationship

$$v = f\lambda \tag{1.4}$$

where v is the velocity of light in the medium, f is the frequency of the signal, and λ is the wavelength in the medium. The velocity when propagating through a material is related to the index of refraction of the material by the following relationship

$$v = \frac{c}{n} \tag{1.5}$$

where c is the speed of light and n is the index of refraction of the material. The speed of light is approximately 299,792,458 m/sec. The wavelength of a signal varies with the index of refraction in the medium.

$$\lambda = \frac{\lambda_o}{n} \tag{1.6}$$

λ_o is referred to as the free-space wavelength, the wavelength that would be measured in a vacuum.

Oftentimes, it is important to look at differences in wavelength, $\Delta\lambda$ or differences in frequency between two signals, Δf. It is important to know how to be able to convert between these two variables:

$$\Delta\lambda = -\frac{v\Delta f}{f^2} = -\frac{\Delta f\lambda^2}{v} \tag{1.7}$$

$$\Delta f = -\frac{v\Delta\lambda}{\lambda^2} = -\frac{f^2\Delta\lambda}{v} \tag{1.8}$$

These equations are very useful since it is often necessary to convert back and forth between these parameters. At the communication band centered at 1300 nm, $\Delta\lambda = 0.1$ nm is equivalent to a Δf of 40 GHz. At the communication band centered at 1550 nm, $\Delta\lambda = 0.1$ nm is equivalent to a Δf of 12 GHz.

REFERENCES

General References on Fiber Optic Communications

1. Palais, Joseph C. *Fiber Optic Communication.* Englewood Cliffs, NJ: Prentice-Hall, 1992.
2. *Fiber Optic Telecommunications II,* edited by Stewart E. Miller and Ivan P. Kaminow, San Diego: Academic Press, 1988.
3. Hentschel, Christian. *Fiber Optics Handbook.* 3rd ed. Boblingen, Germany: Hewlett Packard, 1989.

References on Different Communication Standards

4. Synchronous Optical Network (SONET) Transport Systems: Common Generic Criteria, GR-253-CORE, Bellcore, Piscataway, NJ, Dec. 1994.
5. Digital Hierarchy—Optical Interface Rates and Formats Specifications (SONET), ANSI tl.105–1991, American National Standards Institute, New York, 1992.
6. Optical Interfaces for Equipments and Systems Relating to the Synchronous Digital Hierarchy, ITU G.957, International Telecommunications Union, Geneva, 1990.
7. Benner, Alan F. *Fibre Channel: Gigabit Communications and I/O for Computer Networks.* New York: McGraw-Hill, 1996.
8. Sachs, Martin and Anujan Varma. 1996. Fibre Channel and Related Standards. *IEEE Communications Magazine.* 34, 8: 40–50.
9. Ross, F.E. 1989. An Overview of FDDI: The Fiber Distributed Interface. *IEEE JSAC,* Vol. JSAC-7, 7: 1043–1051.

Wavelength Chirp References

10. Dutta, N.K. and G.P. Agrawal. 1986. *Long Wavelength Semiconductor Lasers.* New York: Van Nostrand Reinhold, 1986.

EDFA References

11. Desivire, E. 1994. *Erbium Doped Fiber Amplifiers: Principles and Applications.* New York: Wiley.
12. *Special Issue on Global Undersea Communication Networks. IEEE Communications Magazine.* 34, 2: Feb. 1996.

WDM References

13. Bellcore GR-2918-Core Document, Generic Criteria for SONET Point to Point Wavelength Division Multiplexed Systems in the 1550 nm Region, Bellcore, Piscataway, NJ, 1996.

Analog Transmission References

14. Darcie, T., et al. 1990. Lightwave Subcarrier CATV Transmission Systems. *IEEE Trans. on Microwave Theory and Techniques.* 38, 5: pp. 524–533.
15. Cox, C.H. 1991. Analog Fiber-Optic Links with Intrinsic Gain. *Microwave Journal,* (Sept.): 90–99.

Jitter References

16. Bures, Kenneth J. 1992. Understanding Timing Recovery and Jitter in Digital Transmission Systems—Part I, *RF Design* (Oct.): 42–53.

17. Trischitta, P. and E. Varma. 1989. *Jitter in Digital Transmission Systems,* Boston: Artech House.

Fiber References

18. Tosco, F., ed., 1990. Fiber Measurements. In *CSELT Fiber Optic Communication Handbook,* Chapter 4, pp. 273–416. "Fiber Measurements," Blue Ridge Summit, PA: McGraw-Hill TAB Books.

19. Marcuse, D. *Principles of Optical Fiber Measurements.* New York: Academic Press, 1981.

Fiber Optic Device References

FP and DFB, VCSEL Lasers

20. Dutta, N.K. and G.P. Agrawal. 1986. *Long Wavelength Semiconductor Lasers.* New York: Van Nostrand Reinhold.

21. Coldren, Larry A. and Scott Corzine. 1995. *Diode Lasers and Photonic Integrated Circuits.* New York: Wiley Interscience.

Mach-Zehnder Modulators

22. Jungerman, Roger, Catherine Johnson, David McQuate, Kari Salomaa, Mark Zurakowski, Robert Bray, Geraldine Conrad, and Donald Cropper. 1990. LiNbO$_s$ modulator for instrumentation applications, *Journal of Lightwave Technology.* 1990. 8, No. 9 (Sept.): 1363–1370.

LED References

23. Saleh, B.A. and M.C. Teich. 1991. *Fundamentals of Photonics.* New York: John Wiley.

Photodetector References

24. Wey, Y.G., K. Giboney, J.E. Bowers, M. Rodwell, P. Silvestre, P. Thiagarajan, and G. Robinson. 1995. 110 GHz InGaAs/InP Double Heterostructure p-i-n Photodetectors, *IEEE Journal of Lightwave Technology* 13, No. 7 (July): 1490–1499.

25. Stillman and Wolfe, *Semiconductors and Semimetals, Vol. 12, Infrared Dectors II,* edited by Williardson and Beer (ed.), Academic Press, 1977.

26. Sloan, Susan. 1994. Photodetectors. In *Photonic Devices and Systems,* ed. Robert G. Hunsperger, 171–246. New York: Marcel Dekker.

CHAPTER

2

Optical Power Measurement

Christian Hentschel

2.1 INTRODUCTION

Optical power measurement is the basis of fiber optic metrology. An optical power detector is found in nearly every lightwave test instrument. Two types of power measurements can be distinguished: absolute and relative power measurement. Relative power measurements are important for the measurement of attenuation, gain, and return loss. (See Chapter 9 for details.) Absolute power measurement is needed in conjunction with optical sources, detectors, and receivers. For example, the absolute power of an optical transmitter or optical amplifier is important for the power margin of a communication system and for eye-safety considerations. The sensitivity of an optical receiver is also specified in units of absolute optical power.

Optical power is generally defined on the basis of electrical power, because electrical power can be precisely measured through current and voltage. Therefore, all optical power measurements should be traceable to electrical power measurements. Most national standards laboratories, such as NIST (U.S.), PTB (Germany) and NPL (U.K.) host substantial activities on this subject.

Two main groups of optical power meters can be identified: power meters with thermal detectors, in which the temperature rise caused by optical radiation is measured, and photodetectors, in which the incident photons generate electron-hole pairs.

Although photodetector-type power meters suffer from a relatively small wavelength coverage and the need for absolute calibration, their astounding sensitivity usually makes them the preferred choice. Nevertheless, power meters with thermal detectors are sometimes preferred in the calibration laboratories because of their wide and flat wavelength characteristics. In addition, thermal detectors can be directly traceable to electrical

Table 2.1 Comparison of thermal power meters and photodetector power meters

Characteristics	Power meters with thermal detectors	Power meters with photodetectors
Wavelength dependence	+ wavelength-independent + wide wavelength range	− wavelength dependent, − wavelength range 2:1
Self-calibration	+ available	− not available (calibration indispensable)
Sensitivity	− very low (typically 10 μW)	+ very high (down to less than 1 pW)
Accuracy	±1% depending on calibration method	±2% depending on calibration method

power measurements. Altogether, there is good reason for the existence of both types of power meters. Table 2.1 compares the two types of power meters.

2.2 POWER METERS WITH THERMAL DETECTORS

Various principles of thermal detectors have been developed. A typical example is shown in Figure 2.1, from Bischoff.[1] This detector uses a method called substitution radiometry, which can be understood as a self-calibration method. In this method, the power meter is first exposed to the optical radiation. Then the radiation is switched off (with a shutter or chopper) and replaced by electrically generated power. This electrical power is controlled

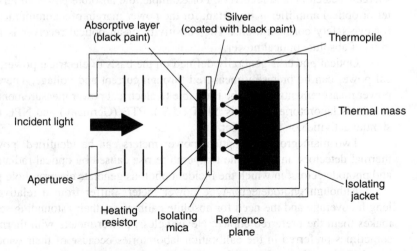

Figure 2.1 Thermal detector with electrical substitution.

so that a time-independent temperature is maintained. Electrical power can be measured very accurately, thereby providing the basis for the accuracy of this method.

In the detector of Figure 2.1, the incident light hits an absorptive layer, for example, one made from black paint. The substitution is accomplished with a heating resistor that is thermally well coupled to the absorptive layer. The back of the heating resistor carries an isolated sheet of silver for the purpose of equalizing any temperature differences. The silver is also coated with black paint. The temperature rise is measured with a thermopile (a series connection of thermocouples) which is brought into close proximity to the silver.

The thermopile produces a voltage proportional to the temperature difference between the absorptive layer and the reference plane, in this case the surface of a relatively large thermal mass. For this application, linearity is irrelevant because the only aim is to achieve equal temperatures for the two types of excitations. The following critical points need to be observed when the goal is highest accuracy:

1. Reference plane on large thermal mass: The thermal mass must be sufficiently large to maintain constant temperature during the relatively long measurement times. Long measurement times are typical because of the long time constants involved.

2. Blocking of background radiation and stray light: The thermal detector not only measures the power of the radiation source, but also changes in room temperature. This effect is reduced by a jacket with thermal isolation. In addition, a series of apertures makes sure that the detector is irradiated by the optical source only.

3. Optimization of heat flow: A negligible thermal resistance between the absorptive layer and the heater would be ideal. The thermal resistance to the jacket—due to convection and radiation—should be as high as possible.

4. High absorptance: Reflected light (both specular and diffuse) does not contribute to the detector's temperature rise and must be corrected for. Therefore, the absorptance should come close to 100%. A reflectance measurement is necessary in any case, as part of the initial calibration.

5. Accurate measurement of the electrical power: The electrical power is measured by current and voltage measurement, where voltage probes near the heating resistor are used to obtain accurate voltage results. A small contribution to the temperature rise is caused by the resistor leads. Bischoff[1] suggests eliminating this effect by an arrangement which includes additional contacts to the resistor and a constant current through the leads, independent of whether the heater is on or not.

More recently, another elegant way of operating this setup was published by the PTB, reference.[2] Instead of sequential exposure to optical radiation and electrical power, the sensor (in this case a thin-film sensor) is continuously heated by electrical power which is slightly larger than the optical power to be measured. The sensor voltage is recorded without the optical power applied. Then the sensor is exposed to the optical power, and a feedback loop reduces the electrical power until the sensor voltage is the same as before (on-line calibration). The desired optical power measurement result is simply the difference of electrical powers between the two steps, without a need of analyzing the sensor voltages.

The biggest problems in using power meters with thermal detectors is their low sensitivity and the correspondent long measurement times. Some improvement is possible on the basis of pyroelectric sensors or thermopiles using semiconductor material. Typical characteristics of thermal power meters are: sensitivity down to 1 μW, uncertainty as low as ±1%, spectral range from ultraviolet to far infrared, and time constant of several seconds to minutes depending on the detector size. These characteristics make thermal power meters well suited for calibration purposes. In other fields of fiber optic measurements they are rarely used, however.

A special form of a thermal power meter is the cryoradiometer.[3] This is a thermal detector that is placed into vacuum and cooled to approximately 6 K using liquid helium. Cryoradiometers are the most precise optical power meters due to the following phenomena:

1. At 6 K, the thermal mass (the energy needed to raise the temperature by 1 K) of the absorbing material is drastically reduced. This reduces the time constants and the measurement times accordingly.

2. Heat loss due to radiation is virtually eliminated because the radiated energy is proportional to T^4 (T in K).

3. Heat contributions from the resistor leads can be eliminated by making them superconducting.

4. Convection losses are eliminated by operating the detector in vacuum.

Based on these properties, cryoradiometers are claimed to have power measurement uncertainties as low as ±0.01%. Cryoradiometers are usually only found in national calibration laboratories because of the high cost of purchasing and operating this equipment.

2.3 POWER METERS WITH PHOTODETECTORS

A big advantage of photodetectors is that they can measure power levels down to less than 1 pW (−90 dBm). High modulation frequency response is another advantage. On the other hand, a relatively strong wavelength dependence is observed, and the spectral band is usually not more than one octave wide, see Section 2.3.2. In contrast to thermal detectors, there is no self-calibration for photodetectors. Some efforts have gone into self-calibration of photodetectors aimed at a quantum efficiency of one.[4] This technique was abandoned because of insurmountable difficulties. Nevertheless, this detector type is the most important today, because of its great sensitivity, fast measurement time, and ease of use.

Photodetector-type power meters are usually categorized into small-area power meters, to be used only when power from a fiber is to be measured, and large-area power meters for open beam *and* fiber applications.

Figure 2.2 shows a cross-section through a commercial large-area optical sensor head based on a photodetector. Important elements are the antireflective coating on the connector adapter, the pinhole and the angled position of the detector, all to avoid multiple reflections. Temperature stabilization using a thermoelectric cooler ensures stable measurement results. The photodetector is operated at zero-bias voltage in order to eliminate any offset currents.

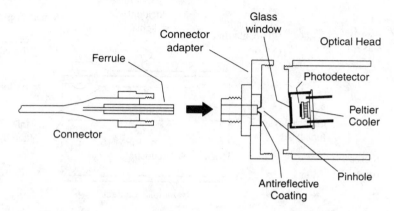

Figure 2.2   Cross-sectional view of a large-area power sensor.

Optical power measurement with photodetector-type power meters often seems as easy as voltage measurements. However, a number of critical points should be checked before a statement on the accuracy can be made. The most important contributions to accurate power measurements are:

Individual correction of wavelength dependence;

Temperature stabilization;

Wide power range with good linearity;

Good spatial homogeneity;

Low polarization dependence;

Low reflections;

Compatibility with different types of fiber.

The following sections discuss these points in more detail.

2.3.1   p-i-n-diode Operation

Figure 2.3 shows a cross-sectional view of a planar InGaAs p-i-n diode as a typical example of a photodetector.

Ideally, each incident photon is absorbed in the intrinsic (i-) layer, and an electron-hole pair is created as long as the photon energy is at least as large as the (material-dependent) bandgap energy. The holes and electrons are swept out of the i-region by the large built-in electric field. This creates the photocurrent. There are two terms which describe the conversion efficiency:

1. Quantum efficiency η, defined as the number of electrons per photon, and

2. Responsivity r, defined as the photocurrent per unit of optical power.

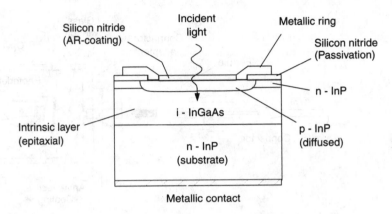

Figure 2.3   Cross-sectional view of an InGaAs photodetector.

In the ideal case of $\eta = 1$, the detector's spectral responsivity is proportional to the wavelength λ. This relation can be derived as follows: the responsivity is defined as the photocurrent, I, per unit of optical power, P:

$$r = \frac{I}{P} \tag{2.1}$$

Let us now calculate the current and optical power which correspond to each photon. Each photon represents the energy E_{ph}:

$$E_{ph} = h\nu = \frac{hc}{\lambda} \tag{2.2}$$

where h = Planck's constant, ν = optical frequency, and c = speed of light in vacuum. Optical power is defined as energy per time span Δt. In our case, the optical power which corresponds to one photon is:

$$P_{ph} = \frac{E_{ph}}{\Delta t} = \frac{hc}{\lambda \Delta t} \tag{2.3}$$

The quantum efficiency η was assumed to be 1. Then the correspondent electrical current is one electron charge q per time span Δt:

$$I_{ph} = \frac{q}{\Delta t} \tag{2.4}$$

Inserting the power and current into Equation 2.1 yields the linear spectral responsivity of an ideal photodetector with $\eta = 1$:

$$r_{(\eta=1)} = \frac{e\lambda}{hc} \tag{2.5}$$

Practical photodetectors deviate from this ideal wavelength dependence in several ways:

1. There is a long wavelength limit (cutoff wavelength) beyond which the photon energy becomes lower than the bandgap energy of the semiconductor material used. The responsivity falls off rapidly after this wavelength. The detector material determines the long wavelength limit, as shown in Figure 2.4.

2. At short wavelengths, some of the photons are absorbed outside of the i-region of the photodetector, and the number of electron-hole pairs is reduced.

3. The responsivity may also be reduced by recombination: when the electrons recombine with the holes before they reach the electrodes, then the photocurrent is reduced: another contribution to $\eta < 1$.

4. Any reflections from the detector surface or from inside the detector reduce the responsivity as well. This is a critical factor because reflections can produce substantial inaccuracies in optical power and insertion loss measurements. Pure InGaAs, for example, has a refractive index of 3.5, leading to a reflectivity of 31%. Antireflective coatings are typically used to reduce the effect. Single-layer, quarter-wavelength coatings are most often used. They reduce the reflectivity down to around 1% within a limited wavelength range. Sometimes, a periodic structure of the responsivity is observed. This indicates optical interference in the diode (see Figure 2.9). Multilayer coatings are applied when low reflectivity over a wider wavelength range is needed.

2.3.2 Spectral Responsivity

Figure 2.4 shows typical responsivity measurement results for three types of photodetectors. All three curves are normalized to a maximum of 1.

Silicon is the appropriate detector for the short-wavelength range between 500 and 1000 nm. For the long wavelength region, both germanium and InGaAs detectors can be

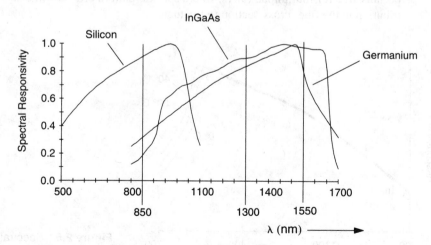

Figure 2.4   Spectral responsivities of three different detector types (all curves normalized to 1).

used. Germanium is presently the lower cost solution. It can be recommended when the sources to be measured are spectrally narrow and the wavelength is well known. This is particularly important around 1550 nm: in this region, a germanium detector produces a 1% error when the power meter's wavelength setting is incorrect by 1 nm.

In contrast, InGaAs detectors are essentially flat around 1550 nm (better than 0.1%/nm wavelength error). This makes them well-suited for optical amplifier (EDFA) applications, as the flat portion of the curve corresponds well to the usable gain region of EDFAs (1525 to 1570 nm). However, InGaAs is a more expensive technology.

2.3.3 Temperature Stabilization

Temperature-stabilized detectors can be expected to generate reproducible measurement results. Figure 2.5 shows that the responsivity of a germanium detector exhibits a relatively small temperature dependence for most of the wavelength range. In contrast, there is a substantial change beyond the cutoff wavelength, for example, at 1550 nm. This change can be most easily described as a shift of cutoff wavelength, in this case approximately 1 nm/K. Nearly the same wavelength shift, 1 nm/K, can be observed in InGaAs detectors at wavelengths around 1650 nm.

2.3.4 Spatial Homogeneity

The responsivity of photodetectors can vary across the detector surface. Figure 2.6 shows an example of the relative responsivity of an InGaAs photodetector of 5 mm diameter at 1550 nm. Wide variations in the homogeneity of commercial detectors—from perfect to marginal (as in Figure 2.6)—are usually observed.

Inhomogeneous photodetector surfaces create measurement uncertainties because the position and diameter of the incident beam cannot be perfectly controlled. This is especially true for multimode fibers, where speckle-pattern effects will cause the power distribution in the fiber cross-section to fluctuate.

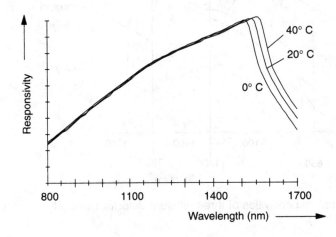

Figure 2.5 Spectral responsivity of a germanium detector at different temperatures.

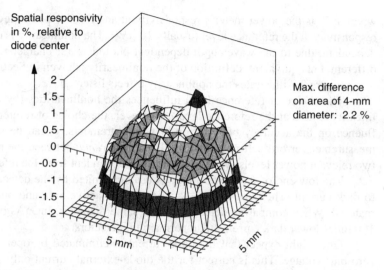

Figure 2.6 Spatial homogeneity of an InGaAs detector, measured at 1550 nm.

2.3.5 Power Range and Nonlinearity

Ideally, optical power meters display correct measurement results over many decades of optical power. However, this capability is often impaired by the meter's nonlinearity. Measurement linearity is important for accurate insertion loss (or gain) measurements. In an optical power meter, nonlinearity contributors can be classified into:

1. The photodetector nonlinearity: Photodetectors are usually thought to be very linear over six or more decades of optical power. However, there are three effects which may cause limitations:
   a. noise at low power levels;
   b. supralinearity at medium power levels;
   c. saturation at high power levels.
2. The electronic nonlinearity which can be split into:
   a. the in-range nonlinearity of the analog amplifier that follows the photodetector, for example offset at low power levels and amplifier saturation at high power levels;
   b. the ranging discontinuity observed by switching from one power range to another, caused by nonmatching amplifier gains.

The nonlinearity is defined as:

$$N(P) = \frac{r(P) - r(P_0)}{r(P_0)} \tag{2.6}$$

where $r(P)$ is the power meter's responsivity at an arbitrary power level, and $r(P_0)$ is the responsivity at the reference level (usually 10 μW). The nonlinearity is usually wavelength-dependent, due to the wavelength-dependent photodetector characteristics. Notice that a different, but equivalent, definition of the nonlinearity is given in Section 2.6.

Figure 2.7 illustrates the nonlinearity effects listed above.

The choice of reference level influences the nonlinearities (by definition, the non-linearity is zero at the reference level). However, the choice of reference level has no influence on the accuracy of insertion-loss measurements. It can be shown that the loss measurement error due to nonlinearity is the *difference* between the nonlinearities at the two relevant power levels. This difference is independent from the reference level.

The low end of the power range is usually limited by the detector's shot noise due to dark current. The dark current depends on the active area and on the semiconductor material. When comparing detectors of the same diameter, InGaAs detectors have up to 100 times lower dark currents than germanium detectors.

One might expect that the dark current is eliminated by operating the detector at zero-bias voltage. This is correct for the diode-external current only. Internally, the dark current is compensated by an equally large diffusion current, where both currents produce shot noise. Accordingly, the shot noise current at zero input power is given by:[5]

$$<i_n^2> \; = 2\,q\,B_n \Sigma I = 2qB_n \times 2I_{\text{dark}} \;\; [A^2] \tag{2.7}$$

where q = electron charge, ΣI = total current, I_{dark} = dark current, and B_n = noise equivalent bandwidth. Time averaging is indicated by brackets $< >$. Note that the shot noise is frequency-independent.

Under light exposure, the photocurrent adds to the total current and produces additional shot noise:

$$<i_n^2> \; = 2\,e\,B_n \Sigma I = 2eB_n(2I_{\text{dark}} + rP_{\text{opt}}) \tag{2.8}$$

where r = responsivity and P_{opt} = received optical power.

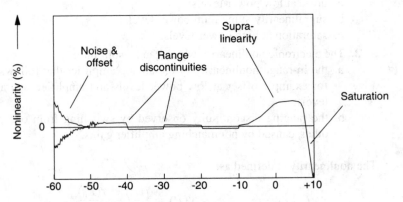

Figure 2.7   Possible nonlinearity effects of an optical power meter.

To relate this to power meter noise, the shot noise can be expressed as noise-equivalent power NEP (this is an RMS value):

$$\text{NEP} = \frac{1}{r}\sqrt{[i_n^2]} = \frac{1}{r}\sqrt{2eB_n(2I_{\text{dark}} + rP_{\text{opt}})}\left[\frac{W}{\sqrt{Hz}}\right] \qquad (2.9)$$

A signal-to-noise ratio, *SNR,* can be calculated using:

$$\text{SNR} = \frac{P_{\text{opt}}}{\text{NEP}} \qquad (2.10)$$

Figure 2.8 shows an example of the power dependence of the SNR. It was calculated using a dark current of 1.5 μA (a typical value for a germanium detector of 5 mm diameter at 25C) and a noise equivalent bandwidth of 100 Hz. At low power levels, the SNR increases by 10 dB for every 10 dB increase in optical power. At higher power levels, where the photocurrent becomes larger than the dark current, the SNR increases by only 5 dB for every 10 dB power increase. This is because the noise level increases with the optical power.

The lowest SNRs exist at low power levels. An improvement is possible by reducing the dark current: either by cooling or by reducing the detector's active area (the dark current is proportional to the active area). A longer averaging time also reduces the noise problem. Variable averaging time should be implemented in a modern power meter.

Range discontinuities are caused by the necessity to switch the gain of the electronic amplifier, depending on the input power level. Otherwise, a power range of more than six decades could not be realized. A range discontinuity means that the power meter does not display exactly the same power level when switching between power ranges. These effects should be substantially lower than 1% in a good power meter.

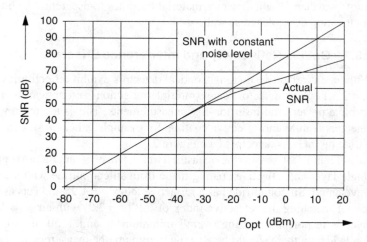

Figure 2.8   Power dependence of the signal-to-noise ratio (example).

A less-known nonlinearity effect is *supralinearity,* an increase in responsivity typically starting at power levels of around 100 μW. To our knowledge, this effect has not been carefully studied. A possible explanation is "traps" in the semiconductor material causing increased recombination at low power levels. When the power reaches higher levels, then these traps become saturated, the recombination decreases, and the responsivity increases. The correspondent nonlinearity can reach several percent. The strongest supralinearity effects are usually observed at the borders of the usable wavelength range of the specific detector. See Stock[6] for more details.

On the high power end, the responsivity drops due to *saturation,* caused by a reduction of the electric field across the pn-junction along with recombination in the active region. This effect starts at input powers of typically a few milliwatts. For many years, such power levels corresponded well to the output power levels of commercial laser diodes. With the advent of optical amplifiers, this situation has changed dramatically: Today's pump lasers produce optical powers in the 100 mW region, and optical amplifiers capable of more than 1 W output power are commercially available. Neutral-density filtering or power splitting can be used to shift the power scale to higher levels. Section 2.4.2 covers the area of high-power measurement.

2.3.6 Polarization Dependence

Crystalline structures in the semiconductor material and in the photodetector's coating or mechanical stress in the detector are the usual causes for polarization dependence in optical detectors. Also, detectors are often tilted against the beam axis to reduce multiple reflections. Tilting always causes additional polarization dependence. The polarization dependencies of modern large-area photodetectors range from a few 0.001 dB peak-to-peak for selected straight detectors to 0.05 dB peak-to-peak in unselected angled detectors. A relatively strong wavelength dependence of the polarization characteristics can also be observed and is usually caused by the quality of the antireflective coating. Beyond the cutoff wavelength, the detector material becomes transparent which tends to increase the polarization dependence as well.

2.3.7 Optical Reflectivity and Interference Effects

Without antireflective coatings, optical detectors exhibit reflectivities up to 30% (see section 2.3.1). Such detectors cause multiple reflection- and optical-interference problems in absolute power- and insertion-loss measurements. This is the reason why all commercial detectors feature antireflective coating, for example, silica on silicon detectors and silicon nitride on InGaAs detectors (see Figure 2.3).

Figure 2.9 shows the measured reflectance of an InGaAs photodetector with a single-layer antireflective coating made from silicon nitride with a thickness of a quarter wavelength. Silicon nitride has a refractive index $n = 1.95$ and acts as an impedance transformer matching the refractive index of InP ($n = 3.2$) with air ($n = 1$). The quarter-wave layer is responsible for the overall minimum around 1250 nm for this specific diode. While ideally there should be a gentle minimum, the measurement shows some additional ripple. This is caused by the upper InP-layer (see Figure 2.3) which forms an additional

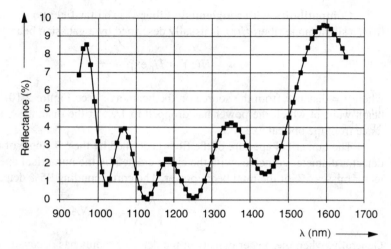

Figure 2.9 Typical reflection pattern of an InGaAs detector with AR coating.

resonator due to the fact that InP has a refractive index of 3.2, in contrast to the refractive index of 3.52 for the intrinsic InGaAs layer.

The result is a wavelength-dependent reflectance between 1% and 10%. Another typical characteristic is that the reflectance varies substantially from detector to detector: It is obvious that a slight thickness change of the InP layer shifts the pattern to a different wavelength.

Reflections and optical interference can also be created by the glass cap which usually covers the detector can. The glass can itself act as a resonator if it is sufficiently flat. Therefore, glass caps with perfect optical quality are to be avoided.

Figure 2.2 demonstrates that reflections from the detector surface *or* the glass cap go back to the connector adapter and the connector. From there, doubly reflected light can strike the detector again. These effects were first mentioned in Gallawa and Li.[7] If the detector is sufficiently large, then the unwanted power fraction on the detector is the product of the photodetector reflectance and the reflectance of the optical interface. To reduce these problems, the adapter of Figure 2.2 is coated with an antireflective coating on the inside, and a pinhole shields the highly reflective connector end.

2.3.8 Compatibility with Different Fibers

A wide range of optical fibers are used in fiber optic communication. There is usually no problem measuring the power from a standard singlemode fiber with its small numerical aperture of 0.1. In contrast, it can be very difficult to measure the output power from a thin-core singlemode fiber with a high numerical aperture of 0.4 (used in optical amplifiers). In some cases, the fiber end may even be angled to reduce optical reflection problems. Then the effective numerical aperture increases even further.

Compatibility with Singlemode Fibers. The far-field power density (irradiance) from a singlemode fiber, $H(z)$, is usually described by a gaussian beam:

$$H(z) = H_0 \exp - \frac{2r^2}{w(z)^2} \qquad (2.11)$$

where z = distance from the source on the beam axis (see Figure 2.10), w = radius of the beam waist at which the power has dropped to $1/e^2$, at the distance z, and r = radial distance from the optical axis.

The numerical aperture of the fiber is defined by the 5% angle of the far field. If the detector diameter coincides with the circle created by the numerical aperture, then the detector misses 5% of the total beam power. The corresponding 95% detector radius is:

$$r_{\text{det}} = z \frac{NA}{\sqrt{1 - NA^2}} \cong zNA \qquad (2.12)$$

Generally, when the power density at the detector radius has decayed to $x\%$, then there is $x\%$ of the total power outside the detector. This is a property of the gaussian beam. The coupling efficiency is given by:

$$\eta = 1 - \exp\left[-\frac{2r_{\text{det}}^2}{w^2} \right] \qquad (2.13)$$

It is advisable to replace w, the $1/e^2$ beam radius, by the 5% beam radius which corresponds to the fiber's numerical aperture. The gaussian beam profile yields:

$$w = 0.817 r_{5\%} = 0.817\, zNA \qquad (2.14)$$

Then the coupling efficiency can be expressed on the basis of the numerical aperture:

$$\eta = 1 - \exp\left[-\left(\frac{1.71 r_{\text{det}}}{zNA}\right)^2 \right] \qquad (2.15)$$

Example: If the detector radius is 2.5 mm, the distance between the fiber end and the detector is 8 mm (as in the power meter of Figure 2.2), and the numerical aperture of the singlemode fiber is 0.3, then the coupling efficiency is 96%, indicating the aperture limitation of this power meter configuration.

Compatibility with Angled Fiber Ends. Special care must be taken to capture light from angled fiber ends. Angled fiber ends are aimed at reducing reflections. Figure 2.10 shows a singlemode fiber, both with straight and angled fiber end. Let us assume that, in the case of the straight fiber end, the detector captures the beam fully, and that in the angled case the detector misses a part of the beam.

In both cases, the numerical aperture of the fiber is defined by the 5%-angle γ of the far field. In the angled case, the tilt of the beam axis, β, can be calculated using Shell's law:

$$\beta = \arcsin(n \sin \alpha) \cong (n - 1)\alpha \qquad (2.16)$$

The effective numerical aperture for the angled case is:

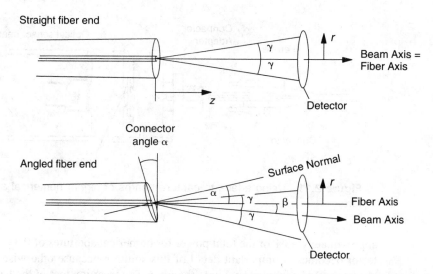

Figure 2.10   Power measurements from straight and angled fiber end.

$$NA_{eff} = \sin(\gamma + \beta) \tag{2.17}$$

in comparison to simply $NA = \sin\gamma$ for the straight case. A shorter distance to the detector would be needed to capture the beam fully. Other possibilities are:

1. Tilting the fiber, so that the beam axis is realigned to hit the center of the detector; or
2. Using a lens to reduce the effective beam diameter (see Figure 2.11).

Compatibility with Fibers of High NA.   Some of the fibers used in conjunction with optical amplifiers are thin-core, high-numerical-aperture fibers. Numerical apertures up to 0.4 and even 0.5 are not unusual. Such numerical apertures can also occur with multimode fibers. Ideally, a power meter should present the same responsivity to all parts of the beam. In situations with high numerical aperture, this requirement is hard to meet. For a given photodetector diameter, often 5 mm, there are three possibilities:

1. Decreasing the distance between fiber end and photodetector. This solution may cause reflection problems because the fiber tends to capture more of the power that is reflected from the detector surface. Another problem can be caused by the fact that the photodetector's responsivity is lower for those parts of the beam that hit the detector at larger angles.
2. Using a lens with high numerical aperture in order to collimate the beam. This solution can also create problems because light emitted at larger angles will be more strongly reflected off the lens than the on-axis beams. Figure 2.11 shows the power meter of Figure 2.2 with a lens inserted into the beam path; this assembly captures

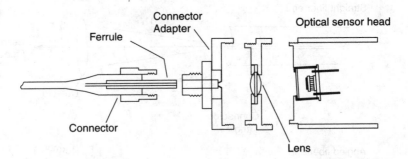

Figure 2.11 Using a lens to capture beams of higher numerical aperture.

approximately 98% of the total power for numerical apertures of 0.3. Tilting the detector becomes an important detail of this solution because otherwise the reflected power would be imaged back into the fiber. The insertion loss of the lens can be calibrated out as discussed below.

3. Using an integrating sphere in combination with the photodetector. Ideally, an integrating sphere should perfectly scatter all incident light so that the power measurement result becomes independent on the fiber's numerical aperture. To accomplish this, the detector should not be exposed to either direct beams from the source or to beams after only one reflection (see Figure 2.12). Even under those conditions, measurements have shown that commercial integrating spheres also have some angle-dependent responsivities depending on their construction. Particularly beams forming large angles (high numerical aperture) against the connector axis go through different attenuations than the near-axis beams.

In addition, some of the materials used to scatter the beam inside the integrating sphere tend to absorb moisture, so that the scattering characteristics change with the relative humidity. Integrating spheres are capable of high-power measurement, because the optical

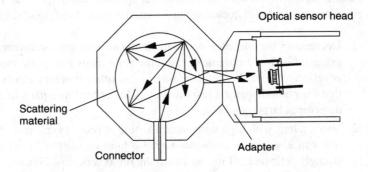

Figure 2.12 Optical power measurement with integrating sphere.

power is attenuated by 30 dB or more before it reaches the detector, and because the absorption takes place over a relatively large area.

Compatibility with Multimode Fibers.   Multimode fibers, particularly graded-index multimode fibers, generate irregular far-field patterns (speckle patterns) which are caused by optical interference between the different fiber modes. This is only a problem when the source is a laser diode; the spectral width of an LED is too large to create optical interference. Speckle patterns go through rapid changes when the fiber is moved, because changing the path lengths of the individual modes by only fractions of the wavelength creates a different speckle pattern.

Along with these rapid changes, speckle patterns usually create additional uncertainties because the photocurrent is a convolution of the speckle pattern with the detector's spatial homogeneity (see Section 2.3.4).

The numerical aperture of multimode fibers ranges from 0.2 to 0.5 depending on the core diameter and refractive-index profile. Therefore, the problems of multimode fibers are essentially the same as for thin-core singlemode fibers.

2.4 ABSOLUTE POWER MEASUREMENT

Absolute optical power, in mW or dBm (decibels relative to 1 mW) is a key parameter for all optical sources. Figure 2.13 shows an example of an absolute power measurement in which the power from a pigtailed laser diode is measured. This is a typical production test, to ensure the appropriate system margin or that the laser meets the specified performance.

While this measurement seems simple, there are a number of questions to be answered before the measurement result can be claimed to be accurate. It is obvious that most of these questions are related to the performance of the power meter used, and that there may be substantial differences in measurement results when different types of power meters are used. See the discussion on the uncertainty of absolute power measurement in Section 2.4.3.

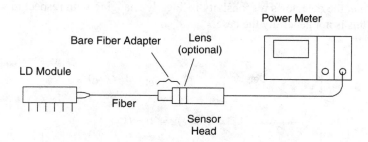

Figure 2.13   Power measurement of a pigtailed laser diode.

Before the discussion of the uncertainties, two specific examples of absolute power measurement are analyzed in more detail below: LED measurement and high-power measurement.

2.4.1 LED-Power Measurement

In contrast to the power from narrow-linewidth laser diodes, LED power is sometimes difficult to measure because of the LED's wide spectral width and the fact that the photodetector's responsivity changes within this spectral range. However, a correction is possible when the detector's spectral responsivity and the LED's spectral power density are known. Figure 2.14 depicts the situation for the example of a 1550 nm LED and a germanium detector.

The symbols in Figure 2.14 are as follows:

$\lambda_0 =$ arbitrarily chosen wavelength (preferably the LED peak wavelength) for which the power meter is corrected;
$r_{rel}K(\lambda) =$ responsivity relative to λ_0, where $r_{rel}(\lambda_0) = 1$;
$p_0 =$ spectral power density of the LED at the wavelength λ_0, in watts/nm;
$f(\lambda) =$ factor describing the LED's spectral emission, where $f(\lambda_0) = 1$.
On this basis, the correct LED power is:

$$P = p_0 \int f(\lambda)\delta\lambda \qquad (2.18)$$

In contrast, the uncorrected measurement result is:

$$P_m = p_0 \int f(\lambda)r_{rel}(\lambda)d\lambda \qquad (2.19)$$

Accordingly, a correction factor can be calculated to be:

$$K = \frac{P}{P_m} = \frac{\int f(\lambda)d\lambda}{\int f(\lambda)r_{rel}(\lambda)d\lambda} \qquad (2.20)$$

Analyzing this equation shows that there is no error if the LED spectrum is symmetrical *and* the detector's responsivity is linearly changing with respect to wavelength. However, this is not generally the case.

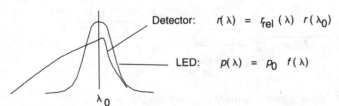

Detector: $r(\lambda) = r_{rel}(\lambda) \, r(\lambda_0)$

LED: $p(\lambda) = p_0 \, f(\lambda)$

λ_0

Figure 2.14 Modeling an LED and a photodetector.

The following measurement procedure is suggested:

1. Determine the LED's center wavelength, for example, from its data sheet.
2. Set the power meter to the LED's wavelength λ_0 and measure the LED power.
3. If the LED spectrum is essentially symmetrical and the photodetector's responsivity is nearly linear within the LED's spectral band, use the measured power as the result.

   If one of these conditions is not met, then calculate the correction factor as in Equation 2.20 and multiply the measured power with the correction factor to obtain the correct power.

2.4.2 High-Power Measurement

Optical power meters based on photodetectors can measure maximum power levels of a few milliwatts. Beyond this power level, the photodetector goes into saturation. For many years, such power levels were sufficient, because they corresponded well to the output power levels of commercial laser diodes.

The advent of optical amplifiers changed this situation. The complication starts with the pump lasers, which produce 100 mW or more. The amplifiers' output powers must be measured as well. Except for preamplifiers, designed to generate a few milliwatts at the most, all optical amplifiers generate power levels exceeding the measurement range of conventional power meters. Today, the highest powers from EDFAs exceed 1 watt. Since the output power is a key parameter, the question is: How can such large power levels be measured with good accuracy?

Figure 2.15 shows a commercial high-power optical head with a 5 mm InGaAs detector and a window made from absorbing glass, to reduce the incident optical power to a suitable level. It can measure up to 500 mW. In order to prevent local overheating of the absorber at power levels exceeding 100 mW, it is recommended to create a spot diameter of not less than 3 mm on the detector (measured at the 5% points). At the given distance of 8 mm between the fiber end and the detector, a 3 mm spot is obtained when the numer-

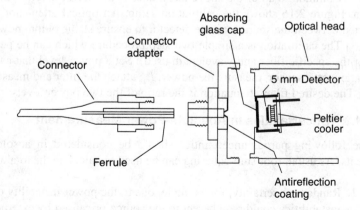

Figure 2.15 High-power measurement with absorber in front of the detector.

ical aperture of the fiber is 0.2. In this case, the coupling ratio is 100%. A beam with a numerical aperture of 0.3 will create a spot diameter of 4.5 mm and a coupling ratio of 96%. A standard singlemode fiber with a numerical aperture of 0.1 yields a spot diameter of only 1.5 mm; in this case a simple spacer between the adapter and the optical head can be used to create a spot diameter of 3 mm.

Alternative solutions for high-power measurement are:

1. Inserting a mesh-type filter consisting of thin wires between the fiber end and the detector. This solution has wide wavelength range and high-power capability, because increased wire temperature will not influence the attenuation.

2. Inserting a scattering filter, for example, a ceramic disc between the fiber end and the detector. This technique also has wide wavelength range and high-power capability. In addition, scattering introduces depolarization, thereby reducing the polarization dependence of the optical head. A polarization dependence of 0.003 dB p-p can be achieved in conjunction with an FP laser diode, and 0.015 dB p-p for a single-line laser, for example, a DFB laser or an external-cavity laser (ECL). A disadvantage is the fact that different beam geometries (fiber types) will cause different attenuations, because this technique splits power away from the detector.

3. Splitting some power away before the measurement, for example, with the help of a fiber coupler. This solution is limited to certain fiber types because the coupler fibers must be of the same type as the fiber to be measured.

4. Inserting an integrating sphere between the fiber end and the detector; see the discussion in Section 2.3.8. This is usually an expensive solution, where some angle dependence and dependence on relative humidity may have to be taken into account.

Several of these techniques can be combined, too. Notice that a collimating lens may have to be inserted before these filters to ensure that beam diameter remains smaller than the detector diameter.

Common to all of the above techniques is the need for calibrating the filter attenuation. Figure 2.16 shows the calibration setup. An optical attenuator may have to be inserted between the source and the detector to ensure stable output power.

The calibration is a simple two-step procedure which can be performed by the user (for the specific fiber and wavelength used): Set a power level that can be handled by the unattenuated sensor, measure the power, P_1, attach the filter and measure the power again, P_2. The desired filter attenuation is the ratio of the two power levels.

2.4.3 Uncertainties in Absolute-Power Measurement

The following partial uncertainties should be considered in absolute-power measurements. As usual, root-sum-squaring can be used to calculate the total uncertainty.

1. Random uncertainty, for example, due to the power instability of the source: Power instabilities could be inherent to the source or caused by external reflections travel-

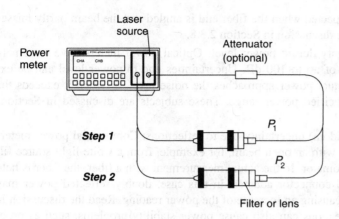

Figure 2.16 Calibrating filtered- or lensed-detectors.

ing back to the source. Most laser sources are sensitive to reflections. This uncertainty depends strongly on the specific situation. It can range from less than 0.1% to several percent.

2. Systematic uncertainty due to power-meter calibration: It is assumed that the power meter is regularly recalibrated following the manufacturer's recommendations and that the wavelength correction is set to the wavelength of the source. The absolute uncertainty and the conditions for which this uncertainty applies should be obtained from the power meter's data sheet (for example, power range, numerical aperture of the fiber, connectors, and wavelength). Absolute uncertainties are ±2% in the best case.

If the actual measurement conditions coincide with the specified conditions, then the uncertainty analysis ends here. If not, then consider the following:

3. Systematic uncertainty due to wavelength: The wavelength of the source (center wavelength) should be accurately known. Otherwise, the partial uncertainty will be the wavelength uncertainty multiplied by the power meter's responsivity versus wavelength slope (%/nm) at that wavelength.

4. Systematic uncertainty due to the spectral width of the source: In the measurement of laser diodes, this uncertainty will usually be negligible. In LED measurement, there will be no error if the spectrum is symmetrical about the center wavelength and the power meter's responsivity is linear within the wavelength range of interest. Otherwise a correction factor or an uncertainty can be calculated on the basis of Section 2.4.1.

5. Systematic uncertainty due to beam geometry: In the best case, the beam is centered on the detector and the beam diameter is about $2/3$ of the detector diameter. If this is not the case, then an appropriate uncertainty may have to be calculated. Particularly,

problems can be expected when the fiber end is angled and the beam partly misses the detector. See the discussion in Section 2.3.8.

6. Systematic uncertainty due to power level: Optical power meters have extremely wide power ranges of up to 100 dB. Uncertainties due to power level can be expected when the actual power approaches the noise level, or when it exceeds the high end of the specified power range. These subjects are discussed in Sections 2.3.5 and 2.4.2.

7. Systematic (and random) uncertainty due to reflections: Commercial power meters are often calibrated with an open beam, for example, from a white light source filtered by a monochromator. In the actual measurement with a fiber, the fiber is held by a connector and connector adapter. In this case, doubly reflected power may strike the detector, causing an increase of the power reading. Read the discussion in Section 2.3.7. Reflections can also cause power stability problems, such as problems described in (1) of this list.

2.5 RESPONSIVITY CALIBRATION

The most important criterion in conjunction with accurate measurement of absolute power is calibration. Generally, all power meters are calibrated through comparison: A test meter and a power measurement standard are exposed to a suitable radiation source, either sequentially or in parallel. If a calibration in fine-wavelength steps over a wide wavelength range is desired, then the source should be a halogen white-light source which is spectrally filtered with a monochromator. A power level of approximately 10 μW and a spectral width of up to 5 nm are desirable. Figure 2.17 shows a typical monochromator-type calibration setup.

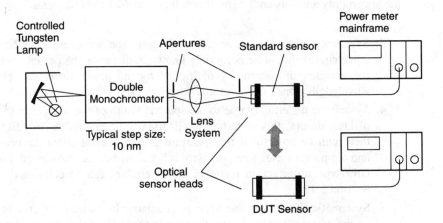

Figure 2.17 Responsivity calibration with monochromator source.

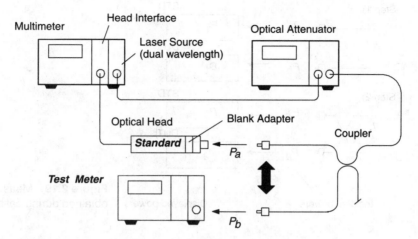

Figure 2.18 Alternative calibration setup using fixed wavelength sources.

Two types of standard sensors are commonly used: thermal detectors and photodetector sensors. Both need regular calibration within an unbroken chain to a national standards laboratory; see Section 2.5.1.

A monochromator-based calibration setup is expensive and difficult to operate and maintain. A more affordable setup is shown in Figure 2.18. A dual-wavelength source (FP laser) generates precisely known wavelengths around 1300 and 1550 nm. The attenuator is used to isolate the source and to set the appropriate power level. The coupler is used to split the power and to provide power monitoring. A specially calibrated optical head is used as the standard. A blank adapter serves as a spacer, to enlarge the spot diameter on the detector to approximately 2.4 mm (at the 5% points).

If dual-wavelength calibration is insufficient, for example, because absolute power measurements over wavelength is required, then a tunable laser source can be used as well. However, some care is recommended to avoid optical inference caused by the narrow linewidth of the tunable laser.

Switching the two coupler arms between the standard and the test meter (DUT) can be used to determine both the split ratio and the correction factor. See Figure 2.19, in which the symbol P is used for the correct power levels from the standards, and D is used for the displayed power of the DUT.

Equations 2.21 and 2.22 show how the coupling ratio and the correction factor can be calculated. Notice that a drift of the source power between the two steps has no influence on either result.

$$\text{Coupling ratio: } c = \frac{P_1}{kD_1} = \frac{kD_2}{P_2} \tag{2.21}$$

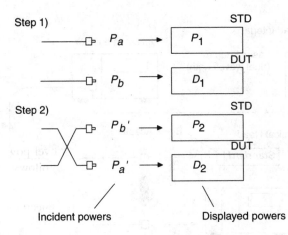

Step 1)

P_a

STD
P_1

DUT
D_1

Step 2)

P_b'

STD
P_2

DUT
P_a'

D_2

Incident powers

Displayed powers

Figure 2.19 Measurement results obtained during calibration.

$$\text{Correction factor: } k = \sqrt{\frac{P_1 P_2}{D_1 D_2}} \qquad (2.22)$$

The correction factor can either be used to correct the test meter or, without correction, as a test result for the calibration certificate.

2.5.1 Traceability and Uncertainty in Responsivity Calibrations

An unbroken chain of comparison to the national laboratory is considered a proof of traceability. Further credibility is usually given by the fact that the national laboratories compare their power scales on a more or less regular basis.[8] Figure 2.20 shows a typical traceability chain for a commercial optical power meter, together with the equipment used

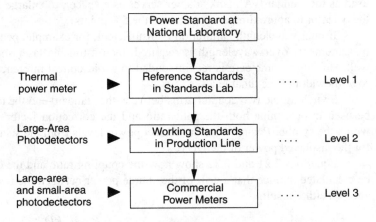

Power Standard at
National Laboratory

Thermal
power meter ▶ Reference Standards
in Standards Lab   . . . .   Level 1

Large-Area
Photodetectors ▶ Working Standards
in Production Line   . . . .   Level 2

Large-area
and small-area
photodectectors ▶ Commercial
Power Meters   . . . .   Level 3

Figure 2.20 Example of a traceability chain.

at each of the levels. Each of these comparisons (indicated by arrows) must be repeated in regular intervals.

For the calculation of the measurement uncertainty of the test meter (the end of the chain), it is important to know the calibration conditions for each of the steps, for example, the measurement instrumentation, the power levels, the wavelengths, and the beam diameters. Only then can the uncertainties for each step and, finally, for the test meter be calculated. Calibration of optical power meters is thoroughly discussed in References 9 and 10. According to Reference 9, each calibration step is accompanied by classes of uncertainties. The term "parent meter" is always used for the higher-level power meter (the standard). For each calibration step, the uncertainty "classes" are as follows:

1. **The parent meter's uncertainty at reference conditions:** This is the uncertainty of the standard power meter for a specific set of conditions, either as stated by the national laboratory, or as calculated along with its own calibration.

2. **The transfer-related uncertainties of the parent meter:** This category of uncertainties is due to differences between the calibration conditions and the "use" conditions of the parent meter causing changes of measurement results of the parent meter. This uncertainty should be accumulated, by root-sum-squaring, from the following effects: aging of the working standard and changes in the wavelength, temperature, reflection conditions, power level (nonlinearity), beam geometry, and spectral width of the source.

3. **The transfer-related uncertainties of the test meter:** This category of uncertainties is due to tolerance bands of test conditions causing changes of test meter results. This uncertainty is strongly influenced by a) how well the test conditions are known, and b) what the stated test conditions (for example, on the calibration certificate) are. It should be accumulated from the following effects and their tolerance bands, all of which are assumed to have an influence on the test meter results: the wavelength, the temperature, the reflection conditions, the power level, the beam geometry, the spectral width of the source, and the state of polarization.

4. **The random uncertainty of the transfer process:** This uncertainty expresses changes of the correction factor obtained in consecutive measurements with the same instruments.

Again, these are the uncertainties for *one* calibration step, to be accumulated to the test meter's "uncertainty at reference conditions." In a traceability chain of, for example, three higher-level instruments, three calibration steps are necessary. Then three recursive uncertainty calculations have to be carried out. On this basis, the calibration uncertainties achieved today for commercial power meters are $\pm 2\%$ in the best case.

The "uncertainty at operating conditions" should be accumulated, by root-sum-squaring, the uncertainty at reference conditions and the additional uncertainty caused by using the instrument at conditions which are different from the calibration conditions; Section 2.4.3, list items 3 to 7.

2.6 LINEARITY CALIBRATION

Power-meter linearity calibration is necessary because of two reasons: first, to extend the calibration of absolute power to the whole power range, and, even more important, to prepare the basis for high-accuracy loss and gain measurements (see Chapter 9). In these measurements, the optical power may have to cover a wide range of six or more decades.

The linearity is expected to be almost wavelength-independent. Therefore, it is sufficient to calibrate at only one or two wavelengths within the detector's spectral responsivity region.

As discussed in Section 2.3.5, photodetectors provide excellent linearity from the noise level to approximately 1 mW. This leads to the fact that often the specifiable linearity is not limited by the linearity of the instrument, but by the performance of the linearity calibration setup. Two calibration methods are discussed below. An overview of the available techniques is presented in Yan and co-workers.[11]

2.6.1 Linearity Calibration Based on Comparison

The easiest way to perform a linearity calibration is to measure an arbitrary attenuation with both the test meter and a standard meter (for example, a meter calibrated by a national laboratory) and to compare the two attenuation results. A possible measurement setup is shown in Figure 2.21.

The first attenuator is used to set the power level, to generate additional fixed attenuations (for example, 10 dB) and to split the power (a power-splitter is built into this specific attenuator model). The second attenuator is used to increase the measurement range: for very high power levels, the second attenuator reduces the power level to the usable range for the standard sensor; for very low power levels, the two sensors can be switched

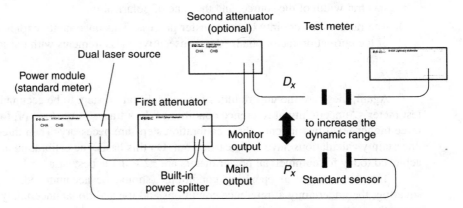

Figure 2.21   Linearity calibration based on comparison with a standard sensor.

and the second attenuator produces the low power levels. Any difference between the two measured attenuations indicates nonlinearity.

The nonlinearity of an optical power meter is internationally defined so that it represents directly the correspondent error in a loss measurement.[9]

$$N(D_x) = \frac{A_m - A}{A} = \frac{D_x/D_0}{P_x/P_0} - 1 \tag{2.23}$$

where A is the true power ratio, A_m is the measured power ratio, D_x/D_0 is the displayed power ratio (of the test meter) and P_x/P_0 are the true power ratio (of the standard meter). See Figure 2.22.

The calibration procedure is as follows:

1. Set the desired reference power on the test meter, D_0 (for example, 10 μW). Record the powers P_0 (standard meter) and D_0.

2. Increase (decrease) the attentuation of the first attenuator (for example, by 10 dB), and record the powers, P_1 $(P_2, ,)$ and D_1 $(D_2, ,)$.

3. Calculate the nonlinearity for the power D_1 $(D_2, ,)$ using Equation 2.23. In these calculations, the reference level is changing from step to step, which is why these nonlinearities are termed "partial."

4. Increase the attenuation further by repeating steps 2. and 3., until the low (high) end of the power range is reached. It is advisable to measure the nonlinearity due to range discontinuities by simply changing the power range and recording the measurement results in both ranges.

5. Decrease the attenuation to obtain the power levels above P_0 and to obtain the correspondent nonlinearity results, by repeating steps 2. and 3.

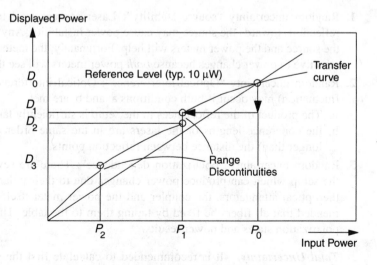

Figure 2.22 Power levels used in linearity calibration.

Finally, the nonlinearity data must be re-calculated on the basis of *one* fixed power level, e.g., 10 μW. This can be accomplished by accumulating the partial nonlinearities. See reference[10] for further details. The results are nonlinearity data for a number of power levels above and below the reference level D_0. By definition, the nonlinearity is zero at D_0.

Notice that the standard meter can operate in a much smaller power range than the test meter. For example, a linearity calibration over 70 dB is possible with a 20 dB calibrated range of the standard meter. This is possible by repetitive adjustment of the power at the standard meter using the second attenuator.

Uncertainty.   Calibrations should aim at the lowest uncertainties. Accordingly, the uncertainties must be well understood. As usual, we distinguish between "systematic" and "random" uncertainties. The following potential uncertainties should be considered.

"Systematic" uncertainties.   This type of uncertainty represents repeatable errors in the measurement data. The most important systematic uncertainty is the (calibrated) nonlinearity of the linearity standard.

Linearity standards are often calibrated using a self-calibrating technique (see Section 2.6.2). The calibration result should include nonlinearity data and their uncertainty. The nonlinearity, if any, can be corrected for. The uncertainty will reflect the performance of the calibration setup and of the meter under test. It will also depend on the power range over which the calibration was carried out.

"Random" uncertainties.   There are two possibilities in the determination of the random uncertainties: Either you evaluate the individual contributions to the random uncertainty, or you evaluate them experimentally as one ensemble. For completeness, random uncertainties can be reduced by averaging.

1. Random uncertainty "source stability": Laser sources tend to drift. Also, back-reflections towards the source may cause power instability. Any attenuation between the source and the power meters will help. Fortunately, the method is more or less insensitive to power changes because *both* power meters will see the same change.
2. Random uncertainty "Optical interference": Optical interference problems (power fluctuation) may occur if both conditions a. and b. are met.
   a. The product of the reflectances in the setup is sufficiently large, and
   b. the coherence lengths of the lasers are in the same order of magnitude as (or longer than) the distance between reflection points.
3. Random uncertainty "polarization dependence": There are several components in the setup which can produce power changes due to their polarization dependence: the optical attenuators, the coupler and the power meter itself. It is highly recommended that all fibers be fixed by taping them to the table. This will ensure stable polarization states and power results.

Total Uncertainty.   It is recommended to calculate first the standard uncertainty representing *one* attenuation step by root-sum-squaring all relevant partial uncertainties (systematic and random) in the form of standard deviations. Then the *total* uncertainty, on the basis of a 95% confidence level, is given by:

$$U = \pm 2\sigma_{\text{single step}}\sqrt{n} \tag{2.24}$$

where $\sigma_{\text{single step}}$ is the standard uncertainty for a single attenuation step and n is the number of steps where counting starts from the reference level. The factor of 2 converts the standard uncertainty to an uncertainty with 95% confidence level.

This calibration method suffers from the fact that a linearity standard is necessary, and that each of the single-step uncertainties includes the uncertainty of the standard; therefore, the uncertainty will rarely be less than ±0.1% for a 10 dB step. These problems are avoided by the superposition method.

2.6.2   Linearity Calibration Based on Superposition

An alternative linearity calibration method is based on power superposition. This is a self-calibrating method which does not need a standard meter. Therefore, a traceability to a national standards laboratory is not mandatory. This principle was first mentioned in Sanders.[12]

A possible measurement setup is shown in Figure 2.23.

In the beginning, the two attenuators are both set to high attenuation and so that each beam separately gives rise to the same powers at the DUT: $D_a \cong D_b$. Each attenuator is equipped with a shutter. The shutter of the respective other attenuator remains closed. Then the two beams are combined by opening both shutters at the same time. This reading should now be the sum of the two preceding individual readings:

$$D_c = D_a + D_b \, (\cong 2D_a) \tag{2.25}$$

Any deviation indicates nonlinearity. Accordingly, the nonlinearity for the first power D_c is:

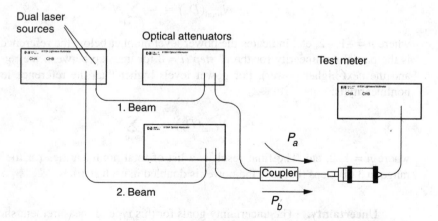

Figure 2.23   Nonlinearity calibration using the superposition method.

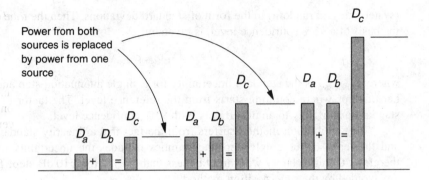

Figure 2.24   Power superposition used in linearity calibration.

$$N_1 = \frac{D_c}{D_a + D_b} - 1 \tag{2.26}$$

The next cycle starts by generating the combined power separately with each of the attenuators, before combining them again. This is indicated in Figure 2.24. Notice that the result of Equation 2.26 should be considered as a partial nonlinearity because it uses a changing power level as the reference. At the end of the measurement, the partial nonlinearities for all steps will be determined.

Finally, the *total* nonlinearity can be calculated, in other words, the nonlinearity with respect to a fixed reference level. Start by choosing a reference level, for example, 10 μW, at which the total nonlinearity is zero by definition. Then use the following equation for power levels lower than the reference level:

$$N_{\text{total}}(D_n) = -\sum_{i=-1}^{n} N_i \tag{2.27}$$

where $n = -1, -2$, etc. indicates the power level number below the reference point and N_i is the partial nonlinearity for the ith *step* ($i = 0$ for the step between the reference power and the next-higher power). For power levels higher than the reference level, the total nonlinearity is:

$$N_{\text{total}}(D_n) = -\sum_{i=0}^{n-1} N_i \tag{2.28}$$

where $n = 1, 2$, etc. The final result is a list of total nonlinearities for the whole power range in 3 dB steps (because the power is doubled in each step).

Uncertainty.   The uncertainty goals for this type of measurement should be better than ±0.1% for a 10 dB power range. This is an aggressive goal. Accordingly, the mea-

surement uncertainties must be well understood. In the following, only the differences between the uncertainties of the comparison method, Section 2.6.1, and this method are discussed.

"Systematic" uncertainties.   The big advantage of this method is that there is no
linearity standard, and consequently no uncertainty due to the linearity standard.

A small systematic uncertainty may be caused by multiple reflections. There is one
main reflection in the setup: the fiber end that stimulates the optical power meter. It usually represents a reflectance $R_a = 3.5\%$ (equivalent to -14.7 dB). If the reflected wave, on
its way back to the source, hits a second reflection point with reflectance R_b, then the incident power level is increased to:

$$P_{\text{total}} = (1 + R_a B_b)P_{\text{inc}}   \text{[watts]} \tag{2.29}$$

This problem is insignificant if the second reflectance is sufficiently small (for example, $R_b \leq 10^{-4}$, equivalent to -40 dB) and remains constant. However, R_b may change
due to activating the optical shutters.

"Random" uncertainties.   The random uncertainties of this method are essentially
the same as the random uncertainties of the comparison method.

Total uncertainty.   Again, it is recommended to calculate first the standard uncertainty representing *one* 3 dB step by root-sum-squaring all relevant partial uncertainties
(systematic and random) in the form of standard deviations. Then the *total* uncertainty, on
the basis of a 95%-confidence level, is given by:

$$U = \pm 2\sigma_{3dB}\sqrt{n} \tag{2.30}$$

where $\sigma_{3\,dB}$ is the standard uncertainty for one 3 dB step and n is the number of 3 dB steps
counted from the reference level. State-of-the-art uncertainties achieved in linearity calibration using the superposition method are $\leq \pm 0.1\%$ for a 10 dB power range.

2.7   SUMMARY

This is a chapter on optical power meters. The range of topics includes all aspects of this
instrument: construction, absolute power measurement, and calibration. In most cases, the
potential uncertainties are also discussed, with the aim of improving the accuracy of
optical-power measurement. The characteristics of state-of-the-art optical-power meters
can be summarized as follows:

±2% uncertainty at calibration conditions,

±3% to 5% uncertainty at operating conditions,

±0.5% nonlinearity for a power range of 50 dB or more,

an approximate power range -90 to 0 dBm,

(attenuation needed for higher power measurement).

ACKNOWLEDGMENTS

The author wishes to thank his colleagues Andreas Gerster, Siegmar Schmidt, and Dennis Derrickson for fruitful discussions and suggestions.

REFERENCES

1. Bischoff, K. 1968/69. *Ein einfacher Absolutempfänger hoher Genauigkeit,* Optik, 28. Band: 183–189.

2. Haars, H. and K. Möstl. 1997. *Dünnschicht-Thermosäule mit Online-Kalibrierung,* PTB Jahresberichte 1996. Physikalisch-Technische Bundesanstalt, Braunschweig, Germany.

3. Stock, K.D. and H. Hofer. 1993. Present state of the PTB primary standard for radiant power based on cryogenic radiometry. *Metrologia* 30: 291–296.

4. Zalewski, E.F. and J. Geist. 1980. *Silicon photodiode absolute spectral response self-calibration. Applied Optics* 19, No. 8: 1214–1216.

5. Hentschel, C. *Fiber Optics Handbook,* 1989. Böblinger, Germany: Hewlett Packard, P/N 5952–9654.

6. Stock, K.D. *Si-photodiode spectral nonlinearity in the infrared. Applied Optics* 25, No. 6, 830–832.

7. Gallawa, R.L. and X. Li. 1987. *Calibration of optical fiber power meters: the effect of connectors. Applied Optics* 26, No. 7: 1170–1174.

8. Gardner, J.L., et al. 1992. *International intercomparison of detector responsivity at 1300 and 1550 nm. Applied Optics* 31, No. 34: 7226–7231.

9. IEC Standard 1315, 1995. *Calibration of fibre-optic power meters,* International Electrotechnical Commission.

10. Hentschel, C. *Setting up a calibration system for fiber optic power meters,* Böblinger, Germany. Hewlett Packard, P/N 5964–9638E.

11. Yan, S., I. Vayshenker, et al. 1994. *Optical detector nonlinearity: a comparison of five methods,* Conference on Precision Electromagnetic Measurements, 455–456.

12. Sanders, C.L. 1962. *A photocell linearity tester, Applied Optics* 1, No. 3: 207–211.

CHAPTER

3

Optical Spectrum Analysis

Joachim Vobis, Dennis Derickson

3.1 INTRODUCTION TO OPTICAL SPECTRUM ANALYSIS

Optical spectrum analysis is the measurement of optical power as a function of wavelength. The spectrum of a light source is an important parameter in fiber-optic communication systems. For example, chromatic dispersion can occur in the fiber and limit the achievable modulation bandwidth of the system. The effect of chromatic dispersion can be seen in the time domain as pulse broadening of a digital waveform. Since chromatic dispersion is a function of the spectral width of the light source, narrow spectral widths are desirable for high-speed communication systems.

The prevalence of wavelength division multiplexed (WDM) systems has stimulated significant activity in the measurement of optical spectra. WDM has also made optical spectrum analysis a key measurement capability that must be embedded inside telecommunication network elements.

Figure 3.1 shows an example measurement made by an optical spectrum analyzer (OSA). It shows the power versus wavelength for a Fabry-Perot (FP) laser. The FP laser shows a series of longitudinal modes that have significant energy over a 20 nm span. The plot shows that the measurement has been made using an instrument filter bandwidth of 0.2 nm with an instrument sensitivity setting of −55 dBm. From the spacing of the modes and power distribution, the laser length and coherence properties of the laser can be determined.

This chapter is the first of three chapters covering the area of optical spectrum analysis. Chapter 3 covers the most common implementation using a diffraction-grating based optical filter. Chapter 4 covers wavelength meters based on Michelson interferometry. Wavelength meters offer high wavelength accuracy. Chapter 5 covers methods of

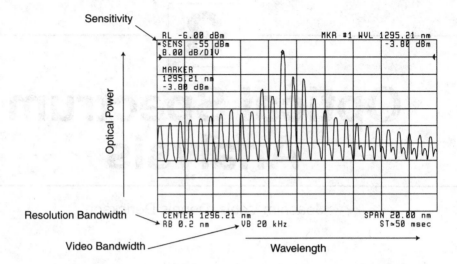

Figure 3.1 Optical spectrum analyzer measurement of a Fabry-Perot laser.

very-high wavelength resolution optical spectrum analysis using homodyne and heterodyne techniques.

3.2 TYPES OF OPTICAL SPECTRUM ANALYZERS

3.2.1 Basic Block Diagram

A simplified OSA block diagram is shown in Figure 3.2. The incoming light passes through a wavelength-tunable optical filter which resolves the individual spectral components. The photodetector then converts the optical signal to an electrical current proportional to the incident optical power.

The current from the photodetector is converted to a voltage by the transimpedance amplifier and is then digitized. Any remaining signal processing, such as applying correction factors, is performed digitally. The signal is then applied to the display as the vertical or power axis. A ramp generator determines the horizontal location of the trace as it sweeps from left to right. The ramp also tunes the optical filter so that its center wavelength is proportional to the horizontal position. A trace of optical power versus wavelength results. The displayed width of each mode of the laser is a function of the spectral resolution of the wavelength-tunable optical filter.

3.2.2 Fabry-Perot Interferometers

The FP interferometer, shown in Figure 3.3, consists of two highly reflective, parallel mirrors that act as a resonant cavity which filters the incoming light. The resolution of FP-interferometer-based OSAs depends on the reflection coefficient of the mirrors and

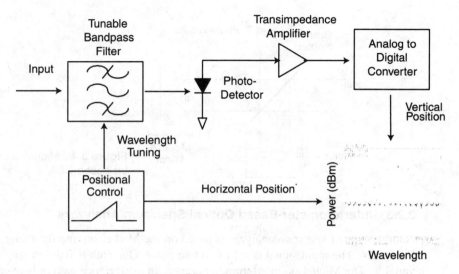

Figure 3.2   Simplified OSA block diagram.

the mirror spacing. Wavelength tuning of the FP interferometer is accomplished by adjusting the mirror spacing or by rotating the interferometer with respect to the input beam.

The advantage of the FP interferometer is its potential for very narrow spectral resolution and its simplicity of construction. The added resolution allows measurements such as laser chirp to be performed. The major disadvantage is that the filters have repeated passbands. The spacing between these passbands is called the free spectral range. If the mirrors are spaced very widely apart, very high resolution can be obtained, but the free spectral range is small. This problem can be solved by placing a second filter in cascade with the FP interferometer to filter out power outside the interferometer's free spectral range. Section 6.1 in Chapter 4 gives a more detailed description of FP-based OSAs.

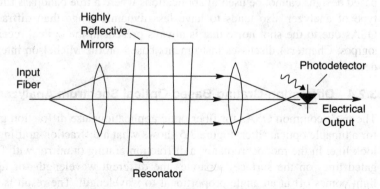

Figure 3.3   FP-interferometer-based OSA.

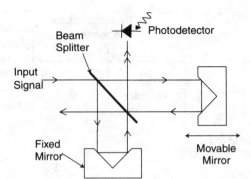

Figure 3.4 Michelson-interferometer-based OSA.

3.2.3 Interferometer-Based Optical Spectrum Analyzers

Another type of spectrum analyzer is based on the Michelson interferometer as shown in Figure 3.4. The input signal is split into two paths. One path is fixed in length and one is variable. The Michelson interferometer creates an interference pattern between the signal and a delayed version of itself at the detector. The resulting waveform is the autocorrelation function of the input signal and is often referred to as an interferogram. Michelson-interferometer-based spectrum analyzers make direct measurements of coherence length. Other types of OSAs cannot make direct coherence-length measurements. If the period of the zero crossings in the interferogram are accurately measured by comparison to a wavelength standard, the wavelength of the unknown signal can be determined with high accuracy. It is the potential for high wavelength accuracy that distinguishes this instrument. A state of the art wavelength meter can measure wavelength to less than 1 part per million. A 1550 nm laser could be measured to ±0.0015 nm.

The Michelson interferometer can also provide displays of power versus wavelength. To determine the power spectra of the input signal, a Fourier transform is performed on the interferogram. The resolution of the instrument is determined by the path-length delay that is used to create the interferogram. Because this instrument does not depend on a tunable bandpass filter for wavelength identification, Michelson-interferometer-based designs cannot be used in applications where a true bandpass filter is required. This type of analyzer also tends to have less dynamic range than diffraction-grating-based OSAs due to the shot noise that is always present in the optical receiver for large input signals. Chapter 4 discusses instruments based on the Michelson-interferometer in more detail.

3.2.4 Diffraction-Grating-Based Optical Spectrum Analyzers

The most common OSAs for fiber optic applications use diffraction gratings as the basis for a tunable optical filter. Figure 3.5 shows what a diffraction-grafting-based OSA might look like. In the monochromator, a diffraction grating (a mirror with finely spaced corrugated lines on the surface) separates the different wavelengths of light. The diffracted light comes off at an angle proportional to wavelength. The result is similar to the rainbow produced by visible light passing through a prism. In the infrared, prisms do not

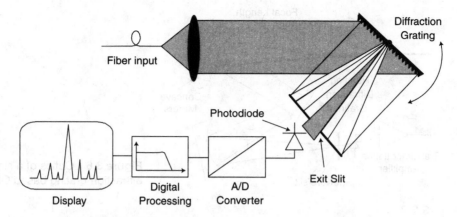

Figure 3.5   Concept of diffraction-grating-based OSA.

work very well because the dispersion (in other words, change of refractive index versus wavelength) of glass in the 1 to 2 μm wavelength range is small. Diffraction gratings are used instead. They provide a greater separation of wavelengths allowing for better wavelength resolution. A diffraction grating is made up of an array of equidistant parallel slits (in the case of a transmissive grating) or reflectors (in the case of a reflective grating). The spacing of the slits or reflectors is on the order of the wavelength of the light for which the grating is intended to be used. The grating separates the different wavelengths of light because the grating lines cause the reflected rays to undergo constructive interference only in very specific directions. Only the wavelength that passes through the aperture reaches the photodetector to be measured. The angle of the grating determines the wavelength to which the OSA is tuned. The size of the input and output apertures together with the size of the beam on the diffraction grating determines the spectral width of the optical filter.

3.3 ANATOMY OF A DIFFRACTION-GRATING-BASED OPTICAL SPECTRUM ANALYZER

In this section, the optical portion of a grating-based OSA will be dissected and an introduction to the function of each of the components will be given. This section will be useful to the reader unfamiliar with the internal workings of monochromators and spectrometers and the terms describing each of the components involved. Section 3.4 will describe some of the parameters that are important in a spectral measurement and relate them back to the workings of the monochromator.

3.3.1 Basic OSA Block Diagram

Figure 3.6 shows the various optical components in a basic OSA.[1] Here one can see that an OSA contains (in the order of the light propagating through the system) an entrance (or input) slit, collimating optics, a diffraction grating, focusing optics, an exit (or output) slit,

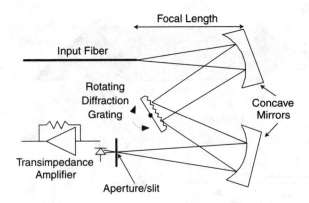

Figure 3.6 Optics of a single pass diffraction-grating based OSA.

and a detector. This optical portion of the OSA is usually referred to as a monochromator or as a spectrometer. Historically, a spectrometer usually refers to an instrument that has some means of separating the input light (for example, the input slit, the diffraction grating, and the output slit) to its various spectral components and a detector so that the spectrum can be recorded. This same instrument without the detector (in other words, having an optical output only), would be called a monochromator. A monochromator can be thought of as a tunable optical filter. One couples light from a source to the monochromator and at the output one has light consisting of a selected portion of the spectrum of that coupled source. Throughout the rest of this chapter, the word monochromator will be used to describe the optical portion of the OSA up to, but not including, the detector.

3.3.2 The Entrance or Input Slit

The light first enters the monochromator through the entrance or input slit. The input slit defines the spatial width of the input image. This input aperture, along with other components in the system, defines the wavelength resolution of the system. The narrower the input slit, the better the possible wavelength resolution. Narrower slits also reduce the optical throughput (number of photons) that the detector sees. In a modern fiber-optic input OSA, the input slit is often a singlemode or multimode fiber. This fiber defines the input image. For the case of a singlemode step-index fiber, the image will approximate a Gaussian amplitude distribution. It is more accurately described by the J_0 Bessel function in the core and the K_0 Bessel function in the cladding.

There are two methods often used to connect the input fiber to the monochromator in fiber-optic input spectrometers. These methods are shown in Figure 3.7.

In Figure 3.7a, the user's fiber directly defines the input aperture. OSAs with an input connection directly to the OSA allow you to connect fibers with different core diameters, and in some cases to apply open beams from optical benches as well. There is no insertion loss and no risk of scratching any internal fiber. This design has several disadvantages. Small particles can fall into the input hole and possibly harm the monochromator. The position of the input spot strongly depends on how an input signal is brought to the OSA. The image quality is defined by the shape of the fiber-end face polish. Scratches

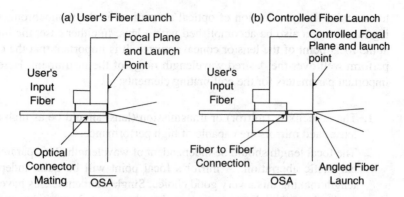

Figure 3.7   (a) Input slit with user's fiber. (b) Input slit using a short connector fiber.

and chips on the fiber core can affect the OSA filter response. Furthermore, a typical connector at the OSA input causes a 14 dB return loss (RL) due to the glass-to-air transition. Such a low RL can have an impact on measurement results. Many semiconductor lasers are sensitive to back reflections.

Figure 3.7b shows the user's fiber coupled to a captive length of fiber that is then used to form the entrance slit to the monochromator. This design has the advantage that the quality of the input image is well defined. The disadvantage of this approach is the insertion loss magnitude and insertion loss uncertainty found when mating together fibers. The use of a captive fiber allows the use of an angled fiber tip launch. The reflectivity of an 8 degree polished singlemode fiber can be greater than 70 dB in return loss. In this case the input reflectivity to the monochromator will be limited by the reflection from the fiber-to-fiber mating at the instrument front panel. The return loss from fiber mating typically exceeds 28 dB depending on the connector style (see Appendix C).

Singlemode versus Multimode Input.   Either singlemode or multimode fibers can be used in the captive fiber design. It is possible to connect singlemode fibers to multimode fiber inputs (typically 50 or 62.5 μm) with very little insertion-loss uncertainty. If the multimode section is very short, the spot size will still be small where the light leaves the multimode fiber. The smaller spot size (9 μm) and the smaller numerical aperture of a singlemode fiber further improves the image quality, allowing tight specifications. However, the absolute power accuracy is affected by the insertion-loss uncertainty of a singlemode to singlemode connection at the instrument's input.

3.3.3 The Collimating Optics

The purpose of the collimating optics is to take the diverging beam from the input slit and collimate this beam to form a plane wave to illuminate the diffraction grating. In Figure 3.6, the collimation is accomplished using a concave mirror. In a reflective system, the collimating optics will be some form of a curved mirror, typically a section of an asphere to minimize the introduction of optical aberrations in the monochromator. The collimating function can also be accomplished with a lens. In either case, the input slit is located at the focal point of the lens or concave mirror. It is important that the collimating optics perform well over the desired wavelength range of the instrument. Here are some of the important parameters for the collimating elements:

1. The reflectivity (mirror) or transmission (lens) should be as high as possible. Both lenses and mirrors are capable of high performance.

2. The focal length should be independent of wavelength. This parameter is also called chromatic aberration. A mirror's focal point will be quite independent of wavelength making this a very good choice. Single-element lenses have significant chromatic aberration. For a wide-wavelength range of operation, a multielement lens would be required to adequately compensate for the dispersion of the glasses used in the lens as well as to minimize the aberrations introduced by the spherical shape of the lens.

3. The size of the collimated beam should be as large as possible to achieve high wavelength resolution. Since the input fiber determines the divergence angle of the light reaching the collimating lens, a long focal length lens will be needed to get a large collimated beam size. It is more economical to make large mirrors as compared to large lenses.

4. Optics should be diffraction limited in their performance. The idea of diffraction-limited optics is illustrated in Figure 3.8. If a collimated beam is incident on a lens, the light will be focused to a small spot. If one thinks of the propagation of light in terms of rays passing through the lens and coming to a spot, one would think that the diameter of the light beam goes to zero at the focal point. In actuality, the beam goes to a minimum beam waist at the focal point of the system. This minimum beam waist is called the diffraction-limited spot size of a lens. Mathematically, the diffraction limited spot size is:[2]

$$w_0 \equiv \frac{2\lambda(fl)}{\pi(\text{diameter})} \tag{3.1}$$

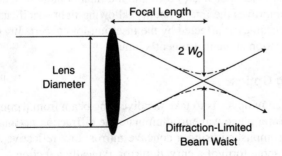

Focal Length

2 W_0

Lens Diameter

Diffraction-Limited Beam Waist

Figure 3.8 Diffraction-limited beam waist for a lens or for a concave mirror.

where w_0 is the spot radius at the $1/e$ power points, λ is the wavelength, fl is the focal length of the lens and *diameter* is the lens diameter. This is the minimum achievable spot size if the entire lens is illuminated by a plane wave. Aberrations in a lens can cause the minimum spot size to be degraded. Let's use this formula to calculate the diffraction-limited spot size of a lens that might be used in a fiber-optic input monochromator. Assume that the lens has a diameter of 5 cm and a focal length of 30 cm. At a wavelength of 1550 nm, the diffraction-limited spot radius would be 6 μm. If the shape of the lens or mirror is not correct, the actual minimum spot size for this lensing system could be significantly larger than 6 μm. It is the quality of this collimating system along with diffraction-grating performance that determines the filter response of the monochromator.

3.3.4 The Diffraction Grating

The diffraction grating is at the heart of the monochromator. The diffraction grating functions to "reflect" light at an angle proportional to wavelength.[3] This is the diffractive element in the system, that is, the light is diffracted according to wavelength. The monochromator can be tuned by changing the angle at which the light is incident on the grating. The diffraction grating is typically a reflective element consisting of a substrate and a reflective coating with periodic perturbations (typically referred to as lines or grooves) that form the grating.

The operation of a diffraction grating begins when light strikes the reflective lines of the grating. Each line of the grating diffracts the light off into a range of wavelet angles. For a given wavelength there will be a certain angle at which the diffracted wavelets will be exactly one wavelength out of phase with one another and will add constructively in a parallel wavefront (see Figure 3.9). The light of a given wavelength leaves the grating at a specific angle. Light of other wavelengths leaves the grating at slightly lower or higher angles. The shape or blaze of each grating line determines the overall efficiency of the diffracted beam with respect to the incident beam power.

The general equation for a diffraction grating is:[1]

$$n\lambda = d(\sin \beta - \sin \alpha) \tag{3.2}$$

where λ is the wavelength of the light (in air for most OSAs), d is the spacing of the lines on the grating, α is the angle of the incident light relative to the grating normal, β is the angle at which light leaves the grating, and n is an integer that is called the order of the spectrum. Here the two angles are defined as being on opposite sides of the grating normal as shown in Figure 3.9. The circled area of the figure illustrates the geometry that defines the grating equation.

When the diffracted rays are one wavelength out of phase with respect to each other, the diffracted beam is called first-order. At another angle where the wavelets are all exactly two wavelengths out of phase and will also add constructively, the spectrum is called a second-order spectrum. Even higher-order diffraction may also be present. Figure 3.10 illustrates diffracted orders. The first reflection is called the zero-order beam ($n = 0$)

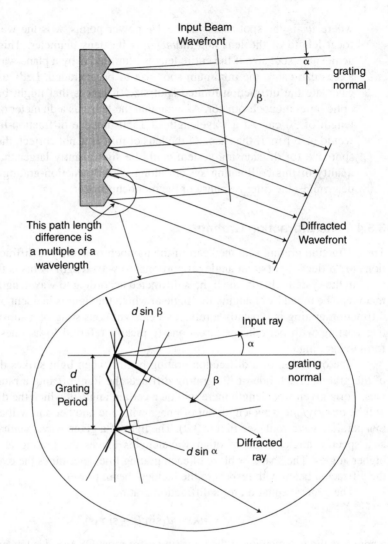

Figure 3.9 Diffraction-grating operation.

where the angle of incidence is equal to the angle of reflection. This zero-order reflected beam is not separated into different wavelengths and is not used by an OSA.

OSAs often use a diffraction grating in a special orientation called the Littrow condition. In this arrangement, the wavelength of interest leaves the diffraction grating and goes directly back along the path of the incident beam as shown in Figure 3.11. For the Littrow condition, the grating equation simplifies to:

$$n\lambda = 2 d \sin\theta \tag{3.3}$$

where $\theta = \alpha - \beta$.

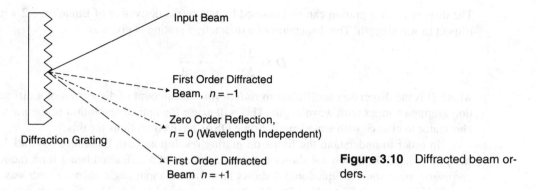

Input Beam

First Order Diffracted
Beam, $n = -1$

Zero Order Reflection,
$n = 0$ (Wavelength Independent)

Diffraction Grating

First Order Diffracted
Beam $n = +1$

Figure 3.10 Diffracted beam orders.

Figure 3.9 shows that a diffracted plane wave is formed at a single angle where constructive interference is occurring between adjacent grooves of the grating. The diffracted beam actually occupies a narrow range of angles. Even for a single-wavelength input beam, this new diffracted wavefront will be slightly diverging. The divergence angle for this diffracted beam is given as:[2]

$$\Delta\beta_{\min} = \frac{\lambda}{Nd\cos\beta} \tag{3.4}$$

where $\Delta\beta_{\min}$ is the divergence angle of the diffracted beam for monochromatic light and N is the number of illuminated lines on the grating. This equation highlights a fundamental limitation on the filter width of a grating-based OSA. The resolution is limited by the diameter of the illuminated grating compared to the wavelength. This is analogous to a phased-array antenna. The more elements in the array, the narrower the beam width of the radiation pattern. The resolution of the overall instrument can be worse than this depending on the size of the input and output apertures and the performance of the collimating optics.

Another important property of the grating is its dispersion. Dispersion is a measure of how many degrees the diffracted beam rotates for a given input wavelength change.

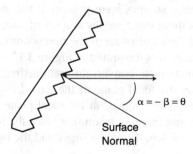

$\alpha = -\beta = \theta$

Surface
Normal

Figure 3.11 The Littrow condition.

The dispersion of a grating can be obtained by taking the derivative of Equation 3.2 with respect to wavelength. The dispersion of a diffraction grating is given as:

$$D = \frac{\Delta\beta}{\Delta\lambda} = \frac{n}{d\cos\beta} \tag{3.5}$$

where D is the dispersion coefficient in *radians/m*. The amount of dispersion of a diffraction grating changes with wavelength. This will cause the optical resolution of the monochromator to change with wavelength. Some OSAs have corrections for this.

In order to understand the limits on grating resolution, both Equations 3.4 and 3.5 must be utilized. Equation 3.4 shows the angular extent of a diffracted beam for a monochromatic input source. Equation 3.4 shows how fast the input angle changes with wavelength. If Equation 3.5 is solved for $\Delta\beta$, it can be equated to 3.4. The minimum achievable wavelength resolution can then be found by solving for $\Delta\lambda$. The resulting equation for minimum resolution is given as:[2]

$$\Delta\lambda_{min} = \frac{\lambda}{Nn} \tag{3.6}$$

where $\Delta\lambda_{min}$ is the minimum-wavelength resolution available from the diffraction grating. This is a very simple result involving only the wavelength, the number of lines illuminated on the grating, N, and the diffracted order, n.

The efficiency of a grating depends on the diffracted angles, the blaze of the grating lines, and the coatings on the grating. For a treatment of this topic, readers are referred to Hutley.[3] An important item to note is that the efficiency of a diffraction grating is polarization dependent. Methods to minimize polarization dependence are shown later in Section 3.4.4.

3.3.5 The Focusing Optics

The purpose of the focusing optics is to take the diffracted light from the grating and image it on the exit or output slit. The focusing optics in a monochromator essentially do in reverse what the collimating optics do. That is, the optics take the diffracted collimated light from the diffraction grating and focus the light on the exit or output slit. In the simplest implementation, these focusing optics could be of the exact same type as the collimating optics, just operated in the reverse direction. The diffraction grating produces a diffracted beam that has an angle proportional to wavelength. Let us consider what would happen if two monochromatic sources were present at the input to the monochromator. The focusing lens converts these input angles to a set of spots at the focal distance from the lens. This line is called the focal plane of the monochromator. The focal plane of the monochromator for this situation is illustrated in Figure 3.12.

The focusing lens functions to convert different diffracted angles from the diffraction grating to a position on the focal plane of the lens. Longer focal-length lenses will cause the spacing between the two signals at the focal plane to be separated by a larger distance. This in itself does not increase resolution since the image size of the DFB laser will be magnified. If the focal lengths of the input and the output collimators are equal,

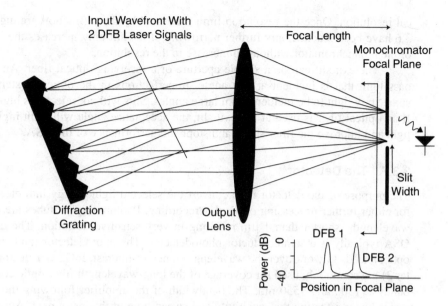

Figure 3.12   The focal plane of the monochromator.

the size of the spot at the output would be equal to the size of the spot at the input. The spot size can be magnified by the ratio of the output lens focal length to the input lens focal length. Notice that in the drawing, the DFB laser images show that the intensity of light away from the peak is greatly attenuated but does not go to zero. This is due to practical issues in a monochromator. Imperfections in the grating can cause light from a DFB laser to be scattered over a wide range of angles. This light will contribute to a low level distribution of light over the entire focal plane. It presents a fundamental limit to the rejection of the optical filter.

3.3.6   The Exit or Output Slit

The exit or output slit along with the entrance or input slit and the diffraction grating affect the resolution of the system. The purpose of the output slit is to spatially filter the light from the diffraction grating. The output slit is put into the focal plane of the monochromator. Because at this point the light is spatially dispersed according to wavelength, any spatial "selection" or filtering of the light will select or filter the spectrum of the light. This is a good place to adjust the optical resolution (in other words, how much of the spectrum of the light to pass) of the instrument. Typically, this slit is realized by an adjustable slit or a series of slits for the desired optical resolutions. A narrow slit will select only a very small portion of the spectrum and thus give high optical resolution. A wider slit will let more of the spectrum through and thus give poorer optical resolution. Typical numbers for optical resolution for a fiber-optic OSA are tenths of nanometers to nanometers of wavelength. One cannot just arbitrarily narrow the slit to give finer and finer optical resolution. Once the resolution limit presented by the diffraction grating in Equation 3.6 have been reached, any further narrowing of the slit only increases the insertion loss of the monochromator without an increase in the resolution.

The exit slit can be a simple aperture or a receiving optical fiber. An exit slit will pass light that is incident at any angle. Imperfections in the optical system may cause some stray light to be incident from large angles. Slits will not reject this high-angle light. If an optical fiber is used at the exit slit, the acceptance angle will limit high-angle stray light and improve the filter shape and stopband performance of the OSA.

3.3.7 The Detector

The purpose of the detector is to convert the selected light energy into electrical energy for either further processing or display/recording. Photomultiplier tubes are often used at wavelengths shorter than 1 μm resulting in very sensitive detection. The detector in an OSA typically is a semiconductor photodetector. The photodetector (and any associated optics) needs to work over the wavelength range of interest. InGaAs detectors (see Chapter 2) are commonly used for coverage of the long-wavelength fiber-optic communication bands at 1300 and 1550 nm. The bandwidth of the amplifier following the detector is a major factor affecting the sensitivity and sweep time of the instrument. Appendix A covers the noise issues of optical detection in detail.

3.3.8 Single Monochromator Summary

A brief overview of a single-pass grating-based monochromator was given in sections 3.3.1 to 3.3.7. The monochromator consists of an entrance (or input) slit, focusing optics, a diffracting element to angularly separate the wavelengths, collimating optics, and an exit (or output) slit. The input slit, the output slit, and the number of illuminated lines on the diffraction grating are the key elements that constrain the achievable width of the monochromater bandpass function. The input slit is a single or multimode fiber for fiber optic applications. The input slit image is then refocussed at the output focal plane. It is important to have a high-quality optical imaging system so that the output spot is as small and aberration-free as possible. The final filter function is a convolution of the output image with the shape of the output slit function. Filter bandwidths of near 0.1 nm at 1550 nm are achievable in monochromaters that will fit into benchtop packages. This filter width is much wider than the linewidth of a typical singlemode laser. The OSA typically traces out the filter shape of the instrument when measuring a DFB-laser mode for example. The high-resolution spectrum analysis techniques of Chapter 5 are needed to resolve the details of a laser-line shape.

Example Calculation of Monochromator Resolution.   To illustrate the resolution capabilities of a single-pass monochromator that might be used for fiber optic applications, consider a monochromator with a singlemode fiber optic input, 5 cm diameter collimating lens with 20 cm long focal length and a diffraction grating with 1000 lines/mm at Littrow angle operating in first order. The wavelength for this example is 1550 nm.

For the Littrow condition, the angle of the input beam with respect to the grating normal would be 51 degrees (from Equation 3.3). The divergence angle for a singlemode fiber is 12 degrees. At 20 cm, the collimated beam diameter would be 4.2 cm. The lenses are just large enough to accommodate the input beam diameter. From Equation 3.6, the minimum available resolution from this system would be 0.037 nm. Using Equation 3.4, the divergence angle for a monochromatic input signal to the monochromator would be 0.00012 rad. A 20 cm long focal-length lens converts this angular change to a 2.4 μm positional change in the focal plane. This means that an aperture of 2.4 μm or slightly larger would be needed to obtain the narrowest instrument resolution. The actual filter shape of the instrument is the convolution of the rectangular slit aperture and the shape of the image in the focal plane.

3.3.9   Single Monochromator Versus Double Monochromator

The monochromator described in the previous section used a single pass off of the diffraction grating to achieve wavelength filtering. The selectivity for this configuration is often not sufficient for measuring side-mode suppression ratio in DFB lasers for telecommunications. One could increase the size of the diffraction grating and the collimated beam size to improve selectivity. The collimated beam size would get too large to be contained in a small benchtop package. In most cases, it is more efficient to cascade two monochromators in series to obtain adequate selectivity. Single monochromators also have a limited stopband performance due to imperfections in the diffraction grating and due to scattered light within the monochromator. Improved stopband performance is obtained with cascaded filters.

3.3.10   Double Monochromator

Double monochromators, such as shown in Figure 3.13, are equivalent to a pair of sweeping filters. While this technique improves dynamic range, double monochromators typically have reduced span widths due to the limitations of monochromator-to-monochromator tuning match; double monochromators also have degraded sensitivity due to losses in the monochromators.

3.3.11   Double-Pass Monochromator

An alternative to the double monochromator is the double-pass monochromator design.[4] The double-pass monochromator provides the dynamic-range advantage of the double monochromator and the sensitivity and size advantages of the single monochromator. Figure 3.14 shows an example double-pass monochromator. It uses the same diffracting grating and collimating optics twice. The grating is used in the Littrow configuration.

The first pass through the double-pass monochromator is similar to conventional single-monochromator systems. The input beam (1) is collimated by the optical element and dispersed by the diffraction grating. This results in an angular distribution of the light, based on wavelength. The diffraction grating is positioned such that the desired wavelength (2) passes through the aperture. The width of the aperture determines the band-

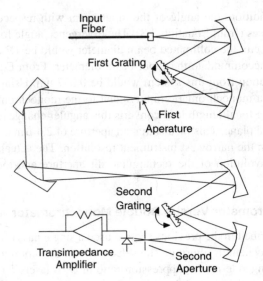

Figure 3.13   Double-monochroma-tor-based OSA.

width of wavelengths allowed to pass to the detector. Various apertures are available to provide a range of resolution bandwidths. The minimum useful aperture size is limited by the diffraction-limited spot size of the optics.

This system shown in Figure 3.14 is unique in that the filtered light (3) is sent through the collimating element and diffraction grating for a second time. During this second pass through the monochromator, the temporal dispersion process is reversed. This means that all of the rays take the same total path length through the monochromator. The small resultant image (4) allows the light to be focused onto a fiber which carries the signal to the detector. This fiber acts as a second aperture in the system. The implementation of this second pass results in the high sensitivity of a single monochromator and the high dynamic range of a double monochromator. The size of the spot at the output of this monochromator is independent of the size of the resolution-determining slit. This allows the use of a small detector for all bandwidth settings. Since the dark current of a detector is proportional to the detector size, better detector sensitivity can be obtained with this double-pass configuration.

3.3.12   Littman Double-Pass Monochromator

In the Littman design (Figure 3.15), the diffraction grating is illuminated at a very shallow angle. The diffraction grating provides a large angular dispersion of wavelengths at shallow angles. The dispersion Equation 3.5 shows that the dispersion increases to a maximum value as β approaches 90 degrees. The diffracted light is retroreflected back to the grating for a second pass in this design. The light is then focused to an exit slit. A major advantage of the Littman configuration is the small size of the monochromator for its resolution. Normally a long focal length lens is required to illuminate a large number of lines

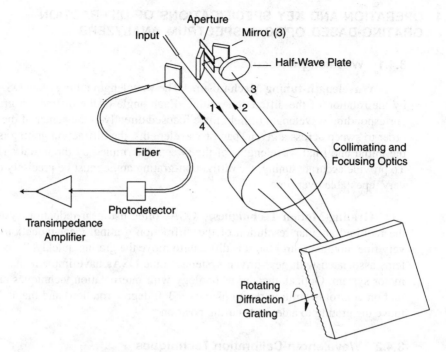

Figure 3.14   Block diagram of double-pass-monochromator OSA.

on the grating. Because the diffraction grating is placed at a shallow angle, only a small collimated beam size is needed for full illumination. If a two-dimensional retroreflector mirror is rotated for wavelength selection, the optical system is very forgiving in terms of optical alignment. A disadvantage of the double-pass Littman monochromator approach is the large amount of polarization sensitivity that is found for a shallow angle grating. The s-polarization (perpendicular to grating lines) is much more efficient than the p-polarization (parallel to grating lines).

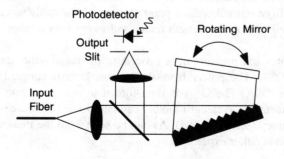

Figure 3.15   Configuration of a double-pass Littman monochromator.

3.4   OPERATION AND KEY SPECIFICATIONS OF DIFFRACTION-GRATING-BASED OPTICAL SPECTRUM ANALYZERS

3.4.1   Wavelength Accuracy

Wavelength-tuning Mechanism.   The wavelength tuning of an OSA is controlled by the rotation of the diffraction grating. Each angle of the diffraction grating causes a corresponding wavelength of light to be focused directly at the center of the output slit. In order to sweep across a given span of wavelengths, the diffraction grating is rotated, with the initial and final wavelengths of the sweep determined by the initial and final angles. To provide accurate tuning, the diffraction-grating angle must be precisely controlled and very repeatable over time.

Grating-Motion Techniques.   OSAs often use gear-reduction systems to obtain the required angular resolution of the diffraction grating. Gear-reduction systems offer very fine motion control but it is difficult to move the grating quickly. To overcome problems associated with gear-driven systems, some OSAs have implemented a direct-drive motor system. Optical encoder technology with interpolation techniques allow very fine motion control (4 million positions over a 360 degree rotation) and the ability to quickly move the grating to a desired starting position.

3.4.2   Wavelength-Calibration Techniques

The OSA determines the wavelength of any data point from the position of the grating. Therefore, any mechanical tolerance has a direct affect on the wavelength accuracy. To compensate for component variations, manufacturers calibrate the wavelength axis. However, shock and vibration as well as temperature changes can cause wavelength shifts on the order of ± 1 nm. Compared to the full wavelength range, this is less than 0.1%—an excellent stability for bearings that have to hold the mass of the grating, position encoder, etc., while still performing many years without noticeable wear.

Calibration with a Well-Known Laser Wavelength.   An OSA can be calibrated by measuring a source with a well-known wavelength.[5] A user calibration is only as accurate as the reference signal. Single-wavelength calibrations can be made at the accurate helium-neon laser wavelengths listed in Table 4.1 in the next chapter. Single-point calibrations allow very accurate measurements near the calibration wavelength but errors start to accumulate away from this calibration point. Alternately, a stable wavelength tunable laser can be calibrated with a wavelength meter and swept over a range of frequencies.

Figure 3.16 illustrates the measurement procedure to transfer the accuracy of a wavelength meter to an OSA. The power from the tunable laser is coupled to both the wavelength meter and the OSA. The OSA is then forced to read the same wavelength value as the wavelength meter. Wavelength meters can determine the source wavelength to 1 part per million accuracy (see chapter 4). As can be seen from the measurement set-up, this calibration method is rather expensive.

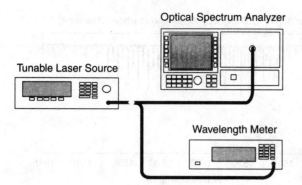

Figure 3.16 Wavelength calibration using a laser signal.

Calibration with Absorption Cells.   Calibrating with gas absorption lines (Figure 3.17) has the advantage that such lines are natural constants.[6–12] The light from a broadband source such as an edge-emitting LED (see Chapter 9) is passed through a glass tube containing a molecular gas. Gas cells absorb radiation near the vibrational and rotational resonances of molecules. The resulting light is collected and passed on to an OSA. The strongest absorption occurs at the fundamental resonance frequency for gas molecules which most often occurs at wavelengths longer than 2 μm. The available absorption lines for the important 1550 nm fiber optic band are limited. The two most promising candidates are acetylene and hydrogen cyanide. The resonances for both of the molecules are overtones of the fundamental vibrational frequencies. Figure 3.18 shows the absorption spectrum for acetylene.[6] There is a set of absorption spikes that are nearly uniformly spaced over the 1510 to 1545 nm range. The length of the evacuated tube for this measurement is 5 cm and the gas pressure is 400 Torr. The magnitude of the absorption lines is less than 3 dB for these conditions. Tables 3.1 and 3.2 list the designations and wavelengths for the absorption lines in acetylene and some of its carbon isotopes.[7] The vacuum wavelengths of these absorption lines in the table have been measured to an accuracy of 0.001 nm. The R prefix in the tables designates resonances in the shorter wavelength lobe and the P prefix is for the longer wavelength absorption lobe. Saka, Sudo, and Ikegami[8] have studied the effect of temperature and pressure for the acetylene absorption band. They concluded that the temperature sensitivity is less than 100 kHz/K and the pressure

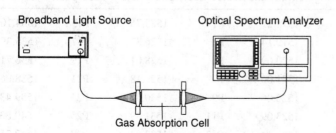

Figure 3.17 Wavelength calibration using natural absorption lines.

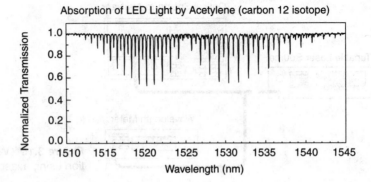

Figure 3.18 Absorption of LED light by acetylene $^{12}C_2H_2$ (source: Sarah Gilbert at NIST).

sensitivity is less than 1.5 kHz/Pascal. An extreme temperature variation would be 100 K resulting in a 1 MHz shift. This corresponds to a 0.000008 nm wavelength shift at 1550 nm and would therefore be unimportant. The absorption level per unit length in the cell can be increased by increasing the gas pressure inside the tube. At higher pressures, the width of the absorption lines will start to broaden as the individual molecules in the gas start interacting with each other.

An alternate absorption cell gas is hydrogen cyanide.[12] The absorption characteristics for a 15 cm long tube at 150 Torr is shown in Figure 3.19. Table 3.3 lists the absorption lines for hydrogen cyanide. Hydrogen cyanide absorption is well centered on the EDFA gain band around 1545 to 1560 nm. For this reason, it is a desirable absorption cell for WDM applications. The only drawback to hydrogen cyanide is its toxicity. Hydrogen cyanide attaches to hemoglobin molecules in the blood rendering them useless to carry oxygen. The amount of gas present in an absorption cell can be made small enough so that calibrations can be accomplished without hazard. DWDM system measurements require high wavelength accuracy giving absorption cell technology significant commercial interest.

Table 3.1 Vacuum wavelengths (nm) of selected acetylene ($^{12}C_2H_2$) absorption lines.

R27	1512.45	R13	1518.21	P1	1525.76	P15	1534.10
R25	1513.20	R11	1519.14	P3	1526.87	P17	1535.39
R23	1513.97	R9	1520.09	P5	1528.01	P19	1536.71
R21	1514.77	R7	1521.06	P7	1529.18	P21	1538.06
R19	1515.59	R5	1522.06	P9	1530.37	P23	1539.43
R17	1516.44	R3	1523.09	P11	1531.59	P25	1540.83
R15	1517.31	R1	1524.14	P13	1532.83	P27	1542.25

Table 3.2 Vacuum wavelengths (nm) of selected acetylene ($^{13}C_2H_2$) absorption lines

R26	1521.20	R12	1526.95	P2	1534.35	P16	1542.39
R24	1521.95	R10	1527.86	P4	1535.43	P18	1543.63
R22	1522.72	R8	1528.80	P6	1536.53	P20	1544.89
R20	1523.52	R6	1529.76	P8	1537.66	P22	1546.18
R18	1524.35	R4	1530.74	P10	1538.81	P24	1547.49
R16	1525.19	R2	1531.75	P12	1539.98	P26	1548.82
R14	1526.06	R0	1533.41	P14	1541.17	P28	1550.18

Air versus Vacuum Wavelengths. All fiber-optic input spectrum analyzers directly measure the wavelength of light in an air environment. No one has undertaken the challenge of building a vacuum chamber into an OSA. Most wavelength measurements are quoted in terms of vacuum wavelengths or optical frequency. This means that the OSA must convert between these display options. The relationships between these variables are:

$$c = f\lambda_{vac} \tag{3.7}$$

where $\lambda_{vac} = \lambda_m n_m$ is the wavelength of light in a vacuum, c = speed of light in vacuum (2.99792458E+8 m/s), f = optical frequency, λ_m = wavelength in medium (for example, air), and n_m = refractive index of the medium.

OSAs directly measure λ_m. The main issue is that the index of refraction depends on temperature, pressure, and humidity. To do a perfect conversion between the vacuum wavelength and air wavelength would require an accurate knowledge of the refractive index. The refractive index of air at sea level for a 15 C temperature and no humidity is 1.000273 (at 1550 nm). Chapter 4 has a section that describes how the index of refraction changes with environmental variables. It is necessary for the user to enter a value of the index of refraction in order to get accurate vacuum wavelength on frequency displays. Before the introduction of DWDM systems, the small errors introduced by uncertainty in the index of refraction were not significant for most measurements. An error of 0.4 nm is made if the index of refraction of air is ignored at a wavelength of 1550 nm. With 0.8 nm-spaced DWDM channels, this is a significant error.

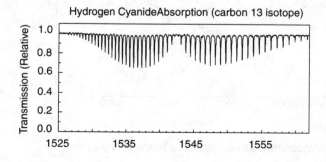

Figure 3.19 Absorption of LED light by hydrogen cyanide ($H^{13}CN$) (source: Sarah Gilbert at NIST).

Table 3.3 Vacuum wavelengths (nm) of selected hydrogen cyanide (H^{13}CN) absorption lines.

R25	1528.05	R12	1534.42	P1	1543.11	P14	1552.93
R24	1528.49	R11	1534.97	P2	1543.81	P15	1553.76
R23	1528.93	R10	1535.54	P3	1544.52	P16	1554.59
R22	1529.38	R9	1536.12	P4	1545.23	P17	1555.44
R21	1529.84	R8	1536.7	P5	1545.96	P18	1556.29
R20	1530.31	R7	1537.3	P6	1546.69	P19	1557.16
R19	1530.79	R6	1537.91	P7	1547.44	P20	1558.03
R18	1531.28	R5	1538.52	P8	1548.19	P21	1558.92
R17	1531.77	R4	1539.15	P9	1548.96	P22	1559.81
R16	1532.28	R3	1539.79	P10	1549.73	P23	1560.72
R15	1532.80	R2	1540.43	P11	1550.52	P24	1561.64
R14	1533.33	R1	1541.09	P12	1551.31	P25	1562.56
R13	1533.87	R0	1541.75	P13	1552.12	P26	1563.50

Let us assume that an OSA is calibrated with an acetylene absorption cell so that a 1550 nm laser reads exactly 1550 nm at sea level. What type of wavelength error would be introduced by a climb into the mountains with this OSA? Figure 3.20 shows how an OSA calibrated at sea level would change its displayed reading as a function of elevation. For a very tall mountain, 5000 m, the instrument would be in error by 0.2 nm. This would not be sufficiently accurate for many applications.

Accuracy After User Calibration. Wavelength reproducibility, as defined for most OSAs, specifies wavelength tuning drift in a 1 min period. This is specified with the OSA in a continuous sweep mode and with no changes made to the tuning. In addition to

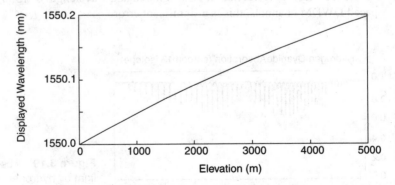

Figure 3.20 Measured wavelength as a function of elevation.

wavelength reproducibility, wavelength repeatability specifies the accuracy to which the OSA can be retuned to a given wavelength after a change in tuning. If a user calibrates the wavelength of the OSA making measurements, then the errors are greatly reduced. The tolerances due to shock and vibrations are compensated, and temperature changes during the measurement usually are less than a few degrees (we assume that the user calibrates the OSA after it has been warmed up). The remaining uncertainties stem from the image quality, the mechanical repeatability, residual nonlinearities, and the limitations of the encoder that determines the grating angle. For further details on OSA calibration, see Reference 5.

3.4.3 Wavelength Resolution and Dynamic Range

Resolution Bandwidth.   The ability of an OSA to display two signals closely spaced in wavelength as two distinct responses is determined by the wavelength resolution. Wavelength resolution is, in turn, determined by the bandwidth of the optical filter. The term resolution bandwidth is often used to describe the width of the optical filter in an OSA. The filter bandwidth is limited by the grating resolution (Equation 3.6) together with the input and output aperture sizes and the quality of the optical components. An example calculation for a single-pass monochromator was given in section 3.3.8. The resolution is also influenced by the number of times that the optical signal impinges upon the diffraction grating. Double monochromators have significantly sharper filter skirts than single monochromators. OSAs have selectable filters of 10 nm down to less than 0.1 nm, that make it possible to select the appropriate resolution for most measurements.

Figure 3.21 shows the filter shape typical for a double-pass monochromator with singlemode fiber at its input for several values of resolution bandwidth. The collimated beam diameter in this example is 2 cm using a 1000 lines/mm diffraction grating in the Littrow configuration. The shape of the filter is important for measuring parameters such as the side-mode suppression ratio in DFB lasers or for measuring the signal-to-noise ratio in amplified DWDM systems. The filter shape also affects noise-level measurements. This figure illustrates graphically how the width of the exit slit and the size of the image in the focal plane combine to determine the filter shape of the instrument. The size

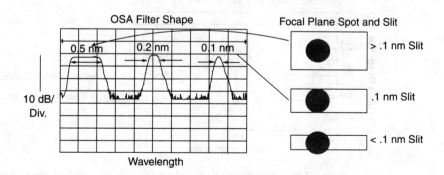

Figure 3.21 Typical filter shape.

of the imaged spot is determined by the number of illuminated lines on the grating and the number of passes off of the grating. The shape of the image can also be influenced by the quality of the collimating optics and the input-image quality. The filter shape of the instrument is the convolution of the image shape and the slit shape. If the exit slit is wider than the image size, the passband is dominated by the slit width. The minimum resolution is obtained when the slit and image size are approximately the same size. If the slit width is decreased further, the insertion loss of the filter will increase for a small narrowing of the filter width.

Noise Bandwidth.   An OSA reads out a nominal resolution bandwidth setting from its front panel control. Most noise measurements are defined for a filter shape with a flat-topped passband and infinitely steep filter skirts. Since OSAs do not achieve this perfect filter shape, an effective noise bandwidth of the filter must be measured and calculated. The effective noise bandwidth of a filter is such that it would pass the same total noise power as a flat-topped filter of the same bandwidth. OSAs have a noise marker function that uses stored filter shape data so that the effective noise bandwidth can be directly read out without the need for the user to measure the filter shape.

Dynamic Range.   Dynamic range refers to the ability of a spectrum analyzer to simultaneously look at large and small signals in the same sweep. The dynamic range of an OSA is primarily determined by the filter shape and stopband performance of the tunable filter. Very sharp filter skirts and deep stopbands in the filter response are desired. Figure 3.22 shows the filter shape of a single monochromator versus a double-pass monochromator instrument. The advantage of double monochromators over single monochromators is that double-monochromator filter skirts are much steeper, and they allow greater dynamic range for the measurement of weak spectral components located very close to a large spectral component. The monochromator parameters for the double-pass case are the same as that for Figure 3.21.

Scattered light within a monochromator limits the ultimate stopband performance of a monochromator. Imperfections in the grating lines are an example mechanism for scat-

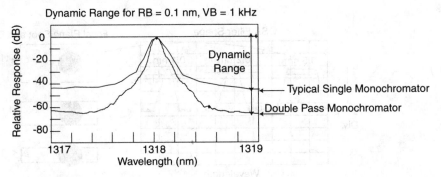

Figure 3.22 Example filter shapes, of a single- and double-monochromator OSA.

tered light. Dynamic range is commonly specified at 0.5 nm and 1.0 nm offsets from the main response. Specifying dynamic range at these offsets is driven by the mode spacings of typical DFB lasers. A −60 dB dynamic-range specification at 1.0 nm and greater indicates that the OSA's response to a purely monochromatic signal will be −60 dBc or less at offsets of 1.0 nm and greater.

Stray-light Limitations on Dynamic Range.   The stopband performance of a monochromator is often limited by stray light. The stray light is caused by scattering of light within the optical system. The stopband performance of the filter can be improved if the detector can distinguish between the stray light and the desired light from the mono-chromator. Improvements in performance can be obtained in several ways. The scattered light in the focal plane of the monochromator can be sensed and the OSA can measure the difference between the received light and the background light level. The improvement in dynamic range performance for such a subtraction scheme is illustrated in Figure 3.23. The main purpose of this chopper mode is to provide stable sensitivity levels for long sweep times, which could otherwise be affected by drift of the electronic circuitry. The desired stability is achieved by automatically chopping the light to stabilize electronic drift in long sweeps. The procedure samples the noise and stray light before each trace point and subtracts them from the trace point reading. Another method to reject stray light is to chop the light in the signal path. The detector can have a bandpass filter at the chop frequency to distinguish stray light from that coming from the desired optical path.

Figure 3.24 shows a DFB laser measured using a single and double-pass monochro-mator. It illustrates the importance of the resolution bandwidth, the filter shape, and dy-namic range of OSAs for DFB laser measurements. Single monochromators often do not have steep enough filter skirts to see the low-level longitudinal modes adjacent to the main peak.

Dense wavelength division multiplexing (DWDM) systems are pushing dynamic range requirements. To characterize channels with 100 GHz (0.8 nm) spacing, the mono-

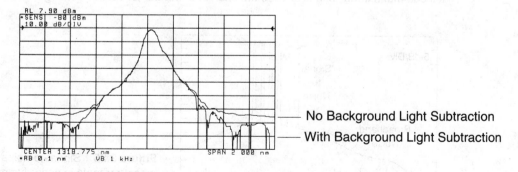

Figure 3.23 Dynamic-range improvement from the background sub-traction (chopper) mode.

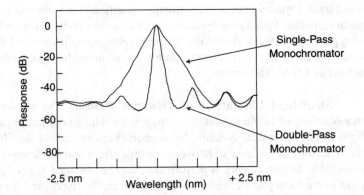

Figure 3.24 Typical dynamic-range limits for single, double, and double-pass monochromators.

chromator must have good stopband performance and steep skirts. Figure 3.25 shows a single-to-noise ratio (SNR) measurement on a four channel DWDM system with 200 GHz (1.6 nm) channel spacing. For signal-to-noise ratio measurements in 0.8 nm spaced channels, the desired dynamic range is at least 40 dB at a 0.4 nm offset.

3.4.4 Sensitivity and Sweep Time

Sensitivity is defined as the minimum detectable signal and is often defined as six times the root-mean-square (rms) noise level of the instrument. The sensitivity of an instrument is determined by the loss in the monochromator, and the sensitivity of the optical detector following the monochromator. The resolution bandwidth of the instrument does not affect the sensitivity as it does in electrical spectrum analyzers.

The dominant loss mechanism for the monochromator is the diffraction-grating efficiency. The blaze (the shape of the lines in the grating) of the diffraction grating can be optimized for the wavelength at which the OSA is intended to operate. For the fiber-optic telecommunications bands at 1300 nm and 1550 nm, the diffraction grating is gold-coated

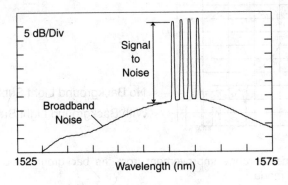

Figure 3.25 Signal-to-noise ratio measurements in optically amplified DWDM systems.

to increase diffraction efficiency. The loss of a single pass through a monochromator is in the 3 to 8 dB range.

Sensitivity is coupled directly to video bandwidth, as illustrated in Figure 3.26. The video bandwidth controls the amount of noise present in the receiving electronics. The sensitivity of the optical receivers is covered in detail in Appendix A. The major considerations are the dark current of the detector and the bandwidth and noise contribution of the amplifying electronics. The dark current of a detector scales with the area of the detector. The minimum size for the detector is determined by the slit widths that are chosen in the image plane. For the special case of the double-pass monochromator shown in Figure 3.14, the detector size can be made very small, independent of the size of the slit width. It is surprising to note that the filter bandwidth does not affect OSA sensitivity. This is in contrast to electrical spectrum analyzers which have a sensitivity proportional to the filter bandwidth.

Sweep-Time Limits.   For fast sweeps and low sensitivity settings, sweep time is limited by the maximum tuning rate of the monochromator. A direct-drive-motor system allows for faster sweep rates when compared with OSAs that use gear-reduction systems to rotate the diffraction grating. As the sensitivity level increases, the video detector bandwidth decreases (the transimpedance amplifier gain increases), resulting in a longer sweep time, since the sweep time is inversely proportional to the video bandwidth.

Continuously variable digital video bandwidths improve the sweep time for high-sensitivity sweeps in two ways. First, the implementation of digital video filtering is faster than the response time required by narrow analog filters during autoranging of amplifier gains. Second, since the video bandwidth can be selected with great resolution, just enough video filtering can be employed, resulting in no unnecessary sweep-time penalty caused by using a narrower video bandwidth than required.

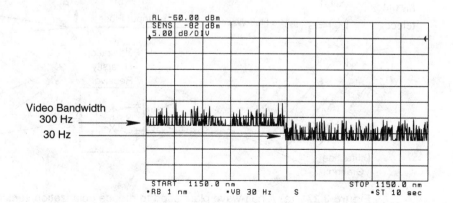

Figure 3.26   Video bandwidth directly affects sensitivity.

3.4.5 Input Polarization Sensitivity

Cause of Polarization Sensitivity.   Polarization sensitivity results from the insertion loss of the diffraction grating being a function of the polarization angle of the light that strikes it. Polarized light can be divided into two components. The component parallel to the direction of the lines on the diffraction grating is labeled p-polarization and the component perpendicular to the direction of the lines on the diffraction grating is labeled s-polarization. The loss of the diffraction grating differs for the two different polarizations, and each loss varies with wavelength. At each wavelength, the loss of p-polarized light and the loss of s-polarized light represent the minimum and maximum losses possible for linearly polarized light. At some wavelengths, the loss experienced by p-polarized light is greater than that of s-polarized light, while at other wavelengths, the situation is reversed. This polarization sensitivity results in an amplitude uncertainty for measurements of polarized light and is specified as polarization dependence.

Solutions to Polarization Sensitivity Problem.   Two solutions for reducing polarization sensitivity in monochromators are shown in Figure 3.27.[4,13,14] In Figure 3.27a, a half-wave plate has been placed in the path of the optical signal between the first and second pass in the double-pass monochromator. A half-wave plate rotates the polarization of an s-polarized input component to a p-polarized component. Similarly a p-polarized component is transformed into an s-polarized component. In this implementation, the total loss of any input polarization receives the same total loss on two passes through the grating. This forms a monochromator that is independent of polarization. The difficulty in this design is finding a waveplate with adequate wavelength coverage.

In Figure 3.27b, the light to the monochromator is separated into separate paths for each polarization with a polarization walk-off crystal. Separate detectors are used at the output to measure the transmitted power in each polarization. If the loss for the monochromator in both polarizations is characterized and stored in the instrument, the dis-

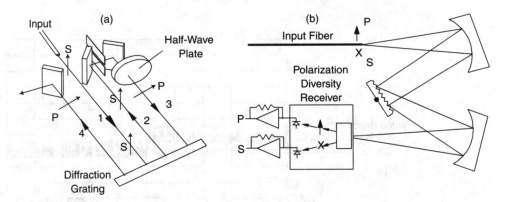

Figure 3.27   (a) A half-wave plate used to reduce polarization sensitivity. (b) Polarization-diversity receiver method to reduce polarization sensitivity.

played results can be made independent of polarization. A disadvantage of this technique is that the instrument sensitivity depends on the input light polarization. This technique does not permit a monochromator fiber output that is polarization insensitive.

3.5 SPECTRAL MEASUREMENTS ON MODULATED SIGNALS

3.5.1 Signal Processing in an OSA

If the light at the OSA input changes with time, then the signal to be measured must be described as a function of wavelength and time. However, the operation of the instrument also depends on time. In order to measure all spectral components, the grating rotates so that different wavelengths pass through the slit. If the modulation rate of the incoming light is very much higher than the rotation rate of the OSA, the correct time averaged spectrum will be measured. If the modulation rate is comparable to the speed of the grating rotation or the speed of the detection electronics, then care must be taken to get correct spectral measurements. Special triggering modes are available in OSAs to facilitate measurements of sources modulated at low repetition rates (for example, <250 kHz).

To understand triggering modes, it is useful to review the data acquisition procedure for an OSA. Figure 3.28 illustrates the standard free-run mode of operation. Refer to Figure 3.2 for the block diagram of an OSA. The instrument initiates a sweep of the diffraction grating. The signal from the photodetector is amplified and applied to the analog-to-digital converter (ADC) for data acquisition samples. The analog-to-digital conversion occurs at a fixed rate (for example, 37.5 μs). In the best case, it takes the trace length (for example, 800 points) times the ADC conversion time to scan a given wavelength range (for example, 30 ms/trace). After the ADC, a digital signal processor (DSP) further processes the data. For example, the video bandwidth function is often implemented in the digital processor. Finally, the data is log-converted and transferred to a display unit. When the sweep has been completed, the grating moves back into the start position. This cycle repeats itself as long as continuous sweep is active.

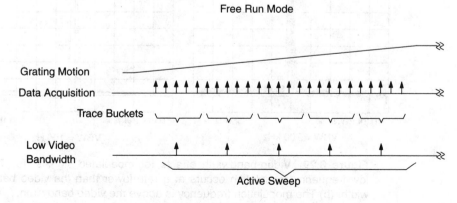

Figure 3.28   Free-run mode.

If the input power and spectrum are constant over time, then only the grating motion and the digital filters in the DSP must be synchronized to generate an accurate trace on the screen. In this case, the grating speed mainly depends on the wavelength range to be covered and the required sensitivity: the slower the grating rotates, the more samples from the ADC can be averaged by the video bandwidth (VBW) function into one trace point on the screen. Sometimes such a trace point is called a "trace bucket" because it actually combines several ADC values.

If the signal is modulated at a high enough frequency, the OSA still can measure the average spectrum without any external synchronization. The VBW must be significantly smaller than the lowest modulation frequency component, otherwise the signal can look very distorted (Figure 3.29).

In many cases the spectrum at a given point within the modulation period is more meaningful than the average spectrum. Some OSAs offer a great variety of triggering modes to characterize test signals from components, sensors, subsystems, or other light sources modulated in the frequency range approximately between 10 Hz and 250 kHz. Again, it is possible to measure only the average spectrum by choosing a low VBW. Even if the analog bandwidth is higher than the modulation of the signal, the VBW function will low-pass filter the samples. A signal at the trigger input of the OSA can synchronize a variety of functions. For example, the trigger signal can (mutually exclusive):

- Initiate measurements with the grating remaining fixed at a specific wavelength (zero span mode)
- Start the grating motion on a trigger signal *(triggered sweep)*

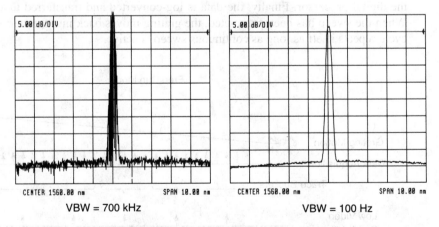

CENTER 1560.00 nm   SPAN 10.00 nm   CENTER 1560.00 nm   SPAN 10.00 nm

VBW = 700 kHz   VBW = 100 Hz

Figure 3.29 Video-bandwidth effects for modulated signals. (a) The low frequency modulation occurs at a rate lower than the video bandwidth. (b) The modulation frequency is above the video bandwidth.

- Sample and A/D convert a data point a specified time after the trigger signal *(ADC trigger mode)*
- Tell the DSP when the optical spectrum is valid *(gated sweep)*

The following sections will describe these methods in more detail.

3.5.2   Zero-Span Mode

If the span is zero (in other words, start wavelength = stop wavelength), then the grating remains at the angular position representing the center wavelength. The optical filter is fixed in wavelength. This measurement will record the power at any particular time after the trigger event that started the measurement. If the time response is successively recorded at many different wavelengths, the spectrum versus time of a pulse signal can be recorded.

Besides looking at a low frequency (< 10 kHz) modulation, this mode has a major speed benefit for an accurate power measurement at one wavelength. Instead of placing a marker at a desired wavelength and then reading its power level, the OSA can be placed in zero span at that wavelength and then *the average power of the whole trace can be read.* Because a trace consists of many points, the values for all of these points can be averaged. To achieve the same noise or modulation suppression for a sweep with span greater than zero, the VBW must be very low, and therefore the sweep time becomes very long.

3.5.3   Triggered Sweep Mode

In this mode, the grating waits in a position according to the start wavelength until it receives a trigger pulse (Figure 3.30). Then the grating starts to move in the same way as in the free-run mode. There is no difference in building trace buckets or averaging the signal by a low VBW. After the sweep, the grating stops at the start position and waits for the next trigger event.

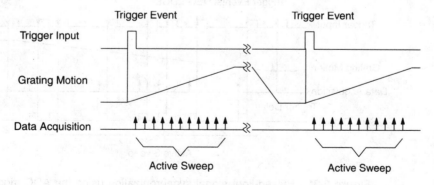

Figure 3.30   Triggered-sweep mode.

Each sweep results in a trace which can be processed further. For example, the max. hold function will take the trace displayed before, compare each trace point with the new data, and then display only the greater value of each point. Often, a swept source, such as a tunable laser, triggers a sweep after each wavelength step. Triggered sweep also works in zero-span mode. In this case, a trigger edge causes the start of the data acquisition for an entire trace.

3.5.4 ADC-Trigger Mode

The ADC-trigger mode samples the raw data at a specified time after a positive or negative edge of the signal is at the trigger input (Figure 3.31). The grating runs continuously but the data acquisition is synchronized. If there is a trigger event, then the OSA will sample the data after the specified delay and digitize it.

Testing an unpackaged source component, such as a laser or LED on a chip, is a common OSA application of this trigger mode. Pulse current to the laser is used to avoid heating effects which alter the spectral shape. Figure 3.31 shows the current versus time and power versus time for a pulsed laser. The ADC-trigger mode allows the spectrum to be sampled during the "on" time of the laser.

3.5.5 ADC-AC Mode

Similar to the ADC-trigger mode, ADC-AC samples the data delayed after a trigger event. While the first event triggers on either a positive or a negative edge, the ADC-AC mode alternates between positive and negative edges (Figure 3.32). In addition, the DSP processes the data differently: It calculates the absolute difference between the samples acquired after the positive trigger edge and the ones acquired after the negative edge. The resulting trace point represents only the modulation amplitude, so that any constant light or light modulated at a different frequency cancels out.

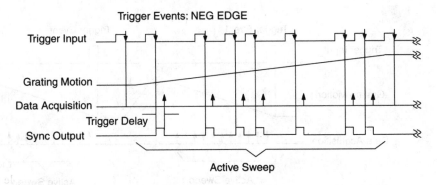

Figure 3.31 Pulsed-light signal synchronization using the ADC-trigger mode.

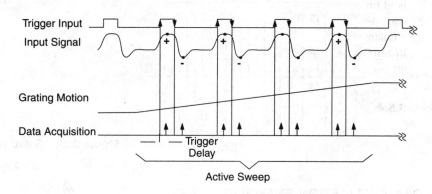

Figure 3.32   ADC-AC trigger.

In this mode, the DSP runs two VBW filters on the raw data from the ADC (one for the positive and one for the negative samples). Therefore, it reduces random noise without affecting the true amplitude of the signal.

The ADC-AC mode is similar to lock-in techniques. It measures the modulation portion of the light only, and it suppresses light that is not modulated. The impact on the effective VBW and the amplitude-range-setting considerations discussed above apply to the ADC-AC mode as well.

Applications for this mode include tests for systems incorporating EDFAs, or open-beam setups which use a 270 Hz modulation in order to suppress ambient light.

3.5.6 Gated-Sweep Mode

The gated-sweep mode tells the DSP when to retain or ignore the data coming from the ADC. Both the grating and the ADC run without synchronization to any external signal. If the trigger input is high, then the DSP takes the ADC value as a valid data point. Otherwise, it replaces the sample by a small value (for example, −200 dBm). In both cases it continues processing according to the functions selected (for example, VBW, max. hold, etc.).

If the time of the low level is longer than the time needed for the grating to move from one trace point to the next one, then the trace will have gaps (Figure 3.33). There are two alternatives to close the gaps: either increase the sweep time to at least 1.2 to 2 times the longest "low-level" period, or activate the max. hold function and let the OSA sweep several times. In the first case, the DSP will have at least one data sample marked valid (high level) per trace point. In the second case, multiple sweeps fill the gaps because the high and low levels of the gating signal occur independently of the grating position.

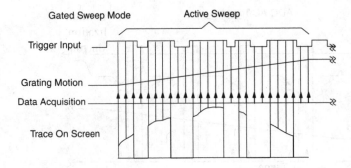

Figure 3.33   Gated-sweep mode.

3.6 OSA APPLICATION EXAMPLES

3.6.1 Light-Emitting Diodes (LEDs)

Light-emitting diodes (LEDs) produce light with a broad spectral width. They can be modulated at frequencies up to about 200 MHz. Figure 3.34 shows the spectrum of an LED. The broad spectral width of the source is often specified by the full-width at half-maximum (half-power points of the spectrum). Typical values for full-width at half-maximum (FWHM) range from 20 nm to 80 nm for LEDs.

There are many parameters of LEDs that are commonly measured. These parameters can be automatically measured as shown in Figure 3.34. Some parameters (such as mean wavelength and spectral width) have two methods by which they can be measured. One method takes into account the entire spectrum, while the other takes into account only a few points of the spectrum. The definition of each parameter is described below.

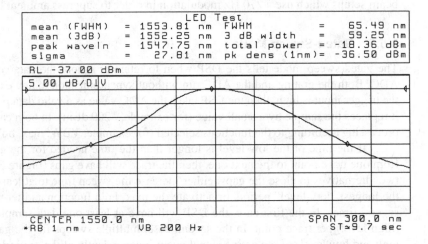

Figure 3.34   Spectrum of a light-emitting diode.

Total Power. The summation of the power at each trace point, normalized by the ratio of the trace-point spacing/resolution bandwidth. This normalization is required because the spectrum of the LED is continuous, rather than containing discrete spectral components (as a laser does).

$$P_{\text{Total}} = \sum_{i=1}^{N} \left(P_i \, \frac{\text{trace point spacing}}{\text{resolution bandwidth}} \right) \tag{3.8}$$

Mean (FWHM). This wavelength represents the center of mass of the trace points. The power and wavelength of each trace point are used to calculate the mean (FWHM) wavelength.

$$\lambda_{\text{Mean}} = \sum_{i=1}^{N} \left(\frac{\lambda_i P_i}{P_{\text{Total}}} \, \frac{\text{trace point spacing}}{\text{resolution bandwidth}} \right) \tag{3.9}$$

Sigma. An rms calculation of the spectral width of the LED based on a Gaussian distribution. The power and wavelength of each trace point are used to calculate sigma.

$$\sigma = \sum_{i=1}^{N} \left[(\lambda_i - \lambda_{\text{Mean}})^2 \, \frac{P_i}{P_{\text{Total}}} \, \frac{\text{trace point spacing}}{\text{resolution bandwidth}} \right] \tag{3.10}$$

FWHM (Full-Width at Half-Maximum). Describes the spectral width of the half-power points of the LED, assuming a continuous, Gaussian power distribution. The half-power points are those where the power-spectral density is one-half that of the peak amplitude.

$$FWHM = 2.355\sigma \tag{3.11}$$

3 dB Width. Used to describe the spectral width of the LED based on the separation of the two wavelengths that each have a spectral density equal to one-half the peak power-spectral density. The 3 dB width is determined by finding the peak of the LED spectrum, and dropping down 3 dB on each side.

Mean (3 dB). The wavelength that is the average of the two wavelengths determined in the 3 dB width measurement.

Peak Wavelength. The wavelength at which the peak of the LED's spectrum occurs.

Density (1 nm). The power-spectral density (normalized to a 1 nm bandwidth) of the LED at the peak wavelength.

Distribution Trace. A trace can be displayed that is based on the total power, power distribution, and mean wavelength of the LED. This trace has a Gaussian spectral distribution and represents a Gaussian approximation to the measured spectrum.

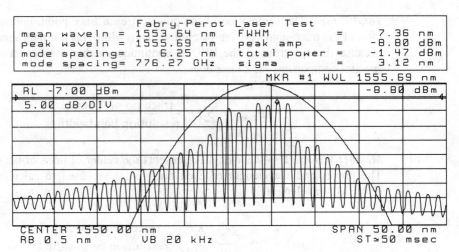

Figure 3.35 Results of an automatic Fabry-Perot laser measurement routine.

3.6.2 Fabry-Perot Lasers

As with the LED, many OSAs have an automatic measurement routine for FP lasers. The results from an FP laser measurement routine are shown in Figure 3.35. The following parameters are of interest.

Total Power. The summation of the power in each of the displayed spectral components, or modes, that satisfy the peak-excursion criteria. (See below for discussion of peak-excursion criteria.)

$$P_{\text{Total}} = \sum_{i=1}^{N} P_i \tag{3.12}$$

Mean Wavelength. Represents the center of mass of the spectral components on screen. The power and wavelength of each spectral component is used to calculate the mean wavelength.

$$\lambda_{\text{Mean}} = \sum_{i=1}^{N} P_i \lambda_i \tag{3.13}$$

Sigma. An rms calculation of the spectral width of the FP laser based on a Gaussian distribution. The power and wavelength of each spectral component is used to calculate the mean wavelength.

$$\sigma = \sqrt{\frac{1}{P_{total}} \sum_{i=1}^{N} P_i (\lambda_i - \lambda_{Mean})^2}$$ (3.14)

FWHM (Full-Width at Half-Maximum). Describes the spectral width of the half-power points of the FP laser, assuming a continuous, Gaussian power distribution. The half-power points are those where the power-spectral density is one-half that of the peak amplitude.

$$FWHM = 2.355\sigma$$ (3.15)

Mode Spacing. The average wavelength spacing between the individual spectral components of the FP laser.

Peak Amplitude. The power level of the peak spectral component of the FP laser.

Peak Wavelength. This is the wavelength at which the peak spectral component of the FP laser occurs.

Peak Excursion. The peak excursion value (in dB) can be set by the user and is used to determine which on screen responses are accepted as discrete spectral responses. To be accepted, each trace peak must rise, and then fall, by at least the peak excursion value about a given spectral component. Setting the value too high will result in failure to include the smaller responses near the noise floor. Setting the value too low will cause all spectral components to be accepted, but unwanted responses, including noise spikes and the second peak of a response with a slight dip, could be erroneously included.

Peaks Function. The peaks function displays a vertical line from the bottom of the grid to each counted spectral component of the signal. This function is useful to determine if an adjustment of the peak excursion value is required.

Distribution Trace. A trace is displayed that is based on the total power, individual wavelengths, mean wavelength, and mode spacing of the laser. This trace has a Gaussian spectral distribution and represents a continuous approximation to the actual, discrete spectrum.

3.6.3 Distributed Feedback (DFB) Lasers

Distributed feedback lasers are similar to FP lasers, except that all but one of their spectral components are significantly reduced as shown in Figure 3.36. Because its spectrum has only one line, the spectral width of a distributed feedback laser is much less than that of an FP laser. This greatly reduces the effect of chromatic dispersion in fiber optic systems, allowing for greater transmission bandwidths.

Most OSAs have an automatic measurement routine for distributed feedback lasers. The results from a DFB-laser measurement routine are shown in Figure 3.36. The following parameters are often of interest and are measured by the automatic routine.

Peak Wavelength. The wavelength at which the main spectral component of the DFB laser occurs.

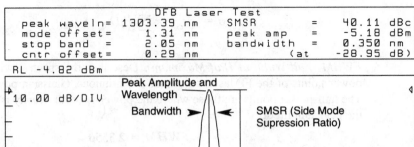

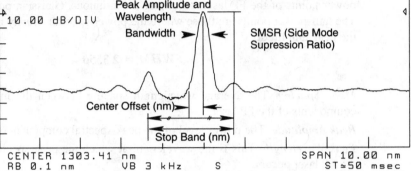

Figure 3.36   Results of automatic DFB-laser measurement routine.

Side Mode Suppression Ratio (SMSR). The amplitude difference between the main spectral component and the largest side mode.

Mode Offset. Wavelength separation (in nanometers) between the main spectral component and the SMSR mode.

Peak Amplitude. The power level of the main spectral component of the DFB laser.

Stopband. Wavelength spacing between the upper and lower side modes adjacent to the main mode.

Center Offset. Indicates how well the main mode is centered in the stopband. This value equals the wavelength of the main spectral component minus the mean of the upper and lower stopband-component wavelengths.

Bandwidth. Measures the displayed bandwidth of the main spectral component of the DFB laser. The amplitude level, relative to the peak, that is used to measure the bandwidth can be set by the user. In Figure 3.36, the amplitude level used is −20 dBc. Due to the narrow line width of lasers, the result of this measurement for an unmodulated laser is strictly dependent upon the resolution-bandwidth filter of the OSA. With modulation applied, the resultant waveform is a convolution of the analyzers filter and the modulated laser's spectrum, causing the measured bandwidth to increase. The combination of the modulated reading and unmodulated reading can be used to determine the bandwidth of the modulated laser and the presence of chirp.

Peak Excursion. The peak excursion value (in dB) can be set by the user and is used to determine which three on-screen responses will be accepted as discrete spectral responses. To be counted, the trace must rise, and then fall, by at least the peak excursion value about a given spectral component. Setting the value too high will result in failure to count small responses near the noise floor.

Peaks Function. The peaks function displays a vertical line from the bottom of the grid to each counted spectral component of the signal. This function is useful to determine if an adjustment of the peak excursion value is required.

3.6.4   Optical Amplifier Measurements

Chapter 13 discusses optical amplifiers and how to characterize them. Therefore, we give here only a quick overview of the most common measurement setup and procedure.[15–17]

Most test setups contain a tunable laser with adjustable output-power level and an OSA (Figure 3.37). The laser drives the amplifier into its saturated gain operating point. The OSA characterizes the signal and noise spectrum before and after amplification. From these two measurements, the gain and noise figure of the optical amplifier can be determined.

The measurement of signal power at the input and output is straightforward. The measurement of the amplifier output noise is more difficult. First, the signal is covering up the noise level at the wavelength of interest. The second consideration is that broadband noise present at the input signal may be amplified and add to the noise output of the optical amplifier. Because the amplified laser signal hides the noise level at the wavelength of interest, the most common measurement technique uses interpolation (Figure 3.38). Markers on either side of the signal are averaged to infer the noise level at the signal wavelength. The accuracy of the noise measurement is very important. The amplitude accuracy of the instrument may need to be compared to a power meter. It is also important that the noise bandwidth marker function be used to measure the noise level. The noise marker takes into account the optical filter shape of the monochromator to define the effective noise bandwidth of the filter. If the noise marker function is not available, it will be necessary to characterize the filter shape of the monochromator as an additional measurement step. To account for the amplified broadband noise of the source, noise subtraction algorithms have been used and are described well in Chapter 13.

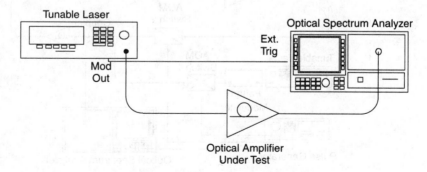

Figure 3.37   Basic optical-amplifier test setup.

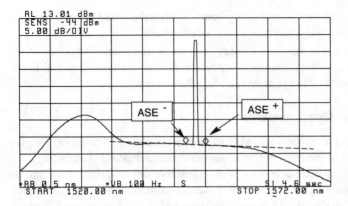

Figure 3.38 Interpolation of the amplified spontaneous emission.

3.6.5 Recirculating Loop

Researchers desire to measure how an optical spectrum shape can change when traveling along thousands of kilometers of fiber and through many optical amplifiers. Since it is impractical and expensive to build very long link lengths, researchers have resorted to methods involving recirculating loops. The setup in Figure 3.39 uses a loop technique. Typically, the loop consists of only a few EDFAs with 30 to 70 km long fibers between them, and one circulation lasts about 0.2 to 2 ms. Several circulations of a signal then simulate how that signal would behave on a very long link. This experimental setup presents significant measurement challenges for OSAs. The spectrum needs to be measured once for each circulation around the loop.

Let us examine how the spectrum can be resolved for each circulation. A pulse generator controls the timing. First, the acousto-optic modulator (AOM) 1 is open and AOM

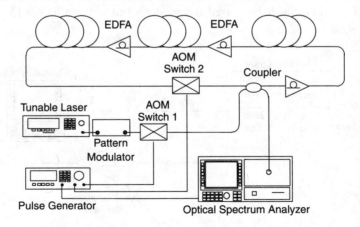

Figure 3.39 Recirculating loop setup.

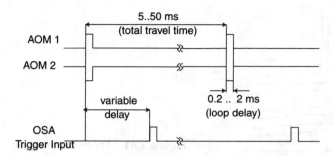

Figure 3.40 Recirculating loop timing.

2 is closed (Figure 3.40). At this time, the TLS in conjunction with an external modulator fills the loop with a pseudorandom modulated bit pattern (see Chapter 8). Second, switch 1 closes and switch 2 opens, allowing the pattern to circulate adequately (about 5 ms per 1000 simulated kilometers). Third, the OSA measures the spectrum at a variable delay (in other words, at a variable simulated distance). The three steps are repeated until the OSA has built a complete trace.

To measure the spectrum at a given distance, two trigger techniques may be used: ADC trigger or gated sweep. When using ADC trigger, the OSA samples only one data point per trigger but the time of the sampling is very well known. The OSA sweep time must exceed the total travel time multiplied by the number of trace points (for example, 50 ms × 800 = 40 s. To allow for processing overhead, choose 50 to 80 s).

When using gated sweep, the OSA keeps data samples as long as the OSA trigger is high. If the sweep time is as long as above, then the trace will be completed within one sweep. Otherwise, use max. hold, so that several sweeps can close the gaps caused during the time the ADC trigger input is low. Gated sweep also provides the advantage that the OSA measures the spectrum of a longer piece of the bit pattern (in other words, over the width of the gating pulse).

Under control of a remote program and by using a spreadsheet or math program, it is possible to create a three-dimensional (3-D) graph showing the signal and ASE amplitude as a function of wavelength and distance. There are two basic methods, each having their own advantages and disadvantages: scanning along the wavelength axis or scanning along the time axis.

Wavelength Scan.   The OSA repetitively measures the spectrum for subsequent variable delays. Each time the trace data is transferred to a computer which finally creates the 3-D plot (Figure 3.41). Because it typically takes about a minute to acquire one trace (see above), the total measurement time is in the order of M minutes (M = number of different delays, typically 10 to 30). The default trace length is 800, so this method provides good wavelength resolution. However, a fine-distance resolution requires a large M and therefore a long total time.

Time-Domain Scan.   When the OSA is set to zero span, the grating behaves as a filter with fixed-center wavelength. If the signal going to AOM 1 triggers a sweep, then

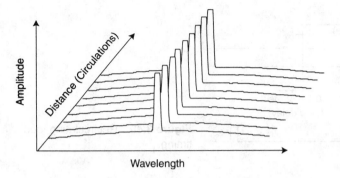

Figure 3.41 Three-dimensional loop spectrum.

subsequent ADC values represent the power versus time of the center wavelength of the OSA. With 50 ms sweep time and 800 points/trace, the distance resolution is about 7.5 km. This measurement has to be repeated for *N* wavelengths (typically 50 to 200) in order to create the 3-D plot (Figure 3.42). *N* and the span to be covered determine the wavelength accuracy. Assuming that a trace transfer to a computer lasts only few seconds, the data acquisition for this 3-D plot takes several minutes.

3.7 SUMMARY

The operation of a diffraction-grating based optical spectrum analyzer was described and example measurements were shown. The resolution and dynamic range of the monochromator filter was found to be determined by the number of illuminated lines on the grating, the input and output slit widths, and the quality of the collimating optics and the number of passes off of the diffraction grating. The wavelength accuracy of a monochromator is very important due to the demands of DWDM telecommunications systems. Absolute

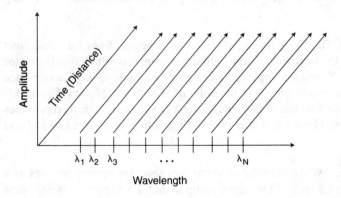

Figure 3.42 3-D scan using the zero span method.

wavelength calibration can be obtained by comparison to a laser wavelength standard, a wavelength meter (Chapter 4) or a gas absorption cell. Care must be taken in OSA design to minimize the polarization dependence of the monochromator design. The measurement of signals with low frequency modulation requires a set of triggering features to be present in the optical spectrum analyzer. Applications for OSAs including measurements of LEDs, FP lasers, DFB lasers, EDFAs, and recirculating loop experiments were given.

ACKNOWLEDGMENTS

The author would like to thank Val Mcomber, Mike LeVernier, Jim Stimple, Kenn Wildnauer and many other colleagues for their excellent material that has been the basis to write this section. Furthermore, many thanks to Anneliese, Kay-Mario, Vanessa and Yasmin for their support and patience while the author was working on his part of the book in his spare time.

REFERENCES

1. *Optical spectrum analysis basics, application note 1550–4.* 1992. HP literature number 5963-7145E. Santa Rosa, CA: Hewlett Packard Lightwave Division.

2. Hecht, E. and A. Zajag. 1979. *Optics.* Reading, MA: Addison-Wesley.

3. Hutley, M.C. 1982. *Diffraction Gratings.* London: Academic Press.

4. Wildnauer, K.R. and Z. Azary. 1993. A double-pass monochromator for wavelength selection in an optical spectrum analyzer. Hewlett Packard Journal, 44, No. 6, pp. 68–74.

5. *Calibration of Optical Spectrum Analyzers.* Geneva, Switzerland IEC Technical Committee No. 86: Fiber Optics, Working Group 4, Subgroup 5.

6. Gilbert, S.L., T.J. Drapela, and D.L. Franzen, 1992. Moderate-accuracy wavelength standards for optical communications, in *Technical Digest—Symposium on Optical Fiber Measurements.* Boulder, CO: NIST Spec. Publ. 839: pp. 191–194.

7. Varanasi, P. and B.R.P. Bangaru. 1975. Intensity and half-width measurements in the 1.525 μm band of acetylene. *J. Quant. Spectrosc. Radiat. Transfer* 15, 267.

8. Sakai, Y., S. Sudo, and T. Ikegami. 1992. Frequency stabilization of laser diodes using 1.51–1.55 μm lines of $^{12}C_2H_2$ and $^{13}C_2H_2$. *IEEE J. Quantum Electron.* 28: 75.

9. Baldacci, A., S. Ghersetti, and K.N. Rao. 1977. Interpretation of the acetylene spectrum at 1.5 μm, *J. Mol. Spectrosc.* 68:183.

10. Chanm, K., H. Ito, and H. Inaba. 1983. Optical remote monitoring of CH_4 gas using low-loss optical fiber link and InGaAsP light-emitting diode in the 1.33 μm region. *Appl. Phys. Lett.* 43: 634.

11. Nakagawa, K. Labachelerie, M. Awaji, Y. Kourogi, M. 1996. Accurate Optical Frequency Atlas of the 1.5 μm bands of acetylene. Journal of the Optical Society of America B, 13, No. 12, pp. 2708–2714.

12. Sasada, H. Yamada, K. Calibration lines of HCN in the 1.5 μm region. Applied Optics, Vol. 29, (1990) pages 3535–3547.

13. Sonobe, Y., S. Ishigaki, T. Kikugawa. Apparatus for measuring spectral power of a light beam. U.S. patent number 4,758,086.

14. Shirasaki, M., R. Yamamoto, Y. Watanabe, and S. Nishina. 1991. Polarization independent grating-type optical spectrum analyzer with fiber interface. San Diego, CA: Optical Fiber Conference, Optical Society of America, Paper WNS, pp. 130.

15. *EDFA testing with the time domain extinction technique.* 1995. HP literature number 5963-7147E. Santa Rosa, CA: Hewlett-Packard Lightwave Division.

16. *EDFA noise gain profile and noise gain peak measurements.* 1995. HP literature number 5963-7148E. Santa Rosa, CA: Hewlett-Packard Lightwave Division.

17. *WDM passive components test guide.* 1996. HP literature number 5965-3124E. Santa Rosa, CA: Hewlett-Packard Lightwave Division.

CHAPTER

4

Wavelength Meters

Dennis Derickson, Loren Stokes

4.1 INTRODUCTION

Chapter 3 covered optical spectrum analysis in general, and then went on to explain the operation of grating-based optical spectrum analyzers (OSAs) in detail. This chapter concentrates on another topic in optical spectrum analysis—wavelength meters. Wavelength meters can measure amplitude versus wavelength, as do grating-based OSAs. Wavelength meters distinguish themselves by making very accurate measurements of wavelength. A grating-based OSA can measure a 1550 nm signal with ±0.1 nm absolute-wavelength accuracy, assuming that the instrument has had a recent user calibration. A wavelength meter can make this measurement with better than ±0.001 nm accuracy. This represents a 100 times improvement.

One could ask why it is important to measure a 1550 nm signal to ± 0.001 nm accuracy. Here are four examples of fiber optic measurements that require very accurate measurement of wavelength.

1. A researcher may want to study long-term wavelength drift of a distributed feedback (DFB) laser for use in a wavelength division multiplexed (WDM) communication system. WDM lasers must be proven to drift less than 0.1 nm over a 25 year lifetime with accelerated aging studies.

2. A wavelength tunable laser is used for wavelength-resolved insertion loss measurements on a fiber-Bragg grating filter (see Chapter 9). Fiber-Bragg gratings have insertion-loss features that can change very rapidly with wavelength on the filter skirts. Tunable lasers have frequency steps of less than 0.001 nm but the step size may not always be linear. One picometer out of a wavelength of 1550 nm requires a

wavelength accuracy of 0.64 parts per million (ppm). Thus the wavelength meter can be used to calibrate the wavelength steps of the tunable laser so that the insertion loss plot has a sufficiently accurate wavelength axis.

3. When making chromatic dispersion measurements (see Chapter 12), it is necessary to measure the slope of the group delay versus wavelength function. Manufacturers need to know the zero-dispersion wavelength for components like dispersion compensators to tenths of a nanometer. Dispersion measurement algorithms require a derivative of the group delay with respect to wavelength. In order to get accurate dispersion values, the relative wavelength steps must be measured very accurately.

4. Accurate measurement of wavelength, power, and signal-to-noise ratio is very important for telecommunications systems using wavelength division multiplexing. A common WDM channel spacing is 100 GHz (0.8 nm at 1550 nm). The laser must be tuned to the precise center of each channel by temperature adjustment. This requires a wavelength settability of less than 0.05 nm. The measured data from the wavelength meter can be used to wavelength and amplitude stabilize the signals in the WDM system.

4.1.1 Wavelength Definition

For accurate measurement of wavelength in instrumentation, it is important to be clear on the medium that the wavelength is measured in. Some OSAs display the wavelength as measured in air. Other sources quote the vacuum wavelength of a signal. The relationship between the wavelength in air and vacuum is the well-known relationship:

$$\lambda_{vac} = n_{air}\lambda_{air} \tag{4.1}$$

where λ_{vac} is the vacuum wavelength, λ_{air} is the air wavelength, and n_{air} is the index of refraction of air. Although the wavelength in air and vacuum are close ($n_{air} \approx 1.00027$), this difference is 270 ppm. A 1550 nm vacuum wavelength source would have a wavelength of 1549.58 nm in standard dry air (15C temperature and 760 Torr pressure). Grating-based OSAs directly measure the wavelength in air. The instrument then calculates the vacuum wavelength based on an assumption for the index of refraction. With WDM channels spacings of 100 GHz, one can no longer ignore the difference between vacuum and air wavelengths. Section 4.4.2 describes models for the index of refraction in air as a function of temperature, pressure, and humidity. With proper attention to environmental variables, the vacuum wavelength to air wavelength conversion can be resolved to better than 1 ppm accuracy.

4.1.2 Methods of Accurate Wavelength Measurement

There are several approaches that are used to accurately measure wavelength.

1. Optical bandpass filter techniques. Chapter 3 has already covered optical spectrum analysis using a grating-based filter. Fabry-Perot (FP) filters are discussed in Section 4.6.1. Wavelength calibrator cells (see Chapter 3) are used to improve the accuracy of these measurements.

2. Interferometric fringe-counting techniques are commonly used for high accuracy applications. This chapter will concentrate on describing this method.

3. Wavelength discriminator techniques. A sloping insertion loss versus wavelength function can be used to determine optical wavelength. Section 4.6.4 will describe one implementation of this technique.

This chapter starts with a description of wavelength meters based on the Michelson interferometer. The principles of operation will be discussed in Section 4.2. The Michelson interferometer measurement technique will be generalized to multiple signal environments using Fourier transform techniques in Section 4.3. Then a discussion of wavelength accuracy considerations will be made in Section 4.4. Section 4.5 discusses other measurement considerations for Michelson interferometer instruments. Section 4.6 will discuss some alternate wavelength meter and spectrum analysis techniques that fall into the general class of wavelength meters. Finally, a comparison of all wavelength meters types will be made in Section 4.7 to show the strengths and weaknesses of each approach.

As a last note before starting wavelength meter discussion, it is appropriate to mention that there are several popular ways to display the results of a laser wavelength measurement.

vacuum wavelength, λ_{vac}
standard dry air wavelength (see section 4.4), λ_{air}
frequency = speed of light/$\lambda_{vac} = f$ or v
wavenumber or spatial frequency = $1/\lambda_{vac} = \sigma$

Vacuum wavelength is the most often used measure in fiber optic measurements. The frequency of a laser is coming into more common usage because the WDM-wavelength grid is established in THz.

4.2 THE MICHELSON INTERFEROMETER WAVELENGTH METER

Figure 4.1 shows a block diagram of a Michelson interferometer.[1-4] Light from a fiber optic input is collimated and directed to the input of the interferometer. Singlemode fibers are most commonly used for fiber optic applications. The input signal is split into two paths with a beamsplitter. Both beams are then incident on 100%-reflecting mirrors that bounce the light back toward the beamsplitter. These mirrors are most often constructed as retroreflectors so that the beams are reflected back at nearly the same angle as they are sent into the mirror. Part of the light reflected from the two arms of the interferometer goes back toward the input beam. The other portion of the light is incident on a photodetector. Since there is no loss assumed in the interferometer, all of the light is directed to either the photodetector or the input port.

If the variable length interferometer mirror is moved, the amount of light reaching the photodetector will oscillate up and down because of constructive and destructive in-

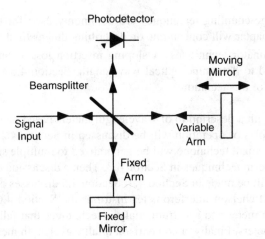

Figure 4.1 Block diagram of a Michelson interferometer.

terference effects between the two paths of the interferometer. Through the analysis of these interference patterns, the wavelength of light can be calculated. There are two alternate but equivalent viewpoints useful in analyzing the interferometer operation. The two beams can be analyzed in terms of light interfering as the path length in the interferometer changes. This will be referred to as the fringe-counting description of wavelength-meter operation. This approach will be discussed first in the next section. Alternately, if one arm of the interferometer is moved at a constant rate, the frequency of the light in the moving arm is Doppler-frequency shifted. The detector then mixes (see Chapter 5) the light from the Doppler-shifted and unshifted arms. The beat frequency between these two signals will be used to calculate the unknown frequency of the input signal.

4.2.1 Fringe-Counting Description of Wavelength Meter Operation

The light reflected from the two mirrors is interfered where it is combined at the detector. In order for interference to occur, the light from the two arms must overlap in space and be identical in polarization. As an example, assume that the light from the input beam has a well-defined input wavelength such as that from a DFB laser. The form for the photocurrent generated from the interferometer detector is:

$$I(\Delta L) = 1 + \cos((2\pi\Delta L)/\lambda_u) + \phi \qquad (4.2)$$

where I is the photodetector photocurrent, ΔL is the optical path length difference between the two interferometer arms, λ_u is the unknown wavelength of the light in the medium of the interferometer, and ϕ is a phase-shift difference for equal path length delays between the two arms. Note that the path-length difference, ΔL, is twice the mirror-movement distance because of the double transit through each interferometer arm. This interference between two light beams coming from a signal source is called homodyne interferometry. Homodyne analysis is studied extensively in Chapter 5.

If the path-length difference, ΔL, is an integer multiple of the wavelength of light in the medium of the interferometer, the light will constructively interfere. All of the input light will be incident on the detector. If the path-length difference results in destructive interference at the photodetector, all of the light will go back out the input port.

In wavelength meter measurement operation, the position of the variable length arm is scanned. Figure 4.2a shows the scanned interferometer output measurements for a 1550 nm DFB laser. The result of a measurement of photocurrent versus mirror position will be referred to as an interferogram. The interferogram shows the detector signal alternating from dark to light as the variable position mirror is scanned. The plot of Figure 4.2a shows only a small segment of the interferogram over a 25 μm window. Figure 4.2b shows the optical spectrum for the same laser. For a narrow spectral width signal (10 MHz linewidth) such as the DFB shown in Figure 4.2b, the interference signal will remain strong for interferometer delays of many meters.

Figure 4.3a shows the interferogram measurement for a 1550 nm LED. Figure 4.3b shows the power versus wavelength for the 1550 nm LED as measured with a grating-based OSA. The interferograms for the LED and DFB are very different. Strong interference is found over a wide path-length difference range for the DFB of Figure 4.2. Interference patterns are found only near zero path-length difference for the LED. The difference between the two interferograms is caused by the difference in coherence properties of a source. For a DFB laser, the two signals arriving at the photodetector have a well-defined phase relationship resulting in a strong interference signal even for wide ranges of the variable mirror position. For the LED source, the phase relationship between the two signals starts to become random as the path-length difference increases. This randomness in the phase relationship is caused by the LED having a less well-defined wavelength due to the source's wide spectral width. For a broadband source with a Gaussian power versus wavelength distribution, the photocurrent interferogram function is given as:

$$I(\Delta L) = 1 + \exp^{-\left(\frac{\pi}{4\sqrt{2}} \cdot \left(\frac{4\Delta L \Delta \lambda_{pulse}}{\lambda^2 - \Delta \lambda_{pulse}^2}\right)^2\right)} \cos\left(\frac{2\pi\Delta L}{\lambda}\right) \tag{4.3}$$

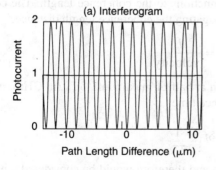

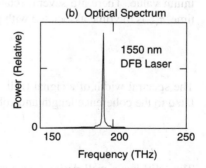

Figure 4.2   (a) Interferogram for a 1550 nm DFB laser. (b) Optical spectrum of the 1550 nm DFB laser.

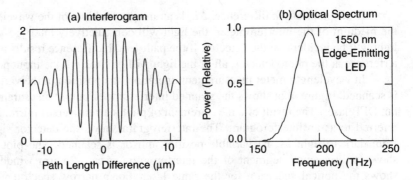

Figure 4.3 (a) Interferogram for a 1550 nm LED. (b) Optical spectrum of the 1550 nm LED.

where $\Delta\lambda_{pulse}$ is the spectral width of the LED at half power points, λ is the center wavelength of the LED, and ΔL is the optical path-length difference.

Figures 4.2 and 4.3 demonstrate that the Michelson interferometer can distinguish between wide and narrow spectral-width signals. The degree of complete constructive and destructive interference in the interferogram is referred to as fringe visibility. The fringe visibility becomes poor for the LED example of Figure 4.3 for path-length differences away from zero. The interferogram for the LED starts to reduce away from zero path-length delay because the phase fluctuations of the two delayed arms are uncorrelated for large path-length differences. Michelson interferometers with low coherence sources inputs can be used for high-resolution reflectometry (Chapter 10).

Let's take a moment to define some of the terminology associated with the spectral width of a signal. The terms linewidth, spectral width, coherence length, and coherence time all refer to the same basic property of a signal. The coherence length of a signal is directly measured in the interferogram examples of Figures 4.2 and 4.3. The coherence length, L_c, is defined as the distance where the coherence function drops to $1/e$ of its maximum value. There are several related functions to the coherence length. The coherence time, τ_C, is the time associated with propagating the coherence length distance.

$$\tau_C = \frac{L_c}{\text{velocity}} \qquad (4.4)$$

The spectral width of a signal (full-width at half of maximum), $\delta f_{1/2}$ (in GHz), is also related to the coherence length and coherence time.

$$\delta f_{\frac{1}{2}} = \frac{1}{\pi \tau_C} \qquad (4.5)$$

The DFB has a well-defined wavelength and therefore would be considered a highly coherent signal. The LED in contrast is more incoherent because it does not have nearly as well defined of center wavelength.

For the example of Figure 4.3, the LED would have an approximate coherence length of 16 μm, and a coherence time of 0.06 ps for a spectral width of 50 nm (6 THz). Let's contrast this to a DFB laser with an unmodulated linewidth of 10 MHz. The coherence length would be 10 meters and the coherence time would be 32 ns. It would not be practical to build a variable length arm into a Michelson interferometer to directly measure the coherence length of an unmodulated DFB laser.

Unknown Wavelength Calculation.   Now that we have spent some time understanding how interferograms are formed in a Michelson-interferometer scan, it is time to see how this information can be used to measure the wavelength of the input signal. The principle of wavelength measurement is very simple. The distance between peaks in the interferograms of Figures 4.3 and 4.4 gives the wavelength. If the interferometer is in a vacuum, λ_{vac} is measured. If the interferometer is in air, λ_{air} is measured.

Let us first assume that one can measure the position of the movable mirror with perfect accuracy and therefore ΔL is a known variable. The measurement of wavelength is then found by moving the variable-length mirror a known amount of distance and then counting the number of interference fringes that appear at the detector. If the mirror movement distance is Δx, the path length change is $\Delta L = 2\Delta x$. The number of counted fringes in the length ΔL is N. The unknown wavelength in the medium of the interferometer, λ_u, will be:

$$\lambda_u = \left(\frac{\Delta L}{N}\right) \tag{4.6}$$

It is important to note that the unknown wavelength is the wavelength measured in the medium of the interferometer. If the interferometer is located in a vacuum, the parameter, λ_u, will be the vacuum wavelength. If the interferometer is located in an air environment, λ_u will be the wavelength for that particular air environment. On first assumption, one can assume that the index of refraction of air is 1.00027. The index of refraction of air will be examined in detail in Section 4.4.2.

So far two limitations have been introduced to the measurement of wavelength:

1. The position of the mirror must be known very accurately.
2. The index of refraction of the interferometer environment has to be known accurately.

These limitations will be addressed in sections 4.2.3 and 4.2.4. Before going on to these discussions, it is useful to view the operation of the interferometer from a Doppler-shift point of view.

4.2.2 Doppler-Shift Approach to Understanding Wavelength Meter Operation

In this approach let us assume that the adjustable mirror is moved at a constant velocity through the zero path-length difference position. The moving mirror will cause a Doppler frequency shift on the light in the moving arm. The Doppler frequency shift will be:

$$\Delta f = (2v_m f_0/(v_i)). \tag{4.7}$$

where Δf is the Doppler frequency shift, v_m is the mirror velocity, f_u is the optical frequency of the signal, and v_i is the speed of light in the medium of the interferometer. What might be a typical Doppler frequency shift? For a mirror velocity of 1.5 m/s and a center frequency of 193.5 THz, (1550 nm light), the Doppler frequency shift would be 1.93 MHz.

The light incident on the photodetector is of two slightly different center frequencies in the moving interferometer case studied here. The detector will measure the beat frequency between the fixed and moving arm paths. If the detector is measuring the beat-frequency signal for the time period, T, the number of zero crossings, N, measured during the period, T, will be:

$$N = \Delta f T \tag{4.8}$$

The unknown frequency of the input signal can then be measured as:

$$f_0 = (v_i N/(2v_m T)) \tag{4.9}$$

In this measurement, accurate knowledge of mirror velocity and a time period are required. This is equivalent to knowing the total distance moved as is analyzed in the fringe-counting method since velocity multiplied by time gives a total distance.

4.2.3 Accurate Measurement of Distance, Velocity, and Time

In Section 4.2.1, it was shown that it is critical to have accurate position measurements of the moveable mirror. One must either know the absolute mirror position accurately, or know the velocity and sample period accurately.

It is interesting to note that Michelson interferometers are often used as accurate distance-measurement tools. If the wavelength of the input laser, λ_{known}, is already a well-known number, Equation 4.4 can be rewritten so that distance is the quantity being measured.

$$\Delta L = \lambda_{known} N \tag{4.10}$$

The key to accurate wavelength measurement therefore is to include a very accurate laser reference in the same interferometer as the unknown input. Figure 4.4 shows a wavelength meter that uses the Michelson interferometer with two input beams. One interferometer uses a well-known laser wavelength standard at its input to measure the mirror motion accurately. The other path measures the interferogram of the unknown input. The implementation shown in Figure 4.4 has the known and unknown signals making equivalent but noncoincident paths through the interferometer. Alternately, the two beams are placed coincident to each other with dichroic (wavelength-separating) filters used to separate the reference and unknown signals at the detector. By comparing the outputs of the reference photodetector and the unknown photodetector, very accurate wavelength

(a) Michelson Interferometer With Reference Laser

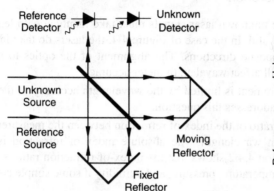

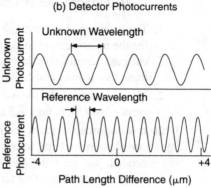

(b) Detector Photocurrents

Figure 4.4   (a) Michelson interferometer with an added wavelength reference input added. (b) Example photocurrent values versus optical path length difference.

measurements can be made. An analysis of the interferometer with the laser wavelength standard will now be made.

4.2.4   Wavelength Measurement with Respect to a Wavelength Standard

The interferogram of the unknown signal is compared to that of the known standard. If the mirror is moved over a specified scan length, the interference fringes in both the reference and unknown arms are counted. The wavelength of the unknown signal can then be calculated by comparing the fringe counts in the unknown and the reference signal paths and taking the ratio of counts.

$$\lambda_u = (N_r/N_u)(n_u/n_r)\lambda_r \qquad (4.11)$$

where λ_u is the unknown wavelength, λ_r is the reference wavelength, n_r is the index of refraction at the reference wavelength, n_u is the index of refraction at the unknown wavelength, N_r is the number of reference counts over a distance L, and N_u is the number of unknown wavelength detector counts over distance L. The equation requires that an accurate ratio of the index of refraction at the reference wavelength to the index at the unknown wavelength be known. Section 4.4.2 explores how accurately this ratio can be determined for an air environment interferometer. It is this relationship between the fringe counting at the reference frequency and the unknown frequency that allows accurate wavelength measurements. Figure 4.4b shows an example of the photocurrents that are measured in the reference arm and the unknown arms for a reference wavelength of 633 nm and an unknown arm wavelength of 1550 nm. Equation 4.11 compares the interferogram period of the two signals and takes into account the index of refraction at both the reference and the unknown wavelengths.

4.2.5 Summary of Michelson-Interferometer Wavelength-Meter Operation

1. The reference laser and the unknown laser signals must take identical path lengths for the measurement to be valid. In the case of Figure 4.4, the lasers do take identical path lengths, but in opposite directions. The alignment of the optics to make these two paths identical will affect wavelength meter accuracy.

2. The accuracy of the measurement is limited by the wavelength accuracy of the reference laser. Section 4.4.3 addresses this question.

3. It is important to know the ratio of the index of refraction between the measurement wavelength and the known wavelength. The absolute index of refraction is not needed, just the ratio. Section 4.4.2 shows that the index-of-refraction ratio is relatively independent of air temperature, pressure, and humidity if some simple corrections are made.

4. The coherence of a source affects the shape of the interferogram. Wide spectral-width sources show interference only over a narrow range of path-length differences in the interferometer.

This section has dealt exclusively with analyzing situations where a single optical source is connected to the measurement instrument. It will be shown that simple fringe counting is of limited usefulness if many different input wavelengths are present simultaneously. The following section will show that by using Fourier transform techniques, multiple signal environments can also be characterized with a Michelson interferometer.

4.3 WAVELENGTH METERS IN MULTIPLE SIGNAL ENVIRONMENTS

Figure 4.2 shows the interferogram that results when a single wavelength signal is applied to a Michelson interferometer. What happens if two or more separate laser sources are combined together into the wavelength meter input? This situation is very common since WDM systems are installed for telecommunications. Figure 4.5 shows the interferogram

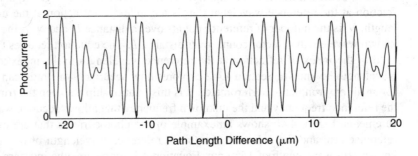

Figure 4.5 The input signal for this interferogram includes a DFB laser at 1300 nm and 1550 nm with equal powers.

that results when 1300 nm and 1550 nm sources are combined and applied to a Michelson-interferometer wavelength meter. The interferogram does not show the regular period of Figure 4.2. In some portions of the interferogram, the interference signal amplitude is also very small. With the simple fringe-counting methods of Section 4.2, it would be difficult to correctly calculate the wavelength of the input signal. The algorithm of Equation 4.11 would produce an average wavelength value between 1300 nm and 1550 nm. This example illustrates the limitation of fringe-counting wavelength meters. With a little more complicated analysis involving Fourier transforms, it will be found that one can display the power versus wavelength for multiple signal environments.

Figure 4.6 illustrates the operation of a Fourier-transform Michelson-interferometer wavelength meter.[5,6] Figure 4.6a shows the interferogram result for an input signal containing a 1300, 1550, and 1650 nm laser with equal powers. This complicated interferogram can be thought of as the sum of three separate interferograms similar to that of Figure 4.2. The single wavelength interferograms of the individual 1300, 1550, and 1650 nm lasers add to produce the complicated pattern of Figure 4.6a.

Let us consider what would result if a Fourier-transform operation were done on the photocurrent versus distance function that makes up the interferogram of Figure 4.6a. The Fourier transform will produce a plot of the photocurrent magnitude versus spatial fre-

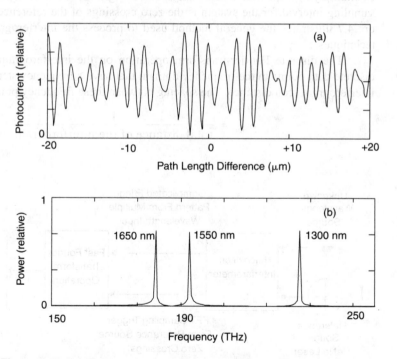

Figure 4.6   (a) The interferogram for a combined 1300, 1550, and 1650 nm signal with equal power. (b) Results of a Fourier transform operation on the interferogram data.

quency, σ (measured in cycles per meter), of the interferogram. Figure 4.6b illustrates the results of performing a Fourier-transform operation on the data of Figure 4.6a. The temporal frequency shown in the plot (cycles per second) is obtained by multiplying by the speed of light in the medium. The Fourier-transform operation on the interferogram allows the wavelength of the signals to be separated and measured individually. Here each of the input signals is individually resolved both in frequency and in power. The Fourier-transform operation does not compromise the wavelength accuracy of the measurement when compared to fringe-counting methods.

In performing the Fourier transform of the data presented in Figure 4.6a, the unknown signal interferogram is sampled at regular distance intervals. The distance spacing between interferogram samples controls the maximum spatial frequency that can be displayed in the result. Using the Nyquist sampling theorem, the maximum spatial frequency that can be measured without aliasing effects is

$$\sigma_{max} = \frac{1}{2\,(\text{distance between samples in the interferogram})} \qquad (4.12)$$

Here σ_{max} is the maximum spatial frequency in cycles per meter. Spatial frequency is also referred to as the wavenumber. The maximum temporal frequency of the signal is found by multiplying Equation 4.12 by the speed of light for the interferometer. A convenient sampling interval for the system is the zero crossings of the reference photocurrent. Figure 4.7 illustrates the general method used to process the interferogram data by Fourier techniques.

The discrete Fourier-transform operation on the interferogram data will produce discrete data points in the frequency domain that have a frequency step related to the total interferogram scan distance. The spacing between these frequency points is

$$\sigma_{step} = \frac{1}{\text{total distance of the interferogram}} \qquad (4.13)$$

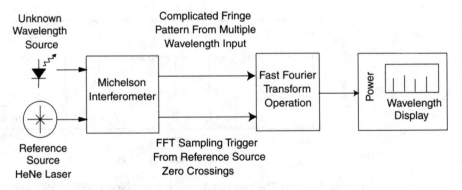

Figure 4.7   The signal processing block diagram of a Fourier-transform wavelength meter.

The temporal frequency step of the signal is found by multiplying Equation 4.13 by the speed of light for the interferometer. For maximum wavelength resolution, it is important to have a small Fourier-transform frequency step. Long interferogram distances are desirable to obtain good wavelength resolution. For ease of processing, the interferometer trace is broken up into 2^n data points so that efficient fast-Fourier-transform algorithms can be used. The raw interferogram data of Figure 4.6a is often multiplied by a windowing function so that the data does not abruptly terminate at the ends of the scan. Abrupt termination of the interferogram data would result in ringing and the introduction of spurious signals in the Fourier transform domain. Specialized digital signal processor chips can be used to accomplish fast-Fourier-transforms quickly. With Fourier-transform capability, the Michelson-interferometer wavelength-meter provides a display very similar to the grating-based OSAs discussed in Chapter 3.

An example calculation is given here to illustrate the use of the equations above. Assume that a helium-neon ($\lambda = 0.633$ μm) reference laser was used and samples were taken at every zero crossing of the HeNe fringe. The maximum frequency that can be displayed without aliasing effects would be 475 THz based on the Nyquist theorem which requires two samples per period at the maximum desired frequency (Equation 4.12). This frequency range is adequate for operation in the fiber-optic communication bands. For a ±30 mm sweep range, the frequency spacing of the data points in the Fourier domain will be 5 GHz (Equation 4.11). The resolution of the system is defined as two Fourier transform data point intervals or in this case 10 GHz. Frequency resolution of 10 GHz is adequate for WDM communication systems spaced at 100 GHz.

Fourier transform wavelength meters are very useful for WDM communication system measurements. Figure 4.8 shows an example measurement of a four-channel WDM source using a Fourier-transform wavelength meter. The frequency spacing between lasers in this example is 100 GHz on average. The Fourier-transform wavelength meter provides accurate measurement of the wavelength, power, and signal-to-noise ratio for WDM communication systems.

4.4 ABSOLUTE WAVELENGTH ACCURACY CONSIDERATIONS FOR MICHELSON-INTERFEROMETER WAVELENGTH METERS

Here are some of the important variables that influence wavelength meter accuracy.

1. Maximum path-length change in the variable arm of the interferometer. The longer the interforometer path-length difference, the better the accuracy. Related to this subject is the ability to count fractional fringes. If one has to round to the nearest integer number of interference fringes and throw away the remainder, valuable wavelength information is being thrown away. There are innovative ways to count down to a fraction of a fringe.

2. Knowledge of the ratio of the index of refraction at the reference wavelength to the index of refraction at the unknown wavelength. The index of refraction in air is a function of humidity, temperature, gas content, etc. How big of an error would re-

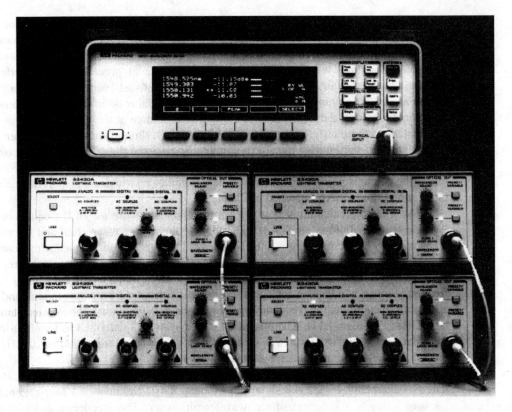

Figure 4.8   A wavelength meter measurement of 4 DFB lasers sources
located on the 100 GHz spacing dense WDM grid.

sult if this were ignored? Should one immerse the interferometer in a known environment (nitrogen or a vacuum)?

3. The wavelength accuracy of the reference source will ultimately limit the accuracy
   of the measurement of the unknown.

4.4.1   The Ability to Count Many Fringes and to Count Them Accurately

The simple wavelength meter as described in Section 4.2 counts the number of fringes
that occur in a given path length for both the unknown signal and the reference signal. In
order to get accurate wavelength measurements, it is desirable to make a very long path-
length difference interferometer to maximize the number of fringe counts. Long sweep
distances require increased measurement time. From Equation 4.11, the calculation of an
unknown wavelength involves ratioing the fringe counts at the reference wavelength to
fringe counts at the unknown wavelength. An entire mirror sweep will probably not result
in an integer number of interference fringes at both the reference and unknown wavelengths. Rounding to the nearest integer will degrade measurement accuracy.

It is common to use electronic zero-crossing counters to measure the number of fringes at both the signal and reference wavelengths. In this implementation of a wavelength meter, the fraction of a fringe found at the beginning and end of sweep will be ignored. If this fractional fringe information is not used, the accuracy of the interferometer measurement will be degraded. Here is an example of the magnitude of error for dropping fractional fringes. An interferometer is scanned over a 3 cm range with an input wavelength of exactly 1550 nm; 38709.68 fringes should be counted. If the 0.68 fringe fraction is not counted, the wavelength meter would measure 1550.027 nm. This represents a 17.4 ppm error which is unacceptable for many applications. Improved accuracy can be obtained by increasing the scanning distance or devising methods for fractional fringe counting. In modern wavelength meter designs, methods of fractional fringe counting have led to dramatic improvements in accuracy. Figure 4.9 illustrates a frequency multiplying method of fractional fringe counting. If the variable length arm of the interferometer is scanned at a uniform rate, the output from the detector will be a sinusoidal signal. This signal is then multiplied by a nonlinear electronic circuit to a harmonic of the input frequency. Frequency multiplication up to 100 times has been achieved.[7-9] The multiplied signal is then counted with an electronic counter. If a 10 times multiplication is used, an effective fringe resolution of 1/10 of the unmultiplied fringe period is achieved. Other methods use phase-locked loop techniques for locking a higher frequency oscillator to the output of the detector.

Folding of the optical path allows the path-length difference to be increased for a given mirror motion.[1] This allows for a more compact optical design and increased measurement accuracy.

4.4.2   Index of Refraction and Dispersion of Air

Equation 4.11 showed that the ratio of the index of refraction at the unknown signal wavelength to the index at the reference wavelength is important for accurate wavelength measurements. The wavelength of a signal in air can be appreciably different from the

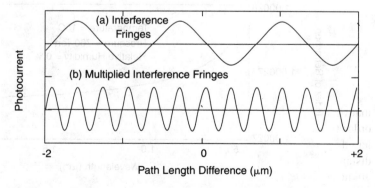

Figure 4.9   The fringe counting capability of a wavelength meter can be increased by using fractional fringe counting techniques: (a) raw photocurrent signal. (b) photocurrent signal multiplied by 4.

index of refraction in a vacuum. Recall that the wavelength in a vacuum and the wavelength in air are related by the equation

$$\lambda_{vac} = n_{air}\lambda_{air}$$ (4.14)

where λ_{vac} is the vacuum wavelength and λ_{air} is the wavelength in air, and n_{air} is the index of refraction of air. Michelson interferometers most often make measurements in an air environment. Because of the variability in the index of refraction of air with respect to environmental conditions, it is important to have a set of standard conditions that can be used to clearly state the wavelength. The two most common ways of stating wavelength are the wavelength in a vacuum and the wavelength in standard dry air. The wavelength in a vacuum is self-explanatory. Standard dry air is defined to have pressure of 760 Torr and temperature of 15 C and no water vapor. The index of refraction of air has been extensively studied and accurate models have been developed.[10,11] The use of these models in wavelength meter calculations can improve the accuracy of the wavelength measurement. The index of refraction is a function of temperature, pressure, and gas composition. A useful result is given by Edlen.[10] The wavelength dependence of the index of refraction for standard dry air is given as:

$$n_s = 1 + 10^{-8}\left(8342.13 + \frac{2406030}{130 - \frac{1}{\lambda^2}} + \frac{15997}{38.9 - \frac{1}{\lambda^2}}\right).$$ (4.15)

Here the wavelength, λ, is in microns. Figure 4.10 shows a plot of how the index of refraction depends on wavelength for the standard dry-air conditions.

A number to remember is 1.000273 for the telecommunication wavelength areas near 1300 nm and 1550 nm. Equation 4.15 can be corrected for temperature and pressure using the following correction equation:

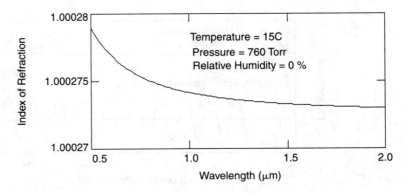

Figure 4.10   The index of refraction versus wavelength for standard dry air.

$$n(T, P) = 1 + \frac{(n_s - 1)(0.00138823)P}{1 + (0.003671)T} \qquad (4.16)$$

Here pressure, P, is in Torr and temperature, T, is in Celcius. A further correction can be made to the index for water vapor content in the air.

$$n(T, P, h) = n(T, P) - h*\left(5.722 - \frac{0.0457}{\lambda^2}\right) 10^{-8} \qquad (4.17)$$

Here h is the partial pressure of water vapor in Torr. Equation 4.17 is most accurate for visible light with degraded accuracy at infrared wavelengths. A more everyday measure of water vapor content is relative humidity. The partial water vapor pressure for a 100% relative-humidity day would be approximately 4.3 Torr, 22.6 Torr, and 112 Torr for a temperature of 0, 25, and 55C respectively.

It is useful to see how the index of refraction varies as a function of the three most common variables: pressure, temperature, and humidity. Figure 4.11a shows how the index of refraction changes as a function of wavelength for three values of temperature.

Figure 4.11b, c, and d show variability of the index of refraction at the wavelength of 1550 nm. Figure 4.11b shows a plot of the index versus temperature with 760 Torr pressure and 0% relative humidity. Figure 4.11c shows a plot of index versus pressure with 25C temperature, and 0% relative humidity. Figure 4.11d shows a plot of index versus relative humidity with 25C temperature and pressure of 760 Torr. From these three graphs, it is seen that pressure variations have the largest effect on the index of refraction for reasonable values of the environmental variables. The average atmospheric pressure can vary from 760 Torr at sea level down to 400 Torr at a height of 5000 m.

It is fortunate that the Michelson-interferometer wavelength-meter measurement does not directly depend on absolute value of the index of refraction in air. The only important parameter is how the ratio of indexes at the reference and unknown wavelengths changes with respect to an atmospheric variable of air (see Equation 4.11). The ratio of the index of refraction at the reference and unknown wavelengths is quite insensitive to atmospheric variables. The conditions that tend to raise the index of refraction at the unknown wavelength also raise the index of refraction at the known wavelength, keeping the ratio very constant. Figure 4.12a shows the index of refraction at 633 nm and 1550 nm as a function of temperature. The two curves track each other very closely.

Figure 4.12b shows how the ratio of the index of refraction between 0.633 (HeNe reference laser) and 1.55 μm (unknown signal wavelength) changes versus temperature with a 760 Torr pressure and 0% relative humidity. It may be possible for a wavelength meter to be exposed to temperatures as low as −40C and as high as 45C in an outdoor environment. This temperature-range excursion would cause the ratio of the index of refractions to change by up to 1 ppm. For laboratory or manufacturing environments where the temperature excursions may only be ±10C, temperature corrections would probably not be necessary. For outdoor measurements in extreme conditions, temperature will affect accuracy.

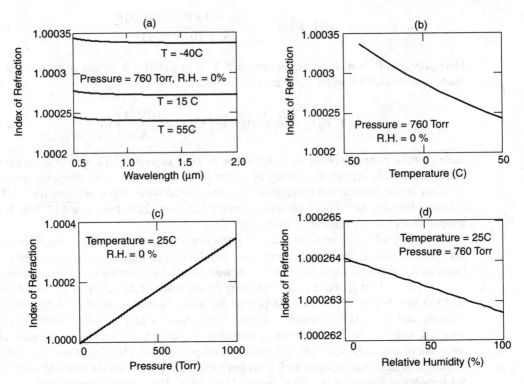

Figure 4.11 (a) The index of refraction as a function of wavelength for 3 different temperatures. (b) Index of refraction as a function of temperature at 1550 nm. (c) Index of refraction as a function of pressure at 1550 nm. (d) Index of refraction as a function of relative humidity at 1550 nm.

Figure 4.12c shows how the ratio of the index of refraction between 0.633 μm to that at 1.55 μm changes versus pressure with a 25C temperature and 0% relative humidity. For a Michelson interferometer in an air environment, a worst case comparison would be to compare a result on top of a 5000 m mountain at a pressure of 400 Torr with that at sea level of 760 Torr. This condition would give a wavelength measurement error of up to 2 ppm. This factor is large enough so that many wavelength meters allow entry of the instrument's air pressure conditions.

Figure 4.12d shows how the ratio of the index of refraction between 0.632 and 1.55 μm changes versus relative humidity with the temperature at 25C and pressure held constant at 760 Torr. In going from 0 to 100% humidity, the ratio changes by only 0.02 ppm. The error will increase for higher temperatures since the water holding content of air increases rapidly with temperature. In general, the effects of humidity may be neglected except that it would be convenient to keep water from condensing on the interferometer optics!

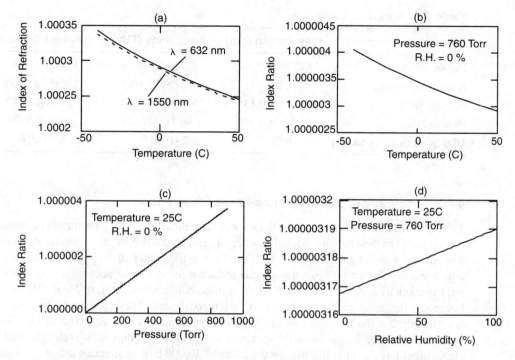

Figure 4.12   (a) Index of refraction as a function of temperature for two different wavelengths. (b) 632 to 1550 nm refractive index ratio as a function of temperature. (c) 632 to 1550 nm refractive index ratio as a function of pressure. (d) 632 to 1550 nm refractive index ratio as a function of relative humidity.

A formula that has been used successfully to take into account the effects of the ratio of the index of refraction on pressure conditions is as follows.

$$f_u = f_{\text{vacuum}} \left[1 + \left(\frac{n_r}{n_u} - 1 \right) \left(1 - \frac{\text{Elevation in m}}{500} \right) (0.05) \right] \qquad (4.18)$$

where n_r is the index of refraction at the reference wavelength and n_u is the index of refraction at the unknown wavelength. This model only takes into account pressure variations and ignores effects of temperature and humidity. The term $(n_r/n_u - 1)$ is put into memory in tabular form. The second term corrects for elevation dependence of the ratio.

With accurate knowledge of the effects of the index of refraction of air on the measurement, extremely accurate wavelength meter measurements can be made without the use of built-in atmospheric monitoring sensors or by immersing the interferometer in a vacuum chamber. When temperature and pressure are corrected for, environmental effects on the index of refraction limit wavelength accuracy only by several tenths of a part per million.

Table 4.1 Reference Laser Data

Parameter	Wavelength (nm)	Frequency (THz)	Wavenumber cm^{-1}
Helium-Neon	632.99076(2)	473.612692(12)	15798.0189(4)
Helium-Neon	730.6805(11)	410.289572(60)	13685.7870(20)
Helium-Neon	1152.59050(13)	260.103184(30)	8676.1083(10)
Helium-Neon	1523.48761(19)	196.780372(25)	6563.8867(8)
DFB Laser (semiconductor)	800–1600	187–1600	6250–12500

4.4.3 Accuracy of the Reference-Laser Wavelength

Table 4.1 lists several possible choices for a reference laser.[12–15] The numbers listed in parentheses are the standard deviation uncertainties in the last digits of the quoted value. The quoted wavelength values are for lasers set to the center of their gain versus wavelength curve. The actual laser design can reduce reference laser accuracy. The speed of light is taken as $2.99792458(1) \times 10^8$ m/s in these conversions. Helium-Neon (HeNe) gas lasers are the most common choice for reference lasers. There are several choices of wavelength, but the 633 nm line is the most common. It is also very useful that HeNe lines are available in the infrared spectrum to check accuracies of wavelength meters. HeNe tubes have a finite lifetime on the order of 30,000 h of continuous usage. The internal filaments deposit metal on the output mirrors causing the reflectivity to drop, eventually stopping the lasing. The DFB laser example provides a flexible alternative that would allow laser lifetimes approaching 10^6 h. If the wavelength stability of the DFB is accurately characterized, it should be an acceptable laser reference for a telecommunications application wavelength meter. Alternately, the semiconductor laser can be locked to other frequency standards to maintain long-term stability.[15]

633 nm HeNe Laser Wavelength Accuracy. The accuracy of the reference laser wavelength depends on laser construction and on knowledge of some of the details of the energy levels in the laser. The gain versus wavelength of the HeNe line has a center-vacuum wavelength of 632.99076 nm. The Doppler-broadened gain versus wavelength function for the HeNe-optical amplifier has a spectral width of 1.5 GHz.

A major consideration for a HeNe laser reference is how close the laser emits to the center of the gain versus wavelength curve. This situation is given pictorially in Figure 4.13. The length of a typical compact FP resonator for a HeNe laser is typically 15 cm. This means that the longitudinal mode spacing (0.5 × velocity/laser length) of the laser will be approximately 1 GHz. Figure 4.13 shows an example situation of the location of the longitudinal modes with respect to the peak of the gain curve. The accuracy of the source for wavelength-meter applications depends on the power-weighted average of the longitudinal mode wavelength locations. The best possible situation would occur if the longitudinal modes are placed symmetrically around the gain center. In the worst case alignment, the average wavelength would be off by half of the longitudinal mode

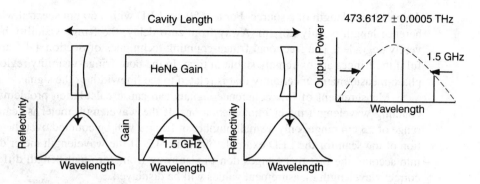

Figure 4.13 Gain peak width and longitudinal mode spacings for Helium Neon lasers.

spacing or 0.5 GHz in this 15 cm cavity length example. For a 15 cm cavity length, this results in a potential 1 ppm error in wavelength. This error will be temperature dependent since the length of the laser resonator is temperature dependent causing the longitudinal mode position to shift. An alternate strategy would be to use a HeNe laser with finely spaced longitudinal modes so that the laser has a multiple longitudinal mode output with the average wavelength being very constant. The accuracy can be significantly improved by using HeNe lasers that have the longitudinal mode position locked to the center of the gain versus wavelength curve. However, this stabilization results in an appreciable increase in the cost of the laser. A gain-center-locked HeNe laser is capable of wavelength errors of less than 0.1 ppm. Further refinements in HeNe laser accuracy are also possible but are not necessary for fiber optic applications.[1]

Other Wavelength References.   HeNe lasers with similar wavelength accuracies are available at 730.6 nm, 1152.6 nm, and 1523.5 nm. The considerations illustrated in Figure 4.13 apply to these wavelengths also except for differences in the width of the gain peak and the spacing of the longitudinal modes. These other wavelengths offer lower gain per unit length requiring a significantly longer laser tube. Without stabilization of the longitudinal modes to the line center, accuracies of less than 1 ppm are achievable at these HeNe wavelengths.

Standards organizations have refined very accurate wavelength standards. The use of these elaborate standards is probably not justified for the accuracies required in fiber-optic communications systems. The commercially available HeNe lasers shown in Table 4.2 can be used to verify accuracy of wavelength meters intended for fiber optic applications.

4.4.4 Dependence on the Signal Spectral Width

Figure 4.3 showed the interferogram for a relatively incoherent source. The broad spectral content of the LED caused the fringe visibility to decrease to zero for offsets far from zero path-length difference. The distance over which fringes are visible is a measurement of the coherence length of a source. For a 1550 nm LED with a 60 nm spectral width, the coherence length is only 20 μm. Away from zero delay, the fringe visibility becomes severely degraded. The fractional fringe-counting techniques of Section 4.4.1 are only useful if the fringes are extremely stable in time. In the poor fringe-visibility region there is a phase-measurement uncertainty that is related to the linewidth of the signal.[1]

Measurement of low coherence signals can cause calculation problems for fringe counting wavelength meter implementations. If the wavelength meter is scanned over a range of ±3 cm range, only a small number of fringes will be counted over the central portion of the scan for the LED example of Figure 4.2. If the wavelength meter doesn't take into account the fact that fringes don't exist away from zero path-length differences, incorrect wavelength measurement values will be displayed.

The Fourier-transform techniques of Section 4.3 are more tolerant to measuring incoherent sources. The Fourier-transform operation will record the characteristic periods that are within the interferogram, even if the fringes are not visible over the entire scan length. Since fewer fringes are involved in the comparison, the center wavelength of the LED will not be measured with a large accuracy. However, this reduced measurement accuracy for an LED is justified because the LED does not have a well-defined center wavelength.

4.4.5 Optical Alignment Issues

It is important that the path lengths be identical between the known and the unknown arms of the interferometer. If the two arms are not well-aligned, the reference laser will not accurately measure the path-length difference for the unknown signal-interference pattern. Alignment errors often enter only second order to the wavelength accuracy. A 1 ppm accuracy requires the input beam to be aligned to a reference beam and/or the instrument optical axis to the order of 10^{-3} radians.[16] Singlemode fiber-optic input-wavelength meters are convenient because the alignment of the input beam to the optical axis of the interferometer is fixed by the fiber launch optics. Wavelength meters that accept open optical beams must be careful so that the input signal is well-aligned to the optical axis of the measurement instrument. A multimode fiber input to a wavelength meter can have an error associated with mode excitation. The distribution of light at the output of multimode fiber can be variable depending on how the fiber optic cable is wiggled. This variability in the excitation conditions leads to wavelength errors due to the uncertainty in the launch conditions into the interferometer as compared to the reference source.

4.4.6 Diffraction Effects

Collimated light slowly diverges according to the mathematics of Gaussian optics. The smaller the diameter of the collimated beam, the wider the divergence cone. Since the two beams of a Michelson interferometer often travel different distances to the receiving detector, the wavefront curvature for these beams will be slightly different. This limitation on fringe counting is illustrated in Figure 4.14. This presents an ultimate limit to the fraction of a fringe that can be counted as described in Section 4.4.1. If the two Gaussian

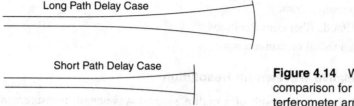

Figure 4.14 Wavefront curvature comparison for non-equal length interferometer arms.

beams travel the same distance, the constant-phase wavefronts perfectly align with each other. If the wavefronts are not perfectly matched, the interference effect will be reduced. The fringe visibility will be reduced as the mirror scan is increased away from equal path delay. For the purpose of fiber-optic communications measurements, this term is very small.[17]

4.4.7 Summary of Wavelength Accuracy Factors

Table 4.2 lists factors that affect wavelength accuracy and comments about each. If all of the factors are taken into account, it is possible to achieve less than 1 ppm accuracy in an air-measurement environment. This would require the use of a HeNe reference laser stabilized to the gain peak, and corrections for temperature and pressure variations.

4.5 MICHELSON WAVELENGTH-METER MEASUREMENT CONSIDERATIONS

In the following section, some of the measurement capabilities of Fourier-transform Michelson-interferometers will be discussed. The areas of discussion are:

1. Wavelength resolution
2. Wavelength coverage
3. Sensitivity, measurement range, dynamic range, and signal to noise

Table 4.2 Comparison of factors that affect wavelength accuracy.

Parameter	Uncertainty Contribution	Comments
Fractional fringe counting error	0.17 ppm	Assumes a 1/10 fringe resolution in 600,000 sample inteferogram
Index of refraction dispersion vs. wavelength	0.2 ppm	Assumes that elevation is known and temp is stabilized ±10C
Reference laser accuracy	0.1 ppm–1 ppm	Depends on wavelength stabilization of HeNe laser

4. Amplitude accuracy

5. Multimode fiber considerations

6. Pulsed signal measurements.

4.5.1 Relative Wavelength Resolution

Figure 4.8 showed the result of a multiple-signal wavelength measurement for a WDM system. This section intends to answer the question, How close can the signals be and still be individually resolved?

The major factor affecting relative wavelength resolution is the length of scan in the Michelson interferometer. This may best be illustrated by an example: An interferometer with a 40.45 mm (±20.225 mm) path delay is taken. The reference fringe period is every $\lambda_{reference}/2$ of mirror translation ($\lambda_{reference} = 0.633$ μm for a HeNe reference laser). This would yield a total of 131072 samples during one scan. If these samples are put into a fast-Fourier-transform operation, there will be 65535 ($2^{17}/2$) real and 65,535 imaginary frequency bins produced. Each frequency domain point will be spaced by 7.227 GHz (see Equation 4.13) in this example. A single, pure optical frequency input will generate a peak that is 2 bins wide (full-width at half of maximum). In this example the resolution would be 14.454 GHz. The resolution is independent of wavelength. At 1310 nm it would be 0.083 nm and at 1550 nm it would be 0.116 nm. The resolution capability of a Fourier-transform Michelson wavelength-meter is thus given as

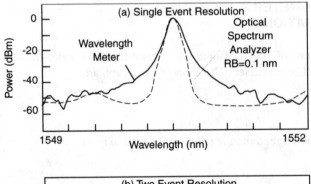

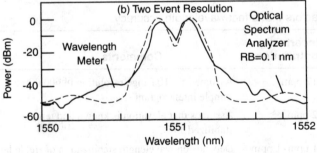

Figure 4.15 (a) Single-event resolution of a DFB laser. (b) Two-event resolution of DFB lasers spaced by 0.2 nm. The wavelength meter results are contrasted with a double-pass grating-based optical spectrum analyzer measurement of the same sources.

$$\text{Resolution (Hz)} = \frac{\text{speed of light}}{\text{total distance of the interferogram}}. \qquad (4.19)$$

Figure 4.15a shows the measurement of single-event resolution for a wavelength meter with a mirror scan length of ±20 mm. The effective resolution for this measurement from Equation 4.19 is about 15.6 GHz (0.13 nm). Figure 4.16a shows a measurement of the two-event resolution for two DFB lasers signals separated by 0.2 nm. The two signals are resolved in this case. The plot also shows the results of these two signals as measured by a double-pass diffraction-grating-based OSA. The resolution capability of both instruments is similar near the signal peaks. The OSA has superior resolution 0.2 nm off of the signal peak.

In a FP laser, it is often desirable to know the relative locations of all of the longitudinal modes. Figure 4.16 shows a plot of the modes of an FP laser as measured with a ±20 mm mirror scan length. The modes are nearly uniformly spaced in frequency step. Dispersion and gain-shape effects in the semiconductor cause the modes to have a slight frequency step nonuniformity. These nonuniformities can be used to measure material dispersion. This measurement requires very accurate relative measurements of wavelength. The spacing between modes is approximately 35 GHz in this example.

4.5.2 Wavelength Coverage

The wavelength coverage is limited fundamentally by the wavelength coverage of the photodetectors. InP/InGaAs photodetectors with thin InP-layers above the absorbing region are capable of operating over the 0.6 μm to 1.7 μm wavelength range. The input fiber also has a specified wavelength range over which it operates in a single transverse mode. Multiple transverse-mode operation of the input fiber will decrease wavelength accuracy.

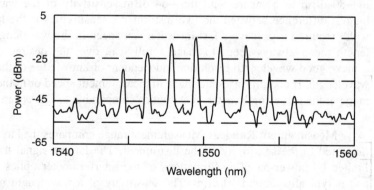

Figure 4.16 A Fabry-Perot laser spectrum measured on a Michelson interferometer wavelength meter.

4.5.3 Sensitivity and Measurement Speed, Measurement Range, and Dynamic Range

Sensitivity and Measurement Speed. Sensitivity refers to the smallest signal that can be measured with a wavelength meter. The sensitivity of a wavelength meter is best demonstrated using a design example.

Example

Assume that a wavelength meter has a mirror scan of ±20 mm and a desired measurement speed of 0.1 s. Let us assume that the wavelength meter has the block diagram as shown in Figure 4.7. The ±20 mm span will produce approximately 2^{17} samples in the 0.1 s interval. This corresponds to a data rate of 1.3 MHz. Let us assume that the optical receiver uses a 10 MHz bandwidth. For a 10 MHz bandwidth, a high-performance optical receiver can use a transimpedance value of 10,000 ohms. The achievable sensitivity of such a receiver can be found from Table A.1 in Appendix A as −80 dBm assuming a photodetector responsivity of 0.5Amp/Watt including Michelson interferometer losses. It takes at least a 15 dB signal-to-noise ratio to make a reasonably accurate wavelength measurement (for example, 3 ppm wavelength accuracy). The resulting sensitivity in this example is therefore −65 dBm.

The example above shows that the sensitivity of a wavelength meter is worse than that of the grating-based OSAs studied in Chapter 3. It is important to compare the two instruments in a similar measurement condition. Let's assume that the goal of a measurement is to find the wavelength and power of a signal known to have a wavelength near 1550 nm. For the OSA, a narrow 10 nm wavelength sweep would be adequate to find the wavelength of the signal. The optical receiver in the OSA can use a narrow detection bandwidth of 100 Hz. From Table A.1 in Appendix A, assuming a 100,000 ohm transimpedance and a detector responsivity of 0.02 Amp/Watt (includes monochromater loss), a sensitivity of −90 dBm should be achievable. The OSA should have a signal-to-noise ratio of 10 dB to make an accurate wavelength measurement. This would result in a sensitivity of −80 dBm to compare with the −64 dBm sensitivity of the wavelength meter. The biggest difference between the two instruments sensitivities is the large detection bandwidth required to achieve a fast measurement speed in the wavelength meter. The wavelength meter always needs to make a full scan over the desired path length delay to achieve good wavelength resolution independent of knowledge of the approximate signal wavelength. If compromises are made in measurement speed or in the desired wavelength range of the wavelength meter, sensitivities comparable to OSAs can be achieved.

Measurement Range. Measurement range compares the largest signal that can be measured to the sensitivity of the instrument. The largest signal that can be handled is limited by power-handling limitations of the interferometer optics or the photodetector and is typically around 10 dBm. The sensitivity of a wavelength meter in the example above was −64 dBm. The resulting measurement range in this example would be 74 dB. This would compare to a grating based OSA measurement range of 90 dB for the above example.

Dynamic Range.   Dynamic range refers to the ability to measure small signals in the presence of large signals. OSAs and wavelength meters have significantly different dynamic-range capabilities. The skirts of the filter in an OSA limit the dynamic range of a double-pass grating OSA to about 70 dB.

The wavelength meter has a dynamic range limited by the large amount of shot noise that is present in the optical receiver. This dynamic range limitation is shown by an example.

Example

The conditions from the wavelength-meter sensitivity calculation above will again be used. Let's assume that an input signal with 1 mW of average power is input to the wavelength meter. For a detector responsivity of 1 Amp/Watt, 1 mA of photocurrent will be produced in the detector. This large input signal will generate a significant amount of shot noise in the receiver. Assuming a 10 MHz receiver bandwidth, Table A.2 in Appendix A shows that the current shot noise value is 57 nA for a 10 MHz bandwidth. This 57 nA of shot noise is 42 dB lower than the average current of 1 mA. The dynamic range of the wavelength meter in this example would be 42 dB.

This dynamic range limitation is illustrated in the FP laser measurement example in Figure 4.16. The wavelength meter used for this measurement has similar parameters as in the above example. There appears to be a noise floor in the measurement that is about 37 dB lower than the peak of the spectrum. This noise floor is due to shot-noise generation in the optical receiver. If the DFB signal were removed from the wavelength meter, the displayed noise floor would drop by over 30 dB. In an OSA, the optical receiver receives a large shot-noise contribution only when the signal falls within the passband of the OSA filter. A typical double-pass grating-based OSA may have a dynamic range of over 70 dB.

Signal-to-Noise Ratio Measurement.   Michelson interferometers with Fourier-transform capability can be used to make signal-to-noise ratio measurements on the signal input. Signal-to-noise measurements are very important for optically amplified WDM systems. The amplified spontaneous emission noise from erbium-doped fiber amplifiers is added to the signal power. If the signal-to-noise ratio become significantly worse than 20 dB (0.1 nm noise bandwidth), bit-error ratio problems will occur.

For wavelength meters, the signal-to-noise ratio measurement capability is limited by the dynamic range of the measurement system as described in the last section. For closely spaced WDM channels, the selectivity of the wavelength meter is another factor that can limit the capability to measure signal-to-noise ratio. The measurement effectiveness is improved by taking long interferometer sweeps and by using a slow measurement speed. The effective filter width of the noise measurement must also be taken into account when calculating noise power. The effective filter width is approximately equal to the resolution given by equation 4.19.

4.5.4 Amplitude Accuracy

Traditional power meters need the wavelength of the light entered before accurate power measurements can be made because of the change in detector responsivity with wavelength. Since wavelength meters automatically calculate wavelength, one does not need to enter wavelength to accurately measure power. The amplitude accuracy of multiple-wavelength Fourier-transform measurements is more complicated to achieve. Amplitude accuracy of the instrument can be calibrated while operating with the data in the frequency domain.

4.5.5 Singlemode and Multimode Fiber Input Considerations for Wavelength Meters

The use of a singlemode fiber optic input for a wavelength meter greatly simplifies the use and the accuracy of the instrument. With a fixed-position fiber input to the interferometer, the alignment of the input signal to the optical axis of the interferometer is well-defined and repeatable. This avoids wavelength errors due to optical axis misalignment.

A wavelength meter using a multimode-fiber input will have reduced performance when compared to a singlemode fiber input in terms of wavelength accuracy. When a narrow spectral width source is coupled into multimode-fiber, a speckle pattern appears in the fiber cross-section because of interference between the multiple paths through the fiber. The optics of the Michelson interferometer images the input fiber-speckle pattern to the detector through each arm of the interferometer. The speckle pattern features must be placed identically on top of each other at the detector in order for interference to occur. This means that the alignment of the interferometer optics to accommodate multimode fiber inputs is as critical as for singlemode fiber inputs. The fringe visibility of the wavelength meter is also degraded compared to a singlemode input. The fringe visibility will also vary with the exact nature of the speckle pattern of the input. This causes the instrument sensitivity to vary with movement of the input fiber.

Multimode fiber also may produce a slightly different wavelength readout depending on how well all of the modes within the fiber are excited. If the waveguide is excited off of the center axis of the multimode fiber, the effective angle of the input beam may be misaligned with respect to the reference laser beam resulting in a path length mismatch and resulting wavelength error.

4.5.6 Measurement of Pulsed Signals

The scanning Michelson-interferometer analysis done up to this point made the assumption that unmodulated signals are present at the input. Wavelength-meter designs based on movement in time of an interferometer arm will not make correct measurements on lasers with very low repetition-rate intensity-modulation or frequency modulation. In fiber optic systems, it is common to measure signals that have intensity modulation. The SONET transmission standard has modulation spectral content that extends down to 8 kHz. Pseudorandom binary-sequence modulated signals can have frequency content at even lower modulation rates. Even cw lasers can have instabilities that can lead to erroneous data in a wavelength measurement.

What are the limitations for measuring modulated or pulsed lasers with a Michelson interferometer? If the modulation period is short compared to the measurement time of the instrument, accurate measurements can be made. If the modulation period is too long, significant wavelength measurement errors can result. Fringe-counting wavelength meters are especially prone to producing erroneous results with low-frequency modulation signals present. However, averaging of the wavelength measurement over many measurements may improve the accuracy of the reading.

Fourier transform wavelength meters are more tolerant of pulsed input waveforms than are fringe-counting wavelength meters. When a periodic low-frequency modulation signal is present on the interferogram, the Fourier transform of that signal will produce modulation sidebands around the actual signal wavelength. The extra sidebands caused by the low frequency modulation will appear as spurious signals in the display that are suppressed from the main signal. The wavelength value of the main signal is still accurately displayed in the pulsed condition.

4.6 ALTERNATE WAVELENGTH METER TECHNIQUES

In this section, some alternate techniques for accurate wavelength measurement are made. The coverage is not complete, but it does discuss some important considerations that may not be well-covered by Michelson-interferometer wavelength-meter designs. The four configurations to be discussed are the FP filter, the static FP interferometer, static Fizeau interferometer, and wavelength discriminator. The static interferometer techniques are excellent for pulsed-laser measurements. Wavelength discriminator techniques have limited accuracy, but have simplicity and low cost.

4.6.1 Fabry-Perot Filters

The first part of this chapter discussed the use of a Michelson interferometer. There are other configurations that can be used to interfere light. The FP interferometer is shown in Figure 4.17. FP filters are also referred to as etalons. In an FP interferometer, light is incident on an optical component with two reflecting surfaces. The light that exits the interferometer consists of a direct path added to a large number of reflected paths through the interferometer.

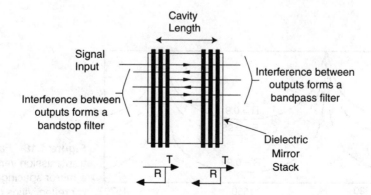

Figure 4.17   A Fabry-Perot interferometer.

The outputs add constructively or destructively depending on the wavelength of light. A similar interference occurs on the reflected signals from the interferometer. The FP interferometer is most commonly used as a tunable filter. The filter applications make use of the fact that the transmission loss through the filter is very wavelength selective. The passband of the filter can be adjusted by changing the angle of the incident light or by varying the spacing between the two reflections. The FP configuration can also be used as a fringe-counting wavelength meter. In the fringe-counting application, a diverging beam is applied to the interferometer. A detector array is then used to measure the distance between interference fringes that are found at the output of the interferometer. This section concentrates on a discussion of the FP interferometer in the filter application. Sections 4.6.2 and 4.6.3 will describe static fringe-counting configurations.

An analysis of the lossless FP interferometer will aid in the understanding of this type of wavelength meter. There are many excellent books that describe in detail the mathematics of an FP filter,[18-20] the results of the analysis are presented here. The filtering function of the interferometer is accomplished by interfering the initially transmitted signal with many delayed versions of the input signal. The maximum tran mission occurs when the delayed signals add-in phase with the first transmitted signal. The transmission function through a FP interferometer is given as[19]

$$\text{Transmission} = \frac{(1 - R)^2}{(1 - R)^2 + 4R \sin^2 \left(\dfrac{2 \pi L n \cos \theta}{\lambda_{vac}} \right)} \cdot \qquad (4.20)$$

In this equation, it is assumed that both mirrors have equal mirror power reflectivity, R. The index of refraction between the mirrors is n, the mirror spacing is L, and the angle of the input light with respect to the perpendicular is θ.

Figure 4.18 shows an example filter function for a mirror spacing of 50 μm and mirror reflectivities of 0.9, 0.95, and 0.99. The most obvious characteristic of the filter is the repeated passband. There are several terms used to describe the characteristics of an FP interferometer passband. The frequency spacing between the repeated transmission peaks of Figure 4.18 is

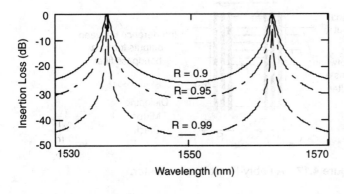

Figure 4.18 Fabry-Perot filter transmission versus wavelength for a mirror spacing of 50 μm and mirror reflectivities of 0.9, 0.95 and 0.99.

$$\Delta f = \frac{c}{2nL \cos \theta}.$$ (4.21)

The free spectral range (FSR) is another term that describes the transmission peak spacing in terms of wavelength spacing:

$$|\Delta \lambda| = \frac{\lambda^2 \cos \theta}{2nL}$$ (4.22)

The width (full-width at half of maximum) of the transmission peak is given as:

$$\delta f_{\frac{1}{2}} = \frac{(1-R)c}{2\pi \, nL \, \sqrt{R} \cos \theta}$$ (4.23)

The width of the transmission peaks in Figure 4.18 would be 100 GHz, 49 GHz, and 9.6 GHz for the 0.9, 0.95, and 0.99 mirror reflectivity cases shown.

The larger the mirror reflectivity, the narrower the transmission response. The term finesse describes the sharpness of the transmission peak in comparison to the width between repeated passbands.

$$\text{Finesse} = \frac{\Delta f}{\delta f_{\frac{1}{2}}}$$ (4.24)

Finesse values of several hundred are easily obtainable with values of up to several thousand being practical.

Repeated Passband Limitation on Narrow Spectral-Width Filters.   An FP filter can have an extremely narrow passband. With a very narrow passband, it is possible to accurately locate the wavelength of a signal in a wavelength-meter implementation where the FP filter is followed by a power meter. Narrow-width passbands can be accomplished with either very high values of mirror reflectivities or with wide mirror spacings. A major limitation of FP filters is the repeated nature of the passband. The repeated passband introduces uncertainty in the actual center wavelength measurement because one does not know which passband is being used in the repeated filter response. The repeated passband effect can be reduced by introducing mirrors that have a wavelength dependent reflectivity.[18] Figure 4.19 shows the measured passband of an FP filter that has many of the repeated passbands eliminated with special mirror coatings. This bandpass filter has a single with a very narrow (1 nm wide) passband. All of the repeated passbands are suppressed in this case out to 1400 nm.

Filter Wavelength Tuning.   The FP filter can be used as a tunable filter to locate the wavelength of an unknown signal in a manner similar to that of the diffraction grating in Chapter 3. FP filters are tuned by adjusting the angle of the incoming light with respect to the mirror or by adjusting the spacing between the mirrors. Equations 4.20 to 4.24 describe how the center-wavelength and filter-width changes with respect to mirror spacing and angle.

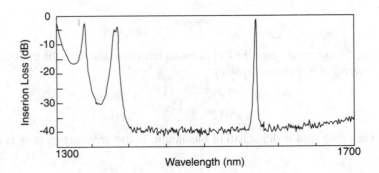

Figure 4.19 Transmission response of a Fabry-Perot filter with mirror coatings to reduce repeated passbands.

As the angle of the filter is changed, the width of the passband degrades. The multiple-reflection light beams do not completely overlap with the initially transmitted beam reducing the magnitude of interference. The maximum tuning range is limited by the range of acceptable filter-passband characteristics. This method of wavelength adjustment has the advantage of stability and simplicity. A disadvantage of angle tuning is that the transmission through the filter becomes polarization sensitive.

Length tuning suffers from the mechanical complexity of keeping the mirrors perpendicular with respect to length change. The free spectral range and filter width are also changing as the FP filter length is adjusted.

4.6.2 Static Fabry-Perot Interferometer Wavelength Meter

The first part of this section emphasized the use of the FP filter as a bandpass filter that can be wavelength tuned. The FP filter can also be used in a static fringe-counting mode of operation.[21] Static FP wavelength meters can measure the wavelength of individual laser pulses as well as cw beams because the measurements are obtained spatially. The word static in the title of this section emphasizes that there are no moving parts in the design. This is in contrast to the Michelson interferometer applications where a mirror is scanned in time to make a wavelength measurement.

The generation of interference fringes in a static FP etalon is illustrated in Figure 4.20. The Fabry-Perot etalon is illuminated with a diverging light beam. A bull's-eye concentric-ring pattern of interference fringes is produced. The spacing between the interference fringes is used to calculate the wavelength of the unknown signal. The wavelength meter is calibrated by comparing the interference pattern of the unknown signal to that produced by an accurate reference wavelength source such as a HeNe laser. Care must be taken so that both the reference and unknown wavelengths illuminate the FP etalon identically. Many of the wavelength accuracy considerations discussed in Section 4.4 of this chapter apply to static FP designs. The wavelength accuracy of the HeNe laser is important, the index of refraction dispersion must be considered, and fractional fringe-counting is important for high accuracy measurements.

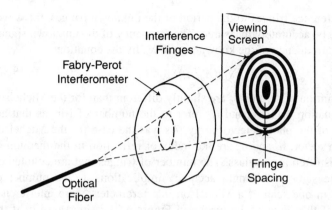

Figure 4.20 Static Fabry-Perot interferometer interference fringes.

4.6.3 Static Fizeau Interferometer Wavelength Meter

The Fizeau interferometer is similar to an FP interferometer except that the two reflecting surfaces form a wedge instead of being parallel. Fizeau wavelength meters are very important for accurate measurements of pulsed and modulated signals. Figure 4.21 shows a block diagram of a Fizeau interferometer[22-25] wavelength meter. The Fizeau interferometer wedge can be considered as a collection of FP interferometers with slightly different mirror spacings. The Fizeau wavelength meter uses reflections off of the front and back surfaces of the wedge to introduce interference. The two surfaces of the wedge have relatively low values of reflectivity so that only two dominant waves are being interfered. The reflected light from the wedge is then imaged on to a detector array where the interference pattern is recorded.

It is the measurement of the period of the interference fringes that allows the unknown wavelength of the signal to be measured. A calibration of the unknown fringe period can be made by comparing it to the fringe period of a reference laser signal. By com-

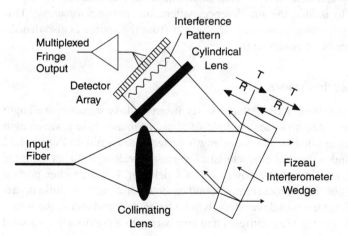

Figure 4.21 Static Fizeau interferometer wavelength meter diagram.

paring the period of the reference fringes to the period of the unknown fringes, the wavelength of the unknown can be accurately measured. The frequency of the unknown signal, λ_u, can be compared to the frequency of the known signal, λ_r, by the equation:

$$\lambda_u = (N_r/N_u)(n_u/n_r)\lambda_r \tag{4.25}$$

In this case, the definition of N_r and N_u are slightly different than for the Michelson interferometer case of Equation 4.11. N_r and N_u refer to the number of fringes that are measured over the length of the photodetector array. As was the case for the Michelson interferometer wavelength meter, n_r and n_u are the index of refraction in the medium of the interferometer which is often air or glass. The number of fringes that are counted on the detector array has the same wavelength accuracy implications as the number of fringes that are measured in one scan of a Michelson interferometer. More interference fringes will be measured if the slope of the wedge in Figure 4.21 is increased or if the length of the detector array is increased. The slope cannot be increased to very large values though because the fringe visibility decreases and the detector array has a finite density of detector elements.

It is the spatial interference pattern on the detector array that is one of the key features of the design. In a scanning Michelson wavelength meter, measurement inaccuracy can result if the amplitude or spectral characteristics of the input signal change over the duration of an interferogram scan. This limitation on measurement of pulsed signals was described in Section 4.5.6 for Michelson wavelength meters. The Fizeau wavelength meter does not require any physical motion to measure wavelength. Instead of adjusting the mirror spacing, the Fizeau interferometer has a built-in scanning function since the signal is simultaneously applied to the entire Fizeau wedge. This makes the Fizeau-wavelength meter design superior for measurements on optical sources with low frequency modulation or instabilities. The use of a detector array is often a cost barrier for use in the telecommunication wavelengths of 1100 to 1700 nm. Low-cost silicon detector arrays are not available at these wavelengths. Arrays of InGaAs/InP detector arrays are available though and the cost should decrease to make commercial products more practical. The Fizeau interferometer also suffers from poor sensitivity. The design of Figure 4.21 uses 4% reflectors to induce the interference pattern on the detector array. This means that 92% of the input power is unused. The power from the source is also distributed among a large number of array elements.

4.6.4 Wavelength Discriminators

A wavelength discriminator uses a filter that has its insertion-loss versus wavelength well-characterized. Figure 4.22 shows an example of a discriminator-based wavelength meter. An input signal is applied to the wavelength meter beam splitter. Figure 4.22 shows the transmission and reflection characteristics versus wavelength for an example discriminator element. A fraction of the power reaches detector 1 and another portion reaches detector 2. If most of the photocurrent is found on detector 1 and very little on detector 2, then it is evident that the signal wavelength must be on the short end of the wavelength range. Thus by ratioing the photocurrents and comparing to a previously measured

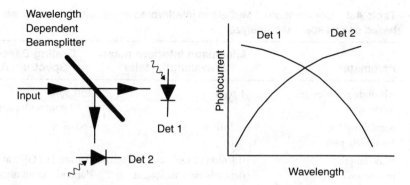

Figure 4.22 Wavelength discriminator.

calibration table, the wavelength of the input as well as the power of the signal can be measured. Absorption filters using doped glass are one method to achieve the change in transmission versus wavelength. This type of wavelength meter is very simple and may be adequate for many applications. Most optical power meters require the user to enter the wavelength of the incoming signal. With the wavelength discriminator, this problem can be solved with simultaneous knowledge of the wavelength and photocurrent.

The method has moderate wavelength accuracy capabilities when compared to interferometer-based wavelength meters. If the discriminator function changes rapidly with wavelength, better wavelength accuracy will be achievable at the expense of wavelength coverage. Typical accuracy for this type of wavelength meter is 650 ppm for a unit that can cover a single telecommunication wavelength band. Wavelength meters of this type cannot distinguish individual signals from a multiple signal environment.

4.7 SUMMARY

Comparison of Wavelength Meter Techniques.   Table 4.3 provides a comparison of the Michelson wavelength meter methods discussed in this chapter with the OSA techniques discussed in Chapter 3. This comparison is made for instruments that are optimized to make measurements with a comparable measurement update rate. The most dramatic difference between the two techniques is the achievable dynamic range for Michelson interferometer wavelength meters. The shot noise generated in the optical receiver is the limiting factor for the Michelson design.

This chapter discussed the accurate measurement of wavelength, concentrating on a Michelson interferometer approach. After reading this chapter, there are several important points that should be highlighted.

1. It is important to state the conditions for a wavelength measurement. Vacuum wavelength numbers and standard dry air wavelengths are most often quoted.

Table 4.3 Comparison of Michelson interferometer wavelength meter with a grating-based optical spectrum analyzer

Parameter	Michelson Interferometer Wavelength Meter	Grating-Based Optical Spectrum Analyzer
Absolute wavelength accuracy	1 ppm	150 ppm < 30 ppm with calibrator
Sensitivity for 1-s update rate	-65 dBm	-80 dBm
Wavelength resolution	0.1 nm (12 GHz at 1550 nm) depends on scan length	0.1 nm (12 GHz at 1550 nm) depends on number of illuminated grating lines
Dunamic range	40 dB limited by shot noise	70 dB limited by filter stopband rejection
Wavelength range	All fiber optic wavelengths	All fiber optic wavelengths

2. The Michelson interferometer is the most common platform used for accurate wavelength measurements. This method compares the wavelength of an unknown signal to that of a highly accurate laser wavelength reference.

3. Wavelength meters for the telecommunication bands can achieve less than ±1 ppm error in absolute wavelength accuracy.

4. Wavelength meters are capable of individually resolving multiple signal inputs simultaneously using Fourier transform techniques.

5. Fabry-Perot filters are useful alternatives to diffraction-grating based OSAs.

6. Static (no moving parts) Fabry-Perot and Fizeau interferometer designs can be optimized to measure the wavelength of sources with low frequency modulation.

7. Wavelength discriminators offer a low-cost approach to wavelength measurement with reduced performance.

REFERENCES

General Michelson Interferometer Wavelength Meter References

1. Obarski, G.E. 1990. Wavelength measurement system for optical fiber communications. Boulder, CO: *NIST Technical Report NIST/TN-1336.*

2. Snyder, J.J. 1982. *Laser wavelength meters. Laser Focus:* 18(1), 55–61.

3. Lawrence, M. 1984. Continuous wave laser wavelength measurement using the travelling michelson interferometer. *Opt. Laser Tech:* 137–140.

4. Monochalin, J.P., M.J. Kelly, J.E. Thomas, N.A. Kumit, A. Szoke, P.H. Lee, and A. Javan. 1981. Accurate laser wavelength measurement with a precision two-beam scanning Michelson interferometer. *Applied Optics* 20, No. 5: 736–757.

Fourier Transform Spectroscopy

5. Thorne, A.P. 1988. *Spectrophysics,* London: Chapman and Hall, pp. 185–201.

6. Junttila, M.L., B. Stahlberg, E. Kyro, T. Veijola, and K. Kauppinen. 1987. Fourier transform wavemeter. *Review of Scientific Instruments* 58, No. 7: 1180–1184.

References on Fractional Fringe Counting

7. Hall, J.L., and S.A. Lee. 1976. Interferometric real-time display of cw dye laser wavelength with sub-doppler accuracy. *Appl. Phys. Lett.* 29 (6): 367–369.

8. Bennett, S.J., and P. Gill. 1980. A digital interferometer for wavelength measurement. *J Phys. E.* 13, 174–177.

9. Ishakawa, J., I. Nobujiko, and K. Tanaka. 1986. Accurate wavelength meter for cw lasers. *Appl. Optics.* 25, No. 5: 639–643.

Reference on Standard Dry Air and the Index of Refraction of Air

10. Edlen, B. 1966. The refractive index of air. *Metrologia* 2, No. 2: 71–80.

11. Peck, E.R., and K. Reeder. 1972. Dispersion of Air. *Journal of the Optical Society of America* 62, No. 8: 958–962.

HeNe Laser Wavelength Accuracy

12. Melles-Griot Product Catalog, 1996. Fundamentals of Helium-Neon Lasers. 1770 Ketterling Street, Irvine, CA., 92714.

13. Jennings, D.A., F.R. Peterson, and K.M. Evenson. 1979. Frequency measurement of the 260-THz (1.15 μm) HeNe laser. *Optics Letters* 4, No. 5: 129–130.

14. Moore, C.E. 1971. *Atomic Energy Levels as Derived from the Analysis of Optical Spectra: Vol 1,* Boulder, CO: NSRDS-NBS 35, Vol 1 (COM-72-51282): 77.

15. Tetu, M. 1997. Absolute Wavelength Stability Ions. Optical Fiber Communications Conference. Dallas, TX: Optical Society of America, Tutorial FE1: 167–220.

Beam Misalignment

16. Kowalski, F.V., R.T. Hawkins, and A.L. Schawlow. 1976. Digital Wavemeter for cw Lasers. *J. Opt. Society of America* 66, No. 9: 965–966.

Diffraction Limits on the Beam

17. Bonsch, G. 1985. Simultaneous wavelength comparison of Iodine-stabilized lasers at 515 nm, 633 nm, and 640 nm. *IEEE Trans. Instrum. Meas.* Vol. IM-34, No. 2: 248–251.

Fabry-Perot Filter References

18. Macleod, H.A. 1989. *Thin-Film Optical Filters* 2nd ed. New York: McGraw-Hill.

19. Haus, Herman. 1984. *Waves and Fields in Optoelectronics.* Englewood Cliffs, NJ: Prentice Hall.

20. Born, M., and E. Wolf. 1970. *Principles of Optics.* Oxford: Pergamon Press.

Static Fabry-Perot Interferometer

21. Byer, R.L., J. Paul, and M.D. Duncan in *Laser Spectroscopy III,* editors J.L. Hall and J.L. Carlsten, Springer-Verlag, Heidelberg, 1977.

Fizeau Interferometer Wavelength Meters

22. Morris, M.B., T.J. McIlrath, and J.J. Snyder. 1984. Fizeau wavemeter for pulsed laser wavelength measurement. *Applied Optics* 23: 3862–3868.

23. Gardner, J.L. 1985. Compact Fizeau wavemeter, *Applied Optics,* Vol. 24: 3570–3573.

24. Gray, D.F., K.A. Smith, and F.B. Dunning. 1986. Simple compact Fizeau wavemeter. *Applied Optics* 25: 1339–1343.

25. Gardner, J.L. 1986. Wavefront curvature compensation in a Fizeau wavemeter. *Applied Optics* 25: 3799–3800.

CHAPTER

5

High Resolution Optical Frequency Analysis

Douglas M. Baney, Wayne V. Sorin

5.1 INTRODUCTION

The intensity and frequency dynamics of optical sources are key characteristics in determining the performance of optical systems. These characteristics determine, for example, the effects of fiber-group-velocity-dispersion on the transmitted signal, channel spacings in wavelength division multiplexed (WDM) systems, and the impact of fiber nonlinearities such as stimulated Brillouin scattering.[1] Measurement of the intensity dynamics is relatively straightforward using a photodetector and an appropriate electronic receiver (see Chapter 6 for more coverage). On the other hand, optical phase noise and frequency chirp, which have substantial impact on the optical power spectrum and the quality of transmitted signals, are not detectable with simple optical power detection. Using the optical mixing and interference techniques described in this chapter, the phase noise and frequency dynamics of optical sources are readily measured. This chapter focuses on the following topics:

- laser linewidth and phase noise (without modulation);
- optical power spectrum of a modulated source;
- time-domain chirp measurement of an optical carrier;
- frequency-domain FM measurement of an optical carrier.

The semiconductor laser is the workhorse optical source for telecommunications. It can be mass-produced in wafer form and efficiently coupled to the singlemode optical fiber that forms high-speed optical links. The light output from these lasers can be modulated to transmit information by varying their injection current. The unmodulated laser exhibits both intensity and phase noise which affects its performance in communications links. When modulation is applied, the optical source spectrum may be broadened (chirped) beyond the limits set by the information bandwidth. The spectral broadening due to chirp combines with wavelength dispersion in optical fiber to erode the shape of the transmitted pulses. This can lead to increased error rates in communications systems. The broadened spectrum of the modulated optical source also limits the proximity of channels in WDM systems. This chapter is aimed at methods to characterize linewidth, power spectrum, and chirping in optical sources. With these tools, a better understanding of the optical sources and ultimately the system performance can be obtained.

The linewidth of a typical single-frequency semiconductor laser is of the order of 10 MHz. To measure this linewidth, a typical grating-based OSA (see Chapter 3) would need about a thousand times improvement in resolution. Other optical filter-based methods such as scanning FP filters and interference filters are not covered in this chapter, the reader is referred to Chapters 3 and 4 for more information on these techniques. In general, these scanning filter methods are not able to achieve the measurement resolutions afforded by the methods presented in this chapter. The measurement techniques discussed in this chapter not only attain the required frequency resolution, but also allow measurement of the frequency response and frequency dynamics of optical signals. As a reference, the following matrix is provided which helps with the task of matching the desired measurement to a specific characterization technique. In Table 5.1, the column labeled *difficulty* refers to the ease with which the measurement can be set up and operated. The section column refers the reader to the appropriate chapter section.

In Section 5.2, basic concepts concerning laser dynamics and interferometry are discussed that will aid in understanding the measurement concepts presented later on. Section 5.3 discusses various methods of linewidth characterization. High-resolution measurement of the power spectrum of a modulated optical signal is examined in Section 5.4.

Table 5.1 Measurement Technique Comparisons

Measurement	Technique/Comments	Difficulty	Section
Linewidth (unmodulated)	Heterodyne: high-resolution, sensitive	low	5.3.1
	Delayed self-heterodyne: ~10 kHz-1 GHz range	low	5.3.2
	Delayed self-homodyne: ~10 kHz-1 GHz range	very low	5.3.3
	Discriminator: phase noise and jitter	moderate	5.3.5
Optical spectrum	Heterodyne: measure any spectrums	low	5.4.1
	Gated delayed self-homodyne: symmetrical spectrum	low	5.4.2
Frequency chirp measurement	Coherent discriminator: oscilloscope	moderate	5.5
Swept FM measurement	Coherent discriminator: network analyzer	moderate	5.6

Sections 5.5 and 5.6 concern the measurement of time-domain laser chirp and frequency domain FM response.

5.2 BASIC CONCEPTS

Measurement Assumptions

Prior to delving deeper into the subject, it is worthwhile to discuss some constraints with respect to the measurement methods presented here. The lasers under test are assumed to operate in a single-longitudinal mode (SLM). This is another way of saying that all the resonant frequencies of the laser cavity are suppressed with the exception of a single-mode. Properly designed DFB and DBR lasers operate in a singlemode. An illustration of a laser spectrum (no modulation) is shown in Figure 5.1. The laser lineshape typically has a Lorentzian-shaped central peak,[2,3] small sidebands caused by relaxation oscillations, and small sidemodes (cavity frequencies) located further away. Many of the measurement techniques presented in this chapter are only valid in the SLM regime. Optical mixing or interference (see Section 5.5.5) plays a key role in the measurement methods presented here. To obtain efficient interference, the following conditions are required between the interfering beams:

- polarization alignment
- spatial overlap

All of the optical fiber in the circuits discussed in this chapter are singlemode. Singlemode fiber insures good spatial overlap of the optical waves that are combined in the measurement setups. Often, the optical fiber used in the measurement setups allow the po-

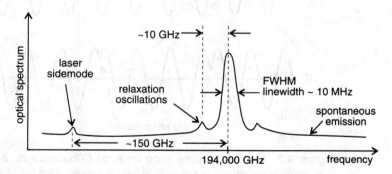

Figure 5.1 Laser spectrum with central peak, relaxation oscillation peaks, laser side-modes, and spontaneous emission.

larization states of the propagating light to freely evolve without any control. To overcome this problem, polarization state controllers are often placed in the measurement circuits to permit polarization alignment.

Coherence Time

The coherence time, τ_c, of a laser is a measure of the spectral purity of the laser frequency over time. In two-path interferometers, the degree to which an optical wave interferes with a time-delayed portion of itself depends on the coherence time of the wave with respect to the optical delay. Coherence time is reduced by random events, such as spontaneous emission in the laser cavity, which alter the phase or frequency of the laser-output field. This is illustrated in Figure 5.2. In Figure 5.2(a), the coherence time is longer since the phase is predictable during the interval of time T_1–T_2. In Figure 5.2b random phase changes cause an uncertainty in the phase relation between time T_1 and time T_2. The coherence time, τ_c, varies inversely with laser linewidth, Δv. It is defined for spectra with Lorentzian lineshapes as

$$\tau_c = \frac{1}{\pi \Delta v} \tag{5.1}$$

Thus as the source linewidth increases, the coherence time decreases. The related concept of coherence length, L_c, is also used in discussions on interferometry. The coherence length is simply the coherence time multipled by the velocity of light: $v_g = c/n_g$, where n_g, the group velocity index, is approximately 1.47 in optical fiber

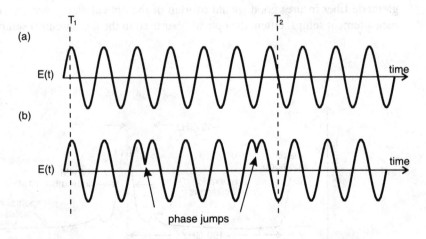

Figure 5.2 Concept of coherence time. (a) Coherent light: across the time interval the phase of the optical field is predictable. (b) Short coherence: random phase jumps cause the phase of the optical field to be uncertain across the time interval.

$$L_c = v_g \tau_c \qquad\qquad (5.2)$$

As an example, consider a laser with a linewidth of 10 MHz. The coherence time and coherence length from Equations 5.1 and 5.2 are 32 ns and 6.5 m respectively.

5.2.1 Linewidth and Chirp

The two dominant causes of spectral broadening in single-longitudinal-mode lasers are phase noise and frequency chirp. Random phase noise is created when spontaneous-emission, originating in the laser cavity gain media, changes the phase of the freerunning laser frequency. This process is magnified by physical effects within the laser cavity. The magnification is quantified by the laser's effective amplitude-phase coupling factor, α_{eff}. A large value of α_{eff} results in increased laser linewidth. More generally, α_{eff} represents the link between power changes in the laser cavity to phase changes of the emitted light. The result is a broadening of the laser spectral linewidth. Another laser process, known as relaxation oscillations, (see Figure 5.1) causes subsidiary peaks centered around the central mode of the laser. These peaks generally lie within 20 GHz of the carrier and are much smaller in amplitude than the main peak.

Laser frequency chirp results in significant spectral broadening when the laser injection current is modulated.[4] The unwanted frequency modulation, or chirp, can broaden the laser spectrum well beyond the freerunning optical linewidth. The magnitude of the chirp is proportional to the amplitude-phase coupling factor α_{eff}. Material and structural properties of the laser contribute to the value of α_{eff}, hence the resulting chirp. Laser chirp (without intensity modulation) is illustrated in Figure. 5.3. The sweeping of the optical phase is due to the presence of frequency modulation or chirp on the optical carrier. A laser undergoing intensity modulation at 2.5 Gbit/s can occupy more than 25 GHz of optical spectrum caused by laser chirp. More detail on laser linewidth, relaxation oscillations, and chirp can be found in Agrawal,[3] Henry,[5] and references therein.

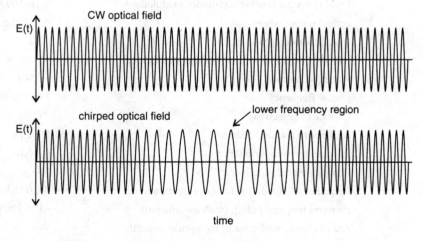

Figure 5.3 Optical field with and without frequency chirp.

Some relations for estimating the laser linewidth, the relaxation resonance frequency, and the chirp of a semiconductor laser are given here. These relations are pertinent to the measurement process since it is important to note relevant experimental conditions when making measurements. The equations below, while not complete, will help indicate some of the dependencies. The variables and constants are defined in Table 5.2.

Static Linewidth

$$\Delta v_1 = \frac{1}{4\pi P} n_{sp} (1 + \alpha_{eff}^2) hv \frac{\log(1/R)}{\tau_p \tau_{rt}} \tag{5.3}$$

Small Signal Chirp

$$\Delta v_{max} = \frac{\alpha_{eff} \, m f_m}{2} \tag{5.4}$$

Large Signal Chirp

$$\Delta v_c = \frac{\alpha_{eff}}{4\pi} \left(\frac{1}{P} \frac{\partial P}{\partial t} \right) \tag{5.5}$$

Table 5.2 Variable Definitions

Symbol	Description	Typical Range of Values
Δv	FWHM optical linewidth (without modulation)	0.01–100 MHz
P	optical output power	0.1–500 mW
n_{sp}	spontaneous emission factor	~2.5
α_{eff}	effective amplitude-phase coupling coefficient	−1–10
h	Planck's constant	6.634×10^{-34} J-s
v	optical frequency	193 THz@$\lambda = 1.55$ μm
R	laser facet reflectivity	~ 0.30
τ_p	cold cavity photon lifetime	1 ~ 2 ps
τ_{rt}	laser cavity round-trip delay	~ 0.5 ps
m	intensity modulation index	0–1
f_m	frequency of sinudoidal intensity modulation	0–20 GHz
Δv_c	transient frequency chirp (with modulation)	~1 to 100 GHz
$\dfrac{\partial P}{\partial t}$	rate of change with time of the optical intensity	W/s

Example

A laser has an effective amplitude-phase coupling factor, $\alpha_{eff} = 5$. If the laser, undergoing 2.5 Gb/s intensity modulation has an intensity slope of 5 mW/30 ps at 4 mW average output, estimate the transient frequency chirp.

Solution

Using Equation 5.5 and $\dfrac{\partial P}{\partial t} = 1.67 \times 10^8 \; W/s$

the chirp $\Delta v \gtrsim_c$ is estimated to be approximately 20 GHz.

5.2.2   Interference between Two Optical Fields

In this section, we examine some of the basics of interference between two optical fields. The concept of interference will be central to the measurement techniques of this chapter. The heterodyne case will be discussed first. This case uses a local oscillator laser as a measurement reference to measure a second signal source with unknown spectral characteristics. Interference of a wave with a delayed version of itself will also be examined. In both cases, we shall see that interference between waves causes intensity variations that are detectable using a photodiode. When the photocurrent is analyzed with electronic instrumentation, information on the optical carrier variations can be obtained. In this chapter, frequency will be denoted by both f and v. f will represent frequencies below ~100 GHz and v will denote frequencies above ~100,000 GHz.

Heterodyne: Interference between Two Fields.   Consider the two optical fields incident on the photodetector after passing through the combiner as shown in Figure 5.4:

$$E_s(t) = \sqrt{P_s(t)} \; e^{\, j(2\pi v_s t + \phi_s(t))} \tag{5.6}$$

$$E_{LO}(t) = \sqrt{P_{LO}} \; e^{\, j(2\pi v_{LO} t + \phi_{LO}(t))} \tag{5.7}$$

These two fields are scaled such that their magnitudes squared are optical powers (in other words, $P(t) = |E(t)|^2$). The optical field frequencies and phases are designated by v and $\phi(t)$. If either field were separately detected on a photodetector, the resulting photocurrent would follow only the power variations, $P(t)$, and all phase information would be lost. The optical phase $\phi(t)$ takes into account any laser-phase noise or optical frequency modulation. The value of v at a wavelength of 1.55 μm is 194,000 GHz. Thus the

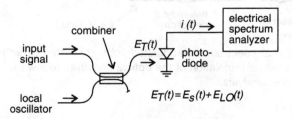

Figure 5.4 Setup for interfering two optical fields. The electrical spectrum analyzer display is proportional to the photocurrent power spectrum.

total phase, $2\pi\nu t + \phi(t)$ of each optical field, changes at a rate much too fast for electronic instrumentation to respond. The optical spectrum corresponding to the two fields is shown in Figure 5.5a. Here the local oscillator has constant power and the signal laser has a small intensity modulation index of m. To obtain the correct spectral display, the local oscillator frequency is set to a lower optical frequency than the signal under study. The optical combiner delivers the spatially overlapped optical fields to the photodetector where interference is detected. The total field $E_T(t)$ at the photodetector is:

$$E_T(t) = E_s(t) + E_{LO}(t) \tag{5.8}$$

Since power is detected (in other words, $P(t) = |E_T(t)|^2$), and not the optical field itself, photodetection is nonlinear with respect to the optical field. This fortunate situation allows us to detect interference between fields. The photocurrent generated in the detector is proportional to the squared magnitude of the field.

$$i(t) = \mathcal{R}|E_T(t)|^2 \tag{5.9}$$

where $\mathcal{R}$ is the detector responsivity given by

$$\mathcal{R} = \frac{\eta_d q}{h\nu} \ [A/W] \tag{5.10}$$

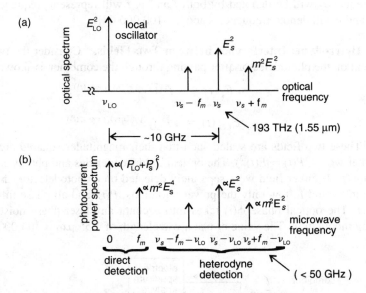

Figure 5.5 (a) Optical spectrum of modulated light. (b) Photocurrent spectrum after heterodyne translation of the optical spectrum to low frequencies for electronic analysis.

where η_d ($0 < \eta_d \le 1$) is the detector quantum efficiency, a measure of the conversion efficiency of incident photons into electrical charge. The parameters q and $h\upsilon$ are electronic charge (1.6021×10^{-19}C) and photon energy ($h = 6.6256 \times 10^{-34}$ J, $\upsilon = c/\lambda$) respectively. Substituting Equation 5.6, 5.7, and 5.8 into 5.9 we obtain using $f_{IF} = \upsilon_s - \upsilon_{LO}$ and $\Delta\phi(t) = \phi_s(t) - \phi_{LO}(t)$:

$$i(t) = \Re[P_s(t) + P_{LO} + 2\sqrt{P_s(t)P_{LO}}\, \cos(2\pi f_{IF} t + \Delta\phi(t))] \qquad (5.11)$$

The first two terms correspond to the direct intensity detection of $E_s(t)$ and $E_{LO}(t)$. The third term is the important heterodyne mixing term. Note that the actual optical frequency is gone and only the difference frequency is left. Thus the heterodyne method is able to shift spectral information from high optical frequencies to frequencies that can be measured with electronics as shown in Figure 5.5b. In the heterodyne method, the local oscillator serves as a reference, with known frequency, amplitude, and phase characteristics. Thus the signal spectrum, including both intensity and frequency contributions can be obtained. Equation 5.11 will be used later on in Section 5.3 in the discussions on optical heterodyne.

Self-Homodyne: Interference between a Field and a Delayed Replica.   Next we will consider the case where one of the two interfering optical fields is a delayed version of the other. This condition can be created by a variety of two-path optical circuits such as the Mach-Zehnder and Michelson interferometers (see Chapter 4) as well as Fabry-Perot interferometers. A Mach-Zehnder interferometer is shown in Figure 5.6a. The input field is split and routed along two paths with unequal lengths. Time τ_o is the differential time delay between the two fields traversing the two arms of the interferometer. The photocurrent generated at the detector is found in a similar way as with the heterodyne case

$$i(t) = \Re[P_1(t) + P_2(t) + 2\sqrt{P_1 P_2}\, \cos(2\pi\upsilon_o \tau_o + \Delta\phi(t, \tau_o))] \qquad (5.12)$$

where $P_1(t)$ and $P_2(t)$ are the powers delivered to the photodetector from each interferometer path. The average phase-setting of the interferometer is given by $2\pi\upsilon_o\tau_o$ and $\Delta\phi(t,\tau_o) = \phi(t) - \phi(t - \tau_o)$ is the time-varying phase difference caused by phase or frequency modulation of the input signal, and the interferometer delay τ_o. The interferometer free-spectral range (FSR) is defined as the change in optical frequency, to obtain a phase shift of 2π between the two combining fields. In other words, it is the frequency difference between the two peaks shown in Figure 5.6b. Obviously, from Equation 5.12, the FSR is the reciprocal of the net interferometer differential delay, τ_o.

Assuming $\Delta\phi(t, \tau_o)$ is small, varying the interferometer delay or the average optical frequency can cause the photocurrent to swing from minimum to maximum as shown in

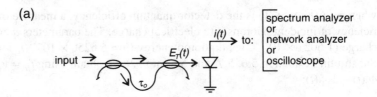

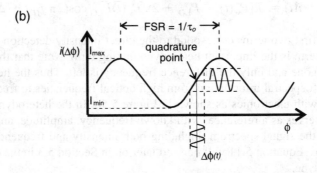

Figure 5.6 (a) Mach-Zehnder interferometer with optical detection and instrumentation for analysis of photocurrent. (b) Dependence of interferometer output on the phase difference between the interfering fields. FSR = free spectral range.

Figure 5.6b. Limitations to the minimum and maximum current swings can be caused by a lack of polarization alignment between the fields, mismatch between path losses through the interferometer, or the limited coherence time of the optical source.

If the average phase $2\pi\nu_o\tau_o$ is equal to $\pi/2$, or more generally, equal to $\pi(2n + 1/2)$, $n = 0, 1, 2, \ldots$, the interferometer is biased in quadrature. This point is indicated in Figure 5.6b. When an interferometer is biased at quadrature, it can linearly transform small optical-phase excursions $\Delta\phi(t,\tau_o)$ into photocurrent variations. This is because the cosine characteristic varies linearly for small changes, $\Delta\phi(t, \tau_o)$ about the quadrature point. Thus the interferometer can function as a frequency discriminator as long as operation is confined to the approximately linear part of the interferometer transfer characteristic. At the quadrature point, Equation 5.12 becomes:

$$i(t) = \Re[P_1(t) + P_2(t) + 2\sqrt{P_1(t)P_2(t)}\sin(\Delta\phi(t,\tau_o))] \qquad (5.13)$$

If $\Delta\phi(t, \tau_o)$ is small such that the approximation $\sin(\Delta\phi(t, \tau_o)) \approx \Delta\phi(t, \tau_o)$ is valid, then the discriminator acts as a linear transducer converting phase or frequency modulation into power variations that can be measured with a photodetector:

$$i(t) = \Re[P_1(t) + P_2(t) + 2\sqrt{P_1(t)P_2(t)}\,\Delta\phi(t,\tau_o)] \qquad (5.14)$$

The first two terms correspond to simple direct detection, the third term is the useful interference signal. In Sections 5.3.5, 5.5, and 5.6, the application of the interferometer as a discriminator to measure laser-phase noise, time-domain chirp, and FM response is presented. In these applications, the interferometer delay, τ_o, must be smaller than the source coherence time τ_c to maintain good interferometer contrast, which is a measure of the difference between $I_{\max}$ and $I_{\min}$ in Figure 6b.

5.3   LASER LINEWIDTH CHARACTERIZATION

In this section, several methods for linewidth characterization of freerunning (unmodulated) singlemode lasers are discussed. Linewidth is often defined in terms of the full-width half-maximum (FWHM) of the optical field power spectrum. Grating-based optical spectrum analyzers (OSAs) don't offer the measurement resolution required for laser linewidth measurement, so alternative characterization methods must be used. The alternative methods brought forth here include the optical heterodyne method, the delayed self-heterodyne method, the delayed self-homodyne method, and an optical discriminator technique. These methods are capable of obtaining the extremely high resolution required for laser linewidth measurements. Understanding the advantages and limitations of each of the methods will aid in deciding which method is best in a particular application. At the end of this section the strengths and weaknesses of the various methods are generalized.

5.3.1   Heterodyne Using a Local Oscillator

Heterodyne analysis is the only technique presented in this chapter capable of characterizing nonsymmetrical spectral lineshapes. Not only will the heterodyne method provide linewidth data, it also is used to measure the optical power spectrum of an unknown optical signal. This method offers exceptional sensitivity and resolution. The key component required for these measurements is a stable, narrow linewidth reference laser.

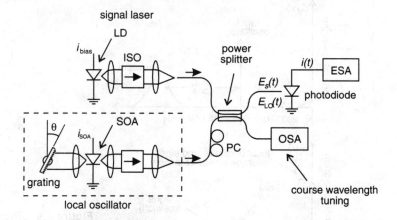

Figure 5.7 Optical heterodyne setup for measuring laser linewidth using an external cavity laser for local oscillator. SOA = semiconductor optical amplifier, ISO = optical isolator, LD = semiconductor laser diode, PC = polarization state controller.

The setup for optical heterodyne discussed here is illustrated in Figure 5.7. In this setup, the reference laser (local oscillator) is tuned appropriately and then its optical frequency is fixed during the measurement. This is possible because of the wide analysis bandwidth offered by electrical spectrum analyzers. An alternative way would be to have a narrow bandwidth electrical detection and a swept local oscillator. This alternative method[6] sets stringent requirements on the tuning fidelity of the local oscillator and will not be discussed further here. In Figure 5.7, light from the local oscillator (LO) is combined with the signal laser under test. A grating-tuned external cavity diode laser is used as the LO in this example. Polarization state converters are placed in the LO path to align the polarization state of the local oscillator to that of the signal under test. The coupler combines the two fields, delivering half the total power to each output port. One port leads to a photodetector which detects the interference beat tone, converting it to an electrical tone. Note that the local oscillator laser frequency must be tuned close to the signal laser frequency to allow the mixing product to fall within the bandwidth of typical detection electronics. A course alignment of the local oscillator wavelength is performed using an OSA or a wavelength meter. The local oscillator frequency is tuned to a frequency just lower than the average frequency of the laser under study. This creates a heterodyne beat tone between the LO and each of the frequency components in the signal spectrum as indicated in Figure 5.8. Thus each frequency component is translated to a low-frequency interference term described by:

$$i(t) = \Re[P_s(t) + P_{LO} + 2\sqrt{P_s(t)P_{LO}} \cos(2\pi(\nu_s - \nu_{LO})t + \Delta\phi(t))] \qquad (5.15)$$

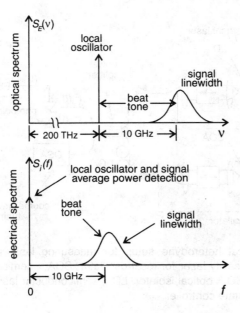

Figure 5.8 The mixing process in terms of beat-tone slices mixed down to low frequencies that can be analyzed with electronic instrumentation.

If the local-oscillator phase noise is small with respect to the test laser, the beat tone will be broadened primarily by the phase noise $\Delta\phi(t) \sim \phi_s(t)$ of the laser under study. The beat frequencies due to signal phase noise are measured using an ESA.

Heterodyne Power Spectrum.   The ESA display is proportional to the power spectrum of the photodetector current which contains products of optical heterodyne mixing as well as direct detection terms[7,8]

$$S_i(f) \approx \mathcal{R}^2 \{ S_d(f) + 2[S_{LO}(\nu) \otimes S_s(-\nu)] \} \qquad (5.16)$$

$$(\text{ESA} \Rightarrow \text{direct detection} + \text{heterodyne spectrum})$$

$S_d(f)$ is the ordinary direct detection that could be measured with just a photodetector and ESA. The second term is the useful heterodyne mixing product which is the convolution of the local oscillator spectrum $S_{LO}(\nu)$ with the signal spectrum $S_s(\nu)$. The convolution originates from the multiplication of the time-varying local oscillator field with the signal field in the photodetector. Multiplication in the time domain is equivalent to convolution in the frequency domain. The lineshape of the laser, including any asymmetries, is replicated at a low frequency set by the optical frequency difference between the two lasers. The convolution given in Equation 5.16 is illustrated in Figure 5.9. As the convolution scans the LO lineshape (Dirac-Δ function as shown) from negative infinity, it passes to zero at ν_{LO}, traces out the test-laser lineshape and continues to positive infinity. The net result is a translation of the test-laser lineshape to the average difference frequency between the LO and the test laser. Note from Equation 5.15 that as the LO linewidth broadens, its linewidth can dominate the photocurrent spectrum and decrease the frequency resolution of the heterodyne measurement.

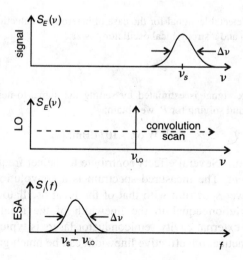

Figure 5.9 Convolution of narrow linewidth laser translates signal spectrum to low frequencies.

If the LO laser linewidth is small compared to the laser under test, the lineshape spectrum of the local oscillator, $S_{LO}(v)$ is approximated with a Dirac-Δ function: P_{LO} $\delta(v - v_{LO})$. Thus from Equation 5.16 the ESA will display:

$$S_i(f) \approx 2\mathcal{R}^2 P_{LO} S_s (v - v_{LO}) \tag{5.17}$$

Thus the ESA gives a measure proportional to the actual laser power spectrum $S_s(v)$ translated to low frequencies accessible to electronics. The key to the exceptional sensitivity of the heterodyne method is evident in Equation 5.17 where the detected spectrum strength increases with local oscillator power, P_{LO}. Large local oscillator power translates to better sensitivity.

Signal-to-Noise Ratio. The measurement signal-to-noise ratio, SNR, is the ratio of the detected heterodyne signal to all of the noise contributions. These noise contributions include the receiver thermal noise, local oscillator RIN, and interference noise. Interference noise is due to optical reflections in the measurement setup that convert laser phase noise into intensity noise which is detected at the receiver (see Section 5.3.3). The best performance is achieved when the dominant noise contribution is the shot noise from the local oscillator. In this case, the detection is said to be quantum-limited or shot-noise limited. The SNR for this case is:

$$\left(\frac{S}{N}\right)_{\text{shot noise limit}} = \frac{\mathcal{R} P_s}{q B_e} \tag{5.18}$$

where P_s is the optical-signal power incident on the photodetector and B_e is the electrical bandwidth of the ESA. Note that this shot-noise limited sensitivity is independent of the local oscillator power and the minimum detectable power is approximately equal to a single photon within the electrical detection response time.

Example

Estimate the minimum detectable signal for the case of heterodyne detection with a $B_e = 100$ kHz electrical bandwidth and a strong local oscillator power.

Solution

The minimum detectable signal is estimated by setting the signal-to-noise ratio to unity. Using $q = 1.602 \times 10^{-19}$ and solving for P_s we obtain:

$$P_s \approx 2 \times 10^{-14} \; W \text{ or } -107 \text{ dBm.}$$

Frequency Resolution. Several effects contribute to the net frequency resolution of the linewidth measurement. The measured spectrum is a convolution (see Equation 5.16) of the signal-field power spectrum with that of the local oscillator laser. This sets the minimum spectral resolution equal to the linewidth of the local oscillator. The linewidth of typical tunable external cavity semiconductor lasers is typically of the order of 100 kHz. However, in practice, the effective linewidth can be much greater due to frequency jitter or 1/f frequency noise. Frequency jitter is the random change in the operating frequency of the laser over time. It is usually caused by environmentally-induced changes in the laser cavity. The time scales for frequency jitter are slow, typically less than a microsecond, but they can still be faster than the integration time of the electronic spectrum analysis. When the measurement averaging time takes a few seconds, the effective linewidth of an external cavity laser can easily be a factor of ten greater (several MHz). Another contributing factor to frequency resolution is the resolution bandwidth of the electrical spectrum analyzer. The resolution bandwidth filter should be set so as not to limit the desired measurement resolution.

Experiment.   The beat spectrum of two semiconductor lasers separated in optical frequency by 8.3 GHz is shown in Figure 5.10. The full-width half-maximum, FWHM, linewidth of a laser is often measured with respect to an assumed Lorentzian spectral shape. The displayed FWHM response is the sum of the laser linewidth and the local oscillator linewidth. Therefore, a narrow LO linewidth is desirable to clearly resolve the laser-under-test linewidth. Often there is frequency jitter as discussed earlier. The effect of frequency jitter is to widen the displayed lineshape. The effects of frequency jitter on the linewidth measurement can be reduced by measuring further down on the displayed lineshape as shown in Figure 5.10. The measured result is then transformed into the correct FWHM linewidth. The advantages of measuring further down on the lineshape can be understood by considering the effect of frequency jitter, on the power spectrum. The Lorentzian-optical power spectrum centered at v_o has a functional form given by:

$$S_E(v) \sim \frac{1}{1 + \left[\dfrac{v - v_o}{\Delta v/2}\right]^2} \tag{5.19}$$

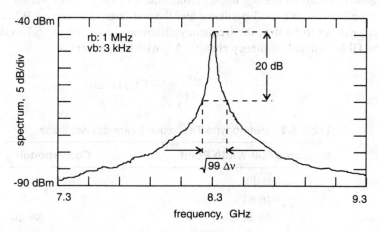

Figure 5.10   Laser linewidth measurement using optical heterodyne method.

Where Δv is the FWHM linewidth and v is optical frequency. When $v - v_o$ is zero, $S_E(v)$ is maximum. As $v - v_o$ increases, the magnitude falls off in a way characteristic of the Lorentzian function. If frequency jitter $\delta v(t)$ is present the spectrum "jitters around" in time:

$$ S_E(v, t) \sim \frac{1}{1 + \left[\dfrac{v - v_o + \delta v(t)}{\Delta v/2} \right]^2} \tag{5.20} $$

When $v - v_o$ is small, on the order of δv, variations in δv cause significant changes in the amplitude of $S_E(v)$ and hence the measured FWHM linewidth. As $v - v_o$ increases, relative to δv, the magnitude of the function depends more on the value of $v - v_o$ and less on the effects of δv. Thus the error due to frequency jitter decreases as the measurement is made further down the skirt of the lineshape, as long as the measurement SNR is adequate and the laser lineshape follows a known functional form.

For the usual case of a Lorentzian-shaped spectrum, the correspondence between the measured full-width at a specific power and the FWHM linewidth is shown in Table 5.3.

Experiment. In this experiment, the dependence of the test laser's optical frequency on bias current is characterized. The measurement setup consists of a wavelength-tunable external cavity laser and the test laser (DFB-LD) as shown in Figure 5.7. The grating-based OSA allows setting of the local oscillator to a slightly longer wavelength than that of the test-laser wavelength. The heterodyne beat frequency was set to a baseband frequency of 1.75 GHz by fine-tuning the local oscillator wavelength. The spectrums for three bias currents are shown in Figure 5.11. The DFB was initially biased at 70 mA. Decreasing the DFB laser current to 62 mA increased the DFB laser frequency, caused by thermal heating in the semiconductor cavity. This resulted in a 9.43 GHz separation between the local oscillator and the test-laser frequencies. When the injection current was set to 54 mA, the frequency difference increased to 16.73 GHz. From this data the DFB's optical frequency change, Δv_o with bias current is:

$$ \frac{\Delta v_o}{\Delta i} \approx -1 \text{ GHz/mA} \tag{5.21} $$

Table 5.3 Heterodyne Technique Linewidth Relations

Measured Full-Width Point	Corresponding Width
−3 dB	Δv
−10 dB	$\sqrt{9} \, \Delta v$
−20 dB	$\sqrt{99} \, \Delta v$
−30 dB	$\sqrt{999} \, \Delta v$

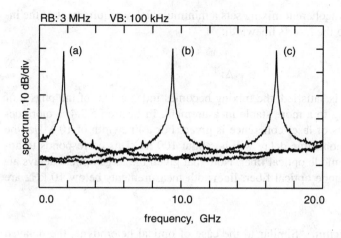

RB: 3 MHz   VB: 100 kHz

Figure 5.11 Optical heterodyne power spectrums of a DFB laser biased at a current of (a) 70 mA, (b) 62 mA, (c) 54 mA.

5.3.2 Delayed Self-Heterodyne

The delayed self-heterodyne technique provides a simple way to perform linewidth measurements without the requirement of a separate local oscillator laser. Taking advantage of the large optical delays attainable with optical fiber, Okoshi and co-workers demonstrated that linewidth measurements could be performed with a simple optical interferometer.[9] The delayed self-heterodyne concept is shown in Figure 5.12a. Incident light is split into two paths by the interferometer. The optical frequency of one arm is offset with respect to the other. If the delay, τ_o of one path exceeds the coherence time, τ_c of the source, the two combining beams interfere as if they originated from two independent lasers offset in frequency by $\delta\nu$ as shown in Figure 5.12b. Thus the system performs similarly to optical heterodyne. The beat tone produced is displaced from 0 Hz by the frequency shift by $\delta\nu$. An electrical spectrum analyzer displays the beat tone which is broadened by the laser linewidth. The translation of linewidth information from optical frequencies to low frequencies where electronics instrumentation operate is shown in Figure 5.13. As with the optical heterodyne case, the spectrum on the ESA is a convolution of the individual power spectrums of the interfering waves.

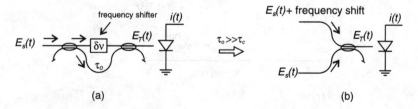

Figure 5.12 Delayed self-heterodyne method: (a) optical measurement setup, (b) equivalent circuit when the interferometer delay time is larger than the signal coherence time.

The requirement for incoherent mixing sets a minimum delay requirement of the interferometer with respect to the laser's linewidth:

$$\tau_o \geq \frac{1}{\Delta\nu} \tag{5.22}$$

When this condition is satisfied, the mixing becomes independent of the phases of the interfering light, leading to a more stable measurement. In Section 5.3.4, a more detailed analysis of the effects of laser coherence is given. For a linewidth of 10 MHz, the minimum required differential time delay will be about 100 ns. This corresponds to approximately 20 m of singlemode optical fiber. Note that given the large optical delays afforded by low-loss singlemode optical fiber, linewidth measurements below 10 kHz are possible.

Photocurrent Spectrum.   Similar to the case of optical heterodyne, the delayed self-heterodyne photocurrent spectrum consists of direct detection as well as the desired mixing product:

$$S_i(f) \approx \mathcal{R}^2 \{ S_d(f) + 2[S_s(\nu - \delta\nu) \otimes S_s(-\nu)] \}$$
$$(\text{ESA} \Rightarrow \text{direct direction} + \text{self-heterodyne spectrum}) \tag{5.23}$$

$\delta\nu$ is the shift frequency applied to the field traversing one arm of the interferometer and $\mathcal{R}$ is the usual detector responsivity. Since the mixing term is essentially the test laser spectrum convolved with itself (see Figure 5.13) and displaced in frequency by $\delta\nu$ the displayed lineshape will always be symmetrical, even if the original lineshape had important asymmetries.

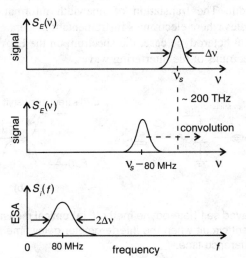

Figure 5.13   The delayed self-heterodyne mixing of the laser field with a frequency shifted replica.

Linewidth Interpretation.   For the case of a Lorentzian-shaped laser field spectrum, the displayed lineshape will be twice that of the actual linewidth as shown in Figure 5.13. Shifting the beat frequency from DC is useful to avoid instrumental limitations such as the DC response, the low-frequency noise, and the local-oscillator feedthrough in electrical spectrum analyzers. The optical lineshapes of semiconductor lasers are typically Lorentzian-shaped. Fortunately the Lorentzian and Gaussian lineshapes retain their form during conversion from the optical spectrum to the electrical domain through the delayed self-heterodyne process. Gaussian lineshapes have a FWHM photocurrent spectrum larger by a factor of $\sqrt{2}$ than the original optical lineshape. The more common Lorentzian lineshape has a FWHM detected lineshape larger by a factor of two.

Laser sources exhibiting frequency jitter or 1/f noise will yield larger measured linewidths in a similar way as the heterodyne technique. In this case, the measured linewidth will vary with the interferometer delay.[10] A larger delay yields a larger linewidth. For Lorentzian lineshapes, the 3 dB linewidth must be inferred from measurements taken further down on the displayed lineshape as with the heterodyne case. Table 5.4 indicates the correspondence between the measured full-width at a specific level and the FWHM linewidth assuming a Lorentzian profile.

Frequency Shifters.   The frequency shift in delayed self-heterodyne linewidth measurements can be obtained with a variety of devices including acousto-optic frequency shifters, phase modulators, and intensity modulators. The use of small signal-injection current modulation on the test laser has also been demonstrated to displace the beat spectrum from 0 Hz.[11] Frequency shifting the light in one arm of the interferometer by 2.37 GHz using a semiconductor amplifier has also been demonstrated.[12] An arrangement using an acousto-optic Bragg-frequency shifter is shown in Figure 5.14. It is important that the frequency shift be larger than the spectral content of the laser under study, if not, foldover effects near zero frequency will distort the observed spectrum.

Experiment.   An acousto-optic frequency shifter provided an 80 MHz shift frequency for a delayed self-heterodyne linewidth measurement as shown in Figure 5.14. An interferometer with a delay, τ_o, of 3.5 μs was used in the experiment. This corresponded to about 715 m of spooled fiber. A polarization controller was used to align the polarizations of the combined fields. The preamplifier following the photodiode provided high gain (~30 dB) to reduce the effects of the ESA noise on the sensitivity of the electronic

Table 5.4   Self-Heterodyne Linewidth Relations

Measured Full-Width Point	Displayed Width
−3 dB	$2\Delta v$
−10 dB	$2\sqrt{9}\,\Delta v$
−20 dB	$2\sqrt{99}\,\Delta v$
−30 dB	$2\sqrt{999}\,\Delta v$

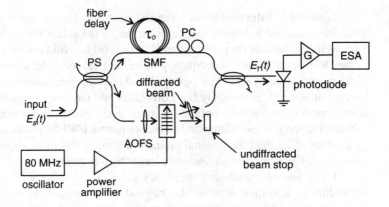

Figure 5.14 Optical self-heterodyne setup for laser linewidth measurement. AOFS = acousto-optic frequency shifter.

spectrum analysis. The optical linewidth measurement is shown in Figure 5.15. A close-in view of the linewidth measurement is shown in the inset. The displayed linewidth is 33.2 MHz which yields a laser linewidth of 16.6 MHz.

5.3.3 Delayed Self-Homodyne

The delayed self-homodyne technique offers a very simple means to measure the linewidth of an unmodulated laser.[9,13,14] Except for the presence of an optical frequency

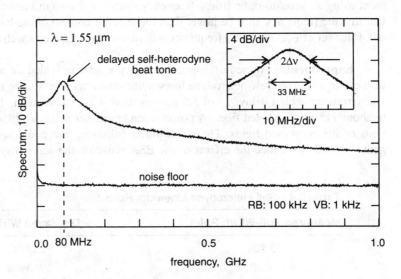

Figure 5.15 Measurement of DFB laser linewidth using optical delayed self-heterodyne technique. Inset shows close-in view of the spectrum.

shifter in the delayed self-heterodyne case, the self-homodyne and self-heterodyne techniques are the same. This method, like the delayed self-heterodyne technique, is well-suited for linewidth measurements because of the extremely high resolution afforded by using optical interferometers with low-loss fiberoptic delays. As with the delayed self-heterodyne method, any frequency jitter of the laser frequency will broaden the measured linewidth. To reduce this effect, the measurements can be performed further down the displayed lineshape. Table 5.4 may be used to infer the correct FWHM linewidth for a Lorentzian lineshape.

Several optical circuit implementations for the delayed self-homodyne method are shown in Figure 5.16. The optical circuit must deliver to the photodetector two fields, one being a delayed replica of the other. The requirement on coherence is satisfied if the differential delay τ_o of the interferometer satisfies Equation 5.22. The reasons for this delay requirement are similar to the delayed self-heterodyne case discussed in Section 5.3.2. More detail on the effect of laser coherence will be presented in Section 5.3.4.

Photocurrent Spectrum. The photocurrent spectrum for the delayed self-homodyne technique consists of direct detection as well as the desired mixing product but without the frequency shift:

$$S_i(f) \approx \mathcal{R}^2 \{S_d(f) + 2[S_s(\nu) \otimes S_s(-\nu)]\} \qquad (5.24)$$

$$(\text{ESA} \Rightarrow \text{direct detection} + \text{self-homodyne spectrum})$$

where $\mathcal{R}$ is the detector responsivity. Since the mixing term is essentially the test-laser spectrum convolved with itself, the displayed lineshape will always be symmetrical, even if the original lineshape had important asymmetries. Note that Equations 5.23 and 5.24 are nearly identical except for the frequency shift used in the delayed self-heterodyne method.

The translation of linewidth information from the optical spectrum to the electrical spectrum is illustrated in Figure 5.17. For the case of laser lineshapes described by Lorentzian or Gaussian functions, the displayed electrical power spectrum will have identical functional shapes to the actual optical spectrum. The reason for this is that the shape of these functions are preserved through the autocorrelation operation. The lineshapes of semiconductor lasers are often approximated by a Lorentzian-shaped profile.

The relationship between the measured self-homodyne FWHM linewidth and the actual optical field linewidth is the same as for the delayed self-heterodyne technique given in Table 5.4. Note that since the delayed self-homodyne method centers the mixing spectrum at 0 Hz, only half of the symmetrical spectrum is viewed (see Figure 5.17). Thus the laser FWHM linewidth corresponds to the measured −3 dB frequency point.

5.3.4 Photocurrent Spectrum: Coherence Effects

In this section, the interplay of the laser coherence time, τ_c and the interferometer differential delay, τ_o is presented in more detail. The reader may jump to the experiment at the end of Section 5.3.4 if a more detailed knowledge of the delay requirements is not desired. This type of analysis, along with actual experiments, provides the basis for the in-

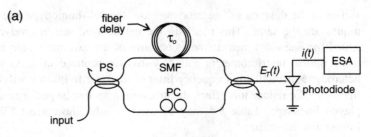

(a)

Mach-Zehnder interferometer

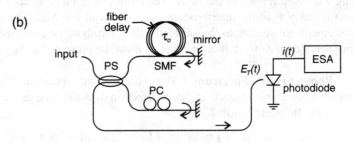

(b)

Micheson interferometer

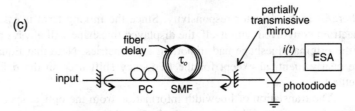

(c)

Low-finess Fabry-Perot interferometer

Figure 5.16 Optical delayed self-homodyne measurement set-up for laser linewidth measurement. (a) Mach-Zehnder interferometer. (b) Michelson interferometer. (c) low-finesse Fabry-Perot filter.

terferometer delay requirements set forth by Equation 5.22. It also applies to the delayed self-heterodyne technique since the physical basis is much the same as with the delayed self-homodyne case.[14] This section is supplemental to the basic linewidth measurement methods. The information provided here, however, can find application in a wide variety of interferometric problems ranging from the linewidth measurement techniques described here to problems of phase noise to intensity noise conversion in optical amplifiers and communications systems.

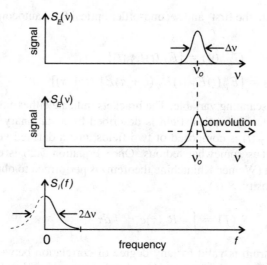

Figure 5.17 The delayed self-homodyne mixing of the laser field with itself.

In this section, we will examine the relationship between the photocurrent power spectrum displayed on the ESA, the laser linewidth, $\Delta\nu$ and the interferometer delay τ_o. The end result is an equation which predicts the photocurrent power spectrum for any degree of correlation between the combined fields. From this equation it is possible to determine the interferometer delay requirement in terms of a linewidth measurement error. The end result is quite broad and can be applicable in a variety of interferometric problems in any regime of coherence. The foregoing premise is that the laser lineshape can be approximated by the usual Lorentzian function (see Equation 5.19).

Analysis. The ESA display is proportional to the photocurrent spectral density by way of the standard resistance, 50 ohms, and any gains or losses in the system. The photocurrent spectrum can be evaluated through a computation of the currents induced by the total incident field at the photodetector. This calculation is quite tedious and is only briefly outlined here to give the reader an idea of the sequences involved. The total complex field at the photodetector, $E_T(t)$ has two contributions, one from each interferometer path.

$$E_T(t) = \sqrt{P_1}\, e^{\, j(2\pi\nu_o t + \phi(t))} + \sqrt{P_2}\, e^{j(2\pi\nu_o(t+\tau_o) + \phi(t+\tau_o))} \tag{5.25}$$

The random-phase noise process of the laser is described by $\phi(t)$. This phase noise leads to a phase jitter, $\phi(t) - \phi(t+\tau)$, that is assumed to have a zero-mean Gaussian probability distribution. By virtue of the Weiner-Khintchine theorem, the photocurrent power spectrum (displayed by the ESA) is the Fourier transform of the photocurrent autocorrelation function, $R_i(\tau)$ defined as:

$$R_i(\tau) = \mathscr{R}q G_E^{(1)}(0)\delta(\tau) + \mathscr{R}^2\, G_E^{(2)}(\tau) \tag{5.26}$$

where $G_E^{(1)}(0)$ and $G_E^{(2)}(\tau)$ are the first- and second-order optical field autocorrelations defined as

$$G_E^{(1)}(0) = [E_T(t)E_T^*(t)] \tag{5.27}$$

$$G_E^{(2)}(\tau) = [E_T(t)E_T^*(t)E_T(t+\tau)E_T^*(t+\tau)] \tag{5.28}$$

and τ is the autocorrelation scanning variable. The brackets indicate either ensemble averaging or time-averaging since the optical field is described by a stationary and random process. Considering that $E_T(t)$ is composed of two fields, one a delayed version of the other, Equation 5.28 becomes somewhat tedious. Once Equation 5.26 is calculated, a Fourier-transform operation (Weiner-Khintchine theorem) is performed to obtain the photocurrent power spectral density, $S_i(f)$

$$S_i(f) = \int_{-\infty}^{\infty} R_i(\tau)e^{-j2\pi f\tau}\,d\tau \tag{5.29}$$

The calculated power spectrum is valid for any degree of correlation between the combined fields. The polarization states of the combined fields are assumed to be aligned to achieve maximum interference. The three contributions (ignoring thermal noise) to the single-sided photocurrent spectrum are the direct detection, $S_{dc}(f)$, the shot noise $S_{shot}(f)$ and the sought after mixing term, $S_{mix}(f)$.[14]

$$S_i(f) = S_{dc}(f) + S_{shot}(f) + S_{mix}(f) \tag{5.30}$$

where $S_{dc}(f) = \delta(f)\mathcal{R}^2[P_1 + P_2 + 2\sqrt{P_1P_2}\cos(2\pi\nu_o\tau_o)e^{-\pi\Delta\nu\tau_o}]^2$

$$S_{shot}(f) = 2q\mathcal{R}[P_1 + P_2 + 2\sqrt{P_1P_2}\cos(2\pi\nu_o\tau_o)e^{-\pi\Delta\nu\tau_o}]$$

$$S_{mix}(f) = \frac{8\mathcal{R}^2 P_1 P_2 \pi^{-1}\Delta\nu^{-1}e^{-2\pi\Delta\nu\tau_o}}{1 + \left(\dfrac{f}{\Delta\nu}\right)^2}[\cosh(2\pi\Delta\nu\tau_o) - \cos(2\pi f\tau_o)$$

$$+ \cos^2(2\pi\Delta\nu\tau_o)\left\{\cos2\pi f\tau_o - \frac{\Delta\nu\sin 2\pi f\tau_o}{f} - e^{-2\pi\Delta\nu\tau_o}\right\}]$$

The equation variables are summarized in Table 5.5.

The important self-homodyne mixing term, $S_{mix}(f)$ appears unwieldy, but simplifies in certain limits. It can be visualized by plotting its value for various settings of $\Delta\nu\tau_o$. This product is proportional to the ratio of the interferometer delay τ_o to the coherence time, τ_c of the laser. Figure 5.18 plots this in terms of relative intensity noise (RIN). Recall that RIN is the ratio of the intensity-noise spectral density to the average power squared. RIN is convenient to use because it normalizes the display with respect to average power. The linewidth was fixed to 30 MHz. When the product $\Delta\nu\tau_o$ is unity, the shape of the spectrum is approximately that of the original Lorentzian laser lineshape but with twice the spectral width. As the product $\Delta\nu\tau_o$ is reduced below 1, the effects of phase between the interfering waves becomes more significant and ripple appears in the spec-

Table 5.5 Variable Definitions

Symbol	Definition
$\delta(f)$	Dirac-Δ function, $\delta(f=0)=1$, $\delta(f\neq 0)=0$
$\mathcal{R}$	detector responsivity [A/W]
P_1	power delivered to detector from path 1 of interferometer [W]
P_2	power delivered to detector from path 2 of interferometer [W]
$\Delta\nu$	laser FWHM optical linewidth [Hz]
ν_o	average optical frequency [Hz]
τ_o	interferometer delay [s]
f	display frequency [Hz]

trum. The spectrum here corresponds to the quadrature case where the efficiency of phase noise to intensity noise conversion is maximum. Also note that the first null in the spectrum appears at a frequency of $1/\tau_o$.

In terms of frequency resolution, these results can be interpreted as follows: For large $\Delta\nu\tau_o$, the resolution is determined by the linewidth of the test laser. In the other extreme, when the product $\Delta\nu\tau_o$ becomes small compared to unity, the resolution is limited by the differential time-delay of the interferometer. In the regime of partial coherence, the resolution is a blend of the interferometer resolution and the laser linewidth. Curve fitting Equation 5.30 with $\Delta\nu$ as a fitting parameter to the measured data can also be used in the regime of partial coherence in order to estimate the laser linewidth.[15] This may be useful

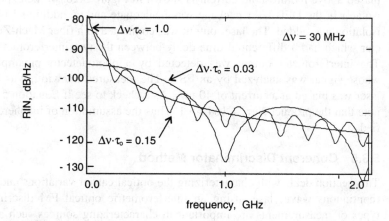

Figure 5.18 Relative intensity noise spectrum for various values of the $\Delta\nu\tau_o$ product.

when this technique is applied to very narrow linewidth lasers and long fiber delays are not available to insure incoherent mixing.

Incoherent Regime: $\Delta\nu\tau_o \geq 1$.   As the product of linewidth and delay, $\Delta\nu\tau_o$ in Equation 5.30 becomes large, the mixing term, $S_{mix}(f)$ tends to a Lorentzian function centered at zero frequency with a FWHM twice that of the original linewidth. In this regime the electrical spectrum is a scaled version of the actual laser lineshape:

$$S_i(f) \approx \mathcal{R}^2 \left[P_1 + P_2\right]^2 \delta(f) + \mathcal{R}^2 P_1 P_2 \frac{4/\pi\Delta\nu}{1 + \left(\dfrac{f}{\Delta\nu}\right)^2} \tag{5.31}$$

The shot-noise term has been dropped, but the DC term was kept since it is useful for the computation of RIN. The effects of interferometric losses and linewidth on the strength of the delayed self-homodyne beat tone are readily determined from Equation 5.31.

Error Due to Partial Coherence: $\Delta\nu\tau_o \leq 1$.   Equation (5.30) provides the theoretical basis for determining the measurement error that would result when there is insufficient delay, τ_o. The theoretical linewidth measurement overestimate, as a function of $\Delta\nu\tau_o$, is shown in Figure 5.19a.[8] Thus the minimum delay, τ_o, can be determined based on the acceptable error caused by the partial coherence. Additionally, Figure 5.19a can be used to correct for measurements performed in the partially coherent regime. In Figure 5.19b the minimum fiber delay length is plotted as a function of laser linewidth. The lower limit for the incoherent range is defined here by the $\Delta\nu\tau_o = 1$ boundary. This corresponds to a linewidth error for the worst-phase condition of approximately 3%. A fiber group index of 1.47 and a Mach-Zehnder topology (see Figure 5.16) were assumed for the calculation.

Experiment.   A delayed self-homodyne linewidth measurement for a DFB laser biased above its threshold current is shown in Figure 5.20. In the experiment, the 30 dB isolator in the DFB laser package was inadequate and an additional isolator with 60 dB isolation was added. The laser output was connected to a fiber Mach-Zehnder interferometer which had a differential time delay between the two interferometer paths of 3.5 μs. The interferometer output was detected by a photodetector-preamplifier combination whose signal was analyzed by an ESA. The measured linewidth was 47 MHz when the laser was biased at a current of 40 mA. As a check to see if Equation 5.22 is satisfied, we note that the product $\Delta\nu\tau_o \approx 164 \gg 1$. Thus the assumption of incoherent mixing is satisfied.

5.3.5 Coherent Discriminator Method

This section deals with characterizing the optical carrier variations and linewidth of a cw (continuous wave) laser using an interferometric optical FM discriminator.[16,17] These types of measurements are important in characterizing sources such as tunable external cavity lasers, local oscillators for coherent detection, and externally modulated FSK (frequency shift keyed) communications lasers. Since we are assuming that the laser runs cw,

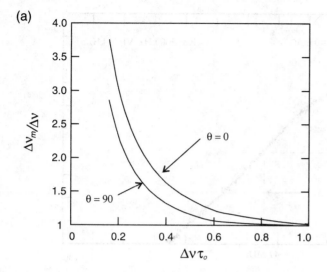

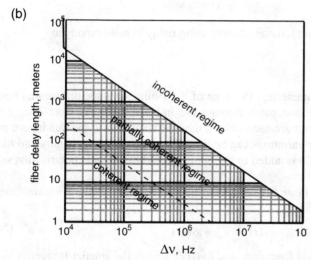

Figure 5.19   (a) Ratio of measured FWHM linewidth to actual laser linewidth for various values of the $\Delta\nu\tau_o$ product. (b) Minimum interferometer delay versus laser linewidth.

its intensity will be constant and variations will occur only in its optical frequency or phase. The case for combined frequency and intensity variations will be covered in Sections 5.5 and 5.6.

The purpose of an optical FM discriminator is to convert optical carrier fluctuations into intensity variations that can then be measured directly. There are many interferometric configurations available to construct such a discriminator, for example, a Michelson interferometer, a Mach-Zehnder interferometer, or an optical resonator could be used (see for example Figure 5.16). This section will use a Mach-Zehnder interferometer to describe the discriminator operation. This type of interferometer is shown in Figure 5.21a.

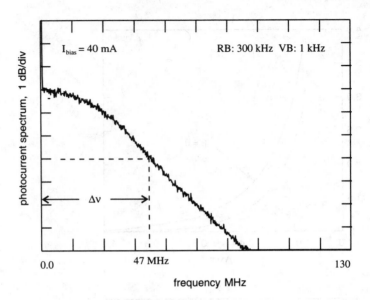

Figure 5.20 Linewidth measurement using delayed self-homodyne technique.

The main parameter that characterizes this type of discriminator is the differential optical time delay, τ_o, between the two paths through the interferometer. A fiber-optic phase modulator and feedback circuit are used to hold the interferometer in the quadrature position where optical frequency variations can be linearly converted into intensity variations. A polarization controller (PC) is added to one arm to avoid orthogonal polarization states for the two interfering signals.

The variations on the laser frequency can be described by using the instantaneous frequency of the optical carrier as

$$v(t) = v_o + \delta v(t) \tag{5.32}$$

where v_o is the average optical frequency and $\delta v(t)$ describes the smaller frequency variations about the large average offset frequency. Without the feedback circuit activated, optical interference results in a photocurrent approximated by

$$i(t) \cong i_o + \Delta i \cos(2\pi v_o \tau_o + 2\pi \delta v(t)\,\tau_o) \tag{5.33}$$

where τ_o represents the differential time delay through the interferometer and Δi is equal to the maximum amplitude of the photocurrent variations due to optical interference. Equation 5.33 is illustrated in Figure 5.21b. In obtaining Equation 5.33, certain constraints are assumed when relating the optical phase to the instantaneous frequency variations. These constraints will be discussed later.

With the feedback circuit activated, the phase modulator is used to stretch one of the fiber arms to keep the interferometer in quadrature. The quadrature position is shown

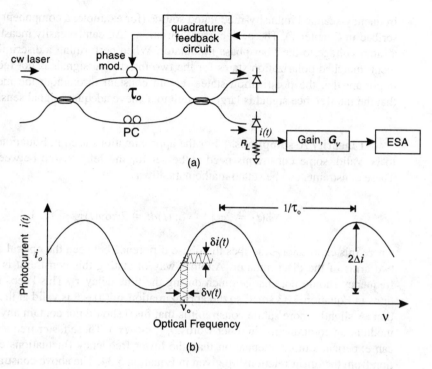

Figure 5.21 Optical frequency discriminator for linewidth measurement.

in Figure 5.21b. The feedback circuit is used to compensate for slow drifts in the average optical frequency and in the fiber lengths of the interferometer. At the quadrature position, small current variations are linearly proportional to the optical frequency fluctuations. Assuming some restrictions on the magnitude of δv compared to τ_o, Equation 5.33 can be approximated to give

$$\frac{\delta i}{\delta v} \cong 2\pi\tau_o\Delta i \tag{5.34}$$

where $\delta i/\delta v$ represents the discriminator slope which describes the conversion of optical frequency variations into photocurrent variations. The factor $2\pi\tau_o\Delta i$ is sometimes referred to as the discriminator constant. The constraints used in obtaining Equation 5.34 will be addressed below. Equation 5.34 gives an important result, it states that we can get a direct measurement of the optical carrier variations even though they are centered at several hundreds of terahertz. It also shows that only two experimental parameters are required to calibrate the FM discriminator. These two parameters are the differential time delay τ_o and the peak current variations, Δi, due to the interference effect. In practice, the differential time delay can be measured in several ways. For example, by using a high-resolution optical reflectometer (see Chapter 10) or by probing the interferometer with a frequency-scanned intensity-modulated source (for example, a component analyzer as described in Chapter 7). The peak current variations, Δi, can be easily measured by sending a ramp voltage to the fiber phase modulator. When performing a discriminator measurement, matched polarization states for the two interfering signals is not required. But it is important that the polarization states remain constant throughout the measurement and that the interference signal is large enough to achieve adequate signal sensitivity.

    Linearity Assumptions.   For the approximations used in Equations 5.33 and 5.34 to be valid, some constraints need to be set for the relationship between $\delta v(t)$ and τ_o. These constraints can be stated mathematically as

$$\Delta\phi(t) = 2\pi\int_{t}^{t+\tau_o} \delta v(t)dt \cong 2\pi\delta v(t)\tau_o << 1 \qquad (5.35)$$

where this expression describes the phase difference between the optical signals from the two arms of the discriminator. Another way of stating this constraint is that the optical frequency should not change much during the time delay τ_o. This keeps the phase variations in Equation 5.33 small so the approximation $\sin(\theta) \sim \theta$ is valid at the quarature position. A slightly more subtle constraint is that $\delta v(t)$ should not contain any high-frequency modulation components (in other words, $f_{max} << 1/\tau_o$). These faster frequency fluctuations can experience more attenuation than the lower frequency fluctuations causing a distortion from the linear relationship given in Equation 5.34. The above constraints can usually be satisfied since the delay, τ_o, can be made arbitrarily small.

    Now we will address some of the details involved in measuring the frequency content of the optical carrier fluctuations. These measurements can be important. For example, external cavity tunable lasers often have mechanical resonances due to the design and dimensions of its physical structure. These resonant frequencies typically occur in the kilohertz range and can be easily identified by sending the output of an FM discriminator into an electrical spectrum analyzer (ESA). It is also possible to measure the fundamental laser linewidth (due to spontaneous emission) from the power spectral density of the output signal, $\delta i(t)$. When discussing frequency domain measurements using an FM discriminator, confusion can occur between the optical frequency and the frequency fluctuations that the optical frequency possesses. In order to minimize this confusion, the symbol for optical frequency will be v and the frequency variations of the optical carrier will be described using f.

    Since the optical frequency variations for a cw laser tend to be somewhat random, it is convenient to measure the power spectral density (PSD), $S_v(f)$, of the optical frequency fluctuations. Using the linear relationship given in Equation 5.34, the PSD of the measured photocurrent, $S_i(f)$, is related to the optical frequency variations by the relation

$$S_i(f) \cong (2\pi\tau_o\Delta i)^2 S_v(f) \qquad (5.36)$$

which is valid for $2\pi\delta v\tau_o << 1$ and $f << 1/\tau_o$. By making a more rigorous analysis, the constraint $f << 1/\tau_o$ can be removed if one uses the more accurate expression of

$$S_i(f) \cong (2\pi\tau_o\Delta i)^2 \, \text{sinc}^2(\tau_o f) \, S_\nu(f) \tag{5.37}$$

where $\text{sinc}(x) = \sin(\pi x)/(\pi x)$. The only remaining constraint required for Equation 5.37 is that $2\pi\delta\nu\tau_o \ll 1$, so that the $\sin\theta \sim \theta$ approximation used for Equation 5.34 remains valid.

As mentioned earlier, the fundamental laser linewidth (linewidth without any external perturbations on the laser) can be determined from the PSD of the optical frequency. For a cw laser with a Lorentzian linewidth of $\Delta\nu$, the PSD of the optical frequency caused by the effects of spontaneous emission is given by

$$S_\nu(f) = \frac{\Delta\nu}{\pi} \tag{5.38}$$

which states that the PSD is a constant, independent of frequency. For actual lasers experiencing external perturbations, this result will usually not be accurate at low frequencies. Depending on the specifics of the laser and the sensitivity of the measurement setup, it is often possible to find a frequency range where the flat spectral characteristics given by Equation 5.38 is valid.

Typically an ESA measures electrical power from which a power spectral density (PSD) can be estimated. The relationship between the output of the ESA and the PSD of the input voltage is given by

$$P_{\text{ESA}}(f) = \frac{R_L^2 \, G_\nu^2 B_e}{R_c} S_i(f) \tag{5.39}$$

where $P_{\text{ESA}}(f)$ is the electrical power measured on the ESA, R_L is the effective load resistance of the receiver that converts the photocurrent into a voltage, G_ν is the voltage gain between the receiver and the ESA, B_e is the effective noise bandwidth for the bandpass filter setting of the ESA and R_c is the characteristic impedance of the ESA (usually $R_c = 50 \, \Omega$). The following example will be used to illustrate the use of some of the above equations.

Experiment.   This experiment will be used to illustrate a measurement of the fundamental linewidth of a DFB laser using a fiber-optic Mach-Zehnder FM discriminator. The experimental setup shown in Figure 5.21 will be used to perform the measurement. First, the time delay of the FM discriminator is measured using a component analyzer (see Chapter 7). By measuring the transmission of optical intensity modulation as a function of the modulation frequency, a sinusoidal pattern is obtained with a period of 1 GHz. Since the interferometer differential time delay is equal to the inverse of this period, we get $\tau_o = 1$ nsec. Next, the DFB laser source is connected to the input of the discriminator and a ramp voltage is sent to the fiber phase modulator. The peak-to-peak variations in the photocurrent due to the interference is measured to be $2\Delta i = 20 \, \mu A$. From the above two measurements, the discriminator constant is calculated to be $2\pi\tau_o\Delta i = 6.28 \times 10^{-2} \, \mu A/MHz$. Next the ramp voltage is removed and the phase modulator is connected to the locking circuit which biases the FM discriminator in quadrature.

Before reaching the spectrum analyzer, the photocurrent first passes through a transimpedance amplifier ($R_L = 10 \, K\Omega$) and then an AC-coupled voltage amplifier ($G_\nu = 100$). A flat spectral density, as predicted by Equation 5.38, is observed on the spec-

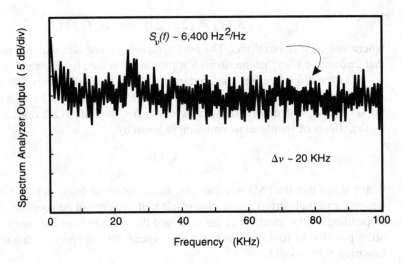

Figure 5.22 Measurement of the optical frequency spectral density of an external cavity laser using an optical discriminator.

trum analyzer. The electrical power in a resolution bandwidth of $B_e = 100$ KHz is measured to be -16 dBm (in other words, $P_{ESA} = 2.5 \times 10^{-5}$ watts). Given that the characteristic impedance of the ESA is $R_c = 50$ Ω, we now have enough information to calculate the PSD for the measured photocurrent. Putting the above values into Equation 5.39 we get a value of $S_i(f) = 1.25 \times 10^{-20}$ A²/W. Using this value and the earlier calculated discriminator constant, Equation 5.36 gives a PSD for the instantaneous frequency of $S_\nu(f) = 3.2 \times 10^6$ Hz²/Hz. From Equation 5.38, we can now calculate the linewidth of the laser, which is $\Delta\nu = 10$ MHz. Although this example yielded only a single laser parameter, the measurement of the power spectral density of the optical frequency provides additional information concerning low frequency noise effects due to electrical pump current noise and the strength of any relaxation oscillations.

Other optical arrangements besides a Mach-Zehnder interferometer can be used to realize an FM discriminator. For example, a fiber ring resonator biased on the edge of a resonance peak can be used to convert optical carrier variations into intensity changes. Figure 5.22 shows an experimental measurement using a fiber ring resonator as a FM discriminator.[16] A feedback circuit is used to lock to the side of the resonator resonance. The discriminator constant (slope of current vs. frequency) is much steeper than the previous example, being approximately 21 μA/MHz. The source being measured was an external cavity laser whose linewidth is much narrower than that of a DFB laser. For this reason, the discriminator constant was chosen to be much larger. For this measurement, the power spectral density for the external cavity laser was determined to be about $S_\nu(f) \sim 6,400$ Hz²/Hz, which corresponds to a fundamental linewidth of 20 KHz. From observing Figure 5.22, the PSD of the carrier frequency is reasonably flat as predicted by theory for a laser with a Lorentzian lineshape. The increase in noise at about 26 KHz is an artifact caused by a resonance in the PZT phase modulator used in the locking circuit. For frequencies below about 5 KHz, the spectral density becomes larger as environmentally induced frequency jitter becomes more dominant.

5.3.6 Comparison of Techniques

In Section 5.3, four different linewidth measurement techniques were described. There is some overlap in capability between the techniques as well as differences in the measurement implementations. The useage of any particular method depends on how well it matches the measurement goals.[17] One aspect of linewidth measurement that quickly differentiates the heterodyne technique from the other methods is asymmetry. If the measurement requires characterization of asymmetries in the lineshape of the laser under study, the heterodyne technique is the only possibility. The delayed self-heterodyne, the delayed self-homodyne, and the discriminator methods do not yield information on linewidth asymmetry. In terms of measurement simplicity, the delayed self-homodyne and delayed self-heterodyne methods are the easiest to implement. The delayed self-heterodyne method is slightly more difficult than the delayed self-homodyne method because of the requirement of an optical frequency shifter. The discriminator method does not directly yield linewidth, therefore the interpretation of the data is more difficult. Another advantage of the delayed self-heterodyne/homodyne methods is the auto-wavelength tracking. Since the local oscillator signal in these measurements is provided by the laser under test, slow drifts in wavelength are tolerated. This is extremely useful when the laser linewidth is characterized with bias current or laser temperature as a parameter. The discriminator method is well-adept to characterizing sources of perturbation such as acoustic resonances that cause frequency jitter. It can also perform narrower linewidth measurements than the delayed self-homodyne and delayed self-heterodyne techniques. The delayed self-heterodyne, the delayed self-homodyne, and the discriminator linewidth measurement methods are based on the usual approximation that the laser exhibits a Lorentzian lineshape. This assumption is usually valid, but should be considered beforehand for any particular laser.

The principle characteristics of the methods covered in this section are outlined here as an aid in determining which method is most suitable to a particular measurement requirement:

Optical Heterodyne

Advantages
- highest sensitivity
- measures extremely narrow linewidths (limited by LO linewidth)
- measures asymmetric lineshape and non-Lorentzian characteristics
- characterizes optical frequency jitter.

Disadvantages

- may require an OSA to match LO wavelength with signal wavelength
- local oscillator needs to have low frequency jitter and linewidth compared to linewidth of laser under test
- requires wavelength tracking if test laser wavelength changes because of changes in temperature, optical feedback, or injection current

Delayed Self-Heterodyne/Self-Homodyne

Advantages

- simple experimental setup
- less sensitive to slow wavelength drift
- measures narrow linewidth lasers $\Delta v \sim 5$ kHz (fiber delay dependent)

Disadvantages

- does not measure asymmetric lineshape
- self-heterodyne: limited on maximum linewidth measurement by frequency shifter
- linewidth overestimate due to frequency jitter (can be corrected)

Coherent Discriminator

Advantages

- measures phase-noise frequency spectrum
- measures laser jitter spectrum
- measures extremely narrow laser linewidths

Disadvantages

- more complicated experimental setup and calibration

5.4 OPTICAL SPECTRAL MEASUREMENT OF A MODULATED LASER

Knowledge of the optical spectrum of a modulated laser is critical to determine the viability of a laser source in a telecommunications system. The broadened spectrum places limits on channel spacings due to cross-talk and creates pulse broadening due to dispersion in the optical fiber. Accurate modeling of laser and modulator performance requires spectral measurements on actual lasers for validation and determination of laser parameters. Typical grating-based OSAs (see Chapter 3) are excellent tools for studying laser side-mode structure but they usually lack the necessary spectral resolution for measurements of the modulation spectrum imparted onto the optical carrier.

Intensity modulation, laser chirp and phase and intensity noise all contribute to the laser's optical spectrum. The spectral broadening during modulation of the laser output power is primarily due to the direct intensity modulation and the resulting frequency chirp. The time-varying optical field is composed of three basic parts: the magnitude, $\sqrt{P(t)}$, which describes the power variation with time, the average operating frequency, v_o and the phase variations of the carrier, $\phi(t)$:

$$E_s(t) = \sqrt{P_s(t)}\, e^{j(2\pi v, t + \phi(t))} \tag{5.40}$$

The corresponding single-sided optical field spectrum is a convolution of the power spectrum of the modulation envelope, $S_m(f)$ with the carrier frequency centered at v_o:

$$S_s(v) = S_m(f_v) \otimes\, <P_s(t)>\delta(v - v_s) \tag{5.41}$$

where the bracket $<>$ denotes a time averaged value.

Ideally, $S_m(f)$ would correspond to the power spectrum of the amplitude modulation used to impart signals onto the optical carrier, however, the effects of laser chirp and linewidth are included in $S_m(f)$.

The measurement methods of this section are geared towards finding the power spectrum of $S_m(f)$. We are less concerned about precise determination of the carrier frequency since this information is readily obtainable from such instruments as wavelength meters or OSAs. In this section two measurement methods will be discussed which allow for high-resolution characterization of $S_m(f)$. The first method, optical heterodyne, allows measurement of arbitrarily shaped power spectrums by using a second laser to serve as a local oscillator. The second approach, the gated delayed self-homodyne (GDSH) technique does not require an additional laser to serve as a local oscillator. It does require access to the modulation input applied to the laser under test. The GDSH technique is most useful in situations where the optical spectrums are approximately symmetric about the optical carrier (predominantly FM broadened). Another method, the scanning FP filter, may also be used for characterization of the optical power spectrum. FP filters are covered in Chapter 4.

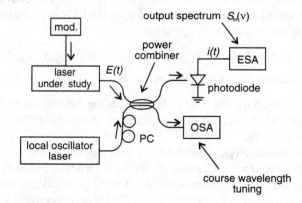

Figure 5.23 Optical heterodyne setup for modulated laser power spectrum measurement. PC = polarization state controller.

5.4.1 Heterodyne Method

The heterodyne approach for measuring the modulated laser spectrum is based on the same principles discussed previously for laser linewidth measurements. The reader is referred to Section 5.3.1 for background information. An optical heterodyne setup is shown in Figure 5.23. The modulated laser signal is combined with the local oscillator signal using a fused fiber directional coupler. The OSA provides course frequency (wavelength) alignment of the local oscillator with respect to the signal laser. The alignment must be precise enough to place the heterodyne beat tone within the bandwidth of the photodiode and ESA. Polarization controllers optimize the signal-to-noise ratio of the displayed spectrum. An electrical preamplifier after the high-speed photodiode improves the system sensitivity by reducing the effect of the thermal noise contributed by the ESA. The desired preamplifier gain is typically of the order of 30 to 40 dB. In this type of heterodyne measurement, the local oscillator is fixed and the optical spectrum is traced out with a broadband ESA. The local oscillator has a narrow linewidth for maximum resolution, therefore for stable measurements, optical reflections must be minimized. Fusion-spliced connections or angled connector interfaces should be used in the measurement setup.

Amplitude and Frequency Scaling. The amplitude scaling displayed on the ESA corresponds to an actual optical spectrum, thus the relative heights of the spectral sidebands are preserved. The display will have a one-to-one frequency correspondence with the actual optical field spectrum, the spectrum is simply translated to a low frequency that can be measured with electronic instrumentation. The bandwidth of the ESA typically places limitations on the extent of the observable power spectrum to less than 100 GHz. The displayed spectrum, which is proportional to the photocurrent spectral density, $S_i(f)$ depends on the local oscillator spectrum $S_{LO}(\nu)$ and the optical spectrum of the modulated laser as well as any direct detection intensity components

$$S_i(f) \approx \mathscr{R}^2 \{S_{dc}(f) + 2S_s(\nu) \otimes S_{LO}(-\nu)\} \tag{5.42}$$

$$(\text{ESA} = \text{direct detection} + \text{mixing})$$

When the local oscillator linewidth is narrow compared to the modulated laser linewidth, its lineshape can be approximated with the Dirac-Δ function. The convolution in Equation 5.42 is straightforward and the observed display on the ESA is the actual optical spectrum frequency translated to within the operation range of the ESA

$$S_i(f) \approx \mathscr{R}^2 \{S_{dc}(f) + 2 <P_s> <P_{LO}> S_m(f) \otimes \delta(f - \nu_s + \nu_{LO})\} \tag{5.43}$$

$$(\text{narrow local oscillator linewidth})$$

A comparison of Equations 5.41 and 5.43 indicates that the heterodyne technique measures the power spectrum of the optical field modulation about the difference frequency between the LO and the signal.

The local oscillator frequency, ν_{LO}, should be set with respect to the signal frequency, such that the average frequency difference places the heterodyne products above the highest spectral extent of the direct intensity detection. This helps avoid confusion between the direct detection spectrum, $S_{dc}(f)$ and the optical field power spectrum, $S_m(f)$.

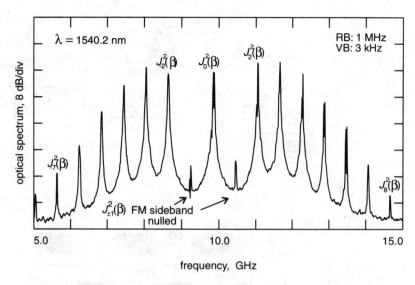

Figure 5.24 Optical heterodyne power spectrum measurement of current modulated DFB laser. Modulation rate is 600 MHz.

Experiment.   An optical heterodyne measurement result is shown in Figure 5.24 using the experimental setup shown in Figure 5.23. The signal source was a DFB semiconductor laser undergoing sinusoidal current modulation at a 600 MHz rate. The local oscillator consisted of a tunable external cavity laser which was set at a frequency approximately 5 GHz less than the DFB optical frequency. The DFB current modulation resulted in frequency chirp yielding a predominantly FM-broadened spectrum as shown. Accurate measurements of the FM index (β) can be obtained by nulling one of the FM sideband pairs. Here, the first-order Bessel function sidebands, $J_{\pm 1}(\beta)$ were nulled yielding an FM index, $\beta = 3.84$. Decreasing the current modulation frequency to 10 MHz resulted in a continuous spectrum as the individual lines were joined due to the spreading effects of the DFB laser linewidth.

5.4.2 Gated-Delayed Self-Homodyne

The optical power spectrum of a modulated laser can be measured using a simple arrangement consisting of an interferometer, and an optical receiver followed by an electrical spectrum analyzer. This measurement method, called the gated-delayed self-homodyne (GDSH) technique[18] makes optical spectrum analysis possible without the requirement of an additional laser to serve as a local oscillator. It is particularly useful for FM-dominated optical spectrums where the optical field spectrum is approximately symmetric with respect to the carrier frequency. An advantage of this method compared to the heterodyne method is the automatic wavelength tracking. For example, if the laser's average wavelength drifts because of temperature changes, the modulated power spectrum will still be displayed since the laser serves also as a local oscillator. For the same reasons, an OSA is

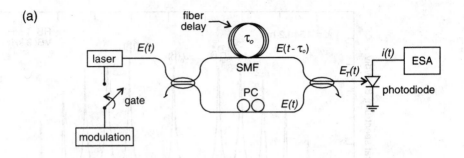

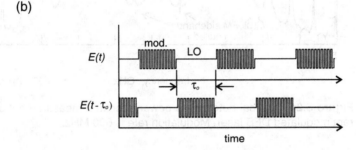

Figure 5.25 Gated-delayed self-homodyne measurement of optical spectrum. (a) Measurement set-up. (b) Timing diagram for signals through interferometer paths.

not required for frequency alignment making the measurement quicker and easier to implement.

The GDSH method setup is shown in Figure 5.25a. As with the optical heterodyne method discussed earlier, the measured spectrum is observed on an ESA. The underlying requirement for this method is that the laser can be operated in two temporally sequential states as shown in Figure 5.25b. In one state, the laser is in its freerunning mode with no modulation applied. This state is called the local oscillator state. The second state is the modulated state with a corresponding optical spectrum to be measured. The differential time delay, τ_o, between the two interferometer paths, results in temporal alignment of a modulated laser state with the local oscillator laser state at the power combiner near the photodetector. To achieve the required temporal alignment, the gate frequency must satisfy:

$$f_g = \frac{1}{2\,\tau_o} \tag{5.44}$$

The system behaves much in the same way as with the heterodyne technique, but with a spectral display about 0 Hz instead of some intermediate frequency. The displayed photocurrent power spectrum on the ESA is the optical spectrum of the modulated laser, as long as the spectrum is symmetric. The optical spectrums will be symmetric if the laser spectral output is dominated by either FM or AM effects.

Photocurrent Power Spectrum.   The analysis of the displayed ESA spectrum and how it relates to the actual optical field spectrum requires calculation of the photocurrent power spectrum.[19] It turns out that the most useful operation is in the incoherent regime where

$$\tau_o > \frac{1}{\Delta v} \tag{5.45}$$

is satisfied. Under this condition, the optical field phases of the two temporal states are uncorrelated allowing the unmodulated state to act as an independent local oscillator. This is the same requirement as with the delayed self-homodyne and delayed self-heterodyne measurement of laser linewidth. The expression for the ESA spectrum in the incoherent case is:

$$S_i(f) \approx R^2\{S_I(f) + 2S_m(v) \otimes S_{LO}(-v)\} \tag{5.46}$$

Thus, as with the heterodyne case, the ESA display has two main components, the filtered direct intensity detection, $S_I(f)$, and the product of the mixing of the laser in its local oscillator state with the modulated state. Since this method is a homodyne technique, it folds the optical spectrum about the LO frequency, therefore it only accurately measures spectras that are symmetrical. A complete theoretical analysis of the technique is given in Baney and Gallion.[19] If there is a shift in the average optical frequency between the local oscillator state and the modulated state, the spectral components below 0 Hz will not align with the positive frequencies when folded about 0 Hz. In practice, this results in the appearance of two FM sidebands in place of one. Using this effect, the GDSH method can accurately measure shifting of the average carrier wavelength between the modulated and unmodulated laser states.

Experiment.   A spectral measurement of a modulated DFB laser using the GDSH technique is shown in Figure 5.26. The measurement setup is shown in Figure 5.25a. The interferometer differential delay, τ_o, was 3.5 µs, thus from Equation 5.44 the required gate frequency was 143 kHz. A separate pulse generator provided a gate signal for the microwave synthesizer. The synthesizer put out 600 MHz cw when the gate signal was high, and nothing when the gate signal was low. This pulsed sinusoidal modulation was applied to the current input of the DFB laser through a bias tee. The DFB was well isolated (~ 60 dB) to reduce the effects of system reflections. The interferometer differential delay, τ_o, was 3.5 µs which along with the DFB laser linewidth (tens of MHz) placed the system in the desirable incoherent regime of operation (see Equation 5.45). Polarization control improved the polarization state matching to increase the displayed SNR. A preamplifier following the photodiode provided approximately 30 dB of gain to overcome the large noise figure of the ESA. The RF power was set to null the first FM sideband (FM index, β, of 3.84) as seen in the figure.

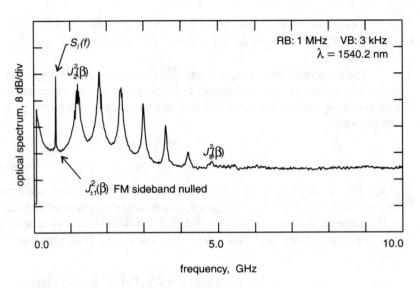

Figure 5.26 Optical power spectrum of current modulated DFB laser using gated-delayed self-homodyne method. Modulation rate = 600 MHz.

Comparison of Heterodyne and GDSH. In Section 5.4, we discussed the heterodyne and GDSH methods for measurement of the power spectrum of a modulated laser. Both techniques offer high resolution characterization of the optical spectrum. In the heterodyne method, frequency resolution is limited by the linewidth of the local oscillator while the GDSH method is limited by the linewidth of the laser under study. The heterodyne method also offers significantly higher sensitivity. This is because of the narrow linewidth local oscillator, more efficient optics and, of course, the heterodyne gain (see Equation 5.17). The heterodyne case is thus preferred over the GDSH technique except when a separate local oscillator laser is not available, or if wavelength tracking between the local oscillator and the laser under test becomes an issue. As a comparison to the scanning FP filter method, suppose we wish to perform high resolution spectrum analysis over the EDFA gain region from 1520 to 1570 nm. This is readily accomplished with either the heterodyne method (since commercially available lasers tune over 100 nm) or the GDSH technique. The FP filter would need a 50 nm free spectral range. To attain the same resolution as the heterodyne technique (assume 1 MHz local oscillator linewidth) the FP filter would require a finesse of 6 million!

5.5 LASER CHIRP MEASUREMENT

In the previous sections, heterodyne and interferometric techniques for characterization of optical source linewidth and power spectrum were discussed. A related measurement is the time domain measurement of the chirp (see Figure 5.3) of a modulated optical source such as a laser or modulator. Measurement of chirp provides valuable information concerning the dynamics of the laser frequency excursions during intensity modulation. Since the velocity of light through optical fiber varies with frequency, and chirped pulse would be expected to spread out over time. Sources exhibiting a large amount of chirp will experience more rapid pulse-spreading, leading to intersymbol interference in high-speed digital links. Generally, the more rapid the optical power variations, the greater the chirp, as predicted in Equation 5.5.

The time dependence of frequency chirp can be characterized using optical discriminators. The purpose of the discriminator is to convert optical frequency variations into detectable intensity variations. Optical slope discriminators may be realized using optical filters, interferometers, or even heterodyne techniques. In principle, any linear optical component with wavelength-dependent transmission characteristics may serve as a discriminator. While we will examine in detail the Mach-Zehnder discriminator,[20] it is noted that other two path interferometers are similar in operation. In fact, a variety of optical circuits have been configured into discriminators; these include birefringent crystals, optical fibers, FP filters, and Michelson interferometers in bulk-optic or fiber form, to name a few.[16,21,22]

Conversion of Chirp to Intensity Variations.   An interferometer circuit for measuring laser chirp is shown in Figure 5.27. Modulated light from the source, in this case a laser diode, enters the Mach-Zehnder interferometer as shown in Figure 5.27a. The differential time of flight between the two interferometer paths is denoted by τ_o. The light is recombined in the output coupler which directs half of the light from each path to the photodetectors. One photodetector provides both average power detection as well as high-speed measurement of intensity variations caused by the discriminated optical frequency variations. The second detector is used for measuring only the average power. When the average powers from each of the two interferometer output ports are equal, the interferometer is in quadrature. This enables conversion of optical frequency chirp into intensity changes via the linear discriminator slope characteristic shown in Figure 5.27b.

A feedback circuit maintains quadrature by adjusting the delay τ_o. The required delay adjustment is small, on the order of an optical wavelength. This delay can be realized using piezo-electric (PZT) devices. The quadrature feedback circuit should respond quickly to environmental factors affecting the interferometer path delay. The measurement apparatus uses a sampling oscilloscope which is triggered by the laser modulation source. This enables the comparison of several measured time records, each for a specific interferometer state, to calculate chirp. The three interferometer states are:

- Quadrature: both shutters open;
- Shutter A closed, shutter B open;
- Shutter A open, shutter B closed.

From an analysis of the photocurrent time records corresponding to the three interferometer states, the chirp is calculated.

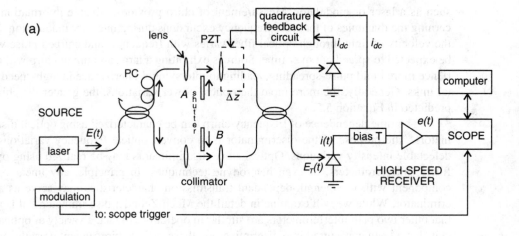

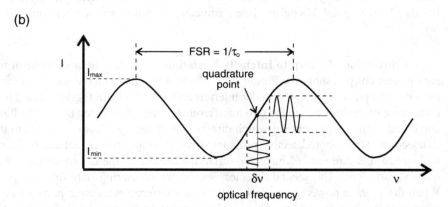

Figure 5.27 (a) Time-domain laser chirp measurement using a Mach-Zehnder interferometer followed by a high-speed receiver. (b) Interferometer discriminator characteristic.

Analysis. For the class of discriminators that consist of interferometers with two optical paths, the average output power from one port of the interferometer is the sum of the path power contributions and an interference term (see Equation 5.12):

$$P_o(t) = P_1(t) + P_2(t) + 2\sqrt{P_1(t)P_2(t)}\cos(\Delta\phi(t, \tau_o) + 2\pi\nu_o\tau_o) \quad (5.47)$$

where ν_o is the average optical carrier frequency and τ_o is the differential interferometer delay. In the absence of any frequency modulation, $\Delta\phi(t)$ is zero. However any changes in the frequency of the signal at the interferometer input will create a change in phase between the combined signals at the interferometer output. We will use the interferometer output intensity in the following analysis, but since the current or voltage in the electronic receiver is proportional to optical intensity, they can be substituted in place of intensity.

To maintain coherent interference between the light beams in Figure 5.27a, the interferometer delay, τ_o, should satisfy:

$$\tau_o \ll \frac{1}{\Delta \nu} \qquad (5.48)$$

Satisfying the quadrature biasing experimentally is equivalent to requiring a relationship between the average optical frequency, ν_o, and the interferometer delay:

$$2 \pi \nu_o \tau_o = \frac{\pi}{2} + 2 n \pi \qquad (5.49)$$

where $n = 0, 1, 2, \ldots$. Note that there are really two quadrature points in Figure 5.27b having opposite slopes. Here we restrict operation to the quadrature point with positive slope.

The instantaneous frequency deviations, $\delta \nu(t)$ of the optical carrier is the time derivative of the phase:

$$\frac{\Delta \phi(t)}{\Delta \tau} \approx 2\pi \delta \nu(t) \qquad (5.50)$$

Equation 5.50 can be approximated when delay τ_o is small compared to the fastest modulation periods:

$$\Delta \phi(t, \tau_o) \approx 2\pi \tau_o \, \delta \nu(t) \qquad (5.51)$$

Substitution of Equations 5.49 and 5.51 into 5.47 leads to

$$P_Q(t) = P_1(t) + P_2(t) + 2 \sqrt{P_1(t) P_2(t)} \sin(2 \pi \tau_o \, \delta \nu(t)) \qquad (5.52)$$

where $P_Q(t)$ is the discriminator output when the interferometer is set at quadrature. To determine $\delta \nu(t)$, we need to measure $P_1(t)$ and $P_2(t)$. These two time records are easily found by alternately blocking each of the interferometer paths.[20] In Figure 5.27a, a shutter is placed in the collimators. Note that the concept of a discriminator constant does not apply here since the discrimination slope depends on the input power which is now changing as a function of time. The direct detection $P_1(t) + P_2(t)$ and the power dependence of the discriminator characteristic, $\sqrt{P_1(t)P_2(t)}$ is used to calculate the chirp $\delta \nu(t)$

$$\delta \nu(t) = \frac{1}{2\pi \tau_o} \arcsin\left(\frac{P_Q(t) - P_1(t) - P_2(t)}{2\sqrt{P_1(t) P_2(t)}} \right) \qquad (5.53)$$

The delay τ_o may be measured by connecting a white light source (ASE from an EELED or EDFA for example) to the interferometer and observing the nulls in the interferometer output spectrum with an OPS. For larger delays, an optical component analyzer or narrow pulse source and high-speed oscilloscope are used to measure τ_o.

Interferometer Delay Requirement.   To maintain unambiguous chirp measurement, the peak frequency excursion $\delta \nu_{peak}$ (see Figure 5.26) must be less than one-fourth the interferometer free-spectral range. Preferably, the maximum chirp is confined to the approximately linear region of the interferometer characteristic to reduce the effects of noise on the chirp measurement

$$\delta \nu_{\text{peak}} \leq \frac{\text{FSR}}{8} = \frac{1}{8\,\tau_o} \qquad (5.54)$$

The approximation given by Equation 5.51 concerns the validity of the derivative approximation as the delay τ_o increases. The delay must be small compared to the inverse of the highest frequency components, f_m, of interest in the measurement.

$$\tau_o \ll \frac{1}{f_m} \qquad (5.55)$$

Experiment.   An example of a time domain measurement of frequency chirping is shown in Figure 5.28.[8] The experimental arrangement was similar to that shown in Figure 5.27. A DFB laser diode was current modulated at a 1 Gb/s rate which resulted in simultaneous power and frequency variation. A high-speed InGaAs photodetector with an integrated traveling-wave GaAs microwave amplifier fed the demodulated signal to a high-speed sampling oscilloscope for analysis. Three measurements were performed as indicated by Equation 5.53. The adjustable delay τ_o was first set to zero using the ASE

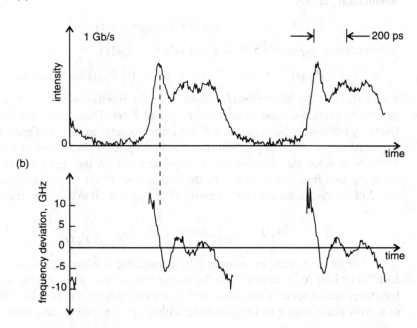

Figure 5.28   Optical discriminator used to measure DFB laser. (a) Intensity modulation. (b) Frequency chirp. Laser current was modulated at a 1 Gb/s rate.

from the DFB laser biased near its threshold current and an OSA to observe the interference nulls. Adjustment of a micrometer controlling the free-space path of one of the collimators fixed the delay τ_o. Polarization controllers were used to align the polarizations of the interfering beams of light. The intensity overshoot in the top trace is due to relaxation oscillations. The rise time was approximately 30 ps. The bottom trace is the measured dynamic frequency chirp. At the beginning of the pulse, the optical frequency chirps over 15 GHz.

Using the technique described in this section, the frequency chirp of a laser or optical modulator can be characterized even in the presence of 100% amplitude modulation. Since the measurement is made in the time domain, both the magnitude and the phase of the amplitude and frequency modulation are measured. In addition to the measurement of the time-domain FM response, this technique can also be applied towards frequency domain measurements of the FM response by characterising laser FM response as a function of modulation frequency applied to the laser.

5.6 FREQUENCY MODULATION MEASUREMENT

This section will discuss a frequency domain technique for characterizing both the AM and FM responses of an optical modulator[8,21,23] using a coherent interferometer. A measurement technique using an interferometer operated in the incoherent regime has been used to measure the FM response versus frequency.[24,25] This method won't be discussed further here since it measures just the magnitude of the FM response and not the phase. The coherent technique presented here consists of a stimulus/response measurement which allows the simultaneous measurement of both the AM and FM imparted onto the optical signal. This type of measurement is useful for characterizing devices such as DFB lasers, $LiNbO_3$ amplitude or phase modulators, and semiconductor electro-absorption modulators.

As discussed in Section 5.5 for the time-domain coherent-discriminator technique, the presence of amplitude modulation can lead to corruption of the measured FM due to the amplitude-dependent discrimination slope of optical interferometers. The direct intensity modulation can be removed from the interference data by making two measurements, each on an opposite slope of the FM discriminator. With these two measurements, the AM and FM responses can be determined. This separation, which requires the measurement of the amplitude and phase at each modulation frequency, can be accomplished using a vector network analyzer.

Figure 5.29 shows a fiber Mach-Zehnder FM discriminator used to perform the frequency domain measurement. The input to the FM discriminator originates from the optical modulator (DFB, $LiNbO_3$, etc.) to be tested. An electrical network analyzer provides the sinusoidal stimulus to the modulator and measures the resulting response (both amplitude and phase) from the discriminator output. Two separate frequency domain measurements are taken, one with the discriminator locked onto the positive slope and the other locked to the negative slope. Since the AM signal is in-phase for both measurements and

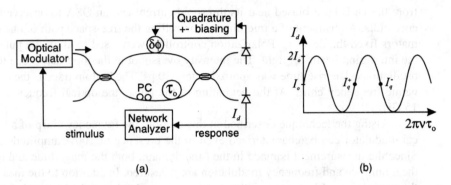

Figure 5.29 Optical FM discriminator used to measure the AM and FM response of an optical modulator.

the FM signal is 180 degrees out-of-phase, by either adding or subtracting the two measurements, the AM and FM responses can be separated. At each frequency, both a magnitude and phase is recorded, therefore the addition and subtraction involves the computation of complex numbers. To better understand this process, the following definitions and equations will be introduced. A more complete analysis can be found in Sorin and co-workers.[23]

Let the complex optical electric field generated by the optical modulator be described by

$$E(t) = \sqrt{P(t)}\, e^{j(2\pi v_o t + \phi(t))} \tag{5.56}$$

where v_o is the average optical frequency and $\phi(t)$ describes the frequency or phase deviations away from the optical carrier. Since the network analyzer delivers a sinusoidal stimulus to the modulator, the response for the optical power can be described by

$$P(t) = P_o[1 + Re\{\tilde{m}(f)e^{j2\pi ft}\}] \tag{5.57}$$

where f is the modulation frequency generated by the network analyzer and P_o is the average optical power. The complex intensity modulation index $\tilde{m}(f)$ represents both the amplitude and phase with respect to the driving stimulus from the network analyzer. The operation $Re\{\}$ denotes taking only the real part of the complex expression.

Assuming that the optical modulator operates in a linear regime, the resulting optical phase modulation can be described by

$$\phi(t) = Re\{\tilde{\phi}(f)e^{j2\pi ft}\} \tag{5.58}$$

where $\tilde{\phi}(f)$ is the complex phase modulation index. As with the intensity modulation, the optical phase modulation has a relative amplitude and phase with respect to the driving signal from the network analyzer.

Sometimes it is more convenient to describe the optical carrier variations in terms of an optical frequency modulation as opposed to a phase modulation. Since optical frequency is related to optical phase through a time derivative, the relationship for sinusoidal modulation can be expressed as

$$\phi(t) = 2\pi \int v(t)dt = Re\left\{\frac{\tilde{v}(f)}{jf} e^{j2\pi ft}\right\} \tag{5.59}$$

where $\tilde{v}(f)$ is the complex optical frequency modulation at the driving frequency f delivered by the network analyzer. $\tilde{v}(f)$ contains both a relative magnitude and phase with respect to the stimulus from the network analyzer.

By sending the modulated optical field described by Equations 5.56 to 5.59 into the FM discriminator, interference signals can be calculated for the two quadrature locking positions. For the positive and negative slope positions, the complex quantities $\tilde{I}_q^+(f)$ and $\tilde{I}_q^-(f)$ represent the measured amplitude and phase of the output photocurrent at the modulation frequency f. The complex intensity modulation index $\tilde{m}(f)$ is related to the sum of these two measurements by

$$\tilde{I}_{AM}(f) = \tilde{I}_q^+(f) + \tilde{I}_q^-(f) = 2I_o \cos(\pi f\tau_o)e^{-j\pi f\tau_o}\tilde{m}(f) \tag{5.60}$$

where I_o is the interference amplitude of the photocurrent and τ_o is the differential time delay through the FM discriminator. The term $\cos(\pi f\tau_o)e^{j\pi f\tau_o}$, describes the filtering effect caused by the Mach-Zehnder interferometer on the input optical intensity. By performing calculations, as described by Equation 5.60, the amplitude and phase of the intensity modulation index can be determined as a function of the stimulus frequency.

The optical phase modulation on the input signal can be calculated in a similar manner by using the difference between the two measured photocurrents. This relationship is given by

$$\tilde{I}_{PM}(f) = \tilde{I}_q^+(f) - \tilde{I}_q^-(f) \approx 4jI_o \sin(\pi f\tau_o)e^{-j\pi f\tau_o}\tilde{\phi}(f) \tag{5.61}$$

where the approximation sign is used to account for the fact that the discriminator slope, see Figure 5.29b, is not perfectly linear. By keeping $|\tilde{\phi}(f)| < 0.1$ rad, this nonlinearity error can be kept below 1%. Another potential error represented by the approximation sign occurs when the input intensity modulation becomes large. This error occurs due to second-order mixing effects between the intensity modulation and the nonlinearly generated higher harmonics of the phase modulation. This error is less than 10% for modulation indices below $|\tilde{m}(f)| \leq 50\%$.[23]

If one wishes to express the optical carrier deviations in terms of a frequency modulation, Equation 5.59 can be inserted into Equation 5.61 to give

$$\tilde{I}_{FM}(f) = \tilde{I}_q^+(f) - \tilde{I}_q^-(f) = 4\pi\tau_o I_o \text{sinc}(f\tau_o)e^{-j\pi f\tau_o}v(f) \tag{5.62}$$

where $\text{sinc}(x) = \sin(\pi/\pi x)/\pi x$. As in Equation 5.61, to keep the approximation reasonably valid, the constraints of $|\tilde{v}(f)| \leq 0.1/\tau_o$ and $|\tilde{m}(f)| < 0.5$ should be observed.

Experimental Results.   To illustrate the use of the above analysis, two experimental examples will be discussed. The first consists of measuring the frequency domain response of a current modulated 1300 nm DFB laser.[23] This measurement was performed by using a HP 8703A optical component analyzer to modulate the pump current of the

DFB laser. With the FM discriminator locked on the positive slope, a frequency measurement from 130 MHz to 10 GHz was performed to generate the data array, $\tilde{I}_q^+$ (f). The polarity to the locking element was then reversed, causing the bias position to change to a negative slope, and the array $\tilde{I}_q^-(f)$ was measured. This data was then inserted into Equations 5.60 and 5.62 to generate the results shown in Figure 5.30. These results were also normalized by the amplitude of the modulated pump current injected into the laser. The differential time delay for the discriminator was set ($1/\tau_o = 17.1$ GHz) to avoid loss of the high-modulation frequencies caused by the filtering effects of the interferometer. The peak response in Figure 5.30, at about 2.7 GHz, is due to relaxation oscillations within the laser. At this modulation frequency, the intensity modulation index was 5%/mA and the optical frequency deviation was about 650 MHz/mA.

The next example consists of measuring the residual phase modulation produced by a LiNbO$_3$ amplitude modulator.[23] Because of its push-pull electrode configuration, the modulator should theoretically impart only intensity modulation and no phase modulation. However, in practice, residual-phase modulation exists due to the nonideal construction of the device. Figure 5.31 shows the photocurrent difference as calculated by Equations 5.60 and 5.61. This measurement is different in nature to that for the DFB laser since the interferometer time delay was made much larger. The inverse of the time delay was $1/\tau_o = 273$ MHz, which is relatively small compared to the 5 GHz measurement range. For this reason the filtering effects of the discriminator are present, as seen by the periodic response in the measured data. It is interesting to note that the peaks in the periodic response correspond to locations where the filtering effects of the discriminator are removed. At these points, the modulation index and phase index can be read directly from

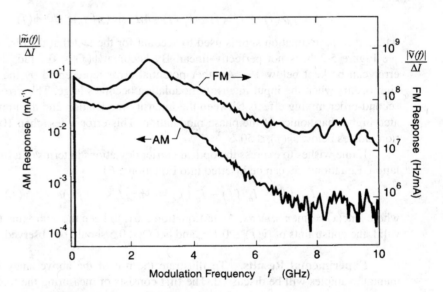

Figure 5.30 AM and FM response versus frequency for a current modulated DFB laser.

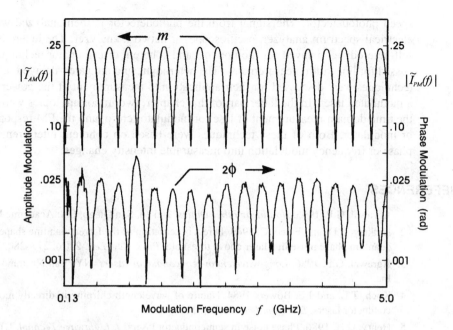

Figure 5.31 Frequency domain measurement of the intensity modulation (m) and phase modulation (2ϕ) response of a $LiNbO_3$ intensity modulator.

the plot in Figure 5.31. For this particular $LiNbO_3$ modulator, the phase modulation was about $2\phi = 0.025$ rad for an intensity modulation of $m = 0.25$. The ratio of these two numbers yields the modulator's amplitude-phase coupling factor α (see Section 5.2.1) of $2\phi/m = 0.1$. This figure of merit represents the ratio of the FM to AM sidebands on the electric field of the modulated optical carrier.

Using the technique described in the above section, the frequency domain response of an optical modulator can be characterized. This technique allows for an accurate measurement of the FM frequency response, even in the presence of some accompanying intensity modulation. Both the magnitude and phase, relative to a sinusoidal stimulus, is measured for the generated intensity and optical frequency modulation. Assuming the modulator is operated in its linear region, this frequency domain data can be used to predict time-domain responses.

5.7 SUMMARY

This chapter covered a variety of techniques for high-resolution measurement of laser linewidth, modulated power spectrum, frequency chirp, and FM response. The measurement techniques have, as a common theme, the concept of interference. The interference allows normally undetectable phase or frequency modulation to be measured with a high-speed photodetector. The signal from the photodetector is then analyzed with either an electrical spectrum analyzer, oscilloscope, or network analyzer. The heterodyne, delayed self-heterodyne, delayed self-homodyne, and coherent discriminator techniques were discussed for linewidth measurements. The heterodyne and the gated-delayed self-homodyne techniques were discussed for their application to measurement of the power spectrum of a modulated laser. In the latter part of this chapter, two important topics were considered: the time domain measurement of laser or modulator chirp, and the FM response of a laser or modulator. Both of these techniques were based on coherent interference to convert phase or frequency modulation into measurable intensity changes.

REFERENCES

1. Agrawal, G.P. 1995. *Nonlinear Fiber Optics.* 2nd ed. San Diego, CA: Academic Press.

2. Hinkley, E.D. and Freed, C. 1969. Direct observation of the Lorentzian line shape as limited by quantum phase noise in a laser above threshold. *Phys. Rev. Lett.* 23(6): 277–280.

3. Agrawal, G.P. 1986. *Long-wavelength semiconductor lasers.* NY: Van Nostrand Reinhold Co. Inc.

4. Koch, T.L. and J.E. Bowers 1984. Nature of wavelength chirping in directly modulated semiconductor lasers. *Electron. Lett.* 20(25): 1038–1039.

5. Henry, C.H. 1986. Phase noise in semiconductor lasers. *J. Lightwave Technol.* LT-4: 298–311.

6. Yoshida, S., Tada, Y., and Nosu, K. 1994. High resolution optical spectrum analysis by coherent detection with multi-electrode DBR-LD's as local oscillators. Conference Proceedings. IMTC/94. Hamamatsu, Japan: Advanced Technologies in I & M. IEEE Instrumentation and Measurement Technology Conference.

7. Nazarathy, M., W.V. Sorin, D.M. Baney, and S.A. Newton. 1989. Spectral analysis of optical mixing measurements. *IEEE J. Lightwave Technol.* 7(7): 1083.

8. Baney, D.M. 1990. Modélization et caractérisation de la réponse FM des lasers à contreréaction distribuée mono et multi-section. Doctoral Thesis, No. 90E015. Ecole Nationale Superiéure des Télécommunications, Paris, France.

9. Okoshi, T., Kikuchi, K., and Nakayama, A. 1980. Novel method for high frequency resolution measurement of laser output spectrum. *Electron. Lett.* 16: 630–631.

10. Mercer, L.B. 1991. 1/f frequency noise effects on self-heterodyne linewidth measurements. *Journal of Lightwave Technology* 9(4): 485–493.

11. Esman, R.D. and L. Goldberg. 1988. Simple measurement of laser diode spectral linewidth using modulation sidebands. *Electron. Lett.* 24(22): 1393–1395.

12. Constable, J.A. and I.H. White. 1994. Laser linewidth measurement using a Mach-Zehnder interferometer and an optical amplifier. *Electronics Lett.* 30(2): 140–142.

13. Baney, D.M. and W.V. Sorin. 1990. Linewidth and Power Spectral Measurements of Single-Frequency Lasers. *Hewlett-Packard Journal* 41(92): 92–96.

14. Gallion, P.B. and G. Debarge. 1984. Quantum phase noise and field correlation in single frequency laser systems. *IEEE J. Quantum Electron.* QE-20: 343–349.

15. Kruger, M.S., L.E. Richter, H.I. Mandelber, and P.A. McGrath. 1986. Linewidth determination from self-heterodyne measurements with sub-coherence delay times. *IEEE J. Quantum Electron.* 22: 2070–2074.

16. Sorin, W.V., S.A. Newton, and M. Nazarathy. 1988. Kilohertz laser linewidth measurements using fiber ring resonators. OFC/OFS '88, paper **WC5,** New Orleans, Louisiana. *Optical Fiber Communications Conference* p. 55.

17. Van Deventer, M.O., Spano, P., and Nielsen, S.K. 1990. Comparison of DFB laser linewidth measurement techniques results from COST 215 round robin. *Electronics Lett.* 26(24): 2018–2020.

18. Baney, D.M. and W.V. Sorin. 1988. Measurement of modulated DFB laser power spectrum using gated delayed self-homodyne technique. *Electron. Lett.* 24(4): 669.

19. Baney, D.M. and P.B. Gallion. 1989. Power spectrum measurement of a modulated semiconductor laser using an interferometric self-homodyne technique: influence of quantum phase noise and field correlation. *IEEE J. Quantum Electron.* 25(10): 2106.

20. Baney, D.M. and P.B. Gallion. 1991. Time domain measurement of chirp suppression for a multisection DFB laser, Optical Fiber Communications Conference, Optical Society of America. Paper WN7. San Diego, CA.

21. Vodhanel, R.S. and S. Tsuji. 1988. 12 Ghz FM bandwidth for a 1530 nm DFB laser. *Electron. Lett.* 24(22): 1359–1361.

22. Bergano, N.S. 1988. Wavelength discriminator method for measuring dynamic chirp in DFB lasers. *Electron. Lett.* 24(20): 1296–1297.

23. Sorin, W.V., K.W. Chang, G.A. Conrad, and P.R. Hernday, 1992. Frequency domain analysis of an optical FM discriminator. *J Lightwave Tech.* 10(6): 787–793.

24. Baney, D.M., P.B. Gallion, and C. Chabran. 1990. Measurement of the swept-frequency carrier-induced FM response of a semiconductor laser using an incoherent interferometric technique. *IEEE Photon. Technol. Lett.* 2(5): 325.

25. Kruger, U. and K. Kruger. 1995. Simultaneous measurement of the linewidth, linewidth enhancement factor and FM and AM response of a semiconductor laser, *J. Lightwave Tech.* 12(4): 592–597.

Polarization Measurements

Paul Hernday

6.1 INTRODUCTION

The concept of lightwave polarization is less familiar to many of us than the concepts of modulation or optical spectra. Many early telecommunications applications did not specify the polarization properties of the signal or the polarization transforming properties of transmission media. A more fundamental reason for our relative unfamiliarity with polarization may be that humans do not have the biological equipment to differentiate between different polarization states. This is not the case with many other organisms. Reflection from surfaces often partially polarizes light, producing patterns which polarization-sensitive vision can detect. The ability of some creatures to see (image) polarization patterns enables them to orient themselves, to recognize elements in their environment, and perhaps to identify members of their own species. The pygmy octopus, the mantis shrimp, and the starfish are among the creatures that can detect polarization patterns.[1] Scarab beetles have the unusual property of converting unpolarized incident light into left-hand circularly polarized light.[2]

Just as some organisms have found an advantage in their ability to recognize polarization patterns, modern telecommunication system designers can benefit from recognizing and controlling the polarization properties of signal light and the polarization conditioning characteristics of transmission media and optical components. In fact, such mastery is critical for high-speed, long-distance fiber optic systems and for high channel-capacity CATV distribution systems. Unlike the conventional waveguides used in the microwave field, optical fiber transforms signal polarization in a random way. Fiber optic components exhibit polarization-dependent insertion loss or gain, and awareness of the polarization relationships within a lightwave system is key to successful design.

The aim of this chapter is to provide an introduction to the concepts and vocabulary of lightwave polarization and its measurement. It also explores several measurement applications. The emphasis is upon providing an interesting introduction to the field, and the reader is referred to other texts for thorough treatments of polarized light. Edward Collett's *Polarized Light: Fundamentals and Applications* provides an extremely interesting and authoritative explanation of polarization concepts and fundamental polarization measurements.[3] Some of Collett's mathematical conventions are followed in this chapter.

Beginning with an introduction to polarization in the context of optical fiber, this chapter will discuss the measurement of retardance and the measurement of polarization cross-talk in polarization-maintaining fiber.

6.2 POLARIZATION CONCEPTS

6.2.1 General Description of Polarized Light

A polarized lightwave signal that is propagating in fiber or in free space is represented by electric and magnetic field vectors that lie at right angles to one another in a transverse plane (a plane perpendicular to the direction of travel). Polarization is defined in terms of the pattern traced out in the transverse plane by the electric field vector as a function of time, as shown in Figure 6.1. These are snapshots in time, showing the electric field as a function of distance. As time passes, the patterns move toward the observer. The polarization ellipse, shown at the left in each example, provides a more convenient description of polarization. By convention, the ellipse is generated by propagating the three-dimensional pattern through a fixed *xy*-plane. The ellipse shows the locus and the direction of rotation—the handedness—of the electric field vector in the fixed plane. In general, fully polarized light is elliptically polarized; linear and circular polarization are simply extremes of ellipticity.

It is not always possible to describe light in terms of a predictable electric field vector. Naturally produced light—sunlight, firelight, light from fireflies—is unpolarized. Unpolarized light of a given intensity can be represented intuitively as an electric field vector that from moment to moment occupies random orientations in the *xy*-plane. The random reorientation need only be fast enough to be beyond observation within the physical context of a particular measurement or application. This qualifier is worth noting, because there are applications in telecommunications and optical sensors in which fully polarized light is polarization-scrambled at a sufficiently high rate for it to appear unpolarized within the lifetime of carriers in an optical amplifier or within the bandwidth of optical instrumentation. It is not possible to systematically predict the electric field orientation of unpolarized light, nor to manipulate the polarization state with retarders (polarization-transforming devices discussed later in this chapter).

The usefulness of a lightwave in a particular application can depend upon its degree of polarization. Light from an ordinary light bulb is entirely unpolarized and is generally adequate for the daily life of non-fiber optics engineers. Light from a diode laser is almost completely polarized, presenting both opportunities and challenges in sensor and telecommunications applications. The light produced by a light emitting diode (LED) may be 10

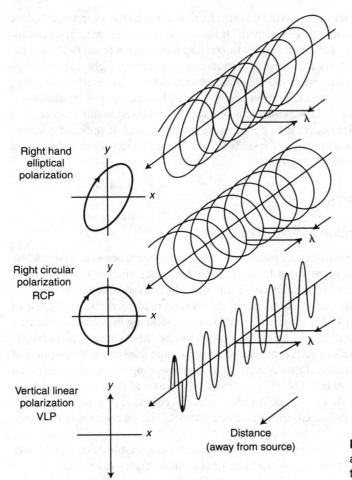

Right hand
elliptical
polarization

Right circular
polarization
RCP

Vertical linear
polarization
VLP

Distance
(away from source)

Figure 6.1 Three-dimensional
and "polarization ellipse" representations of polarized light.

to 20% polarized. Partially polarized light can be modeled as the superposition of a fully
polarized and a completely unpolarized lightwave. The degree of polarization (DOP) is
described by

$$\text{DOP} = \frac{P_{\text{polarized}}}{P_{\text{polarized}} + P_{\text{unpolarized}}} \tag{6.1}$$

involving the amount of power, in linear terms, of the polarized and unpolarized components of the signal. A lightwave traveling through free space will maintain its degree of
polarization indefinitely. However, a nonideal transmission medium or two-port optical
component can change the degree of polarization of the signal in a relationship involving
the spectral width of the signal and the dispersive properties of the transmission path. This
relationship is discussed in Chapter 12, which deals with dispersion measurements.

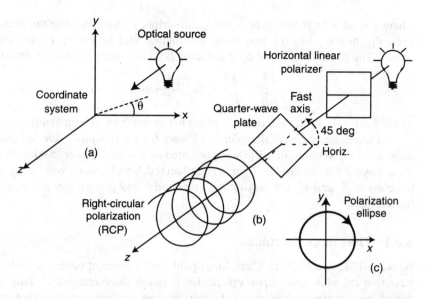

Figure 6.2 (a) A coordinate system for unambiguous description of the state of polarization. (b) Generation of right circular polarization. (c) Corresponding polarization ellipse.

6.2.2   A Polarization Coordinate System

An explicit coordinate system must be specified in order to unambiguously describe a state of polarization. A conventional coordinate system is shown in Figure 6.2a. The z-axis is horizontally oriented in the direction of propagation of the light. The y-axis is vertically oriented pointing up, and the x-axis is horizontally oriented to form a right-handed rectangular coordinate system. Angles are measured with reference to the x-axis, a positive angle indicating the sense of rotation from the x-axis toward the y-axis. In Figure 6.2b, unpolarized light is filtered by a polarizer to produce horizontal linear light. A quarter-wave retarder, which is discussed later in the chapter, is used to resolve the horizontally polarized light into two equal-intensity components that travel at slightly different speeds. For the retarder orientation shown, light emerging from the retarder is right-hand circularly polarized, illustrated in the polarization ellipse of Figure 6.2c.

6.2.3   The Polarization Ellipse

Polarized light can be represented mathematically in terms of the x- and y-axis projections of the electric field vector. Borrowing from Collett, the transverse components are given by [3]

$$E_x(z, t) = E_{0x} \cos(\tau + \delta_x) \tag{6.2}$$

$$E_y(z, t) = E_{0y} \cos(\tau + \delta_y) \tag{6.3}$$

where $\tau = \omega t - kz$ is the "propagator." Subscripts x and y indicate the directions of the axes, E_{0x} and E_{0y} are the maximum amplitudes, and δ_x and δ_y are the corresponding phases. The pattern traced out by the electric field in a fixed xy-plane is described by

$$\frac{E_x^2}{E_{0x}^2} + \frac{E_y^2}{E_{0y}^2} - 2 \frac{E_x E_y}{E_{0x} E_{oy}} \cos \delta = \sin^2 \delta \qquad (6.4)$$

where $\delta = \delta_y - \delta_x$ is the difference in phase between the two components.

The relationships are illustrated in Figure 6.3 for arbitrarily selected amplitudes and phases. By convention, the ellipse is presented as it would appear looking through the xy-plane toward the source. The light is right-handed, which can be verified by inspection of the plots of E_x and E_y. By varying the magnitudes and phases, any polarization state can be described.

6.2.4 The Jones Calculus

Between 1941 and 1948, R. Clark Jones published a series of papers describing a new polarization calculus based upon optical fields rather than intensities.[4] This approach, although more removed from direct observation than previous methods, allowed calculation of interference effects and in some cases provided a simpler description of optical physics.

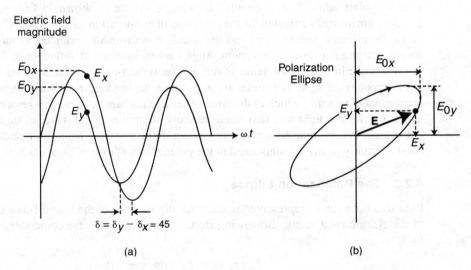

Figure 6.3   The polarization ellipse in relation to the x- and y- components of the electric field.

Polarized light can be represented by a two-element complex vector, the elements of which specify the magnitude and phase of the x- and y-components of the electric field at a particular point in space. The Jones vector has the form

$$\mathbf{E} = \begin{pmatrix} E_x \\ E_y \end{pmatrix} = \begin{pmatrix} E_{0x}\, e^{i\delta_x} \\ E_{0y}\, e^{i\delta_y} \end{pmatrix} \tag{6.5}$$

where real quantities E_{0x} and E_{0y} represent the maximum amplitudes and δ_x and δ_y represent the phases. It is customary to use the Jones vector in normalized form and to simplify the representation of phase. The Jones vector for linear horizontal polarized light

$$\mathbf{E} = \begin{pmatrix} E_{0x} e^{i\delta_x} \\ 0 \end{pmatrix} \quad \text{becomes} \quad \mathbf{E} = \begin{pmatrix} 1 \\ 0 \end{pmatrix}. \tag{6.6}$$

The Jones vector for right-hand circularly polarized light is

$$\mathbf{E} = \frac{1}{\sqrt{2}} \begin{pmatrix} 1 \\ +i \end{pmatrix} \tag{6.7}$$

where the amplitude of the normalized Jones vector is unity and the imaginary bottom element represents the phase relationship $\delta_y - \delta_x = +90$ degrees. The Jones vector is limited to the description of fully polarized light.

The transmission properties of a two-port optical device can be described by the complex two-by-two Jones matrix, which relates the input and output Jones vectors. The Jones matrix representation of an unknown device can be found by measuring three output Jones vectors in response to three known stimulus polarizations, as shown in Figure 6.4. Calculation of the matrix is simplest when the stimuli are linear polarizations oriented at 0, 45, and 90 degrees, but any three distinct stimuli may be used. The matrix calculated in this manner is related to the true Jones matrix by a multiplicative complex con-

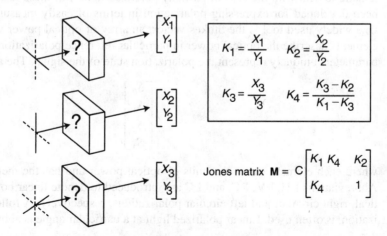

Figure 6.4 Measurement of the Jones matrix of an optical component.

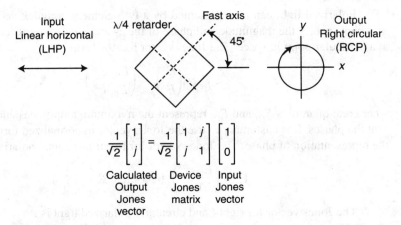

Figure 6.5 Jones calculus description of the effect of a rotated retarder on linear horizontally polarized light.

stant C. The magnitude of this constant can be calculated from intensities measured with the device removed from the optical path, but the phase is relatively difficult to calculate, requiring a stable interferometric measurement. Fortunately, most measurements do not require determination of this constant.

An example of the use of the Jones calculus is given in Figure 6.5. Linear horizontally polarized light (LHP) is passed through a retarder that is tilted to bring the fast axis +45 degrees from the horizontal. The output polarization is right circular (RCP).

6.2.5 The Stokes Parameters

It is not convenient to measure the electric field of a lightwave, and in any case it is problematic to deal with the case of partially polarized light. For these reasons, methods have been developed for expressing polarization in terms of easily measured optical powers. One widely used tool is the Stokes vector, an array of optical power values in which the elements describe the optical power in particular reference polarization states. The Stokes parameters uniquely represent the polarization state of the signal. The array has the form

$$\begin{bmatrix} S_0 \\ S_1 \\ S_2 \\ S_3 \end{bmatrix} \tag{6.8}$$

where each element caries the units of optical power and has the meaning given below. Abbreviations LH, LV, RC, and LC are introduced to denote linear horizontal, linear vertical, right circular, and left circular polarizations, respectively. A following P, for polarization, is often used. Linear polarized light at a particular angle is abbreviated as $L \pm \theta$.

S_0 = Total power (polarized + unpolarized)

S_1 = Power through LH polarizer – power through LV polarizer

S_2 = Power through L +45 polarizer – power through L –45 polarizer

S_3 = Power through RC polarizer – power through LC polarizer

Stokes parameters S_1, S_2, and S_3 can be assigned to an *xyz*-coordinate system as shown in Figure 6.6a. The amount of optical power contained in the polarized part of the lightwave is given by

$$P_{\text{polarized}} = \sqrt{S_1^2 + S_2^2 + S_3^2} \tag{6.9}$$

The *normalized* Stokes parameters are obtained by dividing the Stokes parameters by the total optical power:

$$s1 = \frac{S_1}{S_0} \quad s2 = \frac{S_2}{S_0} \quad s3 = \frac{S_3}{S_0} \tag{6.10}$$

The range of the normalized Stokes parameters is −1 to +1. For example, fully polarized horizontal linear light is expressed by $s1 = +1$, $s2 = s3 = 0$. The normalized Stokes parameters can be assigned to *xyz*-axes as shown in Figure 6.6b.

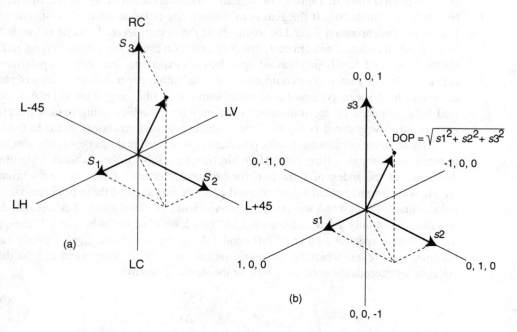

Figure 6.6 Orthogonal representations of (a) the Stokes parameters; (b) the normalized Stokes parameters.

6.2.6 Degree of Polarization

As discussed earlier, the degree to which a lightwave signal is polarized is given by the ratio of polarized light to total light:

$$\text{DOP} = \frac{P_{\text{polarized}}}{P_{\text{polarized}} + P_{\text{unpolarized}}} \qquad (6.11)$$

The degree of polarization can be expressed in terms of the Stokes parameters

$$\text{DOP} = \frac{\sqrt{S_1^2 + S_2^2 + S_3^2}}{S_0} \qquad (6.12)$$

or the normalized Stokes parameters

$$\text{DOP} = \sqrt{s1^2 + s2^2 + s3^2} \qquad (6.13)$$

In the case of fully polarized light, DOP = 1, Equation 6.13 describes a sphere of unit radius. If the light is 50% polarized, Equation 6.13 defines a sphere of radius 1/2.

Depolarization of Lightwave Signals. Depolarization of lightwave signals may be a help or a hindrance. If the goal is to measure the polarization-mode dispersion of a fiber path, depolarization should be avoided. At the system level, the same is true in systems based on coherent detection, where depolarization causes the detected signal to fade. On the other hand, totally unpolarized light does not experience the effects of polarization-dependent loss or gain in fiber components or polarization-dependent sensitivity of photodiodes. In high-speed systems based on erbium-doped fiber amplifiers (EDFAs), polarized light produced by the transmitter laser may be scrambled using optical integrated circuits, effectively depolarizing the light to obtain best performance from the EDFAs.

Under some conditions, a fully polarized lightwave signal can depolarize simply by transmission through a fiber path. High-birefringence (hi-bi) fiber exhibits a significant difference in the index of refraction for orthogonal electric fields. A monochromatic (single-wavelength) optical signal coupled partially into each of these principal states of polarization splits into two wavefronts that experience an entire wave of differential delay in a distance of only a few millimeters (the "beat length" of the hi-bi fiber). If the signal has significant spectral width, the differential delay depolarizes the light. Depolarization becomes most severe when the fast and slow axes are equally illuminated and the differential delay exceeds the coherence time of the signal, given by

$$t_c = \frac{1}{\Delta f} = \frac{\lambda^2}{c\Delta\lambda} \qquad (6.14)$$

where Δf is the source linewidth in Hz, $\Delta\lambda$ is the full-width, half-maximum (FWHM) spectral width of the source in meters, λ is the center wavelength of the source in meters, and c is the speed of light (3×10^8 m/s). Assuming a source with Gaussian spectrum of width $\Delta\lambda$ centered at λ, the lowest degree of polarization which can result from a differential delay $\Delta\tau$ (seconds) is given by

$$\text{DOP} = e^{-\frac{1}{4\,ln2}\left(\frac{\pi c\,\Delta\tau\,\Delta\lambda}{\lambda^2}\right)^2} \qquad (6.15)$$

Depolarization is most severe when fast and slow polarization modes are equally illuminated.

6.2.7 The Poincaré Sphere

The Poincaré sphere,[5] shown in Figure 6.7, is a graphical tool in real, three-dimensional space that allows convenient description of polarized signals and of polarization transformations caused by propagation through devices. Any state of polarization can be uniquely represented by a point on or within a unit sphere centered on a rectangular *xyz*-coordinate system. Circular states are located at the poles, with intermediate elliptical states continuously distributed between the equator and the poles. Right-hand and left-hand elliptical states occupy the northern and southern hemispheres, respectively. The coordinates of a point within or upon this sphere are the normalized Stokes parameters.

Fully polarized light is represented by a point on the surface of the Poincaré sphere. Partially polarized light, which can be considered a superposition of polarized and unpolarized light, is represented by a point within the volume of the Poincaré sphere. The distance of the point from the center of the sphere gives the degree of polarization of the signal, ranging from zero at the origin (unpolarized light) to unity at the sphere surface

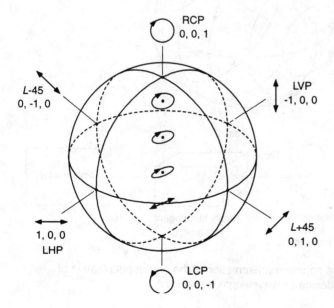

Figure 6.7 The Poincaré sphere representation of polarized light.

(completely polarized light). Points close together on the sphere represent polarizations that are similar, in the sense that the interferometric contrast between two polarizations is related to the distance between the corresponding two points on the sphere. Orthogonal polarizations with zero interferometric contrast are located diametrically opposite one another on the sphere.

As a display device for instrumentation, the Poincaré sphere is generally used to represent fully polarized light or the polarized part of a partially polarized lightwave. In the case of partially polarized light, a ray is extended from the center of the sphere along the normalized Stokes vector to locate a point on the surface of the unit sphere.

Because a state of polarization is represented on the Poincaré sphere as a single point, a continuous evolution of polarization is represented as a continuous path. For example, the evolution of polarization of light emerging from a birefringent device as wavelength is changed is represented by circular traces as illustrated in Figure 6.8. (Birefringence refers to a physical asymmetry in the index of refraction that allows different polarization modes to travel at different speeds.) The amount of rotation is proportional to the change in wavelenth. All of the circles are centered along a diameter of the sphere. The endpoints of the diameter locate the eigenmodes of the device. Eigenmodes are polarization states that propagate unchanged through a device. In linearly birefringent material,

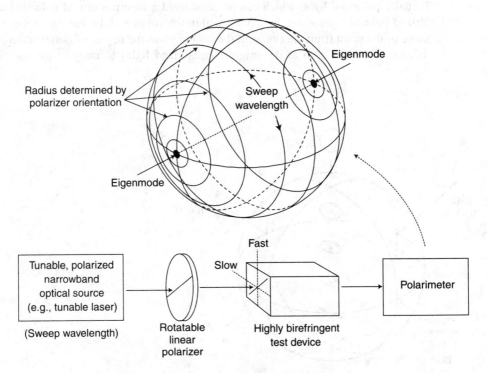

Figure 6.8   Poincaré sphere representation of the output polarization of a highly birefringent device as wavelength is changed.

the eigenmodes are linear and correspond to the fast and slow axes of the device. In Figure 6.8, each circle corresponds to a different rotational angle of the rotatable linear polarizer. The circle is largest when the electric field of the input light resolves equally into the eigenmodes.

A path on the sphere can also record the polarization history of a signal, allowing the operator to view the time-dependent behavior of polarization.

6.2.8   The Polarimeter and Polarization Analyzer

The polarization state of an optical signal can be determined by detecting the optical power transmitted through specially defined polarization filters. The measurement requires splitting the lightwave into samples in space or in time. In the space-division polarimeter, shown in Figure 6.9a, the beam is split into four and the resulting beams are processed in parallel.[6] One beam passes directly to a detector without polarization filtering, to provide a measurement that is proportional to total power. Another beam passes through a linear horizontal polarizer and provides a basis for the measurement of S_1. A third beam passes through a linear +45 degree polarizer and forms the basis for measurement of S_2. The final beam passes through a quarter-wave plate and a linear +45 degree polarizer. Circularly polarized input light is transformed to linear +45 degree by the waveplate and detected after passing through the polarizer.

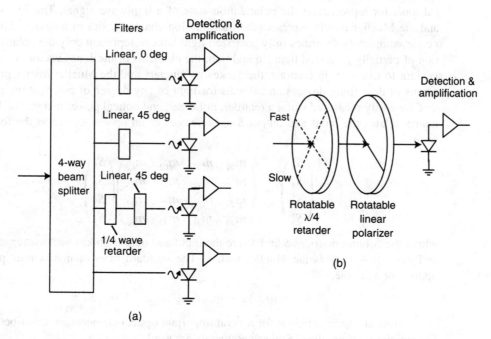

Figure 6.9 Examples of polarimeters based on (a) space division, and (b) time division of the optical beam.

In the time-division polarimeter, shown in Figure 6.9b, the signal passes through an independently rotatable quarter-wave retarder and a polarizer. Angular orientations are orchestrated by hand, or using motors, to sequentially detect the polarization components described above. Total power is determined by analysis of the resulting data.

Space- and time-division polarimeters can be built in a number of configurations. The time-division method is the simplest route for homemade setups and manual measurements. Higher speed measurements require motorization and computer control, and the speed advantage belongs to the space-division polarimeter with its parallel processing and lack of moving parts. All types of polarimeters require a calibration process to overcome such effects as differences in optical path losses and photodiode sensitivities, variations in coupling as optical elements rotate, and imperfections of the optical retarders. In addition, a calibration for total power is needed.

The functionality of a polarimeter can be expanded by adding external polarizers and mathematical algorithms. These enhancements allow the measurement of polarization within a physical reference frame removed from the polarimeter by an arbitrary length of singlemode optical fiber. They also allow the measurement of retardance, polarization-dependent loss, and polarization-mode dispersion in two-port optical devices. This type of instrument is typically referred to as a polarization analyzer.[6]

6.2.9 The Mueller Matrix

The polarization ellipse, Jones vector, Stokes vector, and Poincaré sphere are mathematical tools for representing the polarization state of a lightwave signal. The Jones matrix and the Mueller matrix represent the transmission characteristics of a device. The Jones vector completely describes fully polarized light but can represent only the polarized portion of partially polarized light. In addition, the elements of the Jones vector are not convenient to measure. In contrast, the Stokes parameters and the Mueller matrix provide a means of describing devices and signals for light of any degree of polarization, in terms that are easily measured using a retarder, polarizer, and optical power meter. The Mueller matrix relates the input and output Stokes vectors of an optical device in the following form:

$$
\begin{pmatrix} S_0' \\ S_1' \\ S_2' \\ S_3' \end{pmatrix} = \begin{pmatrix} m_{00} & m_{01} & m_{02} & m_{03} \\ m_{10} & m_{11} & m_{12} & m_{13} \\ m_{20} & m_{21} & m_{22} & m_{23} \\ m_{30} & m_{31} & m_{32} & m_{33} \end{pmatrix} \begin{pmatrix} S_0 \\ S_1 \\ S_2 \\ S_3 \end{pmatrix}
\tag{6.16}
$$

where the column matrices S and S' are the input and output Stokes vectors, respectively, and the 4×4 array is the Mueller matrix. The standard rules of matrix multiplication apply. For example,

$$
S_0' = m_{00} S_0 + m_{01} S_1 + m_{02} S_2 + m_{03} S_3
\tag{6.17}
$$

Mueller matrix relations for several important optical elements are described below. For simplicity, normalized Stokes parameters are used.

   When unpolarized light is incident upon a horizontal linear polarizer, the output light is horizontal linearly polarized and contains half of the incident power. The vertically polarized portion of the light has been filtered out by the polarizer. This relationship is expressed as:

$$\frac{1}{2}\begin{pmatrix} 1 \\ 1 \\ 0 \\ 0 \end{pmatrix} = \frac{1}{2}\begin{pmatrix} 1 & 1 & 0 & 0 \\ 1 & 1 & 0 & 0 \\ 0 & 0 & 0 & 0 \\ 0 & 0 & 0 & 0 \end{pmatrix}\begin{pmatrix} 1 \\ 0 \\ 0 \\ 0 \end{pmatrix} \tag{6.18}$$

Mueller matrices for several other common optical devices are shown below:

Linear vertical polarizer   Linear +45 degree polarizer

$$\frac{1}{2}\begin{pmatrix} 1 & -1 & 0 & 0 \\ -1 & 1 & 0 & 0 \\ 0 & 0 & 0 & 0 \\ 0 & 0 & 0 & 0 \end{pmatrix} \qquad \frac{1}{2}\begin{pmatrix} 1 & 0 & 1 & 0 \\ 0 & 0 & 0 & 0 \\ 1 & 0 & 1 & 0 \\ 0 & 0 & 0 & 0 \end{pmatrix} \tag{6.19}$$

A neutral density filter, or broadband optical attenuator, of strength $1:1/a$ is represented by:

$$\frac{1}{a}\begin{pmatrix} 1 & 0 & 0 & 0 \\ 0 & 1 & 0 & 0 \\ 0 & 0 & 1 & 0 \\ 0 & 0 & 0 & 1 \end{pmatrix} \tag{6.20}$$

   The use of rotatable polarizers is common in experimental work. The Mueller matrix of an ideal linear polarizer with its transmission axis rotated by an arbitrary angle θ from the horizontal, as shown in Figure 6.10, is given by:

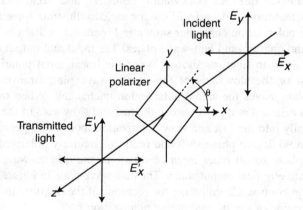

Figure 6.10 Coordinate system for describing the Mueller matrix of a linear polarizer rotated from the horizontal.

$$\frac{1}{2}\begin{pmatrix} 1 & \cos 2\theta & 0 & 0 \\ \cos 2\theta & 1 & 0 & 0 \\ 0 & 0 & \sin 2\theta & 0 \\ 0 & 0 & 0 & \sin 2\theta \end{pmatrix} \qquad (6.21)$$

Chapter 9 describes a method for determining polarization-dependent loss (PDL) from measurements of the Mueller matrix.

6.3 RETARDANCE MEASUREMENT

6.3.1 Introduction

Most optical materials exhibit some degree of refractive index asymmetry that allows light in two orthogonal polarization states to travel at different speeds through the material. This property is called birefringence. The polarization states into which polarized incident light is resolved are defined by the internal structure of the material. For well-defined structures such as a quartz crystal, these states are maintained during transmission through the device and are called the eigenmodes. Birefringence can be linear or circular, although most applications of birefringent devices, and most concerns about birefringence in fiber-optic components, involve linear birefringence. Quartz is linearly birefringent.

Retardance is a measure of the differential phase shift of light in the eigenmodes, more commonly referred to as the fast and slow waves. An example is the retarder or wave plate, a device fabricated to provide a predictable amount of phase shift. Polarized light incident on a typical waveplate decomposes into linear fast and slow waves. In general, the output polarization is different from the input state, depending upon the relationship of the incident polarization to the internal structure, and corresponding eigenmodes, of the waveplate. Retardance is typically expressed in waves or in degrees of phase shift at a specified wavelength.

Polarization Controllers. Because polarization adjusters and controllers are widely used in fiber-optics laboratory work, we will use the topic to illustrate some applications of retardance. In the polarization controller shown in Figure 6.11a, light is transmitted through a polarizer and quarter- and half-wave plates. The input and output of the controller may be in open air or in fiber, as shown. When the linear input polarizer is aligned with either the fast or the slow axis of the quarter-wave plate, transmission through the quarter-wave plate leaves the signal polarization unchanged. When the linearly polarized input light is oriented midway between the fast and slow axes of the wave plate, it is decomposed equally into the fast and slow polarization modes or eigenmodes. The two waves experience a 90 degree phase shift and produce circularly polarized light at the output of the wave plate. At all other input orientations, the quarter-wave plate transforms linear input light to elliptical output states. The half-wave plate in Figure 6.11a allows additional polarization control. Coordinating the rotations of the quarter- and half-wave plates allows the generation of any desired output polarization state.

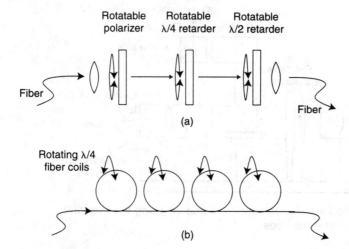

Rotatable polarizer   Rotatable λ/4 retarder   Rotatable λ/2 retarder

Fiber

Fiber

(a)

Rotating λ/4 fiber coils

(b)

Figure 6.11 Examples of polarization manipulators. (a) Polarization controller using polarizer and wave plates. (b) Polarization adjuster based on fiber coils.

Although retarders are typically implemented in the form of discrete optical wave-plates, retarders can also be made from short lengths of highly birefringent fiber (hi-bi, or polarization-maintaining fiber) or from coils of singlemode fiber.[7] A 4-coil polarization controller is depicted in Figure 6.11b. Bending produces an asymmetric stress field which induces a difference in index of refraction between the stressed and nonstressed axes. Retardance increases with the number of turns. The fiber coils are typically arranged so that they can rotate about a common tangent, along which the interconnecting fiber is routed. Rotation of a coil changes the way the input electric field decomposes into the fast and slow axes of the coil, changing the polarization transformation produced by the coil. Several combinations of quarter- and half-wave fiber coils are commercially available and the devices are readily constructed in the laboratory.

6.3.2 Measurement of Retardance

Over the years, many retardance measurement methods have been developed. Some of this work has achieved great accuracy.[8] The two methods described here are not the most accurate, but they do give insight into measurement technique and the concept of birefringence, and they are quite convenient. Both methods make use of a polarimeter or polarization analyzer. The Poincaré sphere method is most intuitive, although the second method, based on the Jones matrix measurement, is faster and more easily automated. The setup for both measurements is shown in Figure 6.12. We will assume that the retarder under test is a quartz fractional waveplate. The fiber path between the output lens and the polarimeter must be fixed in position to prevent spurious polarization shifts during the measurement.

Both measurement methods require the establishment of a physical reference frame, that is, a set of axes within which polarization can be defined. For polarimeters with fiber inputs, this involves fixing the fiber in position to stabilize the polarization transformation it produces. A rotatable linear polarizer is coupled into the fiber end through a lens. Using the

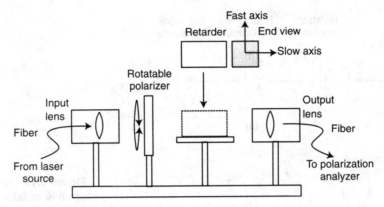

Figure 6.12 Setup for measurement of retardance using the Poincare sphere and Jones matrix methods.

polarizer and an algorithm based on the Jones matrix, the birefringence and polarization-dependent loss of the path connecting the setup to the polarimeter is measured and removed. Polarizer angles of 0, 60, and 120 degrees are commonly used. The first state is typically assumed by the instrument to represent horizontal linear polarization. Once the resulting reference frame is activated, rotation of the polarizer to the 0 degree or horizontal orientation causes a point to appear at 1, 0, 0 (LHP) on the Poincaré sphere, and rotation of the polarizer by 180 degrees traces out the entire equator.

6.3.3 The Poincaré Sphere Method

This method determines retardance from the pattern traced out on the Poincaré sphere when the retarder under test is rotated in a linearly polarized beam. The retarder may be supported in a rotatable mount. The polarization reference frame is not strictly necessary for this method, but it can simplify interpretation of the graphical results. After the reference frame is created and activated, the polarizer is returned to horizontal and the retarder is placed in the beam following the polarizer. The retarder is rotated about its optical axis until the Poincaré display again indicates horizontal linearly polarized light. In this condition, the stimulus polarization is coupled entirely into one of the eigenmodes. Because the electric-field vector is ideally aligned with one of the crystalline axes and experiences only the index of refraction associated with that axis, no polarization transformation occurs. The corresponding polarization state is marked on the Poincaré sphere. Next, the retarder is rotated +45 degrees to allow the input polarization to decompose equally into the two eigenmodes. This condition produces the greatest polarization shift, and the corresponding state is marked on the sphere. The angle formed at the center of the sphere by rays to the two marked points is the retardance of the waveplate. The pattern traced out on the Poincaré sphere for a slightly less-than-quarter-wave retarder is shown in Figure 6.13. If it is known that the retarder exhibits linear birefringence, the retarder may be measured without a reference frame. The figure-8 pattern will appear in an arbitrary orientation on the sphere and the retardance is again determined by marking the central intersection and one extreme of the pattern.

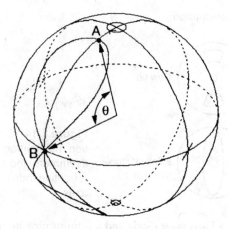

Figure 6.13 Measurement of the retardance θ of a near-quarter-wave retarder using a polarimeter with Poincaré sphere display.

6.3.4   The Jones Matrix Method

The retardance of a two-port device can be determined from the Jones matrix. The measurement setup is identical to that of the Poincaré sphere method, shown in Figure 6.12, and generally involves the use of a polarization analyzer.[6] A reference frame is established at the polarizer, as described above. The polarizer is returned to horizontal orientation and the retarder is placed in the beam following the polarizer. The retarder is rotated about its optical axis until the state of polarization of light emerging from the retarder is linear horizontal. In this orientation, linearly polarized light is propagating solely in one eigenmode. The polarization state of the emerging light is measured at polarizer orientations of 0 degrees (LHP), +45 degrees (L + 45), and +90 degrees (LVP). The Jones matrix is computed from these three measurements. For a quarter-wave plate with its fast axis oriented horizontally,

$$\mathbf{J}\left(\frac{\lambda}{4}\right) = \begin{pmatrix} e^{i\frac{\pi}{4}} & 0 \\ 0 & e^{-i\frac{\pi}{4}} \end{pmatrix} \tag{6.22}$$

The retardance is found by comparing the arguments of the exponentials as follows

$$\frac{\pi}{4} - \left(-\frac{\pi}{4}\right) = \frac{\pi}{2} \tag{6.23}$$

resulting in a retardance of 90 degrees.

6.4   MEASUREMENT OF CROSS-TALK IN POLARIZATION-MAINTAINING FIBER

6.4.1   Introduction

The term polarization-maintaining (PM) refers to a class of highly linearly birefringent singlemode fiber. PM fiber is typically used to guide linearly polarized light from point to

Examples of PM Fiber Construction

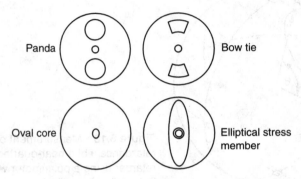

Figure 6.14   Examples of polarization maintaining (PM) fiber. Physical asymmetry induces a difference in refractive index between orthogonal axes.

point, for example between a DFB laser diode and a lithium-niobate modulator in a high-speed telecommunication system. It also finds many specialized applications in lightwave communication and optical sensor research. Birefringence may stress-induced by placing the core between or within glass elements of different physical composition, or may originate with a purposeful asymmetry in the core geometry (form birefringence). In all cases, the result is a difference in the index of refraction between orthogonal axes. Several examples of PM fiber are shown in Figure 6.14.

The birefringence of PM fiber is much larger and more uniform than the residual birefringence of ordinary singlemode fiber. Because the birefringence is associated with a systematic, physical asymmetry of the fiber cross-section, PM fiber exhibits distinct fast and slow principal optical axes. Light coupled into a length of PM fiber resolves into two orthogonal, linearly polarized modes, or waves, according to how the input electric field projects onto the fast and slow axes of the fiber. In most applications, linearly polarized light is aligned with one of the axes, commonly the slow axis.

Only when the electric field of the light is entirely aligned with the slow or fast axis is PM fiber actually polarization maintaining, as indicated in Figure 6.15. Because of the difference in index of refraction between the fast and slow axes, electric fields in the two

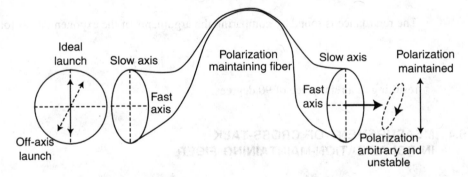

Figure 6.15   Polarization-maintaining (PM) fiber maintains polarization only if the launch polarization is aligned with the fast or slow axis.

axes are phase-shifted relative to one another in proportion to the distance traveled. If electric field components exist in both axes, the polarization state of the propagating light evolves as it travels along the fiber's length and exits at an arbitrary polarization state. Under these conditions, the output polarization of the fiber is readily influenced by temperature and fiber movement.

There are several requirements for guiding linearly polarized light in PM fiber. Light must be linearly polarized to a high extinction before coupling to the core, and birefringence in lenses or optical connectors must be minimized. Light can be scattered between principal axes by impurities and structural flaws, and can also couple between axes where the fiber is sharply stressed by outside forces. In a path made up of several connected PM fibers, light can couple between fast and slow axes at fiber interfaces. This section presents two methods for measuring polarization cross-talk, the figure of merit for the confinement of light to a single axis of PM fiber.

6.4.2   The Crossed-Polarizer Cross-Talk Measurement

The setup for this method is shown in Figure 6.16. To avoid interferometric effects, a spectrally broad optical source is used. As a guide, the coherence time of the source should be much shorter than the differential propagation time along the fast and slow axes of the test fiber. An unpolarized source gives the advantage that coupling to the input of the test fiber is independent of rotation of the input polarizer. The output of the test fiber is detected by a power meter, through a second rotatable polarizer. Input and output optics should be strain-free to avoid influencing the measurement with their own birefringence. Following adjustment of the coupling optics, the two polarizers are iteratively adjusted to minimize the detected signal. In this state, the input polarizer is aligned with one of the principal axes and the output polarizer with the other. The resulting power level P_{min} indicates the amount of light that has coupled into the unintended axis of the PM fiber. Next, the output polarizer is rotated by +90 degrees. The corresponding power level P_{max} indicates the amount of light in the intended axis. Polarization cross-talk is computed according to

$$\text{Polarization cross-talk} = 10 \log(P_{min}/P_{max}) \tag{6.24}$$

Measurement results can be affected by source instability and by variation of the optical coupling with rotation of the polarizers. Iterative adjustment of the coupling and the polarizers can overcome the coupling variation. Polarizers must be strong enough that

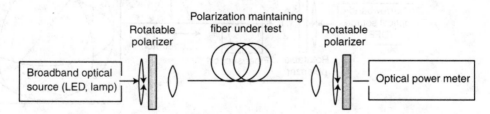

Figure 6.16   Setup for measuring cross-talk in polarization maintaining fiber by the crossed-polarizer method.

with the test fiber removed and the polarizers crossed, the attenuation is at least 10dB greater than the cross-talk to be measured. The strength of a polarizer is given by the extinction ratio

$$\text{Polarization extinction ratio} = 10 \log \frac{P_{\text{block axis}}}{P_{\text{pass axis}}} \tag{6.25}$$

where the numerator and denominator represent the optical power transmitted with the electric field of linearly polarized input light aligned with the block axis and the pass axis, respectively, of the polarizer under test. For example, in order to measure a polarization cross-talk value of −30 dB, the polarizers should have an extinction ratio of at least −40 dB.

6.4.3 The Polarimetric Cross-Talk Measurement

As indicated earlier, PM fiber maintains linear polarization only if the electric field is confined to a single principal axis. If the field has components along both axes, the polarization state within the PM fiber evolves with distance along the fiber. For the same physical reasons, the output polarization of the fiber evolves as the wavelength of a narrowband source is tuned. In the discussion of the Poincaré sphere, it was shown that this polarization evolution traces out circular arcs on the sphere. Polarization cross-talk can be calculated from the diameter of these circular arcs.

The polarimeter-based cross-talk measurement is an example of a *relative* polarization measurement. The cross-talk information is taken from the diameter of the circle, and the specific position of the circle on the sphere is not significant. The PM fiber under test may even be connected to the polarimeter through other sections of fiber.

A setup for measurement of cross-talk is shown in Figure 6.17. The optical source must be highly polarized, exceeding the cross-talk to be measured by at least 10 dB.

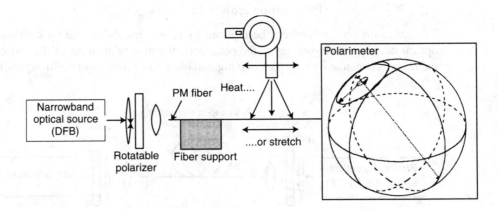

Figure 6.17 Setup for measuring cross-talk in polarization maintaining fiber by the polarimetric method.

In addition, the source spectrum must be narrow enough that the source coherence time is much larger than the difference in propagation time between the fast and slow axes of the fiber under test. This assures that light propagating in the orthogonal fiber axes can interfere at the output. The differential propagation delay of the transmission path can be estimated from the beat length and the overall physical length of the PM fiber. A beat length of 2 mm means that the fast and slow modes experience a 360 degree phase shift in each 2 mm of fiber length. At 1550 nm, a cycle of phase shift corresponds to a differential time delay of about 0.005 ps, so a 2 m segment of PM fiber would have a differential group delay of 5 ps. Assuming a Gaussian spectral shape, the source coherence time is given by

$$t_c = \frac{\lambda^2}{c\,\Delta\lambda} \tag{6.26}$$

where λ is the center wavelength of the optical source and $\Delta\lambda$ is its spectral width. Wavelengths are in meters and c is the velocity of light in free space in m/s. For example, at 1550 nm, an optical source with 0.1 nm spectral width has a coherence time of about 80 ps, sufficiently long to be used in testing the 2 m length of PM fiber. The spectral width of FP lasers and unfiltered LED sources are generally too broad to support this method.

The method makes use of the fact that as polarization-maintaining fiber is stretched or heated, the resulting change in length causes a phase shift between the fast and slow polarization modes of the fiber. A change of phase between orthogonal polarization states always describes a circle on the Poincaré sphere. From the diameter of this circle, the polarization cross-talk can be calculated. To see why this its true, consider the front and side-cutaway views of the Poincaré sphere shown in Figure 6.18. Points A and B correspond to the orthogonal fast and slow polarization modes of the polarization-maintaining fiber. The circular data trace of radius r is shown in cross section as a vertical line of length $2r$. The angle between the AB axis and a ray to a point on the circular trace is given by θ.

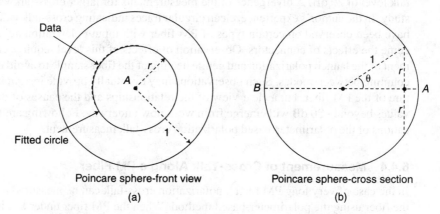

Poincare sphere-front view
(a)

Poincare sphere-cross section
(b)

Figure 6.18 Basis for the calculation of polarization crosstalk from the circular data trace. (a) Poincaré sphere front view. (b) Side-cutaway view.

As discussed earlier, any polarization state can be decomposed into orthogonal pairs of polarization states. In this application, the projection of the data point onto the principal states axis divides the line AB into segments of length $1 + \cos\theta$ and $1 - \cos\theta$. The ratio of optical power in states B and A is given by

$$\frac{P_B}{P_A} = \frac{1 - \cos\theta}{1 + \cos\theta} \tag{6.27}$$

and the polarization cross-talk is given by

$$\text{Polarization cross-talk} = 10 \log\left(\frac{1 - \cos\theta}{1 + \cos\theta}\right). \tag{6.28}$$

To express the cross-talk in terms of the radius of the data trace, note that

$$\theta = \sin^{-1}r \quad \text{and} \quad \cos\theta = \sqrt{1 - r^2} \tag{6.29}$$

leading to the relationship

$$\text{Polarization cross-talk} = 10 \log\left[\frac{1 - \sqrt{1 - r^2}}{1 + \sqrt{1 - r^2}}\right] \tag{6.30}$$

The largest possible circle has unity radius, corresponding to 50% of the light propagating in each of the polarization modes. In the case in which light is very well confined to a single principal axis, the circle shrinks, converging toward a point on the Poincaré sphere that represents, in a relative way, the output eigenmode of the fiber. (The representation is relative because a polarization reference frame has not been set up at the output of the PM fiber).

At the time of writing, careful comparisons of the two measurement methods described above have been undertaken in the measurement community. In an initial, limited interlab comparison, the two methods were found to agree to within a few dB up to a cross-talk level of 20 dB. A divergence of the measurements for larger cross-talk values is under study. In the author's experience, clean circular traces indicating cross-talk values of −50 dB have been observed on certain types of PM fiber with input polarization adjusted to overcome the effects of connectors. Observation of effects of this level requires careful adjustment of the launch polarization and gentle heating of the fiber sample to avoid increasing the coupling between modes. Such observations may eventually provide insight into the structure of the PM fiber, but a clear view of the relationships and the causes of divergence for values beyond −20 dB will emerge from work now underway. Following are several applications of the polarimeter-based polarization cross-talk measurement.

6.4.4 Measurement of Cross-Talk Along a PM Fiber

In the case of very long PM fibers, polarization cross-talk can be measured at points along the fiber using the polarimeter-based method. When the PM fiber under test is stretched or heated, the circle on the Poincaré reveals the polarization cross-talk *in the region of the fiber that is stretched or heated.* The following fiber can be viewed as a fixed birefringence which affects the position of the circle on the sphere but does not affect its diameter. For example, consider a 100 m length of PM fiber. Light is linearly polarized and coupled into the fiber at an initial orientation. The cross-talk is measured by gently heating a short length of fiber immediately following the launch. The polarizer or fiber is rotated as needed to optimize the launch. Once the launch is established (and physically protected from movement of the rest of the fiber), the cross-talk at other points can be measured by simply applying the heat at the point of interest. In each case, about 1/2 m of fiber is heated or carefully stretched to produce an arc and a corresponding cross-talk value. When stretching the fiber, it is important to avoid bending the fiber at the point where it is being gripped, for these bends can cause light to couple between polarization modes.

6.4.5 Cross-Talk Measurement of PM Fiber Interfaces

PM fibers are interconnected by fusion splicing, by proximity coupling with or without index-matching fluid, and by means of various commercial connectors. In each case, maintenance of linear polarization across the fiber interfaces requires that the fiber axes are well aligned and that the fiber ends are prepared and supported in such a way as to avoid cross-talk between polarization modes.

Degradation of cross-talk due to a PM fiber interconnection is best determined by launching linearly polarized light into a single polarization mode of the first segment and then measuring the cross-talk on either side of the interconnection by heating the corresponding fiber.

6.4.6 Measurement of the Polarization Stability of Cascaded PM Fibers

As discussed earlier, if polarized light is not perfectly confined to a single axis, slight changes in fiber length caused by heating or movement can cause the output state to wander. This problem is compounded when several PM fibers are cascaded. Consider the case of a connectorized, linearly polarized source driving a cascade of three connectorized PM fiber jumpers. Typically, the launch into the first fiber is imperfect and the interconnections are slightly misaligned. We are interested in the stability of the final output polarization. The Poincaré sphere measurement in Figure 6.19a shows the pattern resulting from gentle stretches of each of the PM fiber jumpers; each circle indicates the polarization cross-talk in the jumper being stretched. Figure 6.19b was generated for a two-section cascade by slowly stretching one fiber while rapidly stretching and releasing the second. The shaded pattern indicates the range of possible output polarization states. The worst-case cross-talk for the combination of fibers is computed from the diameter of the outer perimeter of the pattern.

6.5 SUMMARY

Polarization measurements play a major role in the design and characterization of modern telecommunication systems, and familiarity with the concepts and vocabulary is extremely useful. In this chapter, we have examined several representations of polarized light:

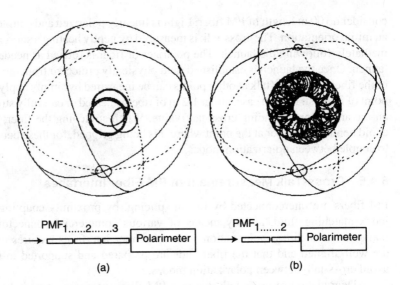

Figure 6.19 Output polarization of (a) three cascaded PM fibers stretched individually, and (b) two cascaded PM fibers stretched simultaneously but randomly.

- The polarization ellipse;
- The Jones vector;
- The Stokes vector;
- The Poincaré sphere.

Partially polarized light can be modeled as the superposition of polarized and unpolarized light. The polarization ellipse, the surface of the Jones vector and Poincaré sphere are used to represent polarized light. The Stokes vector and the interior of the Poincaré sphere represent light of any degree of polarization.

In addition, we have introduced tools for description of the polarization-transforming behavior of two-port devices and fibers:

- The Jones matrix;
- The Mueller matrix.

The Jones matrix is useful for describing the effect of a device on a fully polarized signal. The Mueller matrix is capable of describing the effect of a device on a signal of any degree of polarization.

Several measurements were described, mainly involving the polarimeter, a tool that supports fast and intuitive analysis of polarization behavior. A polarization analyzer extends the functionality of the polarimeter with the addition of insertable polarizers and algorithms that measure and extract information from the Jones matrix. The polarization analyzer allows the creation of a polarization reference frame removed from the polarimeter by a length of singlemode fiber.

Two retardance measurement methods were considered, the Poincaré sphere and Jones matrix measurements. The first offers the advantage of speed and intuitiveness, and the second the benefit of speed. Measurements of polarization cross-talk on polarization-maintaining fiber were also discussed. Numerical agreement of the crossed polarizer and polarimetric methods is under study at the time of this writing.

The reader is referred to Edward Collett's *Polarized Light: Fundamentals and Applications* for an extremely interesting, in-depth discussion of polarization concepts and fundamental polarization movements.[3]

REFERENCES

Polarization in the Biosphere

1. Cronin, T.W., N. Shashar, L. Wolff. 1995. Imaging technology reveals the polarized light fields that exist in nature. *Biophotonics International,* March/April: 38–41.

2. Kattawar, G.W. 1994. A search for circular polarization in nature. *Optics and Photonics News,* September: 42–43.

Polarization References

3. Collett, Edward. 1993. *Polarized light: fundamentals and applications.* New York, NY: Marcel Dekker, Inc.

4. Jones, R.C. 1947. A new calculus for the treatment of optical systems. VI: Experimental determination of the matrix. *Journal of the Optical Society of America,* 37: 110–112.

5. Born, M. and E. Wolf. 1980. *Principles of optics.* 6th ed. Pergamon.

A Polarization Analyzer

6. Heffner, B.L. and P.R. Hernday. 1995. Measurement of polarization-mode dispersion. *Hewlett-Packard Journal.* February: 27–33.

Polarization Controllers

7. Lefevre, H.C. 1980. Singlemode fibre fractional wave devices and polarization contollers. *Electronics Letters* Sept., 16 (20): 778–780.

Retardance Measurement

8. Rochford, K.B., P.A. Williams, A.H. Rose, I.G. Clarke, P.D. Hale, and G.W. Day. 1994. Standard polarization components: progress toward an optical retardance standard, SPIE 2265. *Polarization Analysis and Measurement II:*2–8.

7

Intensity Modulation and Noise Characterization of Optical Signals

Christopher M. Miller

7.1 MODULATION DOMAIN ANALYSIS

The low propagation loss and extremely broad bandwidth of singlemode optical fiber have contributed to the emergence of high-capacity lightwave digital transmission systems and analog-modulated RF and microwave-frequency optical systems. New lightwave components have been developed to meet the performance requirements of these systems. Most notable among these components are high-power single-frequency lasers such as distributed-feedback (DFB) semiconductor lasers, optical intensity modulators, erbium-doped fiber amplifiers (EDFAs), and broad bandwidth p-i-n photodetectors. Modulation frequency domain characterization of these lightwave components, as well as the systems they go into, is required to achieve the desired performance requirements of transmission bandwidth, signal fidelity, and signal-to-noise ratio (SNR).

This chapter starts with an explanation of intensity modulation, how it is used in optical transmission systems, and how it can be characterized. It is important to have a clear understanding of the differences between measurements on the optical carrier signal (as studied in Chapters 3 and 4) and measurements involving the modulation on the optical carrier. Section 7.2 discusses the measurement of the modulation transfer function for components such as DFB lasers and p-i-n photodetectors. Modulation transfer function measurements are made using instrumentation based on an electrical vector network analyzer in conjunction with calibrated optical to electrical (O/E) and electrical to optical (E/O) converters. Section 7.3 describes measurements made on the modulated optical signal. Parameters such as optical modulation index and distortion can be made with instrumentation consisting of a calibrated O/E converter in conjunction with an electrical spectrum analyzer. Section 7.4 deals with the measurement of intensity noise on optical signals. Finally, Section 7.5 reviews the topic of modulation domain calibration techniques. All of the measurements discussed in this chapter require accurate characterization of reference optical transmitters and reference optical receivers, at times in conjunction with the appropriate measurement instrumentation. It is the accuracy of these calibrations that ultimately determines the quality of the measurements described in this chapter.

7.1.1   Simplified Transmission Systems

A basic intensity-modulated transmission system is shown in simplified form in Figure 7.1. Information is transmitted by varying the power of the optical carrier by means of an electrical drive signal. This simplified system can either model a digital or analog lightwave transmission system depending on the type of electrical modulation applied to the optical carrier, and the characteristics of the electrical and optical components used. In a digital lightwave system, a binary electrical-data waveform is used to modulate the optical carrier. In these types of systems, described in detail in Chapter 8, the bit-error ratio as well as the characteristics of the modulated optical waveform are important. In an analog lightwave system, an electrical signal is used to linearly modulate the optical carrier. This is typical of routing radio frequency (RF) or microwave signals using fiber optics such as in the case of satellite antenna remoting, radars, or in various shipboard or avionics systems. Analog fiber-optic links are very attractive because of their light weight and low loss over long distances. Because fiber is a dielectric, these optical links are free from the effects of electromagnetic interference (EMI). Many electrical signals can be multiplexed simultaneously, as in the distribution of cable television (CATV) signals using fiber optics, where 60 or more analog-modulated, cable-television-frequency subcarriers are placed on the optical carrier.

7.1.2   Lightwave Transmission Components

Measurement examples will be given throughout this chapter to illustrate some of the measurement concepts. Three lightwave components will be used as devices under test for these measurements: a directly modulated DFB laser, a Mach-Zehnder optical modu-

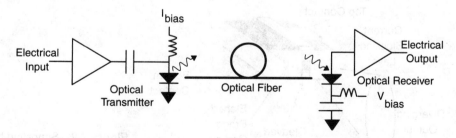

Figure 7.1   Simplified intensity-modulated fiber-optic transmission link.

lator, and a p-i-n photodetector. This section will describe these three components in some detail to make the measurement examples more informative.

Direct and External Modulation.   There are two basic approaches to modulating the laser transmitter in intensity-modulated transmission systems. The first technique is to directly modulate the bias current to the laser which causes a corresponding change in the optical output power of the laser. The second approach is to operate the laser at a constant optical output power and externally modulate the optical signal. Mach-Zehnder modulators fabricated in either lithium niobate or gallium arsenide are often used for this purpose. In many lightwave systems, the modulation bandwidth of the laser or the external modulator will limit the maximum transmitted data rate or modulation frequency.

DFB Semiconductor Laser.   A common type of laser deployed in direct modulated systems is the distributed feedback (DFB) semiconductor laser as shown in Figure 7.2. This laser uses a double-heterostructure design, in which the active region is placed between layers of larger bandgap material. This bandgap difference confines the electrons and holes to a narrow region and provides the high carrier density necessary for laser gain. The InGaAsP/InP material system is used for semiconductor lasers designed to lase at a wavelength of 1300 nm or 1550 nm. The narrower band gap InGaAsP material has a higher refractive index than InP, forming an optical waveguide for the laser in the plane perpendicular to the wafer surface. The optical waveguide for the plane parallel to the wafer surface is provided in this case by a regrown InP layer. This regrown InP layer is often made semi-insulating to increase lateral current confinement and to reduce parasitic capacitance, which can limit the laser's modulation bandwidth.

A periodic Bragg grating is etched near the InGaAsP guiding region to provide frequency selective feedback to allow oscillation at one dominant mode in the laser cavity. When the bias current to the laser is modulated above the lasing threshold, a proportional modulation of the output optical power is induced.

Mach-Zehnder Modulator.   Externally modulated laser transmitters are often deployed where high power, wavelength stability, and a large modulation bandwidth are required. An optical intensity modulator based on the Mach-Zehnder interferometer configuration is shown in Figure 7.3. An optical waveguide is formed in LiNbO$_3$ by diffusing a

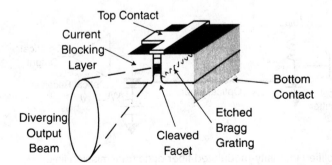

Figure 7.2   Simplified buried-heterostructure DFB laser structure.

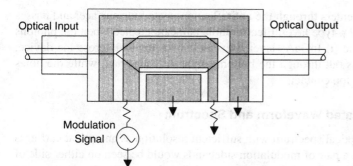

Optical Input

Optical Output

Modulation
Signal

Figure 7.3   Integrated Ti:LiNbO$_3$ Mach-Zehnder interferometer optical modulator.

strip of titanium into the wafer at an elevated temperature. Electrodes are deposited onto the substrate to form a coplanar microwave waveguide. The input light is split in a Y branch. The two parallel optical waveguide arms form two phase modulators, which operate in a push-pull fashion. The phase modulation is made possible by the electro-optic effect in LiNbO$_3$ in which the velocity of light is proportional to the applied electric field. The light from the upper and lower waveguide arms are recombined in the output Y branch. If the two optical waves arrive in phase, the light will be guided on the output waveguide. If the two signals arrive out of phase, then the optical wave will not be guided and will slowly be dispersed into the substrate. Thus, the Mach-Zehnder modulator can produce a transmission loss that is dependent on the electrical modulation signal. This transmission loss through the device is polarization dependent, so the polarization of the optical signal at the input of the device must be well controlled.

p-i-n Photodetector.   Receivers deployed in many high-performance lightwave transmission systems employ p-i-n photodetectors because of their broad detection bandwidth, excellent linearity, and temperature stability. A cross-section of a mesa p-i-n photodetector is shown in Figure 7.4. InP/InGaAs/InP heterojunction p-i-n photodetectors operate by absorbing infrared light in the 1.0 to 1.6 μm wavelength range and converting

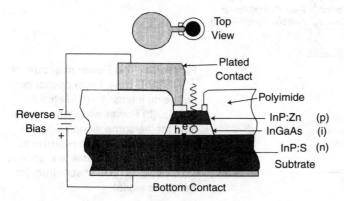

Top
View

Plated
Contact

Polyimide

Reverse
Bias

InP:Zn   (p)

InGaAs   (i)

InP:S   (n)

Subtrate

Bottom Contact

Figure 7.4   Mesa p-i-n photodetector cross-section and top view.

it to a photocurrent. Light enters through the antireflection-coated top surface, and passes through the transparent InP p-type layer. Electron/hole pairs are created when the photons are absorbed in the intrinsic (i) InGaAs layer. Reverse bias is applied across the device, which sweeps the electrons out through the bottom n-type InP substrate, while the holes are collected by the p-type top contact.

7.1.3 Intensity-Modulated Waveform and Spectrum

If one could observe the optical spectrum with sufficient resolution as an optical carrier is being intensity modulated, a pair of modulation sidebands would be seen on either side of the optical carrier, offset from the carrier by the modulation frequency. This is shown in Figure 7.5a for the case of an optical carrier at 193 THz being modulated at 1 GHz. Unfortunately, the resolution of commercial diffraction-grating based OSA is approximately 0.1 nm, which roughly corresponds to 12 GHz at a carrier wavelength of 1550 nm. This makes observation of modulation sidebands difficult. However, heterodyne signal analysis techniques described in Chapter 5 could make this measurement. It is impossible to directly observe an optical carrier waveform being modulated in the time domain, as such measurement instrumentation does not exist. In principle, it would look like the waveform

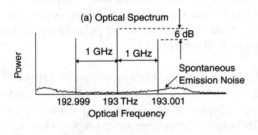

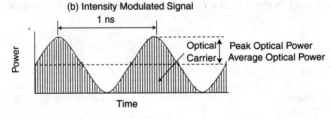

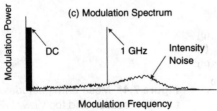

Figure 7.5 (a) Power spectrum of a 193 THz (1554.4 nm) optical carrier being intensity modulated at 1 GHz. (b) Power waveform versus time for the same signal. Both the power envelope and a representation of the optical carrier are shown. (c) Modulation power spectrum for the same signal.

in Figure 7.5b. The vertical lines under the modulation envelope are only symbolic of the optical carrier in this illustration. A laser with a center frequency of 193 THz being modulated at 1 GHz would have 193,000 optical cycles for each 1 ns modulation period.

Because the optical carrier cannot be observed directly in the time domain, nor the modulated optical spectrum be easily resolved, direct detection of the modulation envelope is used to make measurements. A photodetector such as a p-i-n photodiode is used to convert photons of light into an output current. The resulting signal consists of both a modulated component and a time-averaged or dc component of the lightwave signal. This process is essentially baseband demodulation of the optical signal. Assuming the photodetection process is a linear one, the spectrum or waveform of the modulation on the optical carrier can now be measured using conventional electronic measurement instrumentation such as oscilloscopes and RF spectrum analyzers. Notice also in Figure 7.5b, the lightwave waveform is truly being intensity modulated, as there will always be some optical power at the optical carrier frequency which in direct detection corresponds to the time-averaged optical power.

The example in Figure 7.5 corresponds to sinusoidal amplitude modulation with a 100%-modulation depth. In Figure 7.5a, the optical power in each modulated sideband is one-half that of the optical carrier. In Figure 7.5b, the peak modulated power is equal in amplitude to the average optical power. Figure 7.5c, shows the direct-detected optical modulation spectrum, where both amplitude-modulated sidebands fall on one another and sum to a peak-modulated optical power equal to the optical power in the optical carrier, which appears as the dc component.

It is also possible to frequency modulate an optical carrier although this is not typically done. However, frequency modulation or chirping of a semiconductor laser is a common artifact of direct current modulation. This amplitude-phase coupling is often referred as to the alpha- or linewidth-enhancement factor. At low modulation frequencies, this effect is dominated by thermal heating inside the laser cavity changing the effective cavity length and thus the lasing frequency. At higher modulation frequencies, the optical index of the laser cavity varies due to its dependence on the carrier density. This changes the optical transit time in the cavity, causing the lasing frequency to fluctuate.[1] This chirping or spectral broadening of the optical carrier is an important consideration when transmitting signals over long distances where chromatic dispersion in the fiber can cause the signal to become distorted. Chirp and its measurement is discussed more fully in Chapter 5.

7.1.4 Modulation-Frequency-Domain Measurements

In order to optimize the performance of any high-speed transmission system, a number of modulation-frequency-domain measurements must be made on the components and devices that are integrated into the system. Generally, these frequency domain measurements can be grouped into three general categories:

- Modulation transfer function (Section 7.2)
- Modulation signal analysis (Section 7.3)
- Intensity noise characterization (Section 7.4)

Modulation transfer function measurements are necessary to determine if the optical components have the required modulation bandwidth, with appropriate amplitude and phase characteristics, to allow adequate transmission and reception. Modulation signal analysis is useful to analyze the modulation spectrum, to determine the linearity of the modulation function, or conversely to measure any distortion. Maintaining a minimum SNR is a requirement of any transmission system. Typical noise sources include shot noise, receiver noise, and intensity noise. The intensity noise of the semiconductor laser is often the primary limiting factor in the noise performance of many analog lightwave transmission systems, making its measurement extremely important. See Appendix A for a detailed theoretical description of these various noise sources.

     Each of these measurement categories will be examined in greater detail in the following sections.

7.2 MODULATION TRANSFER FUNCTION

A fundamental requirement of any component or device in a transmission system is that it have sufficient modulation bandwidth to allow the transmission and reception of the intended information. For digital systems using a non-return-to-zero (NRZ) binary format, the modulation bandwidth needs to be greater than one-half the bit rate. For example, at least 500 MHz of bandwidth is required for a 1 Gbit/s NRZ transmission rate. In addition, the phase response over this bandwidth should vary linearly with modulation frequency in order to avoid the effects of intersymbol interference (ISI). For a return-to-zero (RZ) binary format, the modulation bandwidth needs to be at least greater than the bit rate. In analog systems, such as those used in fiber-optic CATV transmission, a modulation bandwidth approaching 1 GHz is required to carry the multiple television subcarriers.

     This modulation transfer function of the component or system as a function of the radian frequency, w, can be expressed simply as

$$H(jw) = \frac{O(jw)}{I(jw)} \qquad (7.1)$$

where $H(jw)$ is the modulation transfer function, $O(jw)$ is the output response, and $I(jw)$ is the input stimulus.

     The most common technique to measure the modulation transfer function is to use a lightwave component analyzer. The complex frequency notation, jw, is used to indicate that both the magnitude and the phase function of the transfer function are being measured. Figure 7.6 illustrates the modulation transfer functions that will be examined in this section. A brief description of each is given here:

- E/O measurement: A small signal electrical current is applied to an optical source such as a laser and the resulting output optical-modulation power is measured.
- O/E measurement: A small modulation-depth optical signal is applied to a receiver such as a photodetector and the resulting output electrical current is measured.

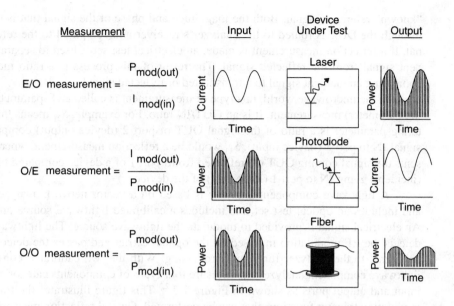

Figure 7.6 Modulation transfer functions for optical components. (a) E/O measurements. (b) O/E measurements. (c) O/O measurements.

- O/O measurement: A small modulation-depth optical signal is applied to a two-port optical device such as optical fiber and the resulting output optical modulation is measured.
- E/E measurement: The modulation response of the entire link of Figure 7.1 from the electrical input to the electrical output can also be characterized. This measurement is not illustrated in Figure 7.6. Since it is strictly an electrical measurement, it will not be further discussed in this section.

7.2.1 Lightwave Component Analyzer

A lightwave component analyzer is used primarily to measure the small-signal linear transmission and reflection characteristics of a lightwave component or system as a function of modulation frequency. It is a measurement system that operates by injecting a modulated signal into a test device and comparing the modulated input signal to the signal that is transmitted or reflected by the test device. This comparison of the transmitted signal to the incident signal results in a ratio measurement. The concept of making ratio measurements to test the response of electrical devices and systems has traditionally been used in the RF and microwave industries. The instrument used to make these types of transmission measurements for high-frequency electrical networks is called a vector-network analyzer.

Electrical vector-network analyzers use a swept-RF frequency source as a stimulus signal to the device under test (DUT). This incident sinewave signal is also used as a "known" reference signal. Both the magnitude and phase of the signal that is transmitted through the DUT is routed to the analyzer's receiver and compared to the reference signal. If a reflection measurement is made, an electrical test set is used to separate the incident signal from the reflected signal. The result of this process is a ratio measurement comparing an incident signal to a transmitted or reflected signal.

In the microwave world, this type of measurement is called an S-parameter (scattering parameter) measurement. It is an OUT/IN ratio. For example, S_{21} means that the scattering parameter is a ratio of the signal OUT of port 2 (device output) compared to the signal IN to port 1 (device input). S_{11} would be a reflection measurement, where 11 represents the signal coming OUT of (reflected from) port 1 of a device compared to the signal (incident) going IN to port 1 (same port) of the device.

A lightwave component analyzer is based on a vector network analyzer platform and includes an optical test set that includes a calibrated lightwave source and receiver. An electrical signal is provided to modulate the lightwave source. The lightwave receiver detects the RF modulation imposed on the optical carrier and passes the detected electrical signal to the analyzer for signal processing. With the two calibrated converters, the lightwave component analyzer can measure four types of components categorized by their input and output ports as shown in Figure 7.7.[2,3] This figure illustrates the transmission-modulation transfer functions that can be performed. Optical reflection measurements can also be made with the addition of optical directional couplers.

7.2.2 E/O Transfer Function Measurements

The modulation transfer function for an E/O transducer such as a semiconductor laser can be measured in the following manner. The lightwave component analyzer measures the input modulating current to the laser as well as the laser's output optical modulation power. The lightwave component analyzer then displays the ratio of the two measurements in watts/amp. Although responsivity is often used to describe a static or dc parame-

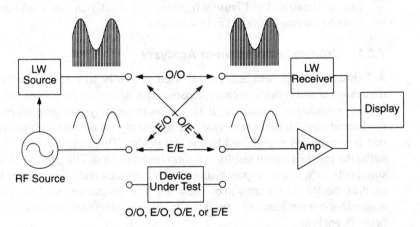

Figure 7.7 Lightwave component analyzer block diagram and measurement modes.

ter, the conversion efficiency of an optical source is a dynamic characteristic and can be referred to as "slope responsivity," R_s. (For lasers this is often referred to as the laser efficiency.) Slope responsivity is used to describe how a change in input current produces a change in optical power. Graphically, this is shown in Figure 7.8. As one may expect, the slope responsivity is a function of the frequency of the modulating current. As the frequency of modulation increases, the conversion efficiency will eventually degrade or "roll off." The frequency at which the conversion efficiency drops to one-half of the maximum value is the "−3 dB point" when displayed logarithmically, and indicates a laser's modulation bandwidth.

Example: DFB-Semiconductor-Laser Intensity-Modulation Response

The frequency response of a laser is dependent on biasing conditions. As the dc bias current of the laser is increased, the bandwidth generally increases. This is caused by a shift in the "relaxation oscillation" characteristic, which is a natural resonance frequency of the laser. This relaxation oscillation phenomenon creates an observable resonance in the intensity modulation response of the laser.

Measurement Procedure

This is basically an E/O transfer measurement well-suited for the lightwave component analyzer. First, an initial calibration is required to perform an accurate measurement. This calibration allows the lightwave component analyzer to remove the response of the test system, including the electrical cables, optical fiber, and the instrument itself. Prior to the calibration of the instrument, the following instrument settings need to be made:

- start and stop frequencies,
- sweep type (linear or logarithmic),
- number of measurement points,
- measurement sweep time,
- source power level.

To perform a simple frequency-response calibration, the connections shown in Figure 7.9 must be made. The analyzer measures the appropriate paths so the frequency and phase response of the "unknown" paths are characterized. The analyzer then uses this information in conjunction with the internal calibration data to generate an error matrix. The lightwave source and receiver characteristics are predetermined during a factory calibration and stored

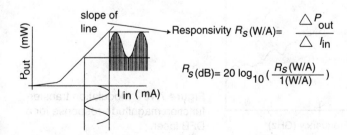

$$R_s(\text{W/A}) = \frac{\triangle P_{out}}{\triangle I_{in}}$$

$$R_s(\text{dB}) = 20 \log_{10}\left(\frac{R_s(\text{W/A})}{1(\text{W/A})}\right)$$

Figure 7.8   E/O slope responsivity.

Electrical Vector Network Analyzer

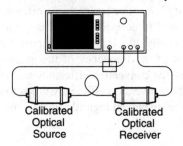

Calibrated
Optical
Source

Calibrated
Optical
Receiver

Figure 7.9 E/O calibration configuration.

in memory. The accuracy of this measurement is dependent on accurate calibration of the magnitude and phase response of the lightwave source and receiver used with the vector network analyzer. Section 7.5 discusses how this calibration may be performed. Once the setup and calibration have been completed, the laser under test is connected between the electrical and optical measurement interface connections, substituting for the calibrated lightwave source.

Figure 7.10 shows a measurement of the conversion efficiency of a DFB semiconductor laser as a function of modulation frequency. The horizontal axis in this case covers the frequency range from 130 MHz to 10 GHz. The vertical axis is in units of watts/amp, displayed logarithmically where O dB represents 1 W/A. This particular laser has a −3 dB modulation bandwidth of approximately 3.5 GHz. By varying the external bias supply to the laser, one can make a number of measurements at different bias conditions to determine the optimum bias for maximum frequency response. Typically, as the bias is increased, both the responsivity and bandwidth will increase until the bias reaches a certain point where the high-end frequency response begins to degrade.

Figure 7.11 shows the measured phase response of the laser. A laser's modulation response would ideally exhibit a linear phase response versus modulation frequency. The derivative of the phase response with respect to modulation frequency is referred to as the group delay. A linear phase response is equivalent to a constant group delay. If the group delay is not constant with the modulation frequency, the modulated waveform will be dis-

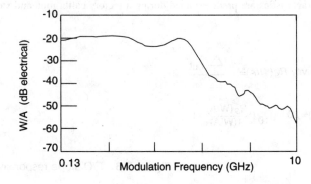

Figure 7.10 Modulation transfer-
function magnitude-response for a
DFB laser.

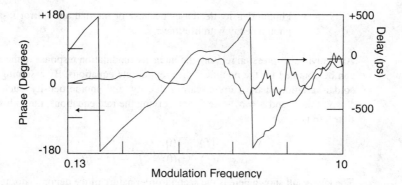

Figure 7.11 Modulation transfer-function phase-response for the DFB laser of Figure 7.10.

torted. The frequency range where the phase response begins to deviate from linear often occurs near the frequency where the magnitude of the modulation response begins to roll off.

Measurement Interpretation

The measurement of a semiconductor laser's modulation response conveys a significant amount of information. There are several factors that limit the modulation bandwidth of semiconductor lasers. The measured modulation response can be compared to the predicted theoretical response based on semiconductor laser dynamics. An analysis utilizing the rate equations can be used to determine the intrinsic intensity modulation response. The rate equations comprise a set of coupled equations that describe the relationship between the laser's bias current, I, electron density, N, and photon density, S. A simplified version of these equations for single longitudinal mode operation is shown:

$$\frac{dN}{dt} = -g(N - N_t)S + \frac{I}{qV} - \frac{N}{\tau_n} \tag{7.2}$$

$$\frac{dS}{dt} = \Gamma g(N - N_t)S - \frac{S}{\tau_p} + \frac{\Gamma \beta N}{\tau_n} \tag{7.3}$$

where N_t is the carrier density for transparency, g is the differential gain, q is the electron charge, V is the volume of the active region, τ_n is the photon lifetime, Γ is the optical confinement factor (which is the fraction of the optical power in the active region), and β is the fraction of spontaneous emission coupled into the active region.[4]

Analyzing the rate equations illustrates four points:

- Electron density increases because of the injection of a current I, into a volume V.
- Electron density decreases because of both spontaneous emission with a lifetime τ_n and stimulated emission, whose rate is dependent on the differential gain, g.
- Photon density increases because of stimulated and spontaneous emissions confined inside the active layer.

- Photon density decreases because of internal and mirror losses, which cause a photon decay with lifetime τ_p.

An analytical expression relating the intensity modulation response to the modulating current can be obtained for the nonlinear differential rate equations. By assuming small signal sinusoidal variation of the current, carrier density, and photon density; such that $I = I_0 + ie^{jwt}$, $S = S_0 + se^{jwt}$, and $N = N_0 + ne^{jwt}$, and solving the rate equations, the following transfer function results:

$$\frac{s(f)}{i(f)} = \frac{s(0)}{i(0)} \times \frac{f_0^2}{f_0^2 - f^2 + jff_d} \tag{7.4}$$

The key result shown here is the second-order nature of the denominator term, which predicts a damped resonance in the laser's modulation frequency response. Two important terms, f_0 and f_d, appear in the equation. f_0 results from a resonance between the electrons and photons in the active region and is related to the geometric mean of the stimulated electron lifetime and photon lifetime.[5,6] f_d is the damping frequency and is dependent on the photon lifetime and ϵ, which characterizes several nonlinear mechanisms in the differential gain. The relaxation oscillation frequency, f_p, can be calculated from f_0 and f_d as shown:

$$f_0 = \frac{1}{2\pi\sqrt{\tau_n\tau_p}} \tag{7.5}$$

$$f_d = \frac{\epsilon S}{2\pi\tau_p} \tag{7.6}$$

$$f_p^2 = f_0^2 - \frac{f_d^2}{4} \tag{7.7}$$

Theory predicts that the intensity modulation response of semiconductor lasers is peaked at f_p, the relaxation oscillation frequency.[7] The amount of peaking or, conversely, damping is a strong function of ϵ. Beyond the relaxation oscillation frequency, a roll off of 12 dB per octave increase in modulation frequency is predicted. In addition, theory predicts that the relaxation oscillation frequency increases with the square root of the bias current, and that the peaking in the response becomes more damped until it is critically damped.

Unfortunately, achieving a high relaxation oscillation frequency does not guarantee a wide modulation bandwidth. Both device structure and package parasitics can dramatically alter the drive current passing through the active region. Parasitics requiring consideration include the bond wire inductance, the bonding pad capacitance, any device capacitance, and contact resistance.

Example: External Modulator Intensity Modulation Response

The optical transmission response of a Mach-Zehnder modulator is a function of the applied dc bias voltage as shown in Figure 7.12. The transfer curve highlights four parameters:

1. The insertion loss, which is the optical loss at the maximum transmission point of the curve.

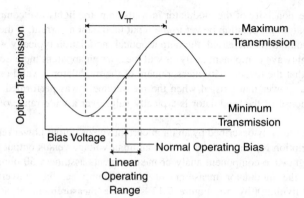

Figure 7.12 Mach-Zehnder interferometer transmission transfer curve versus bias.

2. The switching voltage, V_π, which is the difference in bias voltages at the minimum and maximum transmission points.
3. The extinction ratio, which is the ratio of the maximum to the minimum optical transmission levels.
4. The nominal operating point, which is the bias voltage that results in optical transmission halfway between the minimum and maximum transmission levels.

The modulator response is linear for small deviations from the nominal operating bias point. When a sinusoid with radian frequency w is applied, the transmitted light power $I(t)$ can be written as:

$$I(t) = \frac{I_0}{2}\left[1 + \cos\left(\pi\,\frac{V_b + E(w)V_p\cos(wt)}{V_\pi}\right)\right] \tag{7.8}$$

where I_0 is the maximum transmitted light power, $E(w)$ is the frequency-dependent modulation efficiency, V_p is the peak signal voltage applied to the modulator, and V_π is the modulator's characteristic switching voltage. The operating point on the modulator's transfer curve can be adjusted by varying the dc bias, V_b.[8]

Measurement Procedure

The modulation response of external intensity modulators can be characterized using a lightwave component analyzer in much the same way as laser sources. This is another class of E/O measurements where the stimulus is a swept-frequency electrical signal and the response out of the modulator is intensity-modulated light. The modulation bandwidth and phase measurements of external intensity modulators are made with the lightwave component analyzer in the same configuration that is used for laser measurements. However, a significant difference exists in this measurement because the modulator is a three-port device. While the frequency response of a modulator is often independent of input optical power, the responsivity is not. The conversion efficiency of the modulator is not only a function of the electrical input, but also the optical input level.

To compute the responsivity of the modulator in watts/amp, the lightwave component analyzer compares the output modulation power to the input modulation current. If the input optical power to the modulator is increased, the output optical modulation typically will increase also. Thus, the lightwave component analyzer will measure an apparent increase in responsivity. This means that the responsivity measurement of the modulator is valid only for the specific optical input power that existed when the measurement was performed. However, the frequency response of the modulator is typically valid over a wide range of input optical powers.

Though lasers are typically described by an input current versus output power relationship, the preferred description for a modulator is often an input voltage versus output power relationship. Because lightwave component analyzer measurements assume a 50 ohm measurement environment, the modulator measurement in watts/amp can be converted to watts/volt by scaling (dividing) by 50. Figure 7.13 shows a measurement of a wide-bandwidth LiNbO$_3$ Mach-Zehnder external modulator. The unusual response in the low frequency range is caused by electrical impedance matching circuitry in the modulator. This is caused by the high dielectric constant of LiNbO$_3$ which results in a low characteristic impedance of 22 ohms for the coplanar microwave transmission line.

Measurement Interpretation

Several factors limit the modulation efficiency of optical-intensity modulators at high-modulation frequencies. The main limitation is a mismatch of the velocities of the copropagating optical and microwave traveling waves as shown in Figure 7.14. The optical signal travels about twice as fast as the modulating microwave signal. If the difference in time of propagation through the modulator is great enough, the microwave signal will get out of phase with the modulation impressed on the optical signal and begin to reverse the effect. Velocity mismatch can be reduced and broad bandwidth operation achieved by keeping the device length short. Microwave losses in the coplanar electrodes also contribute to a reduced modulation efficiency at high frequencies. If the electrodes are sufficiently thick, the loss will be mainly due to the skin effect and will increase as the square root of frequency.

7.2.3 O/E Transfer Function Measurements

The measurement process for O/E devices, such as lightwave receivers and photodiodes, is similar to that for E/O devices. The slope responsivity, R_r, for these O/E devices describes how a change in input optical-modulation power produces a change in output

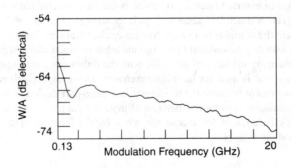

Figure 7.13 Measurement of a Mach-Zehnder modulator modulation bandwidth.

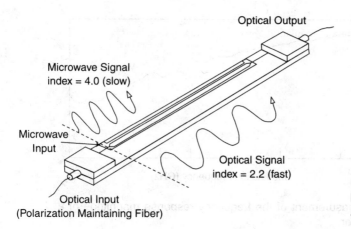

Figure 7.14 Velocity mismatch between the microwave modulation and the optical carrier.

electrical-modulation current. Graphically this is shown in Figure 7.15. The lightwave component analyzer measures the input optical-modulation power and output modulation current. It then displays the ratio of the two in amps/watt.

Example: p-i-n Photodetector Frequency Response

The frequency response of a p-i-n photodetector is determined by a combination of the RC time constant of the photodiode and the transit time for the generated electrical carriers to travel through the i-region. Both a low capacitance and a short transit time are desirable. However, these two parameters are interconnected. A thin i-region can result in a short transit time, but it also increases the capacitance. Therefore, both parameters must be optimized in relation to one another in the design of high-speed p-i-n photodetectors.[9]

Measurement Procedure

The measurement process for a p-i-n photodetector is virtually identical to the laser measurement. The configuration and user calibration of the lightwave-component analyzer is the

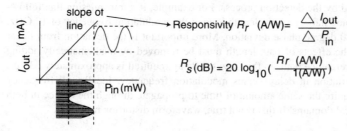

$$Responsivity\ R_r\ (A/W)= \frac{\triangle\ I_{out}}{\triangle\ P_{in}}$$

$$R_s(dB) = 20\ \log_{10}(\frac{R_r\ (A/W)}{1(A/W)})$$

Figure 7.15   O/E slope responsivity.

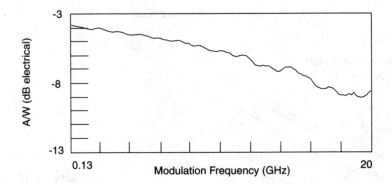

Figure 7.16 Measurement of the frequency response magnitude of a p-i-n photodetector.

same. Once the setup and calibration have been completed, the p-i-n photodetector is placed in the measurement path.

Figure 7.16 shows a measurement of the conversion efficiency of a high-speed p-i-n photodetector as a function of modulation frequency. The vertical axis display units are in amps/watt and the horizontal axis is the modulation frequency. In this case, the vertical axis is in a logarithmic format where 0 db represents 1 A/W. This photodiode under test has a modulation bandwidth of approximately 15 GHz.

Measurement Interpretation

The frequency response of lightwave receivers can show some distinct resonances that will impact the time-domain performance. There can be several reasons for this behavior. Once the photodiode has converted the modulated light to a proportional electrical current, the next task is to efficiently transfer the demodulated signal to a following electrical circuit such as a preamplifier. High-speed systems usually require this transfer over 50 or 75 ohm transmission lines. However the output impedance of a photodiode is usually much higher than 50 or 75 ohms. This can result in signals being reflected and a degradation in conversion efficiency if the signal transmitted from the photodiode encounters another impedance mismatch along the transmission path. Reflections can occur resulting in "mismatch loss" and ripple in the measured frequency response.

The phase response of a lightwave receiver is also an important parameter. If the phase response of the receiver is not linear, the detected waveform will be distorted. Figure 7.17 shows the phase response of a p-i-n photodetector assembly. When components have significant length in either the optical or electrical paths, the relative phase will have a large variation. This is not caused by the detection process. For example, if a transmission line following the photodiode is one-half of a wavelength long at the modulation frequency of 10 GHz, this will result in a 180 degree phase deviation. More important is the deviation from linear phase. To view this, the effects of path length must be removed by mathematically adding delay to the analyzer reference path. In this case the delay required is approximately 630 ps. Also shown is the variation in delay versus modulation frequency. Ideally, all frequency components would require the same amount of time to propagate through the device in both the optical and electrical domains. If this is not true, waveform distortion results.

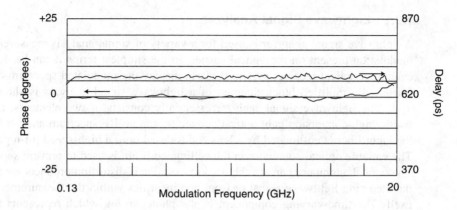

Figure 7.17   Measurement of the phase response of a p-i-n photodetector.

7.2.4   O/O Transfer Function Measurements

The modulation response for two-port optical measurements can also be made. One of the major applications of this mode is in the measurement of chromatic dispersion of optical fiber. The group delay of the fiber is calculated from the measured phase response. If the optical frequency for the measurement is adjusted and the group delay recalculated, the dispersion of the optical fiber can be determined. Chapter 12 covers this measurement area in detail.

7.3   MODULATION SIGNAL ANALYSIS

As lightwave transmission systems and their components have evolved, designers have become interested in the measurement and characterization of the optical modulation depth, signal strength, distortion, and intensity noise of optical signals. Optical modulation depth and distortion are very important in analog systems employing multiple microwave subcarriers such as CATV optical-distribution systems and microwave links for antenna remoting. These CATV systems typically employ high-power DFB lasers or linearized external modulators. They can transmit 60 or more subcarriers spaced in a bandwidth of up to 750 MHz or higher. In antenna-remoting applications, the microwave subcarrier frequencies modulating the optical carrier can reach 15 GHz or more. A recent development is fiber optic links designed for expanding coverage in cellular networks to microcells. These employ subcarriers in the 800 MHz to 1 GHz frequency range.

The most common technique to make these measurements utilizes a lightwave signal analyzer.

7.3.1 Lightwave Signal Analyzer

A lightwave signal analyzer is used for a variety of signal analysis measurements on the modulation present on the optical carrier. In the simplest terms it can be described as a calibrated optical receiver system that can display the modulated spectrum on an optical carrier. A simplified block diagram of a lightwave signal analyzer is shown in Figure 7.18. The lightwave signal analyzer essentially consists of two elements: a wide-bandwidth, high-conversion gain optical converter, and an RF spectrum analyzer. Light from the input fiber is collimated by a lens and focused onto a high-speed p-i-n photodetector. The variable optical attenuator in the collimated beam is used to prevent signal overload. The optical attenuator can also be used to examine if distortion products are generated by the incoming lightwave signal or from nonlinearities within the measurement instrument itself. The time-varying component of the photocurrent, which represents the demodulated signal, is fed through a preamplifier and then to the input of the RF spectrum analyzer. The optical converter in front of the RF spectrum analyzer must have high-conversion gain to overcome the relatively high-noise figure of the RF spectrum analyzer. RF spectrum analyzers are swept-tuned heterodyne electrical receivers employing RF mixers to down-convert the electrical signal to an intermediate frequency (IF) so it can be further processed and detected. The DC portion of the photocurrent is fed to a power meter circuit to allow the display of the average optical power. A major contribution of a lightwave signal analyzer is its optical calibration. It is calibrated in both absolute and relative optical power.[10] Various calibration techniques are described in Section 7.5

Electrical dB and Optical dB Displays.   In this section, the lightwave signal analyzer examples will be displayed in either electrical dB or optical dB units. Since the photodetector is a power-to-current converter, a 1 dB change in optical power corresponds to a 2 dB change in electrical power, thus the dB optical and dB electrical terminology. For some measurements such as modulation-depth characterization, it is useful to use dB optical displays on the lightwave signal analyzer. This allows an easy comparison of the average optical power to the modulated optical power. In other applications such as relative intensity noise characterization, it is more useful to have a display in dB electrical units.

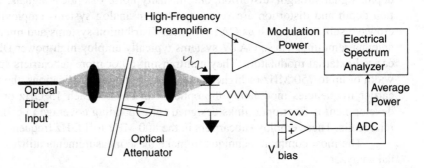

Figure 7.18   Block diagram of lightwave signal analyzer.

The lightwave signal analyzer is often confused with an OSA which is discussed in Chapter 3. Although both instruments have frequency-domain displays, the information they provide is quite different. The OSA shows the spectral distribution of average optical power as a function of optical wavelength. It is suitable for observing the modes of a multimode laser or the sidelobe rejection of a single-frequency laser. The measurement resolution is typically about 0.1 nm or approximately 18 GHz at a wavelength of 1300 nm. The lightwave signal analyzer displays the total average optical power as well as the power of the intensity modulation on the optical carrier. It does not provide any information about the wavelength of the optical signal. This distinction was shown in Figures 7.5a and 7.5c.

7.3.2 Intensity Modulation and Modulation Depth

The measurement of intensity modulation on an optical carrier requires some type of analyzer. If there is only one modulating signal such as in a digital transmission system, the intensity modulation can be measured on a digital communications analyzer, which is based on an oscilloscope platform, as described in Chapter 8. However, if there are multiple modulating signals, it is much more convenient to use a lightwave signal analyzer. Figure 7.19 shows multiple modulating signals that are present on an optical carrier. The horizontal axis is the modulation frequency, indicating the measurement is being made over a wide range of frequencies. The vertical axis is displayed logarithmically in units of dBm where 0 dBm is equal to 1 mW. The amplitude and frequency of each modulating signal can be easily measured using built-in marker functions common in lightwave signal analyzers and spectrum analyzers.

An important parameter in optical CATV and other analog-modulated optical transmission systems is the optical modulation depth (OMD). Sometimes it is referred to as the optical modulation index (OMI). They represent the same ratio which is defined as:

$$OMD = \frac{P(f_{\mathrm{mod}})}{P_{\mathrm{AVG}}} \qquad (7.9)$$

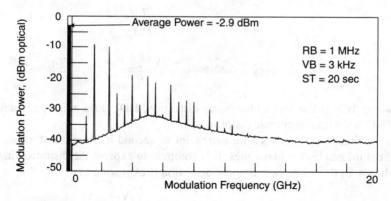

Figure 7.19 Lightwave signal analyzer display of multiple-modulating signals.

where OMD is the optical modulation depth, $P(f_{mod})$ is the peak optical modulated signal power as shown in Figure 7.5b, and P_{AVG} is the average optical power.

In many CATV systems, the OMD of each modulated subcarrier is only a couple of percent. This is because there are typically many subcarriers (TV channels) being modulated simultaneously on the optical carrier, and the cumulative modulation depth is much higher. If the modulation depth becomes too large, signal distortion or clipping can result. This measurement can be made on the lightwave signal analyzer using a specific marker function that determines the ratio of the modulated signal power at a given frequency to the average optical power.

7.3.3 Distortion

The goal of any transmission system is to transfer information as accurately as possible. To the first order, both lasers and p-i-n photodetectors are linear devices. A laser translates an electrical current into a proportional optical power. A p-i-n photodetector translates optical power into a proportional electrical current. This is equivalent to saying their slope responsivities, which were discussed in Section 7.2, are close to being linear. However, neither device is perfectly linear, particularly when large modulation levels are involved. Several different types of distortion products are common from these components such as harmonic distortion and intermodulation distortion.

When a device such as a laser is intensity modulated there will be modulated signal power at the modulating frequency; and depending on the linearity of the device, there will also be some modulating power at harmonics of the modulating frequency. Some modulated power at the second harmonic and third harmonic of the modulating frequency is very common. Harmonic distortion is defined as the ratio of modulated power in a harmonic of the modulating frequency to the power at the modulating frequency. For example:

$$HD_2 = \frac{P(2f_{mod})}{P(f_{mod})} \tag{7.10}$$

$$HD_3 = \frac{P(3f_{mod})}{P(f_{mod})} \tag{7.11}$$

$$THD = \frac{P(2f_{mod}) + P(3f_{mod}) + ... + P(nf_{mod})}{P(f_{mod})} \tag{7.12}$$

where HD_2 is the second harmonic distortion, HD_3 is the third harmonic distortion, and THD is the total harmonic distortion.

Figure 7.20 shows a measurement of second harmonic distortion. Also shown are the third and higher harmonics. It is common to express the distortion in decibels (dB) as shown. In this case, the harmonic distortion is expressed as:

$$HD_x = 10 \log \left(\frac{P(xf_{mod})}{P(f_{mod})} \right) \tag{7.13}$$

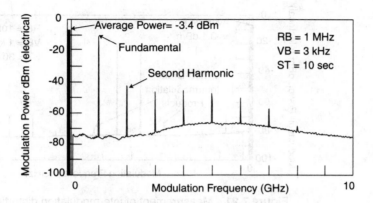

Figure 7.20   Measurement of harmonic distortion.

Intermodulation distortion (ID) occurs when two or more modulating signals are present. In this case, device nonlinearities cause the two modulating signals to interact, producing new signals at the sum and difference frequencies. Second-order intermodulation distortion for two signals can be measured as:

$$ID2(f_i \pm f_j) = \frac{P(f_i \pm f_j)}{P(f_i)} \tag{7.14}$$

where $P(f_i) = P(f_j)$.

For three modulating signals, third-order intermodulation can be expressed as:

$$ID3(f_i \pm f_j \pm f_k) = \frac{P(f_i \pm f_j \pm f_k)}{P(f_i)} \tag{7.15}$$

where $P(f_i) = P(f_j) = P(f_k)$.

A particularly distressing form of intermodulation distortion is third-order intermodulation (IMD) for two closely spaced signals. This is because the IMD signals fall close to the original modulating frequencies.

$$IMD(2f_i - f_j) = \frac{P(2f_i - f_j)}{P(f_i)} \tag{7.16}$$

Figure 7.21 shows an intermodulation distortion measurement.

Care must be taken to make accurate distortion measurements. A measurement system consisting of a lightwave signal analyzer or RF spectrum analyzer can contribute its own distortion. Usually, the instrument distortion is specified either in terms of harmonic distortion at a given input power level or in terms of a second-order intercept (SOI) or a third-order intercept (TOI). The intercept refers to the power level at which the distortion term, either second order or third order, is equal in amplitude to the power of the fundamental signal. Both SOI and TOI are extrapolated quantities used as figures of merit, since equal levels of signal power and harmonics are not likely to occur in practice. How-

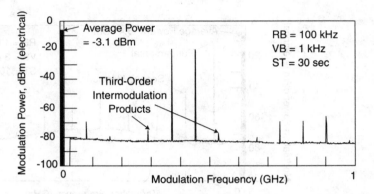

Figure 7.21 Measurement of intermodulation distortion.

ever, modest ratios of peak power to average power can cause measurement distortion by overdriving the input to the instrumentation. This is common with pulse waveforms with low-duty cycles or when a number of modulated subcarriers are present. A technique to check for instrumentation linearity is to vary the input optical attenuation value and observe if any change in the displayed spectrum shape or amplitude occurs. If an optical attenuator change does not affect the harmonic distortion level, then the displayed distortion is generated outside of the measurement instrumentation.

Distortion in Lightwave CATV Systems. As previously stated, lightwave CATV systems employ subcarrier modulation. The video information is vestigial-sideband amplitude-modulated (AM-VSB) on a TV frequency channel or carrier. These TV signals are combined and are used to modulate the optical carrier. Typically, there will be a number of second-order and third-order distortion products generated. For N TV signals or subcarriers there will be the following second-order products: N harmonics at $2 f_i$, and approximately N^2 order terms at $f_i \pm f_j$. The latter tends to dominate the second-order distortion performance of these systems. The summation of the second-order distortion terms is referred to as the composite second-order (CSO) distortion. For N-subcarriers there are the following third-order products: N harmonics at $3 f_i$, approximately N^2 two-tone intermodulation terms at $2 f_i \pm f_j$, and approximately N^3 three-tone intermodulation terms at $f_i \pm f_j \pm f_k$. The last term tends to dominate the third-order distortion performance of these systems. The summation of the third-order distortion terms is referred to as the composite triple-beat (CTB) distortion. When 60 CATV subcarriers are placed in the 50 MHz to the 450 MHz frequency band, it can be shown that the CSO distortion is worse at the low-band edge where approximately 50 terms contribute, whereas it is at a minimum in the middle of the frequency band. However, the CTB peaks in around 200 MHz where over 1100 third-order terms fall on one another.

Distortion in lightwave CATV systems can come from many sources including the laser transmitter and the receiver. DFB semiconductor lasers have several distortion mechanisms including a frequency-dependent distortion related to the relaxation oscillation,[11,12] and frequency-independent distortion caused by gain saturation[13] and spatial hole burning.[14] In addition, the frequency chirp inherent in the direct modulation of DFB lasers causes distortions due to the nonflat gain profile of fiber amplifiers and the chromatic dispersion of optical fiber. External intensity modulators have low or zero chirp, however these devices are only linear for small-signal modulation. This is because the optical intensity transfer function is sinusoidal with applied voltage as was discussed in Section 7.2.2. When the modulator is biased at the half-power point, second-order distortions are suppressed, however third-order distortions must be reduced through some type of linearization. In addition, p-i-n-based photoreceivers have linearity limitations at high power levels and operating frequencies due to internal electric field perturbations that impact the carrier velocity and generate harmonics and intermodulation products in the output current.[15,16]

7.4 INTENSITY NOISE CHARACTERIZATION

In many applications, the intensity noise spectrum is very important. First and foremost, the intensity noise spectrum affects the SNR in a transmission system. Many CATV systems employ low-noise Nd:YAG lasers with external Mach-Zehnder intensity modulators to achieve both high-power and low-noise transmission. However, much work has been done to improve both the output power and noise performance of DFB semiconductor lasers to make them an attractive lower-cost alternative in these applications. In addition, the noise spectrum can provide valuable information about a semiconductor laser's relaxation oscillation frequency.

7.4.1 Intensity Noise Measurement Techniques

The intensity noise from an optical source can be measured in a number of ways. A lightwave signal analyzer described in Section 7.3.1 is the most convenient measurement tool. Both the intensity modulation and the intensity noise are displayed simultaneously. In principle, a broad-bandwidth photoreceiver in conjunction with an electrical spectrum analyzer with the appropriate calibrations can also serve the purpose.

Alternative approaches to measure laser intensity noise would be to use a photoreceiver with a noise figure meter or electrical power meter. Noise figure meters typically offer good noise figures (6 to 8 dB) because of a built-in preamplifier, and can automatically correct for their internally generated thermal noise when a 50 ohm termination is placed at the input. However, there can be significant measurement uncertainty due to mismatch loss if the photoreceiver is not reverse terminated in 50 ohms. Electrical power meters offer good sensitivity over a very wide bandwidth. To use them effectively, a bandpass filter is required to select the noise measurement range of interest. Also, the noise power is effectively integrated over this bandwidth, which in some cases can be desirable. However the spectral characteristics of the noise are not measured.

Example: DFB Semiconductor Laser Intensity Noise

Intensity fluctuations, as well as phase fluctuations, are inherent characteristics of semiconductor laser devices. These fluctuations are due to the quantum nature of the transitions of the lasing process, which gives rise to the intensity noise. Dependent on the structural parameters of the device, the intensity noise is attributed to both the spontaneous emission and carrier-recombination processes.

Measurement Procedure

Prior to measurement, the following lightwave signal analyzer instrument settings need to be made:

- start and stop frequencies,
- resolution bandwidth,
- video bandwidth,
- reference level,
- noise marker.

The start and stop frequencies are set for the desired measurement frequency range. A wide span is useful for observing the full laser-intensity noise spectrum. The resolution bandwidth sets the effective measurement resolution and should be set to a value approximately one-hundredth of the measurement span to maintain a reasonably fast measurement sweep time. The video bandwidth which serves the same function as trace averaging is also set at a value low enough to maintain reasonable sweep time. The reference level should nominally be placed high enough to allow the display of any modulated signals. However, in the case of measuring noise, the reference level can be set to a sufficiently low level such that the internal attenuator auto-sets itself to zero attenuation. This value results in the best instrument sensitivity setting.

The intensity noise spectrum of a DFB semiconductor laser at several bias settings is shown in Figure 7.22. Typically, analyses and measurements have shown that the intensity noise increases with rising current and optical power, reaches a maximum at threshold, and then decreases above threshold.

Measurement Interpretation

The frequency dependence of the intensity noise can be analyzed by adding Langevin sources to the rate equations as described in Section 7.2.2. These Langevin terms mathematically model the spontaneous emission and the carrier shot noise. The photonic shot noise caused by the particulate nature of photons is not included in the rate equations. The small signal solution below shows that the frequency dependence of the intensity noise, $n(f)$, has exactly the same denominator term as the intrinsic modulation response (Equation 7.4):[17,18,19]

$$n(f) = \frac{f_0^2}{a} \times \frac{1 + b^2 f^2}{f_0^2 - f^2 + jff_d} \tag{7.17}$$

where a and b are terms dependent on the Langevin noise sources. Once again, the denominator term generates the damped resonance at the relaxation oscillation frequency. The intensity noise spectrum has the same general shape as the intrinsic intensity modulation response, except for the frequency-squared term in the numerator that indicates an increased noise con-

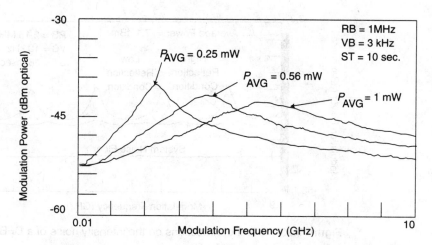

Figure 7.22   Measurement of a DFB laser intensity noise at several bias settings.

tribution at high frequencies. Theoretically, the SNR is relatively constant out to the relaxation oscillation frequency and decreases monotonically at higher frequencies. Typically, as the bias current increases, the resonance peaks of both the intensity-modulation response and intensity-noise spectrum track one another until the modulation response becomes limited by parasitics. Since the relaxation oscillation peak in the intensity-noise response is not affected by device capacitance, it is often used as an indicator of the intrinsic laser modulation response.

Example: Effect of Reflections on DFB Semiconductor Laser Intensity Noise

The laser intensity noise spectrum can be affected by both the magnitude and polarization of optical power that is reflected back to the laser. Reflections from connectors or splices can have a dramatic effect on laser intensity noise.[20]

Measurement Procedure

The measurement settings for the lightwave signal analyzer are basically the same as before. Two intensity-noise traces of the DFB semiconductor laser are shown in Figure 7.23. One is with the reflections minimized indicating the nominal intensity noise spectrum. The other trace is with reflections present showing their effect in enhancing the noise level. Markers can be used to determine the frequency separation between the noise peaks. This frequency separation, Δf, can be used to determine the distance, D, between the reflections by using the following relationship:

$$D = \frac{c}{2n\Delta f} \tag{7.18}$$

where c is the speed of light in a vacuum, 3×10^8 m/s, and n is the index of refraction in glass (approximately 1.5).

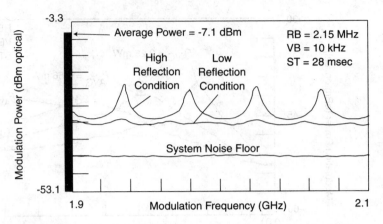

Figure 7.23   Effect of reflections on the intensity noise of a DFB laser.

Measurement Interpretation

An optical resonant cavity formed between the reflection and the back-facet mirror of the laser can enhance the noise of the laser diode. Interference between the forward transmitted wave and the reflected wave in the laser cavity offsets the gain of the laser, producing a sinusoidal variation of the noise spectrum that is dependent on the resonance path length. The reflected power upsets the dynamic equilibrium of the lasing process and typically increases the amplitude of the intensity noise particularly around the relaxation oscillation frequency. It can also induce a ripple on the noise spectrum with a frequency that is inversely proportional to the round-trip time from the laser to the reflection. To avoid this noise increase caused by reflected power, isolators are often employed. It has been reported that from 30 to 60 dB or more, isolation may be required to eliminate this effect. In addition, an increase in the intensity-noise spectrum can be observed at low frequencies in system measurements using DFB lasers. This effect has been attributed to interferometric conversion of the laser phase noise or linewidth-to-intensity noise because of reflections at connectors and splices. See Appendix A for a more detailed discussion on the phase-noise to intensity-noise conversion process.

7.4.2   Relative Intensity Noise

Relative intensity noise (RIN) is an important quantity related to SNR. RIN is defined as:[6,18,21]

$$RIN = \frac{(\Delta P)^2}{(P_{AVG})^2} \tag{7.19}$$

where $(\Delta P)^2$ is the mean square intensity-fluctuation spectral-density of an optical signal and P_{AVG} is the average optical power. A very low value of RIN is desirable. The quantity can be used to determine the maximum realizable SNR in a lightwave transmission system where the dominant noise source is laser intensity noise. The following relationship shows the theoretical relationship between the signal-to-noise level and the RIN:

$$\frac{S}{N} = \frac{m^2}{2B \times \text{RIN}} \qquad (7.20)$$

where m is the optical modulation index and B is the noise bandwidth. A rate equation analysis shows that over a large operating range, the RIN of semiconductor lasers decreases (improves) with increasing laser power.[19]

Traditionally, RIN has been measured by using a photodetector to detect the laser output power. The electrical output of the photodetector is often amplified, and then displayed, on an electrical spectrum analyzer. The measurement on the electrical spectrum analyzer is intended to correspond to an amplified electrical equivalent of $(\Delta P)^2$ in the equation defining RIN. Simultaneously, the dc photocurrent must be monitored. This dc photocurrent, when squared and multiplied by 50 ohms and the amplifier gain, results in the corresponding electrical-equivalent term for $(P_{\text{AVG}})^2$. Here 50 ohm impedances are assumed for the amplifier and the spectrum analyzer input.

Care must be used in this measurement technique to obtain accurate results. The photodetector, amplifier, and spectrum analyzer frequency responses, as well as the mismatch losses between them, must all be well characterized. In addition, the amplifier gain must be large enough, and its noise figure low enough in relation to the spectrum analyzer noise figure to provide adequate sensitivity to make RIN measurements.

Unfortunately, the spectrum analyzer measures other noise components in addition to the laser RIN. These include the shot noise power caused by the quantum nature of the photodetection process and the receiver's total measured thermal noise power. By dividing by the calibrated amplifier gain, the noise power contributions can be conveniently referenced to the photodetector output and expressed as:

$$P_n(\text{measured}) = P_n(\text{laser}) + P_n(\text{shot}) + P_n(\text{thermal}) \qquad (7.21)$$

$P_n(\text{laser})$ is the detected laser intensity noise. $P_n(\text{shot})$ is the detected shot noise power, which can be expressed as:

$$P_n(\text{shot}) = (i_n)^2 \times 50\Omega \qquad (7.22)$$

where i_n is the shot noise current, and $(i_n)^2 = 2qI_{\text{dc}}$ for a 1 Hz bandwidth. q is the electronic charge equal to 1.60×10^{-19} coulomb. $P_n(\text{thermal})$ is the receiver noise, which can be expressed as:

$$P_n(\text{thermal}) = \frac{(F_A G_A + F_{SA} - 1)kT_0 B}{G_A} \qquad (7.23)$$

where F_A is the amplifier noise figure, F_{SA} is the spectrum analyzer noise figure, and G_A is the amplifier gain, all expressed in linear power units. k is Boltzman's constant and B is the noise bandwidth. For a 1 Hz bandwidth, $kT_0 = 4.00 \times 10^{-21}$ at the standard reference temperature of 290 K. By dividing by the detected average power, $P_{\text{AVG}}(\text{electrical}) = (I_{DC})^2 \times 50\Omega$, one can determine the RIN:

$$\text{RIN(measured)} = \text{RIN} + \frac{2q}{I_{dc}} + \frac{P_n(\text{thermal})}{P_{AVG}(\text{electrical})} \qquad (7.24)$$

For shot noise limited detection, the measurable RIN can be shown to be equal to $2q/I_{dc}$, which indicates the minimum measurable RIN decreases linearly as the detected photocurrent increases. RIN is often expressed in dB. Thus, RIN decreases 10 dB per decade increase in detected photocurrent. If the responsivity of the photodetector is 1mA/mW, then a minimum RIN of -155 dB/Hz could be measured at 1 mW of average optical signal power, but only -145 dB/Hz could be measured at 100 μW of average optical power. This relationship is shown in Figure 7.24. Also shown for comparison are the thermal noise contributions of the electrical spectrum analyzer, with and without a low-noise preamplifier, relative to the room temperature 1 Hz kT_oB noise floor of approximately -174 dBm. Both the receiver thermal noise and shot-noise contributions can limit the accuracy of RIN measurements.

RIN can also be measured on a lightwave signal analyzer, which has the capability to measure both the average optical signal power as well as the detected intensity noise fluctuations. In addition, the lightwave signal analyzer has the ability to measure both the detected shot noise and thermal noise levels, and correct for these contributions to make RIN measurements below the shot-noise limit.[22] A measurement of a low-noise DFB semiconductor laser is shown in Figure 7.25. RIN (system) is the value that the lightwave signal analyzer directly measures with all noise terms included. RIN (laser) refers to the desired laser intensity noise value with the thermal and shot-noise contributions removed. The typical RIN measurement range of the lightwave signal analyzer as a function of input optical power is shown in Figure 7.26. Subtraction of the thermal noise and shot noise from RIN (system) provides approximately an additional 16 dB measurement range for the determination of RIN (laser).

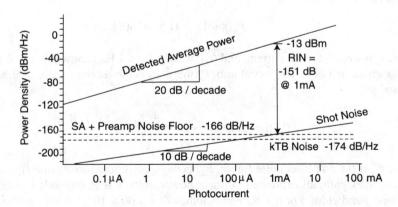

Figure 7.24 Noise considerations when making an RIN measurement on an electrical spectrum analyzer.

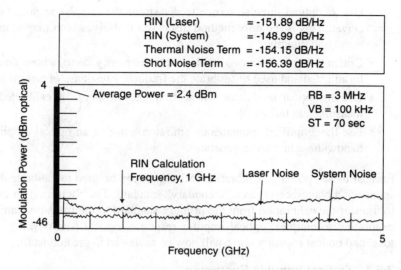

Figure 7.25   RIN measurement on a lightwave signal analyzer.

7.5   MODULATION DOMAIN CALIBRATION TECHNIQUES

Optical frequency response calibration is required in order to make accurate measurements of the modulation transfer function of optical devices as well as intensity modulation and intensity noise measurements. There are several techniques that employ a variety of optical sources that can be used to calibrate the frequency response of measurement instrumentation:

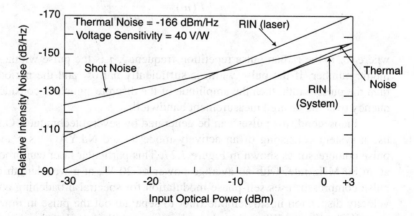

Figure 7.26   RIN measurement range on a lightwave signal analyzer.

- Use an optical impulse source to determine the impulse response of an optical receiver. This is the only method discussed that gives both magnitude and phase information directly.
- Optically heterodyne two stable single-frequency lasers whose beat frequency can be adjusted and used to calibrate the frequency response of an optical receiver.
- Use an optical modulator whose frequency response is calibrated by a two-tone measurement technique.
- Use the amplified spontaneous emission noise of an optical amplifier as a broadbandwidth white noise generator.

Each of these "primary" calibration techniques can be used to calibrate the frequency response of a photoreceiver as a "secondary" standard. The photoreceiver can then use used to transfer the calibration from the primary standard to the measurement system.[23] These calibration techniques: optical impulse response, optical heterodyning, modulator two-tone, and optical intensity noise will now be reviewed in greater detail.

7.5.1 Optical Impulse Response

The frequency response of a measurement system can be computed from the Fourier transform of the time-domain impulse response. This approach has the advantage that it provides both amplitude and phase information about the measurement system which is important for lightwave component analysis and other vector-network-analysis-based measurements. An optical impulse generator can be approximated by a sufficiently narrow laser pulse with a low repetition rate. When viewed in the frequency domain, this type of optical signal generates a frequency domain spectrum consisting of a comb of discrete signals, at harmonics of the pulse repetition frequency, whose amplitude follows the response:

$$P(w) = \frac{\sin\left(\frac{nw_0\tau}{2}\right)}{\frac{nw_0\tau}{2}} \qquad (7.25)$$

where w_0 is the radian pulse repetition frequency, τ is the pulse width, and n is the harmonic number. If the pulse width is sufficiently narrow and the period is much longer than the pulse width, then the amplitude of the discrete signals is considered flat with frequency over the desired measurement bandwidth.

Picosecond (ps) pulses can be generated by several techniques. One technique is to use a system consisting of an actively mode-locked Nd:YAG laser and a fiber-grating pulse compressor as shown in Figure 7.27. This particular laser can produce 80 ps pulses at an 80 MHz rate, with an average power of 20 W, at a wavelength of 1.06 μm. The pulse compressor uses self-phase modulation for spectral broadening and positive group velocity dispersion in singlemode fiber to separate out the pulse in time as it propagates through the fiber.[24] When this chirped and spectrally broadened pulse is passed through the diffraction grating pair, a time delay proportional to wavelength is introduced, which

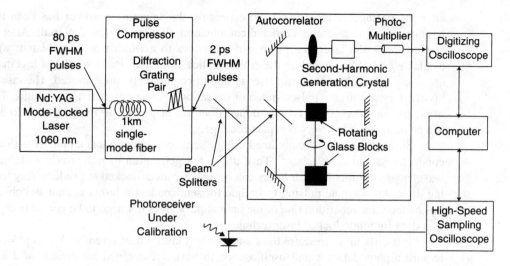

Figure 7.27 Optical impulse response measurement system.

compresses the pulse to about 2 ps full-width at half-maximum (FWHM). Other solid-state lasers such as Ti:Sapphire lasers are available to cover wavelengths shorter than 1 μm. Also shown in Figure 7.27 is an autocorrelator which is often used to characterize the width of very short pulses.

Another technique to generate short optical pulses is to use a colliding-pulse pas-sively mode-locked erbium-doped fiber ring laser as shown in Figure 7.28. This source is convenient because it can be easily packaged and does not require the use of a large opti-cal table with operator adjustments.[25] The key to the operation of the laser is the saturable absorber. The leading edge of the incoming pulse is absorbed. After absorbing a set num-ber of photons, the absorber becomes transparent. This causes a narrowing of the pulse on each pass through the absorber. The absorption is quickly reset to a high absorption value

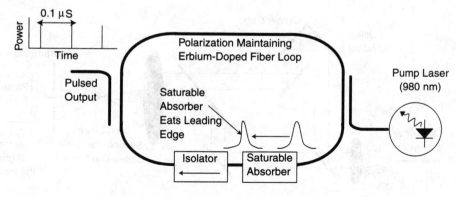

Figure 7.28 Erbium-doped fiber ring laser.

after the passage of the optical pulse because the saturable absorber has been proton-bombarded to reduce the carrier-recombination lifetime in the material. After many round-trips in the laser, the pulse width narrows to a width that is consistent with the available optical bandwidth in the erbium optical amplifier. Pulse widths of less than 1 ps are easily achievable. This ring laser is said to be passively mode locked. The laser starts up into the repetitively pulsed mode of operation without any external stimulus. The use of polarization-preserving erbium-doped fiber has made this design very easy to use and can be designed into a very compact package.

Similar results can be obtained using a semiconductor optical amplifier with a semiconductor saturable absorber.[26] This allows the generation of pulses at a wide range of wavelengths. Semiconductor lasers can be actively mode-locked to produce very low timing jitter. An alternate pulsing technique for semiconductor lasers is gain-switching.[26] In this method, the repetition rate of the optical pulse does not need to be perfectly repetitive as it does for mode-locked laser techniques.

It is difficult to measure the shape of short laser pulses directly. A 2 ps pulse would require a photodetector and oscilloscope to have a combined bandwidth of 250 GHz. Therefore, the autocorrelation function is often measured, from which the power spectral density and width of the pulse can be inferred. The autocorrelator shown in Figure 7.29 operates in the following manner. First, the input beam is split into two paths, one of which is varied in length via a moveable corner cube to introduce a swept differential delay. The two beams are then combined in a crystal with nonlinear characteristics, which generates a second harmonic with an amplitude proportional to the product of the intensities of the two beams overlapped in the crystal. The second harmonic light is detected by the photomultiplier tube. The output signal can then be displayed on an oscilloscope to trace out the optical pulse's autocorrelation function as the corner cube is swept. The Fourier transform of this time record is the power spectral density of the optical pulse.[27] Once the optical impulse has been characterized, it can be used to determine the frequency response of the measurement instrumentation. Depending on the bandwidth of the instrumentation, it may be necessary to deconvolve the contribution of the optical impulse.

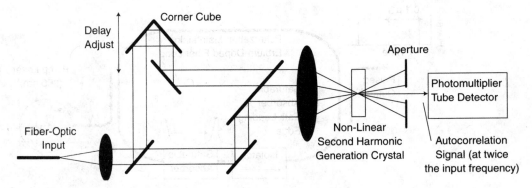

Figure 7.29   Autocorrelator.

The autocorrelation technique gives information on the magnitude of the modulation spectrum from the optical impulse source. It does not directly give information about the optical phase of the source. An alternate measurement technique is based on a nose-to-nose high-frequency oscilloscope measurement.[28] In this method, the magnitude and phase response of a high frequency sampling oscilloscope is made. After the oscilloscope is calibrated, the magnitude and phase of an optical receiver under test can be made with the impulse response method.

7.5.2 Optical Heterodyning

An optical heterodyne source can be built using diode-pumped Nd:YAG ring cavity lasers whose nominal wavelength of 1.32 μm can be varied by changing the laser crystal temperature. Figure 7.30 shows an example of such a calibration source. Three lasers are shown. One acts as a master laser oscillator and the other two as slave laser oscillators. The frequency of each of the slave lasers is temperature-adjusted to a known frequency away from the master laser with a synthesizer/phase-locked loop. The short term optical linewidth of either the master or slave oscillators is less than 10 kHz. By adjusting the frequencies of the slave lasers, a beat frequency between dc and approximately 50 GHz can be produced.[29] This beat frequency can be varied by adjusting the temperature of one slave laser relative to the other. The master-slave configuration is used because it allows small difference frequencies, or a beat frequency near dc, but avoids the problem of the slave lasers injection-locking one another. The use of open-beam combiners in this optical heterodyne source is preferable to fiber-coupled versions.

Let the difference between the laser slave frequencies be f, and the individual output powers be P_1 and P_2. If the laser outputs are linearly polarized and aligned, the signal at the combined output is:

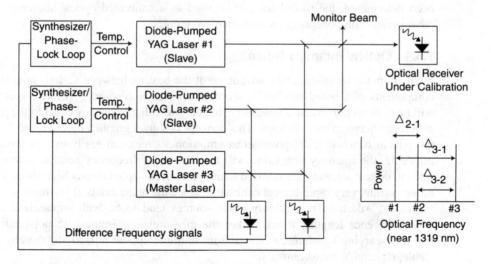

Figure 7.30 Dual YAG heterodyne frequency response measurement system.

$$P_O = P_1 + P_2 + 2\sqrt{P_1 P_2} \cos(2\pi ft) \tag{7.26}$$

If $P_1 = P_2$, the resulting output signal is 100% modulated at the difference frequency. In a test system one of the coupler/splitter's outputs is connected to the DUT or measurement system to be calibrated. The other output is used for monitoring the laser power. Optical receiver calibrations accurate to within 0.06 dB have been performed using this configuration.[23,27] Additional information on heterodyne theory is found in Chapter 5.

7.5.3 Two-Tone Technique

A Mach-Zehnder interferometer, whose use as an optical intensity modulator has already been described in Section 7.2.2, can be used as a calibrated source. To be used as a calibrated source, the modulator is biased for minimum signal transmission. When two modulation signals are applied with radian frequencies w_1 and w_2, having peak voltages V_{p1} and V_{p2}, one of the signals present at the modulator output will be a difference frequency component:

$$I(\Delta f) = I_0 J_1\left(\pi \frac{E(w_1)V_{p1}}{V_\pi}\right) J_1\left(\pi \frac{E(w_2)V_{p2}}{V_\pi}\right) \tag{7.27}$$

This expression can be used to measure the modulator's frequency-dependent modulation efficiency. If the two modulation signals are very close in frequency, then $E(w_1)$ is essentially the same as $E(w_2)$. By measuring I_0, V_π, the applied voltages V_{p1} and V_{p2}, and the optical power at the difference frequency, $I(\Delta f)$, the expression can be solved numerically for $E(w)$. By keeping the difference frequency small and constant, $E(w)$ can be determined by stepping w_1 and w_2 over the range of modulation frequencies. Once $E(w)$ has been determined, the modulator can be used as a calibrated optical source to measure other optical components or measurement systems.[27]

7.5.4 Optical Intensity Noise

Intensity-noise techniques take advantage of the beating between various optical spectral components of a broad-bandwidth spontaneous emission source. Any two spectral lines will beat, or mix, to create an intensity fluctuation with a frequency equal to the frequency difference between the two lines. This concept is shown graphically in Figure 7.31. Since the optical bandwidth of spontaneous emission sources can easily exceed thousands of gigahertz, the intensity beat noise will have a similar frequency content. These fluctuations in optical intensity are referred to as spontaneous-spontaneous beat noise. This technique permits very rapid optical calibration since the noise exists at all frequencies simultaneously. Additionally, intensity noise sources tend to be both unpolarized and have short-coherence lengths, which makes the measurements immune to polarization drifts and time-varying interference effects from multiple optical reflections, thereby allowing stable, repeatable measurements.

There are many sources of broad-bandwidth spontaneous emission. Hot surfaces (such as tungsten light bulbs) can emit optical radiation ranging from the visible to the far

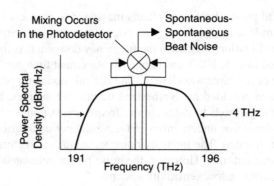

Figure 7.31 Spontaneous-spontaneous beat noise arising from an optical noise source.

infrared. Semiconductor sources such as edge emitting light-emitting diodes (EELEDs) provide increased power densities and a wavelength range of about 100 nm. Still higher power densities can be obtained from solid-state sources such as fiber optic amplifiers.

The ability to couple these sources of broad-bandwidth light efficiently into single-mode fiber is also important. Coupled-power densities can range from about 500 pW/nm for a light bulb to greater than 1 mW/nm for amplified spontaneous emission (ASE) from a fiber optic amplifier. The high-power densities from fiber optic amplifiers are particularly well-suited for intensity noise generation.[30] One must be careful in this method not to use power levels high enough to cause nonlinear responses in the optical receiver.

7.5.5   Comparison of Calibration Techniques

Table 7.1 shows a comparison of the calibration techniques for modulation domain measurements. Each of these calibration techniques have advantages and disadvantages. The optical impulse response technique allows direct measurement of the impulse response of photoreceiver-oscilloscope system. The measurement can be fast. However, depending on

Table 7.1   Comparison of the Calibration Techniques for Modulation Domain Measurements

Calibration Method	Wavelength Range	Phase Determined?	Modulation Frequencies	Comments
Optical impulse	1060 nm with Nd:YAG 1550 nm with EDFA Many other types	Yes	MHz- >50 GHz	Need autocorrelator, may require deconvolution of optical pulse
Optical heterodyne	1320 nm with YAG, others with DFBs	No	kHz- 50 GHz	Requires an optical bench setup
Two-tone technique with modulator	1300 or 1550 nm	No	kHz- 30 GHz	Relies on accurate electrical power measurements
ASE detection	1550 nm with EDFA	No	kHz- >50 GHz	Detector saturation issues

the pulse width of the optical pulse, the source effects may need to be deconvolved from the measurement. The heterodyne system is easy to extend to very high frequencies and very useful for performing calibration directly in the frequency domain. It is the technique that is most easily referenced to U.S. NIST standards. Measurement time can be long because of the thermal tuning of the laser wavelength. The optical modulator technique has very fine frequency resolution provided by synthesized microwave sources, but requires careful calibration in order to be used to calibrate the frequency response of devices or measurement systems. Any errors or uncertainties in the modulator calibration are transferred to the system being calibrated. The intensity noise technique also offers high resolution as well as fast measurement time. However, the noise power generated may not be always sufficiently high to calibrate low sensitivity systems.

REFERENCES

1. Henry, C.H. 1986. Phase noise in semiconductor lasers. *J. Lightwave Tech.* LT-4: 298–311.

2. Wong, R.W., P.R. Hernday, M.G. Hart, and G.A. Conrad 1989. High-speed lightwave component analysis. *Hewlett-Packard J.* 40 (3): 35–51.

3. Wong, R.W., P.R. Hernday, and D.R. Harkins. 1991. High-speed lightwave component analysis to 20 GHz. *Hewlett-Packard J.* 43 (1): 6–12.

4. Ikegami, T. and Y. Suematsu. 1968. Direct modulation of semiconductor junction laser. *Elect. Comm. Japan* 51-B: 51–58.

5. Lau, K.Y., C. Harder, and A. Yariv. 1981. Ultimate frequency response of GaAs injection laser. *Optics Comm.* 36: 472–474.

6. Lau, K.Y. and A. Yariv. 1985. Ultra high-speed semiconductor lasers. *IEEE J. Quantun Elect.* QE-21: 121–138.

7. Bowers, J.E. 1986. High-speed semiconductor laser design and performance. *Solid State Elect.* 30: 1–11.

8. Jungerman, R.L. et al. 1990. High-speed optical modulator for application in instrumentation. *J. Lightwave Tech.* 8: 1363–1370.

9. Bowers, J.E. et al. 1986. Millimetre-waveguide-mounted InGaAs photodetectors. *Elect. Lett.* 22: 633.

10. Miller, C.M. 1990. High-speed lightwave signal analysis. *Hewlett-Packard J.* 41 (1): 80–91.

11. Lau, K.Y. and A. Yariv. 1984. Intermodulation distortion in a directly modulated semiconductor injection laser. *Appl. Phys. Lett.* 45: 1034–1036.

12. Darcie, T.E., R.S. Tucker, and G.J. Sullivan. 1985. Intermodulation and harmonic distortion in InGaAsP lasers. *Electron. Lett.* 21: 665–666.

13. Tucker, R.S. and D.J. Pope. 1983. Circuit modeling of the effect of diffusion on damping in a narrow-stripe semiconductor. *IEEE J. Quantum Elect.* QE-19: 1179.

14. Takemoto, A. et al. 1990. Distributed feedback laser diode and module for CATV systems. *IEEE J. Selected Areas in Commun.* 8: 1365.

15. Dentan, M. and B. De Cremoux. 1990. Numerical simulation of the nonlinear response of a p-i-n photodiode under high illumination. *J. Lightwave Tech.* 8: 1137–1144.

16. Hayes, R.R. and D.L. Persechini. 1993. Nonlinearity of p-i-n photodetectors. *IEEE Photonics Tech. Lett.* 5: 70–72.

17. Bowers, J.E. and M.A. Pollack. 1988. Semiconductor lasers for telecommunications. In *Optical Fiber Telecommunication II,* ch. 13, eds. S.E. Miller and I.P. Kaminow, San Diego, CA: Academic Press.

18. Huag, M. 1969. Quantum mechanical rate equations for semiconductor lasers. *Phys. Rev.* 184: 338–348.

19. Agrawal, G.P. and N.K. Dutta. 1986. *Long-wavelength semiconductor lasers,* ch. 6, New York: Van Nostrand Reinhold.

20. Gimlett, J.L. and N.K. Cheung. 1989. Effects of phase-to-intensity noise conversion by multiple reflections on gigabit-per-second DFB laser transmission systems. *J. Lightwave Tech.* 7: 888–895.

21. Yamamoto, Y. 1983. AM and FM quantum noise in semiconductor lasers—Part I: Theoretical analysis. *IEEE J. Quantum Elect.* QE-19: 34–36.

22. Miller, C.M. and L.F. Stokes. 1990. Measurement of laser diode intensity noise below the shot noise limit. Boulder, CO: *NIST Symp. on Optical Fiber Measurements.*

23. Hale, P.D., C.M. Wang, R. Park, and W.Y. Lau. 1996. Photoreceiver frequency response transfer standard: Calibration using a swept heterodyne method. *J. Lightwave Tech.* 14 (11): 2457–2466.

24. Sieman, A.E. 1986. *Lasers.* Mill Valley, CA: University Science Books.

25. Lin, H., D.K. Donald, K.W. Chang, and S.A. Newton. 1995. *Colliding pulse mode-locked lasers using erbium-doped fiber and a semiconductor saturable absorber.* Baltimore, MD: 1995 Conference on Lasers and Electrooptics, Optical Society of America, paper JTuE1.

26. Derickson, D.J., R.J. Helkey, A. Mar, J. Karin, J. Wasserbauer, and J. Bowers. 1992. Short pulse generation using multisegment mode-locked lasers. *IEEE J. Quant. Elect.* 28 (10): 2186–2201.

27. McQuate, D.J., K.W. Chang, C.J. Madden. 1993. Calibration of lightwave detectors to 50 GHz. *Hewlett-Packard J.* 44 (1): 87–92

28. Verspecht, J. and K. Rush. 1994. Individual characterization of broadband sampling oscilloscopes with a nose-to-nose calibration procedure. *IEEE Trans on Instrumentation and Measurement* 43 (2): 347–354.

29. Tan, T.S., R.L. Jungerman, and S.S. Elliot. 1988. Calibration of optical receivers and modulators using an optical heterodyne technique. *IEEE-MTT-S International Microwave Symp. Digest:* 1067–1070.

30. Baney, D.M. and W.V. Sorin. 1995. Broadband frequency characterization of optical receivers using intensity noise. *Hewlett-Packard J.* 46 (1): 6–12.

CHAPTER

8

Analysis of Digital Modulation on Optical Carriers

Stephen W. Hinch, Christopher M. Miller

This chapter describes methods for analyzing the physical characteristics of digitally modulated optical carriers. Three categories of measurements are discussed: bit-error-ratio testing, waveform analysis, and jitter testing. In each case, the theory and importance of the measurement is presented, followed by descriptions of measurement equipment and techniques. Where more than one approach exists, the benefits and limitations of each are examined. Measurement examples are presented to illustrate typical uses of the techniques. The chapter begins with a brief review of fiber-optic transmission systems and the standards that drive physical-layer measurement requirements.

8.1 DIGITAL FIBER-OPTIC COMMUNICATIONS SYSTEMS

Digital fiber-optic communications systems have gained widespread acceptance throughout the world for both telecommunications and data communications applications. Telecom systems typically operate over singlemode fiber at distances from 10 km to over 100 km and employ lasers emitting in either the 1310 nm or 150 nm wavelength range. Data communications systems typically cover shorter distances of up to a few kilometers, often over multimode fiber. They can employ either lasers or LEDs, typically in the 650 to 850 nm range.

Figure 8.1 illustrates a simplified fiber-optic telecommunications system. Individual 64 kilobit-per-second (kb/s) voice channels are time-division multiplexed into a primary rate serial data stream. In North America and Japan, 24 voice channels are multiplexed

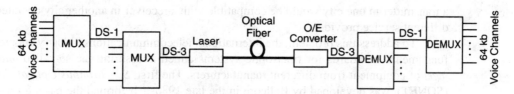

Figure 8.1   Simplified fiber optic telecommunications system.

onto a 1.544 megabit-per-second (Mb/s) data stream known as a DS-1 signal. (The DS-1 signal is sometimes referred to as a T-1 signal, although this more properly refers to the older analog version at a similar rate.) In Europe and many other parts of the world, the primary rate signal combines 30 voice channels onto a 2.048 Mb/s data stream. Multiple primary rate signals can be multiplexed to form higher rates, further increasing capacity. Table 8.1 summarizes the various standard electrical transmission rates in common use around the world.

The serial electronic data stream intensity-modulates the laser transmitter, whose output is a series of corresponding light pulses launched into the fiber optic cable. As it propagates along the cable, the light is subject to attenuation and distortion before eventually reaching the receiver. There, it is converted back to an electrical signal, amplified, and retimed against a recovered clock signal. The low-rate electronic tributary signals can then be demultiplexed eventually back to the individual 64 kb/s voice channels.

8.1.1   SONET/SDH Standards

The earliest fiber optic transmission systems did not conform to any standards. Most manufacturers used proprietary transmission rates and protocols, making it impossible for equipment from different manufacturers to work together. With the deregulation of telecommunications in the U.S. and the increasing importance of international communications, this lack of standardization became a major problem. It was no longer certain that

Table 8.1   Standard Telecommunications Electrical Transmission Rates

Hierarchy Level	Standard Transmission Rates (Mb/s)		
	North America	**Japan**	**Europe**
1	1.544	1.544	2.048
2	6.312	6.312	8.448
3	44.736	32.064	34.368
4	274.176	97.728	139.264
5		396.200	

a transmitter in one city would be compatible with a receiver in another city operated by a different service provider.

To address this problem, the international telecommunications industry adopted two fundamental standards for fiber-optic systems aimed at assuring the so-called *mid-span meet* of equipment from different manufacturers. The first, *Synchronous Optical Network* (SONET), was developed by Bellcore in the late 1980s.[1] It formed the basis for a subsequent ANSI standard.[2] In Europe, a similar standard, *Synchronous Digital Hierarchy* (SDH), was developed by the International Telecommunications Union (ITU).[3] Because of the close cooperation between these standards bodies, in most respects the two documents are functionally equivalent.

SONET/SDH standards define a variety of requirements ranging from such physical parameters as bit rates, optical power levels, wavelengths, and waveform shapes, to higher level functionality such as protocol structures and coding formats. The standards prescribe that the laser be modulated in a simple binary *non-return-to-zero* (NRZ) format. In this format, the laser is turned on for the full duration of a logic one pulse and turned off (or nearly off) for the full duration of a logic zero pulse (Figure 8.2). NRZ coding is the predominant format for all types of fiber optic transmission.

The basic SONET protocol can be viewed as a frame 90 bytes wide by 9 bytes high (Figure 8.3), transmitted over a period of 125 μsec. Since each byte consists of 8 bits, a total of 6480 bits are transmitted over this period to produce the basic SONET data rate of 51.84 Mb/s. This is called the *OC-1* rate. Higher rates are defined as integer multiples of the basic OC-1 rate. The first three columns in the frame are reserved for various overhead functions while the remainder is available to carry the payload. The size of the available OC-1 payload area is sufficient to carry a DS-3 tributary signal.

SDH does not use the 51.84 Mb/s rate but starts with a basic rate three times faster, at 155.52 Mb/s. Called *STM-1,* this rate allows mapping of a 139.264 Mb/s tributary. The

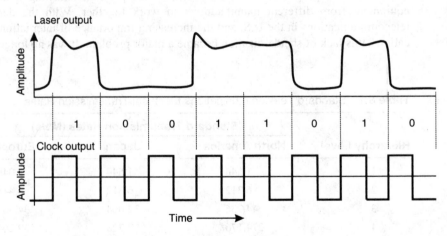

Figure 8.2   NRZ coding format.

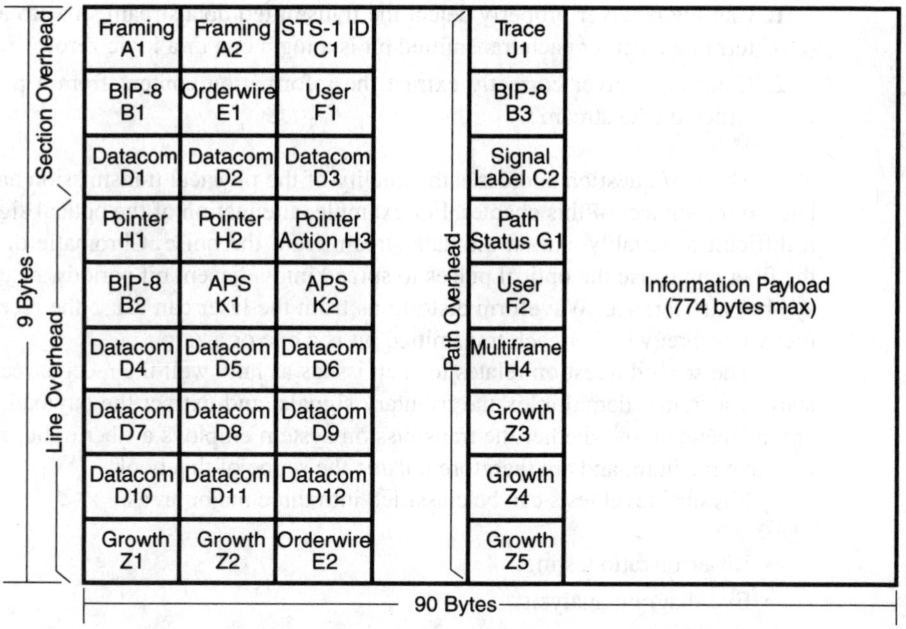

Figure 8.3 Basic SONET frame. Although represented for convenience as a series of rows and columns, the data is actually transmitted serially by row.

frame structure and payload mapping are identical to the equivalent SONET rate. Table 8.2 lists the commonly used rates for both SONET and SDH.

8.1.2 Performance Analysis of Fiber Optic Systems

As with any communications system, the basic performance measure for a fiber optic system is the accuracy with which it can transmit information from one point to another. This involves answering two questions:

Table 8.2 SONET/SDH Transmission Rates

Transmission Rate (Mb/s)	SONET designation	SDH designation
51.84	OC-1	—
155.52	OC-3	STM-1
622.08	OC-12	STM-4
1244.16	OC-24	—
2488.32	OC-48	STM-16
9953.28	OC-192	STM-64

1. Can the receiver properly detect the transmitted data stream so as to accurately determine whether each transmitted bit is a logic one or a logic zero?

2. Can the receiver correctly extract the information content from a properly reconstructed data stream?

The first question addresses the quality of the physical transmission and directly relates to the subject of this chapter. For example, attenuation of the optical signal can make it difficult to reliably extract the data stream from the noise. Chromatic distortion along the fiber can cause the optical pulses to spread into adjacent bit periods, resulting in inter-symbol interference. Waveform distortions from the laser can cause the receiver to incorrectly interpret whether each transmitted bit is a one or a zero.

The second question relates to such issues as how well the receiver can identify the start of a frame, demultiplex the tributary signals, and extract the payload. These issues are independent of whether the transmission system employs a fiber optic, copper, or microwave medium, and are therefore outside the scope of this book.

Physical level tests can be classified into three major areas:

- Bit-error-ratio testing,
- Eye-diagram analysis,
- Jitter analysis.

The following sections describe each category in detail.

8.2 BIT-ERROR RATIO

In the simplest sense, the fundamental measure of performance for a digital communications system is how accurately the receiver can determine the logic state of each transmitted bit. This figure of merit is called *bit-error ratio*, defined as

$$\text{BER} = \frac{E(t)}{N(t)} \tag{8.1}$$

where BER is the bit error ratio, $E(t)$ is the number of bits received in error over time t, and $N(t)$ is the total number of bits transmitted in time t.

Bit error ratio is a statistical parameter. The measured value depends on the *gating time, t,* over which the data is collected and on the processes causing the errors. For errors due primarily to Gaussian noise, relatively stable results are obtained when the gating time is sufficient to capture on the order of 50 to 100 errors. If the errors occur in bursts caused by nonrandom effects such as channel crosstalk or external interference, this simple measure of BER might not adequately represent system performance.

To address the limitations of a simple BER metric, several alternative measures have been defined. The *error-free second* is defined as a second in time that is free of errors. The converse of this is the *errored second*. A list of common BER metrics is presented in Table 8.3. While these are useful metrics for voice communications, they may

Table 8.3 Common BER Metrics (from CCITT Rec. G.821)

Metric	Definition
Error ratio	Ratio of counted errors to total number of bits in the gating period.
Percent unavailability	The error ratio is calculated over 1 s gating periods. An unavailable period begins when the error ratio is worse than 1×10^{-3} for 10 consecutive seconds. The unavailable period ends when the error ratio is better than 1×10^{-3} for 10 consecutive periods. These 10 s are considered part of the available time. Percent unavailability is the ratio of the unavailable seconds to the total gating period expressed as a percentage.
Percent errored seconds	The ratio of the errored seconds in the available time to the total number of seconds in the available time, expressed as a percentage.
Percent severely errored seconds	The ratio of the total number of available seconds with an error ratio worse than 1×10^{-3} to the total number of available seconds, expressed as a percentage.
Percent degraded minutes	Severely errored seconds are discarded from the available time and the remaining seconds are grouped into blocks of 60 s. Blocks having an error ratio worse than 1×10^{-6} are called degraded minutes. Percent degraded minutes is the ratio of the total number of degraded minutes to the total number of 60 s blocks in the available time expressed as a percentage. Incomplete blocks of less than 60 s are treated as complete blocks of 60 s.

not be adequate for more demanding data communications applications where even a single error is unacceptable. In these cases a BER of 10^{-15} or better is often necessary. At 100 Mb/s this corresponds to less than 1 error every 2700 operating hours.

8.2.1 BER Measurement

Two types of bit error ratio measurements can be conducted. *In-service testing* is performed on the system during actual operation to give early warning of problems. In one approach, a single 64 kb/s line is taken out of service and a known test pattern injected onto the line. The error performance of this line can be considered representative of all other lines on the system. Other approaches involve monitoring line-coding violations or trace signals. For example, SONET/SDH standards include a parity byte in the overhead structure that allows frame-based error detection without the need to remove revenue-producing lines from service.

   Out-of-service testing involves injecting a known test pattern onto the serial line. The system cannot carry live traffic during the test, so it is best suited for research and development or manufacturing test environments. The equipment used for out-of-service testing is known as a *bit-error-ratio tester,* or *BERT,* described in the following section.

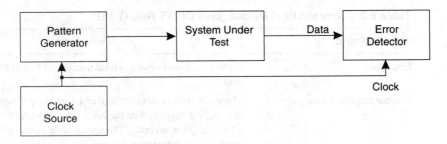

Figure 8.4 Basic bit-error-ratio tester.

8.2.2 BERT Design

The concept behind bit-error-ratio testing is illustrated in Figure 8.4. The BERT consists of two sections: a *pattern generator* and an *error detector*. The pattern generator creates the test pattern together with a separate clock signal at the selected data rate. This pattern is injected into the system under test and received at the error detector's data input. The error detector includes its own pattern generator that produces an exact replica of the known test pattern and a comparator that checks every received bit against this internally generated pattern. Each time the received bit differs from the known transmitted bit, an error is logged.

The pattern generator and the error detector must operate at identical clock rates and the phase relationship between them must be stable. The easiest way to ensure this is to use the pattern generator's clock as the clock source for the error detector. This is easy enough when the two units are in close physical proximity—a direct electrical connection can be made between them. When they are physically separated, for example at opposite ends of a transmission link, a direct connection may not be possible. In this case, the error detector's clock signal must be recovered directly from the data. Refer to Section 8.2.4 for additional details.

In general, the exact time delay through the system under test is not known. Therefore, the error detector must have some provision for automatically synchronizing its internally generated pattern to the incoming data. This synchronization process makes use of the fact that when the two patterns are out of sync, the likelihood of a received bit matching the expected bit is pure chance and the error ratio is approximately 0.5. So whenever the measured BER exceeds a predefined threshold, typically set at 0.2, the error detector attempts to resynchronize to the data. This is done by stepping the phase relationship between the two patterns until a BER minimum is achieved.

Some pattern generators also include a separate trigger output which produces only a single pulse at the start of every repetition of the pattern. The trigger output is not used in BER testing but can be important in eye-diagram analysis (Section 8.3.7).

BER tests may be performed on an entire system, as shown in Figure 8.4, or on individual network elements. In the latter case, additional components may be needed to perform the test. For example, to measure the sensitivity of an optical receiver, the electrical output of the pattern generator must be applied to a test laser; to test laser performance, an optical receiver must be inserted in front of the error detector. Several representative test configurations are described in Section 8.2.5.

8.2.3   Test Patterns for Out-of-Service Testing

The most common test pattern for out-of-service BER testing is the *pseudo-random binary sequence,* or *PRBS*. This is a repetitive pattern whose pattern length is of the form 2^N-1, where N is an integer. (This choice assures the pattern repetition rate is not harmonically related to the data rate.) Typical values of N are 7, 10, 15, 20, 23, and 31. Within the pattern, the bit sequence is designed to approximate the characteristics of truly random data. The approximation is especially close for longer pattern lengths.

A PRBS pattern is generated by using a train of shift registers with feedback as shown in Figure 8.5. The patterns with N = 15 and N = 23 have been adopted in international standards.[4] The shift register combinations for these patterns are described in Table 8.4.

A PRBS frequency spectrum is shown in Figure 8.6. It consists of a series of discrete lines whose envelope follows a sin x/x shape. The spacing between lines is given by

$$\Delta f = \frac{f_b}{2^N - 1} \tag{8.2}$$

where Δf = frequency spacing between spectral lines, and f_b = bit rate.

Other patterns are often used for specialized tests. *Mark density* patterns are similar to PRBS patterns but allow the user to vary the ratio of logic ones to logic zeros. They are often used to test the sensitivity of clock recovery circuits to unequal one-zero distributions. To obtain a true 50%-mark density they must be of the form 2^N rather than $2^N - 1$. *Zero substitution* patterns are similar but allow portions of the pattern to be replaced with

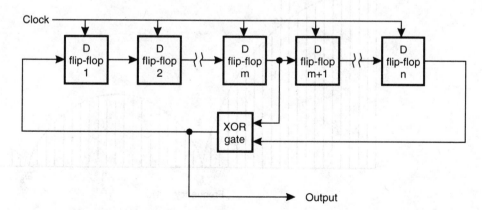

Figure 8.5   PRBS generator using shift registers with feedback.

Table 8.4   PRBS Shift-Register Combinations

Sequence Length	Shift-Register Configuration
2^7-1	$D^7 + D^6 + 1 = 0$, inverted
$2^{10}-1$	$D^{10} + D^7 + 1 = 0$, inverted
$2^{15}-1$	$D^{15} + D^{14} + 1 = 0$, inverted
$2^{20}-1$	$D^{20} + D^{17} + 1 = 0$, inverted
$2^{23}-1$	$D^{23} + D^{18} + 1 = 0$, inverted
$2^{31}-1$	$D^{31} + D^{28} + 1 = 0$, inverted

specified runs of logic zeroes. They can test the ability of the clock recovery circuit to remain synchronized in the absence of data transitions over extended periods.

System-level SONET/SDH testing is often done with a standardized SONET test pattern. This consists of a header simulating the actual SONET overhead structure followed by a $2^7 - 1$ PRBS simulating the payload. The system BER measured with this pattern is generally considered more representative of actual operation than the BER obtained with a simple PRBS test pattern.

Some pattern generators permit the user to create a wide variety of custom patterns. This makes it possible to create stress patterns that exercise the extremes the DUT is likely to see. A pattern generator that allows these patterns to be created on a computer and downloaded into memory is much easier to program than one that requires the user to manually enter the pattern bit-by-bit from the instrument front panel.

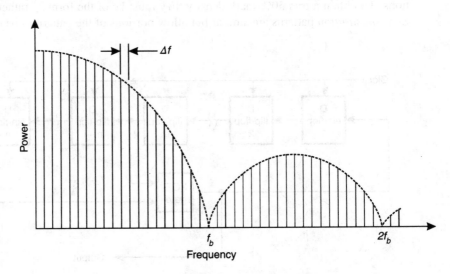

Figure 8.6   PRBS frequency spectrum.

8.2.4   Clock Recovery

To measure BER accurately, the phase relationship between the clock and data inputs of the error detector must be stable. In many cases, this can be achieved by connecting the pattern generator clock output directly to the error detector clock input as was shown in Figure 8.4. In some situations, however, this is not possible. For example, when testing performance of an entire system including a long run of fiber, the pattern generator and error detector may not be located in close physical proximity. In this case, the error detector's clock input must be derived from the data signal. This is accomplished by using a clock recovery circuit as shown in Figure 8.7.

The frequency doubler is required because, as seen from Figure 8.6, the spectrum of an NRZ signal has a null at the clock frequency. The doubler alters this spectrum to include a strong clock signal. This is usually stabilized by phase-locking it to a low-noise voltage controlled oscillator. At data rates above a few gigabits per second, high-Q passive filters centered at the clock frequency are preferred because of the difficulty of designing a suitable phase-locked-loop.

8.2.5   Example Measurements

BER tests can be used in a variety of ways. The following examples illustrate methods for characterizing the performance of systems, components, and fibers.

Example: Measurement of Fiber Dispersion Power Penalty

The index of refraction of optical fiber varies as a function of wavelength. This effect, called *chromatic dispersion,* causes light at different wavelengths to travel through the fiber at slightly different velocities. Since all lasers have a finite linewidth, the different wavelength components will not all arrive at the end of a long length of fiber at the same time. This effect causes fast risetime pulses to flatten and spread out over time, leading to intersymbol interference (Figure 8.8). The result is known as dispersion power penalty, defined as the difference in minimum detectable power level before and after dispersion through the length of fiber.

The test setup to measure the dispersion power penalty of a length of optical fiber is shown in Figure 8.9. The major components include the BERT, a laser source, a variable optical attenuator, an optical coupler, a lightwave receiver, and an optical power meter. Specific characteristics of each component are:

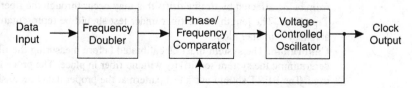

Figure 8.7   Phase-locked-loop clock recovery circuit.

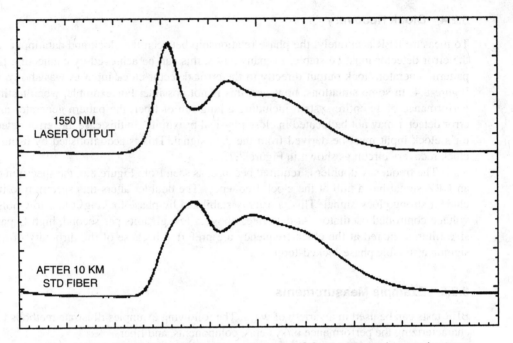

1550 NM
LASER OUTPUT

AFTER 10 KM
STD FIBER

Figure 8.8 Effect of chromatic dispersion on an OC-48 1550 nm laser pulse transmitted through standard fiber whose zero-dispersion point is 1310 nm. Note increased pulse width and distortion after only 10 km of fiber. A directly modulated laser exhibits greatest frequency chirp as it turns on, so chromatic dispersion most affects the front end of the pulse.

- **Laser.** Since dispersion power penalty depends on the spectral characteristics of the laser source, the test laser must be representative of the actual lasers used in the operating system.
- **Optical attenuator.** The optical attenuator must have the ability to vary attenuation in at least 0.1 dB increments over a range of at least 40 dB. It must have excellent repeatability, low chromatic dispersion, and excellent return loss.
- **Optical coupler.** The optical coupler should have approximately a 50:50 coupling ratio.
- **Lightwave receiver.** The lightwave receiver must have high sensitivity. For SONET/SDH testing up to 2.5 Gb/s, a sensitivity of at least −27 dBm is recommended. Because of this, it should use an APD rather than p-i-n photodiode. It also must include clock recovery and data retiming circuits to prevent the measurement from being affected by timing drifts that may occur through the fiber.
- **Test fiber.** The length of the fiber under test should be representative of that used in actual system operation.
- **Calibration.** The system must be calibrated before measuring the fiber. This involves determining the system sensitivity with no fiber in place. The procedure below may be used. The BERT should use a test pattern at the proper data rate to simulate actual operation. In the absence of another specification, a $2^{23}-1$ PRBS pattern is recommended. All power levels should be recorded in dBm.

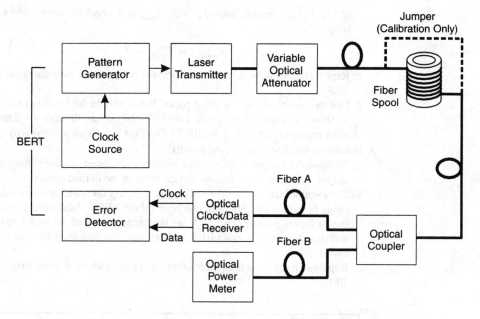

Figure 8.9   Test setup for measuring chromatic dispersion power penalty.

1. Calibrate the loss through the optical splitter:
    a. Connect the jumper as shown in Figure 8.9 and remove input fiber B from the optical power meter.
    b. Remove input fiber A from the lightwave receiver and connect it to the optical power meter. Provide a good optical match for fiber B by connecting it to the lightwave receiver.
    c. Record the attenuator setting necessary to achieve −27.0 dBm optical power as measured on the optical power meter.
    d. Reconnect fiber A to the lightwave receiver and fiber B to the optical power meter without changing the optical attenuator setting.
    e. Record the reading, P_m, in dBm from the optical power meter. Since the power level applied to the lightwave receiver is known to be −27.0 dBm from step (c), the calibration factor to convert from the measured power meter reading in dBm to the optical power at the lightwave receiver is:

$$P_{cal} = -27 - P_m$$

2. Determine system sensitivity without the fiber spool in place.
    a. With the equipment connected as shown in Figure 8.9, connect the optical jumper between the optical attenuator output and the optical coupler input.
    b. Set the optical attenuator to the level necessary to achieve a BER of about 10^{-10}. Set the gating period of the BERT as necessary to capture 100 errors. Record the exact BER and the optical power level into the power meter. Determine power at the re-

ceiver, $P_{receiver}$, from measured power, P_{meas}, at the optical power meter using the relation:

$$P_{receiver} = P_{meas} + P_{cal}$$

  c. Repeat step (b) for at least three other input power levels over the range 10^{-8} to 10^{-4} BER.
  d. Plot the results on log-log graph paper. Points should fall roughly on a straight line as shown in Figure 8.10. Draw a best-fit straight line through the data and determine the power level, P_{ref}, for 10^{-10} BER. This is the system sensitivity.
3. Determine the dispersion power penalty.
  a. Remove the jumper as shown in Figure 8.9 and insert the fiber spool between the output of the optical attenuator and the input of the optical coupler.
  b. Do a preliminary scan of sensitivity by stepping the variable attenuator through a range of values and observing the BER on the BERT. Adjust the attenuator so that the BER is approximately 10^{-10}. Set the gating period of the BERT to capture 100 errors and record the exact BER and the optical power level into the lightwave receiver.
  c. Repeat step (b) for at least three other input power levels over the range 10^{-8} to 10^{-4} BER.

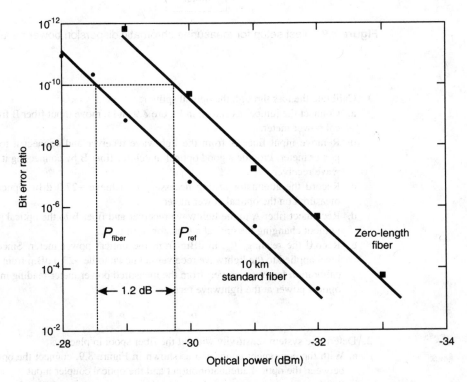

Figure 8.10 Plot of BER vs. optical power level before and after 10 km of fiber.

d. Plot the results on log-log graph paper. Draw a best-fit straight line through the data and determine the power level, P_{fiber}, for 10^{-10} BER.

e. Calculate the dispersion power penalty from:

$$D = P_{\text{fiber}} - P_{\text{ref}}$$

Example: Optimizing Laser Extinction Ratio

The optimum extinction ratio setting for a laser is the bias point that produces the lowest BER. This is determined by adjusting the laser bias for the minimum BER while monitoring real-time error performance on the BERT. The following procedure may be used (equipment requirements are the same as in Example 1):

1. Connect the equipment as shown in Figure 8.11. Set the BERT to the appropriate data rate and select a data pattern that simulates expected operation (a $2^{23}-1$ PRBS pattern is recommended).

2. Set the optical attenuator to an attenuation level high enough to result in a BER in the range of 10^{-6} to 10^{-9}.

3. While observing the instantaneous BER on the BERT, slowly adjust the laser bias in the direction that causes the BER to improve. If the BER goes to zero, increase the attenuation to again obtain a measurable BER. Continue the adjustment until no further improvement is noted. This is the point of optimum bias.

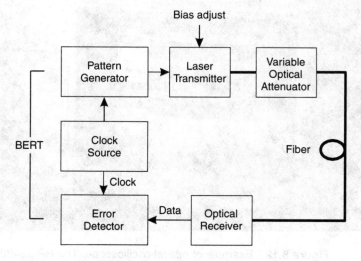

Figure 8.11   Test setup for optimizing laser extinction ratio.

8.3 EYE-DIAGRAM ANALYSIS

BER testing is useful in determining the quality of a transmission link, but it is less suitable for troubleshooting any problems that arise. For this, it is helpful to examine on an oscilloscope the actual shape of the optical waveform. A traditional oscilloscope input is electrical, so a separate optical-to-electrical converter (*O/E converter*) must be used at its vertical channel input. Such systems can provide excellent qualitative representations of the waveform shape, but are difficult to accurately calibrate. Some products now on the market simplify the measurement process by incorporating a built-in, calibrated O/E converter (Figure 8.12).

8.3.1 Eye-Diagram Generation

Oscilloscopes have traditionally been used to display repetitive waveforms such as sine, square, or triangle waves. These are known as *single-value* displays because each point in the time axis has only a single voltage value associated with it.

When analyzing a digital telecommunication waveform, single-value displays are not very useful. Real communications signals are not repetitive but rather consist of random or pseudorandom patterns of ones and zeroes. A single-value display can only show

Figure 8.12 Example of optical oscilloscope. The HP 83480A includes calibrated optical channels and extensive measurement capabilities. (Hewlett-Packard Company.)

a few of the many different possible one-zero combinations. Pattern-dependent problems such as slow rise time or excessive overshoot will be overlooked if they don't occur in the small segment of the pattern appearing on the display. For example, the amount of overshoot on the zero-to-one transition of a SONET laser transmitter depends on the exact pattern preceding it (Figure 8.13).

An *eye diagram* (sometimes called an *eye pattern*) overcomes the limitations of a single-value display by overlapping all of the possible one-zero combinations on the oscilloscope screen (Figure 8.14). Eye diagrams are known as *multi-valued* displays because each point in the time axis has multiple voltage values associated with it.

An eye diagram is generated as shown in Figure 8.15. The data signal (either live traffic or a test signal such as a PRBS pattern) is applied to the oscilloscope's vertical input. A separate trigger signal at the data rate is applied to its trigger input. Ideally, the trigger signal is a sine or square wave at the clock rate of the data, although the data signal itself can be used when only qualitative analysis of the eye diagram is necessary.

The oscilloscope triggers on the first clock transition after its trigger circuit is armed. Upon triggering, it captures whatever data waveform is present at the vertical input and displays it on the screen. The scope is set for infinite persistence so that the waveform remains on the screen and subsequent waveforms continue to add to the display.

After a short deadtime, the trigger circuitry re-arms and triggers on the following clock transition. The data pattern at this instant will most probably be different from the previous pattern, so the display will now be a combination of the two patterns. This

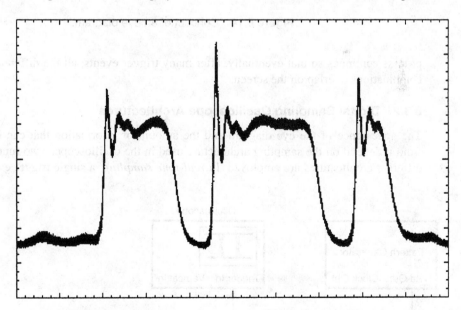

Figure 8.13 Overshoot on a directly modulated laser. The amount of overshoot depends on the data pattern preceding each pulse.

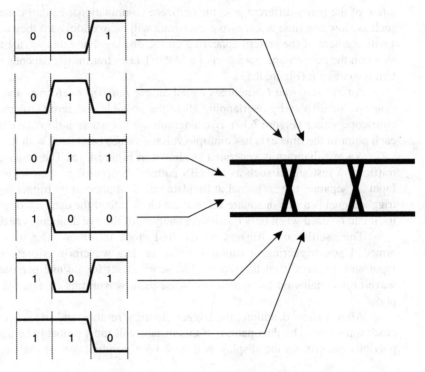

Figure 8.14   Concept of an eye diagram.

process continues so that eventually, after many trigger events, all the different one-zero combinations overlap on the screen.

8.3.2   Digital Sampling Oscilloscope Architectures

The appearance of the eye diagram and the amount of information that can be gleaned from it depend on the sampling architecture used in the oscilloscope. Two fundamentally different architectures are employed. In *real-time sampling*, a single trigger event causes

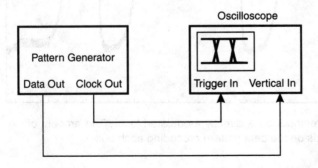

Figure 8.15   Setup for generating an eye diagram.

the entire waveform to be digitized by a fast analog-to-digital (A/D) converter and displayed on the oscilloscope screen. This approach displays a true representation of the actual waveform, but its bandwidth is limited to only a few gigahertz by the availability of high-speed A/D converters. To achieve higher bandwidths, the oscilloscope must use an *equivalent-time sampling* architecture. In this approach, the signal is reconstructed by sampling multiple repetitions of the waveform at a low rate.

Both real-time and equivalent-time oscilloscopes use digital oscilloscope technology as shown in Figure 8.16. The signal at the vertical channel input is conditioned by an optional attenuator and preamplifier, then sampled at discrete instants in time. The sampled voltages are amplified, then digitized by an A/D converter and stored in memory. A microprocessor CPU processes the vertical channel data relative to the trigger signal and presents the result on the display. The details of the sampling, digitizing, and triggering processes depend on whether real-time or equivalent-time sampling is used.

8.3.3   Real-Time Sampling

In real-time sampling, the signal is digitized continuously at a rate fast enough to accurately reproduce the waveform. Shannon's sampling theorem states that to unambiguously digitize a sine wave, the sampling frequency must be at least twice the signal frequency:[5]

$$f_s \geq 2 \times f_{\text{sig}} \tag{8.3}$$

where f_s is the sampling frequency, and f_{sig} is the signal frequency.

The minimum sampling frequency defined by Equation 8.3 is known as the *Nyquist frequency*. At this rate, a sine wave is guaranteed to include two samples per cycle. In reality, observed signals are rarely simple sine waves, so the sampling rate must be higher

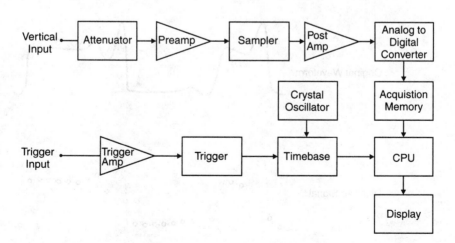

Figure 8.16   Basic digital oscilloscope architecture.

than the Nyquist frequency. Practical real-time sampling oscilloscopes use a sample rate two to five times higher than the Nyquist frequency.

In operation, the real-time sampling scope continuously acquires data at the sampling frequency and stores it in memory in a first-in-first-out sequence. When a trigger event is received, the scope reads the memory and displays the points on screen (Figure 8.17). This approach allows waveforms preceding the trigger event to be displayed. For example, by setting the scope to trigger on a specified failure condition, events leading up to the failure can be observed. The amount of time that can be displayed preceding the trigger is determined by the scope's memory depth.

8.3.4 Equivalent-Time Sampling

The limitations of Shannon's sampling theorem can be overcome through equivalent-time sampling. The tradeoff is that to accurately display the waveform, two conditions must be met:

1. The waveform must be repetitive.
2. A stable trigger signal at the waveform frequency must be available.

Although both conditions must be met to display a single-value representation of the waveform, as we shall see, only the second condition is required to display an eye diagram.

Two different equivalent-time sampling techniques are used: *sequential sampling* and *random repetitive sampling. Microwave transition analysis,* a third technique is essentially a specialized version of sequential sampling with the advantage of not requiring a separate trigger signal.

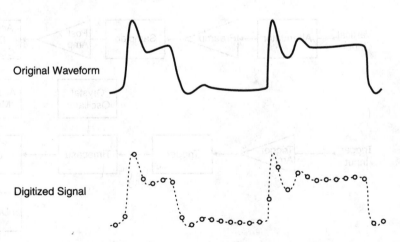

Original Waveform

Digitized Signal

Figure 8.17   Principle of real-time sampling.

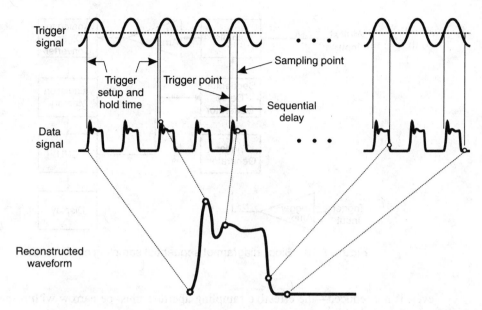

Figure 8.18   Principle of sequential sampling.

Sequential Sampling.   In this approach the signal is reconstructed by sampling sequential repetitions of the waveform. On each trigger, the sampling scope measures only the instantaneous amplitude of the waveform at the sampling instant (Figure 8.18). On the first trigger event, the scope samples the waveform immediately and displays the value at the very left edge of the screen. On the next trigger event, it delays slightly before sampling the data. The amount of this delay depends on the number of horizontal data points on the screen and on the selected sweep speed, and is given by:

$$D = \frac{t}{N - 1} \qquad (8.4)$$

where D is the amount of delay, t is the full-screen sweep time, and N is the number of horizontal points across the screen.

For example, if a full-screen sweep time of 10 nsec is selected and the display has 451 horizontal points, then the delay is $(1/450) \times 10$ ns, or 22.2 ps. The amplitude sampled at this instant is displayed one point to the right of the original sample. On each subsequent trigger event, the scope adds an ever-increasing delay before sampling, so that the trace builds up from left to right across the screen. The block diagram of a sequential sampling oscilloscope is shown in Figure 8.19.

Sequential sampling allows a slow A/D converter to be used, permitting much higher bandwidths to be obtained than with a real-time sampler. Sequential sampling scopes with up to 50 GHz bandwidth are commercially available using A/D converters running at only 40 kilosamples/s. The bandwidth requirement on the sampler itself, how-

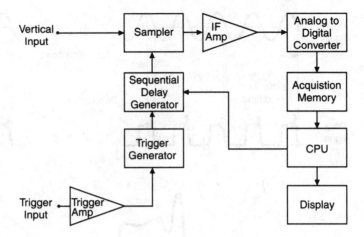

Figure 8.19   Block diagram of sequential sampling oscilloscope.

ever, is not reduced—the effective sampling aperture must be narrow with respect to the signal bandwidth.

Since sequential sampling commences upon receiving a trigger event, it is difficult to obtain information about the waveform prior to the trigger. In principle, a delay line can be inserted in the vertical channel path so that the trigger event occurs before the signal of interest arrives at the scope input. However, the delay line introduces attenuation and distortion, so it is practical only in noncritical applications.

Random Repetitive Sampling.   In this approach, the signal is continuously sampled at a rate independent of the trigger. The amplitude of each sampled point is stored in memory together with a record of the exact time the sample was taken. When a trigger event is received, the scope determines the time difference between the trigger and each data point to determine where on the screen to place the data point (Figure 8.20). Using this approach, it is possible to view pre-trigger events.

Unlike sequential sampling, random repetitive sampling can allow more than one data point to be acquired on each trigger event. For example, if the sampling rate of the scope is 20 megasamples/s (50 ns between successive samples) and the timebase is set to 100 ns full scale, then on average the scope will acquire two points on each trigger.

The main limitation with random repetitive sampling is that it does not work well at microwave frequencies, where sweep times can be very short. For example, a SONET OC-48 signal at 2.48832 Gb/s has a bit period of 402 ps. A typical full-screen sweep time to view this signal would be 1 ns. If the scope samples at 50 ns intervals, then the probability that any given point is within 1 ns of the trigger (and therefore would appear on screen) is only 1/50. At this rate it would take a very long time to fill the screen with data.

Microwave Transition Analysis.   A disadvantage of a sampling oscilloscope architecture is that it requires a separate trigger signal to start the acquisition process. If a

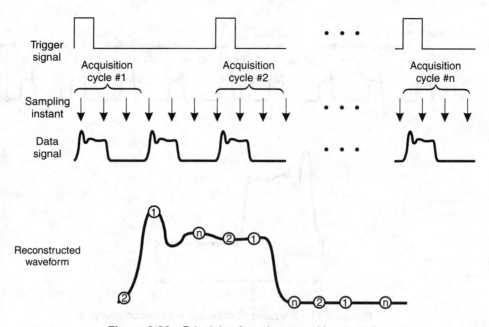

Figure 8.20   Principle of random repetitive sampling.

trigger is not available, it must be derived from the data signal either by direct triggering or by recovering a clock signal from it. This generally adds complexity or reduces the measurement accuracy, or both.

Like a digital sampling oscilloscope, *microwave transition analysis* acquires a waveform by repetitively sampling the input, in other words, one or more cycles of the input signal occur between consecutive sample points.[6] However, unlike an oscilloscope, the sampling frequency is synthesized, based on the frequency of the input signal and the desired time scale. A synthesized sampling rate is an attribute that a microwave transition analyzer shares with dynamic signal analyzers. Also in common is an abundance of digital signal processing capability such as fast-Fourier-transforms (FFTs) that allow simultaneous viewing of the time waveform and frequency spectrum.

This sampling technique is referred to as *harmonic repetitive sampling*. The sample rate is set so that successive sample points step through the measured waveform at a specified time step. The sampling period, T_s, is computed using the fundamental signal period, the time span, and the number of trace points. T_s is set such that an integer number of signal periods plus a small time increment occur between successive sample points:

$$T_s = (N \times T_{signal}) + \Delta T \tag{8.5}$$

where T_s is the sampling period, N is the integer number of signal periods between successive samples, T_{signal} is the fundamental signal period, and ΔT is the time resolution which equals time span/number of trace points.

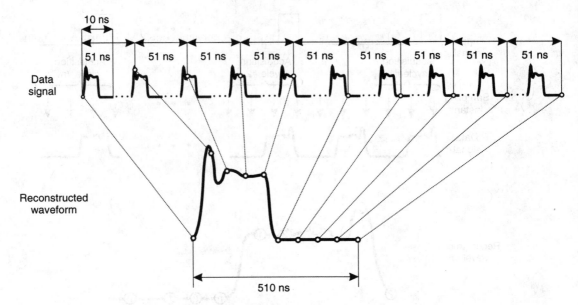

Figure 8.21 Principle of microwave transition analysis.

For example, suppose that the input signal is a 100 MHz square wave as shown in Figure 8.21. Given a minimum sampling period of 50 ns, five cycles of the input waveform are skipped between samples. However, if the sample period were exactly 50 ns, sampling would occur at the same point on the waveform at every fifth cycle, and no new information would be obtained. For a time span of 10 ns and a trace having 10 points, a time resolution, ΔT, of 1 ns is required. For this time resolution, the actual sample period is 50 ns + 1 ns = 51 ns. Thus, at every fifth cycle of the input waveform the sampling point moves forward 1 ns, and after 50 cycles of the input waveform, one complete trace of the input signal would be displayed.

8.3.5 Oscilloscopes for Eye-Diagram Analysis

Although either real-time or equivalent-time sampling oscilloscopes can be used for eye-diagram analysis, equivalent-time scopes have several advantages. They achieve much higher bandwidths at lower cost, and above a few gigahertz are the only available choice. They also generally have better vertical resolution because it is easier to build high-resolution A/D converters at low frequencies. Equivalent-time scopes often have 12 bit resolution compared to 8 bit resolution for fast real-time scopes. Finally, equivalent-time scopes tend to be much less expensive than real-time scopes with similar bandwidths. For these reasons they are preferred for all but the lowest bit rates.

The tradeoff in using an equivalent-time scope is a loss of information about the exact waveform characteristics. When sampling a repetitive waveform such as a sine wave, this doesn't usually pose a problem; the screen displays a sine wave that is a sampled representation of the original waveform. When sampling a random data pattern, however, the eye diagram appears as a series of disconnected dots. All information about the exact nature of the individual waveform is lost—if the eye diagram shows excessive overshoot or slow rise time the exact data pattern causing the problem can't be identified. (In limited cases, this can be overcome by using a display technique called *eyeline* mode, which reconstructs the waveforms comprising the eye, as described in Section 8.3.7.)

Another limitation of equivalent-time scopes is that they are not well-suited for capturing random glitches or noise spikes on the signal. While real-time scopes can be set to trigger on isolated glitches that are then shown on the display, equivalent-time scopes can only show glitches if they occur repeatedly at the same point in the pattern. Fortunately, random glitches are not usually important in optical eye-diagram analysis, so this is not a serious limitation.

Histograms.   Many measurements on an eye diagram are best performed using statistics. For this, an oscilloscope that can perform histogram analysis on the displayed waveforms is necessary. One way to implement histograms is shown in Figure 8.22. A

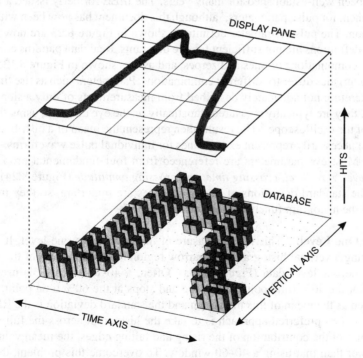

Figure 8.22   Concept of histograms in a digital oscilloscope. Each sampled data point is mapped onto a 3-dimensional database array. The oscilloscope display consists of the 2-dimensional X-Y representation of the database. The full 3-dimensional array is available for statistical analysis.

database array is established within the scope's internal memory, with each screen pixel location mapping into a corresponding database location. Each time a pixel is hit by a waveform, the number stored in its database location is incremented by one. The maximum number of hits that can be counted is limited by the memory depth of the database. For example, a 16 bit memory depth allows 65,536 hits to be recorded.

Statistical analysis is performed by windowing a selected region of the database to find the mean and standard deviation over the windowed region. Vertical histograms indicate the mean and standard deviation of voltage (or optical power) values. Horizontal histograms represent distributions in time.

8.3.6 Eye-Parameter Analysis

Two types of analysis can be performed on eye diagrams. *Eye-parameter analysis* examines fundamental waveform characteristics such as rise time, overshoot, and jitter. *Mask measurements* compare the overall shape of the eye against a predefined mask.

Eye parameters are similar to pulse parameters for single-value waveforms, which have been well-established for many years. The IEEE formerly issued a formal definition document for pulse parameters,[7] although this document has now been withdrawn from circulation. The pulse-parameter definitions shown in Figure 8.23 are now widely used, but these definitions are not sufficient for eye diagrams. Real data patterns consist of many different combinations of ones and zeroes, and as was shown in Figure 8.13, exact waveform characteristics differ for different combinations. Parameters such as rise time and overshoot are therefore not adequately described from measurements of only a single waveform. Instead, they are typically measured statistically on the eye diagram, using the histogram feature of the oscilloscope.[8] The results then represent the mean of a distribution of values and do not necessarily represent exact values for individual pulse waveforms.

Most eye parameters are referenced from four fundamental properties of the eye: *one level, zero level, crossing time,* and *crossing amplitude* (Figure 8.24). Both the means and the standard deviations of these properties are important, so they must be measured using the histogram function of the oscilloscope.

One Level.   This is the measure of the mean logic one-level. It is found by constructing a vertical histogram in a narrow region about the center of the eye that includes only the one-level data (Figure 8.25a). Often, a 40–60 window is used: the histogram starts at the 40% time point of the eye and stops at the 60% time point. The one level is defined as the mean of this histogram and the standard deviation is the RMS noise.

A less preferred approach is to take the histogram across the full width of the eye. Because of the contribution of the rising and falling edges, the mean value found this way is lower than that using a 40–60 window. To overcome this problem, the one level must be defined as the histogram *peak* rather than its mean. The problem is that the peak value does not necessarily stabilize as additional data is collected, so it is not very repeatable. However it does have the advantage of being less sensitive to any waveform aberrations occurring at the center of the eye.

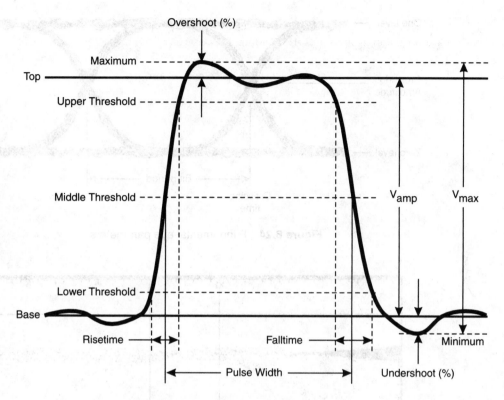

Figure 8.23   Pulse parameter definitions.

Zero Level.   The zero level is the mean logic zero-value. Measurement techniques are identical to those used in finding one level.

Eye Crossing.   Eye crossing consists of two parts: *crossing time* and *crossing amplitude.* By convention, the start of the bit period is defined as the point in time where the rising and falling edges of the eye intersect. To find this point, a horizontal histogram is constructed using only data from the rising and falling edges (Figure 8.25b). The mean of this histogram is the crossing time, and the standard deviation is the jitter. To find the crossing amplitude, a vertical histogram is constructed on the same data window.

More complex eye parameters begin with these three properties as references. In some cases, additional histograms must be constructed over other regions of the eye. The following discussion of rise/fall time and overshoot measurements illustrate the concepts involved. Definitions for a number of important eye parameters are shown in Table 8.5. Measurement of extinction ratio warrants additional consideration and is discussed in Section 8.3.8.

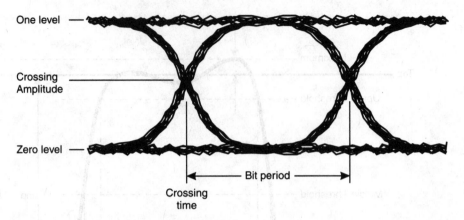

Figure 8.24   Fundamental eye parameters.

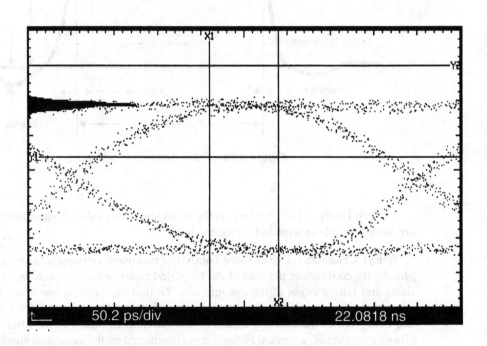

histogram

scale	207 hits/div		offset	0 hits	
mean	1.167061 mW		median	1.167745 mW	$\mu\pm1\sigma$ 80.4%
std dev	24.2726 μW		hits	3.765 khits	$\mu\pm2\sigma$ 97.6%
p-p	180.39 μW		peak	518 hits	$\mu\pm3\sigma$ 99.6%

(a)

Figure 8.25   Histogram windows for finding eye parameters. (a) Voltage histogram window for one level. *(continued)*

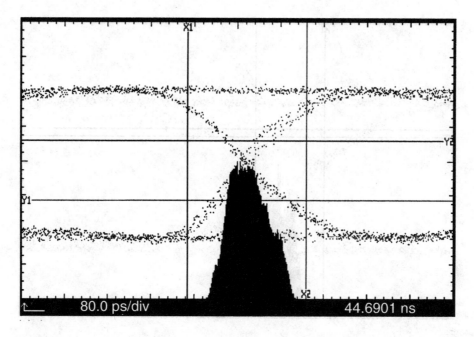

histogram

scale	237 hits/div	offset	0 hits
mean	45.11172 ns	median	45.10890 ns
std dev	34.6512 ps	hits	50.61 knits
p-p	208.0 ps	peak	949 hits

μ±1s	65.5%
μ±2s	97.7%
μ±3s	100%

(b)

Figure 8.25 (b) Time histogram window for eye crossing time. *(continued)*

Rise Time/Fall Time. Traditionally, rise time is defined as the time from when the rising edge reaches 10% of its final amplitude to the time it reaches 90% of its final amplitude. Fall time is similarly defined for the falling edge. For optical signals, however, these points are often obscured by the noise and jitter on the one and zero levels, so 20% and 80% thresholds are normally used. To convert from 20–80 rise time to 10–90 rise time, assuming a fourth-order Bessel-Thomson filter response, use the approximate relationship:

$$T_{10-90} = 1.25 \times T_{20-80} \tag{8.6}$$

The rise-time measurement begins by determining the one and zero levels of the eye. Once these levels are known, the time locations of the 20% and 80% points can be found by constructing narrow horizontal histograms about these amplitudes as shown in Figure 8.25c. A similar approach is used to determine fall time.

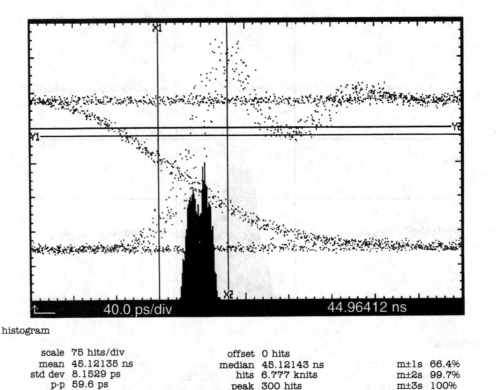

histogram

scale	75 hits/div	offset	0 hits		m±1s	66.4%
mean	45.12135 ns	median	45.12143 ns		m±2s	99.7%
std dev	8.1529 ps	hits	6.777 knits		m±3s	100%
p-p	59.6 ps	peak	300 hits			

(c)

Figure 8.25 (c) Time histogram for 80% threshold point of eye diagram rising edge.

Overshoot. The level of overshoot on a directly modulated laser often varies by more than a factor of three for different one-zero combinations. Moreover, amplitude noise can create an occasional spike well above normal. The definition of overshoot on an eye is therefore somewhat subjective. One choice is to define it as the peak value of the overshoot, but the measured peak depends greatly on the length of time over which the data is collected and it rarely repeats on subsequent measurements. Another possibility is to define it as the mean value of the overshoot. While this value is somewhat more stable, it seriously understates the true overshoot level, for fully half the bit sequences exceed this level.

Empirical studies have shown that most people select a point at about the 95th percentile of data above the one level as their intuitive choice for eye overshoot. While this point is not easily found from the mean and standard deviation of the histogram, some oscilloscopes include an internal algorithm to measure it. It can also be approximated by adding two standard deviations to the mean overshoot.

Table 8.5 Eye-Parameter Definitions

Eye Parameter	Definition
Eye height	$(V_{top} - 3\sigma_{top}) - (V_{base} + 3\sigma_{base})$
Eye width	$(t_{crossing2} - 3\sigma_{crossing}) - (t_{crossing1} + 3\sigma_{crossing})$
RMS jitter	$\sigma_{crossing}$
Q factor	$\dfrac{V_{top} - V_{base}}{\sigma_{top} + \sigma_{base}}$
Duty-cycle distortion	$\dfrac{|t_{r50\%} - t_{f50\%}|}{t_{crossing2} - t_{crossing1}} \times 100\%$
Overshoot	$\dfrac{V_{top} + V_{95}}{V_{top} - V_{base}} \times 100\%$

V_{top} = mean logic one-level
V_{base} = mean logic zero-level
$\sigma_{crossing}$ = standard deviation of time histogram about eye-crossing level
σ_{top} = standard deviation of amplitude histogram about logic one-level
σ_{base} = standard deviation of amplitude histogram about logic zero-level
$t_{r50\%}$ = time location of 50% amplitude level of rising data edge
$t_{f50\%}$ = time location of 50% amplitude level of falling data edge
$t_{crossing1}$ = time location of first eye-crossing
$t_{crossing2}$ = time location of second eye-crossing
V_{95} = 95th percentile of amplitude distribution extending upward from mean logic one-level to maximum data value

8.3.7 Eyeline Diagrams

Sequential sampling oscilloscopes display eye diagrams as a series of disconnected points on the screen. While these points accurately represent all the combinations of digitized bit patterns, each point is obtained from a separate trigger event, so there is no way to determine exact characteristics of any specific bit combination (Figure 8.26a).

When sampling live data there is no alternative, but when sampling repetitive waveforms such as PRBS patterns it is sometimes possible to show the individual bit sequences. This is done by synchronizing the oscilloscope trigger to the pattern repetition rate. In this case, the oscilloscope repeatedly triggers at the same point in the pattern, so the display is a sampled representation of that segment of the pattern.

This display mode, called *pattern triggering,* is commonly used in situations where the device under test can be stimulated with the digital pattern generator from a BERT. Pattern generators typically include a trigger output that can be set to produce only a single trigger pulse at the start of each pattern. The problem in using this technique to generate eye diagrams is that portions of the pattern far from the trigger point cannot accurately be displayed. While it is theoretically possible to show all parts of the pattern by increasing the oscilloscope delay, in reality this is impractical because of accuracy and jitter limitations in the oscilloscope timebase.

This problem can be overcome by adjusting the pattern trigger output rather than the oscilloscope delay. Some pattern generators allow the pattern trigger point to be adjusted bit-by-bit within the pattern, and certain lightwave oscilloscopes have been designed to take advantage of this capability.[9] Using this feature it is possible to generate a type of display known as an *eyeline* display. A computer program run from the oscilloscope automatically sets the pattern trigger location in the data pattern. The oscilloscope samples one data point for each repetition of the pattern; the full data pattern is transmitted between successive triggers. After an entire waveform record is taken (typically 500 to 4000 points, depending on the number of horizontal points across the display), the computer program delays the trigger point by one bit and the oscilloscope repeats the process. Eventually, the trigger point moves through the entire pattern, and the eye diagram shows all possible bit combinations. An eyeline display is shown in Figure 8.26b.

One advantage of the eyeline display mode is that signal averaging can be used to reduce the effects of noise. (Averaging is not possible on an ordinary sampled eye because the result is the average between the two logic levels, which causes the eye to collapse.) Signals too small to be seen without averaging can be readily identified using the eyeline mode as shown in Figure 8.27. Another advantage is that it can aid troubleshooting by showing the bit sequence leading up to a mask test violation, as seen in Figure 8.28. This can be used to identify the cause of pattern-dependent errors.

While the eyeline display mode is a powerful analysis tool, it is not the solution to all problems. It can only be used with pattern generators having a programmable trigger output, so it is not suitable for analyzing live traffic. And because it relies on multiple repetitions of the pattern to generate the eye, it is most suitable for short pattern sequences that repeat rapidly. At an OC-48 data rate (2.48832 Gb/s), for instance, a complete eye showing all bit combinations of a $2^7 - 1$ PRBS pattern takes less than 2 seconds to generate. A complete $2^{23} - 1$ PRBS pattern at OC-3 (155.52 Mb/s), however, would take 7.3 years!

Eyeline diagrams can also be generated through the use of microwave transition analysis. In fact, this was the method by which the technique originated.[10] Using the process of harmonic repetitive sampling as described in Section 8.3.4, conventional eye diagrams are generated when the signal frequency is set equal to the clock frequency. Eyeline diagrams are generated when the signal frequency is set equal to the pattern repetition frequency. For PRBS patterns the signal frequency is thus the clock rate divided by the pattern length. An advantage of this approach is that it can be used with pattern generators that do not have the ability to reprogram the pattern trigger point within the pattern sequence.

The technique is illustrated in Figure 8.29. In this example, the clock rate is 1 GHz and the pattern length is 15 bits, so the pattern repetition period, T_{signal}, is 15 ns. Using Equation 8.5, for a time span of 15 ns, minimum sample period of 50 ns, and a number of trace points equal to 15, ΔT is 1 ns and T_s is found to be 61 ns. Thus, every fourth cycle of

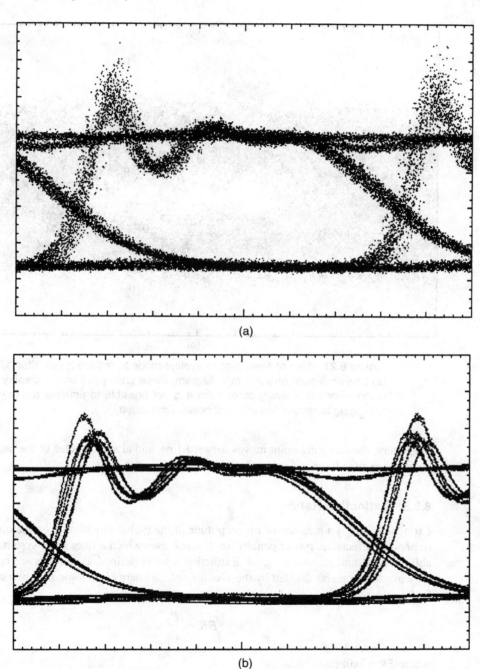

(a)

(b)

Figure 8.26 Comparison of ordinary eye diagram and eyeline diagram displayed on sequential sampling oscilloscope. (a) Ordinary eye diagram. (b) Eyeline diagram. Unlike (a), the individual waveforms comprising the eye can readily be observed.

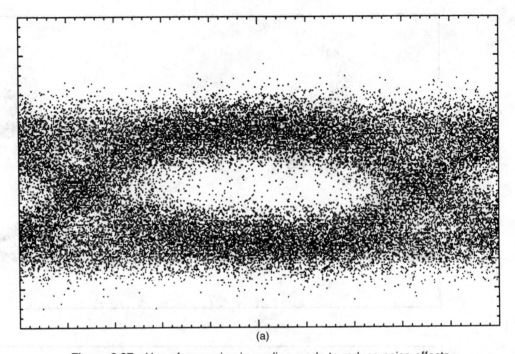

(a)

Figure 8.27 Use of averaging in eyeline mode to reduce noise effects.
(a) Low amplitude ordinary eye diagram. Since each point on the display
is independent of every other point it is not possible to activate display
averaging to reduce the effect of noise. *(continued)*

the pattern, the sampling point moves forward 1 ns, and after 60 cycles of the pattern, one
complete eyeline trace is displayed.

8.3.8 Extinction Ratio

Extinction ratio is a measure of the amplitude of the digital modulation on the optical carrier and so affects the power penalty, or distance over which a fiber optic system can reliably transmit and receive a signal. Extinction ratio is defined as the average optical energy in a logic one bit divided by the average optical energy in a logic zero bit (see Figure 8.30):

$$ER = \frac{P_1}{P_0} \qquad (8.7)$$

where ER = extinction ratio
$\quad P_1$ = average optical power in logic one pulse
$\quad P_0$ = average optical power in logic zero pulse

Extinction ratio is often expressed as a dB value:

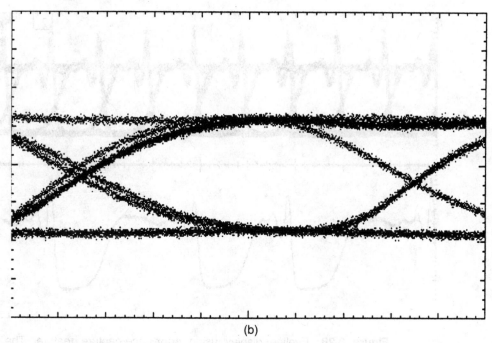

(b)

Figure 8.27   (b) Eyeline diagram of same signal in (a), with 64-point averaging activated.

$$EX = 10 \log_{10}(ER) \tag{8.8}$$

In certain situations it is convenient to to express extinction ratio as a percentage defined by:

$$XR = \frac{100}{ER} \tag{8.9}$$

While it would seem that an infinite extinction ratio value is best because it produces maximum signal swing, this is not usually practical for laser transmitters. As was shown in Figure 8.13, when a directly modulated laser that has been biased near the completely "off" state (in other words, when transmitting a logic zero) is turned on to transmit a logic one, it suffers from turn-on delays and waveform distortions that can cause transmission errors. Therefore, the laser must be biased so that a small amount of optical power is transmitted even when sending a logic zero pulse. This reduces extinction ratio, so the optimum bias point for the laser becomes a compromise between best waveform fidelity and best extinction ratio. Externally modulated lasers do not suffer this problem, but even the best lithium-niobate modulator is not completely opaque to light when biased off.

The difference between the theoretical system sensitivity with infinite extinction ratio and that with a finite extinction ratio is called *extinction-ratio power penalty*. The relationship between extinction ratio and power penalty is shown in Figure 8.31.

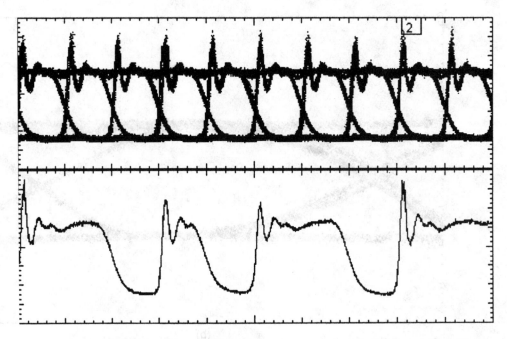

Figure 8.28 Eyeline display using error-trace-capture feature. The upper trace shows the entire eyeline diagram. A custom mask labeled "2" has been created on this trace to capture waveforms having excessive overshoot. Patterns that violate this mask are displayed on the lower trace.

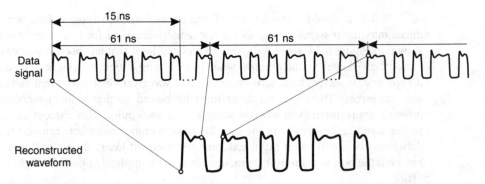

Figure 8.29 Principle of eyeline mode using microwave transition analysis. In this example the bit rate is 1 Gb/s, the clock rate is 1 GHz, and the pattern length is 15 bits. The displayed time span is set to 15 ns.

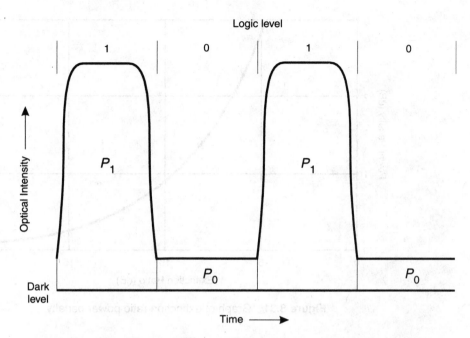

Figure 8.30   Definition of extinction ratio. Note that extinction ratio is defined as the ratio of the *average* logic one and zero powers, not the ratio of the peak powers. It is therefore necessary to integrate the instantaneous logic one and zero level powers across the full bit interval to calculate extinction ratio.

Extinction ratio is defined in terms of average optical one and zero powers (rather than instantaneous one and zero voltages) because this accurately reflects the performance impact on a real transmission system. The decision circuit in most system receivers determines whether each bit is a one or a zero through a technique called *integrate-and-dump* detection.[11] This approach minimizes the effect of noise fluctuations and produces the receiver with the best possible sensitivity.

Integrate-and-dump detection uses a receiver with heavily filtered bandwidth followed by a fast sampler (Figure 8.32). The filter serves as an integrator which averages the instantaneous voltage over the period of the filter time constant. In the ideal case, the filter bandwidth is set so that the averaging period extends over one entire bit period. A threshold detector then samples the voltage at the very end of the period and determines whether it is above or below the predefined threshold. Averaging in this way minimizes noise effects. In contrast, a receiver that made a decision based only on an instantaneous value at the center of the bit period would be very sensitive to noise variations. (Note that the actual sampling interval should still be as short as practical.)

Standards for communications systems such as SONET/SDH and Fibre Channel specify minimum extinction ratio requirements for laser transmitters. It is therefore im-

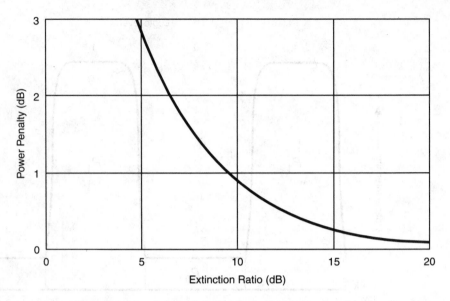

Figure 8.31 Graph of extinction ratio power penalty.

portant that any given transmitter, when measured on different test systems, yields a similar extinction ratio value.

Extinction Ratio Measurements. While extinction ratio is simple in concept, it is notoriously difficult to accurately measure. The most common method uses an O/E converter and electrical oscilloscope to measure the one and zero levels on an eye diagram. This method is described in TIA/EIA Recommendation OFSTP-4.[12] The histogram capabilities of the oscilloscope are used to determine the one and zero level amplitudes. To accurately represent average energies in the one and zero bits, OFSTP-4 recommends the measurement be made using an optical reference receiver bandwidth as defined in SONET/SDH standards: 4th-order Bessel-Thomson response with a 3 dB frequency at 3/4 of the bit rate. This filter approximately integrates the input voltage over a single bit inter-

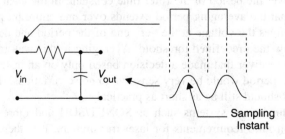

Figure 8.32 Concept of integrate-and-dump detection. The low-pass filter serves as an integrator, so that the instantaneous voltage at its output at the end of the bit period corresponds to the average power over the entire bit period.

val, and since it is also specified for mask measurements, allows both mask and extinction ratio measurements to be made simultaneously.

The design of the reference receiver is critical to the accuracy of the measurement. Two aspects are particularly important: frequency response and output offset voltage. SONET/SDH standards require that the frequency response fall within a very narrow window around ideal (Figure 8.33). The response must be well-controlled all the way down to DC. Low-frequency gain shifts are particularly troublesome, as even a 0.5 dB increase in gain below 1 MHz can result in as much as a 3 dB error in measured extinction ratio.[13]

Much of the measurement error caused by frequency response can be removed using a technique reported by Andersson and Akermark.[14] When extinction ratio is expressed in percent according to Equation 8.9, errors caused by frequency response variations appear as a constant percentage offset that is independent of the actual extinction ratio. To calibrate this offset, a signal with known extinction ratio is applied to the reference receiver input. The difference between the true extinction ratio and the measured extinction ratio in percent is the calibration factor, $C_\%$. Once this factor is known, true extinction ratio, XR, can be found from measured extinction ratio, XR_{meas}, using the relationship:

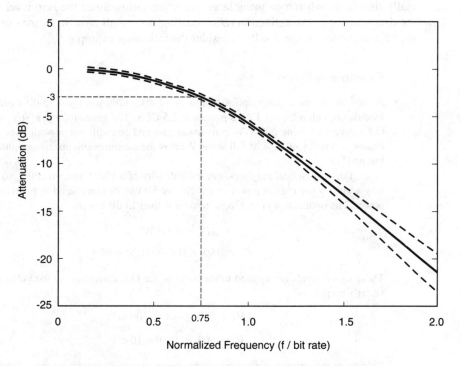

Figure 8.33   SONET/SDH filter tolerance window for data rates up to 622 Mb/s/ The tolerance is ±0.3 dB up to the –3 dB frequency, expanding to ±2.0 dB at twice the –3 dB frequency.

$$XR = XR_{meas} - C_\% \tag{8.10}$$

where XR and XR_{meas} are expressed in percent according to Equation 8.9.

Output offset voltage is defined as any residual voltage at the reference receiver's output when no light is present at its input. Small offsets, as long as they are stable, can be subtracted from the measured amplitudes using the relationship,

$$ER = \frac{P_1 - P_{offset}}{P_0 - P_{offset}} \tag{8.11}$$

where P_1 = measured one level, P_0 = measured zero level, and P_{offset} = measured offset level.

Offsets that drift with time or temperature cannot reliably be removed. It is also difficult to compensate for offsets that are large compared to the one level and zero level.

Other parameters that can affect measured results include data pattern, data rate, and oscilloscope accuracy. The data pattern and data rate should reasonably simulate the expected application. For example, the extinction ratio measured on a laser transmitter modulated with a $2^{23} - 1$ PRBS pattern at 2.5 Gb/s will be quite different from that measured on the same laser using a 1010 pattern at 50 Mb/s.

Oscilloscope accuracy is often overlooked as an error source, but it can be especially significant when measuring large extinction ratios. Since the zero level amplitude is the denominator of the extinction ratio equation, any small error in its measurement has a significant impact on the results. Consider the following example:

Example

A laser having an average optical power of −7 dBm (200 μW) and 13 dB extinction ratio is modulated with a $2^{23} - 1$ PRBS pattern at 2.5 Gb/s. The measurement system consists of an O/E converter having 500 V/W conversion gain and an oscilloscope with a specified vertical channel accuracy of ±1% of full scale. What is the measurement uncertainty due to the oscilloscope?

For all practical purposes, the mark density of a PRBS pattern is 0.5, so to the necessary accuracy the optical power in a logic one bit can be considered to be twice the average power. The optical power in a logic zero bit is then 13 dB lower:

$$P_1 = 0.4\ mW$$

$$P_0 = 0.05 \times 0.4 = 0.020\ mW$$

These signal levels are applied to the input of the O/E converter, so the voltages at the converter's output are:

$$V_1 = 500 \times 0.4 = 200\ mV$$

$$V_0 = 500 \times .020 = 10\ mV$$

To display the entire eye diagram on the screen, the oscilloscope is set to 250 mV full scale. The uncertainty in the measurement is then ±2.5 mV. The worst case measured extinction ratio is then found from:

$$ER_{max} = \frac{(200 + 2.5)}{(10 - 2.5)} = 14.3 \text{ dB}$$

$$\Delta_{ER} = 1.3 \text{ dB}$$

This amounts to a 35% measurement error!

One final word on measuring extinction ratio on an oscilloscope is that it should not be measured on an unfiltered eye diagram. Recall from Equation 8.7 that extinction ratio is defined as the ratio of two *average* optical powers. The unfiltered eye, however, displays *instantaneous* power, so the value at the center of the eye is not necessarily an accurate representation of average power. As was shown in Figure 8.32, the center of the filtered eye corresponds to the average optical power over the full bit period and so should be used for determining extinction ratio.

Power Meter Measurement Method.   The limitations inherent in using an eye diagram to measure extinction ratio have led to the development of a new measurement technique based on direct measurements of optical power.[15] Briefly, as shown in Figure 8.34, two optical powers are measured at the output of a high-speed O/E converter. One path uses a microwave power meter to measure the power in the modulating signal, P_{mod}. The other path uses a voltmeter to measure the average photodiode current, which is proportional to overall average optical power, P_{av}. Extinction ratio is found from the relationship,

$$ER = \frac{P_{av} + P_{mod}}{P_{av} - P_{mod}} \tag{8.12}$$

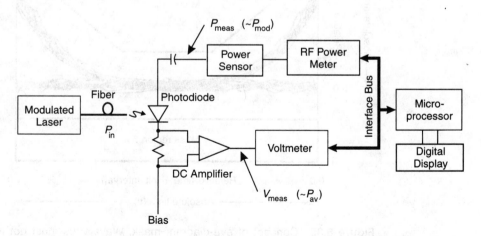

Figure 8.34   Power meter method for measuring extinction ratio.

8.4 MASK MEASUREMENTS

Mask tests are often used in production environments as an alternative to eye-parameter analysis. By comparing an eye diagram against a predefined mask, the overall quality of the waveform can be assessed in one quick measurement.

8.4.1 Mask Definition

Technically, a *mask* defines a set of keep-out regions within which a waveform must not intrude. The converse is known as a *template*, which defines an envelope within which the waveform must remain. Masks are used for eye diagrams while templates are used for pulse waveforms. The principles of mask testing apply equally to templates, so in this chapter the general term "mask" is used to describe both templates and true masks.

A mask consists of two parts, as shown in Figure 8.35:

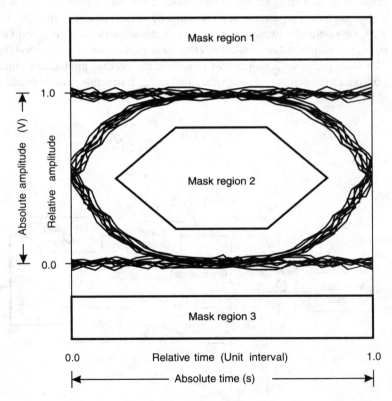

Figure 8.35 Concept of eye-diagram mask. Waveforms must not intrude into the shaded regions.

- A set of regions, or polygons, on the oscilloscope screen that define keep-out areas for the waveform. Waveforms that intrude into these polygons are counted as mask violations.
- Definitions of the time and amplitude scales for the mask.

The earliest masks were simply drawn on the oscilloscope screen with a grease pencil. Later oscilloscopes included rudimentary built-in mask drawing features. A limitation of these instruments was that the masks were drawn in screen coordinates that did not relate to the scale of the waveform. As the user adjusted the horizontal or vertical scales, the mask remained fixed on the screen. More recent oscilloscopes reference masks to true time and amplitude coordinates so that as the user changes the oscilloscope settings, the mask follows the waveform. This also makes it easy to rescale the mask for different data rates or amplitude levels.

It is easiest to first define the mask in terms of a relative coordinate system, then apply actual time and amplitude scales as a second step. The relative coordinate system is scaled to the eye diagram: the point (0, 0) is the amplitude of the zero level at the left-eye crossing and the point (1, 1) is the one level of the right-eye crossing. The time from one eye crossing to the next is the bit period, known as the *unit interval* (UI). (Not all masks use an amplitude scale relative to the waveform—some templates require fixed voltage levels independent of measured signal levels.)

This approach permits the entire mask definition to be easily compressed or expanded in either axis by assigning new amplitude or time values to the (0, 0) and (1, 1) coordinates. It also makes it easy to keep the mask aligned to the signal when the timebase or amplitude settings of the oscilloscope are adjusted.

8.4.2 Mask Margins

In a manufacturing environment it is often desirable to add a test-line margin to industry standard masks. At other times, it is useful to reduce the size of the mask to determine by how much a waveform fails the test. Some oscilloscopes allow this to be accomplished by adding a mask margin to the standard mask. The margin is defined as a percentage above or below the size of the standard mask. Since there are no industry standards that define mask margins this feature is more useful as a qualitiative rather than quantitative measure of waveform performance. An example of a mask margin is shown in Figure 8.36.

8.4.3 Mask Alignment

Before conducting a test, the mask must be properly aligned to the waveform. This is done by locating three points on the eye: one level, zero level, and crossing time. The mask's relative coordinate system can be positioned against the eye using these points as references.

For masks requiring fixed-voltage one and zero levels it is only necessary to locate the start of the bit interval. On a pulse waveform this is defined, by convention, as the time at which the rising edge reaches the 50%-point between the zero and one levels. In defining the voltage levels for such masks it is important to account for any external atten-

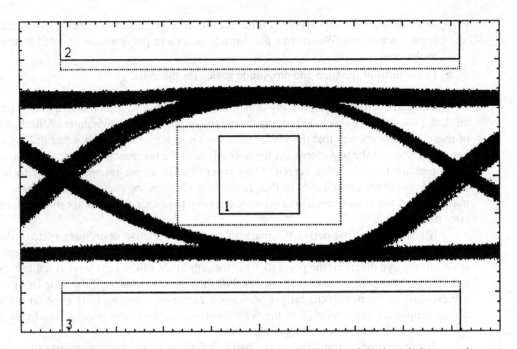

Figure 8.36 Mask with 25% margin added. Both the original and the margin masks are shown.

uation in the signal path. For example, the European PDH 2.048 Mb/s mask requires a zero level of 0.00 volts and a one level of 2.37 volts, defined in a 75 ohm system. When measured on an oscilloscope with a 50 ohm input impedance a 75 to 50 ohm minimum loss pad must be used. The 5.7 dB insertion loss of this pad must be taken into account when measuring voltage levels on the oscilloscope.

Automatic mask alignment algorithms in many oscilloscopes operate by centering both the start of the bit period and the mean one and zero levels inside the mask limits as shown in Figure 8.37a. However, some mask standards allow the waveform to pass if it can be adjusted to fit anywhere within the mask window, as shown in Figure 8.37b. For such masks, if the oscilloscope does not include an automated "best fit" routine it should at least allow manual adjustment of the mask with respect to the waveform.

8.5 JITTER TESTING

8.5.1 Introduction

Jitter is defined as short-term phase variations of the significant instants of a digital signal from their ideal positions in time. The significant instant can be any convenient, easily identifiable point on the signal such as the rising or falling edge of a pulse. A second para-

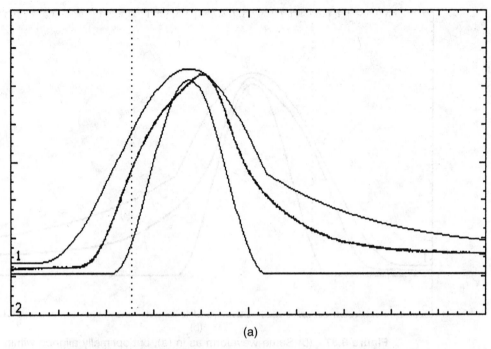

(a)

Figure 8.37 Results of different automated mask alignment algorithms. (a) Algorithm that centers the middle of the rising edge within the template. In this case, the waveform appears to fail the test. *(continued)*

meter closely related to jitter is wander. Wander similarly refers to long-term variations in the significant instants. Although there is no standardized boundary separating jitter from wander, phase variations below 10 Hz are normally called wander.

Figure 8.38 compares an ideal clock signal to a clock signal having timing jitter. By plotting the relative displacement in the significant instants, T_n, versus time, the jitter time function is obtained. In reality, jitter time functions are typically not sinusoidal, but as we will see shortly, it is often useful to characterize jitter performance by applying a controlled level of sinusoidal jitter to the transmission system.

Jitter amplitude is often expressed in terms of UI, as defined in Section 8.4.1. Figure 8.39 illustrates an eye diagram with 0.5 UI of jitter applied to the data. Notice that with jitter applied, the width of the eye opening is reduced such that the peak-to-peak jitter at a crossing point covers half of the bit period (nominal time between the crossing points). By specifying jitter in unit intervals rather than absolute time units, the result is independent of the actual data rate. This makes it possible to easily compare jitter amplitude at different hierarchical levels in a digital transmission system.

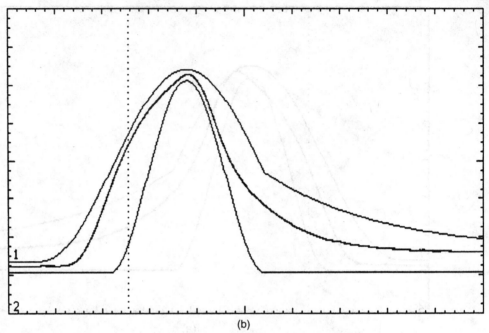

(b)

Figure 8.37   (b) Same waveform as in (a), but optimally aligned within the template. The waveform is now seen to pass the test.

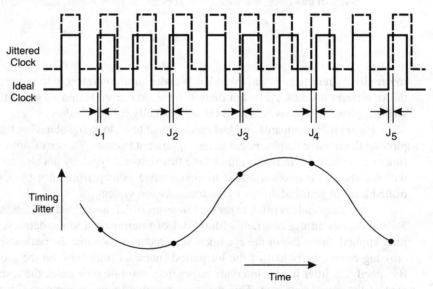

Figure 8.38   Jitter time function derived from comparing a jittered clock with an ideal clock. (Hewlett-Packard Journal. © Copyright 1995 Hewlett-Packard Company. Reproduced with permission.)

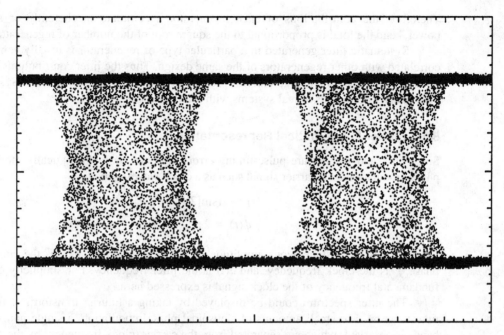

Figure 8.39 Eye diagram with 0.5 UI of jitter intentionally impressed upon the data. (Hewlett-Packard Journal. © Copyright 1995 Hewlett-Packard Company. Reproduced with permission.)

8.5.2 Jitter Issues

It is important to control jitter because it can degrade the performance of a transmission system. Jitter causes bit errors by preventing the clock recovery circuit in the receiver from sampling the digital signal at the optimum instant in time. To accurately determine whether a given bit is a one or zero, the signal should be sampled at the instant in time where the vertical eye opening is maximum. This decision point is set by the recovered clock signal. If jitter on the data causes this point to move away from the optimized location, the decision margin decreases and the system BER increases. In addition, jitter can accumulate in a transmission network depending on the jitter generation and transfer characteristics of the interconnected equipment.

Several types of jitter can arise in a transmission system. *Random jitter* is independent of the transmitted data pattern. It originates primarily from noise generated by the regenerator electronic components. *Systematic jitter* is pattern dependent and results from the finite Q of the clock recovery circuit and its relationship to the transmitted data spectrum. Systematic jitter can also result from waveform distortion and amplitude-to-phase-noise conversions.

Since transmission systems often employ a number of regenerators between terminals, it is useful to consider how jitter accumulates as a function of the number of regenerators in a system. Random jitter generated internally by a particular regenerator is uncorrelated with the random jitter generated by all the other regenerators in the system.

Therefore, the contributions of each regenerator accumulate incoherently or add "in power," and the total is proportional to the square root of the number of regenerators.

Systematic jitter generated in a particular type of regenerator is usually completely correlated with other regenerators of the same design. Thus the jitter contributions add "in voltage" and accumulate coherently in proportion to number of regenerators. Systematic jitter usually dominates in real systems with multiple regenerators.

8.5.3 Jitter Mathematical Representation

Since jitter and wander are pulse-timing errors they can be mathematically modeled as phase modulation on a carrier signal such as a clock waveform.

$$V(t) = A\sin[2\pi f_c + J(t)]$$
$$\phi(t) = 2\pi f_c + J(t) \tag{8.13}$$

where $V(t)$ is the instantaneous amplitude including modulation, $\phi(t)$ is the cumulative phase, f_c is the clock frequency, and $J(t)$ is the jitter frequency. For simplicity only the fundamental frequency of the clock signal is expressed here.

The jitter spectrum could be displayed by taking a Fourier transform of the jitter time function. For small amounts of sinusoidal phase modulation, a single pair of sidebands is observed, which are separated from the carrier (clock frequency) by the modulation frequency. For small values of modulation index, the magnitude of the sideband is linearly proportional to the modulation index. As the modulation index increases, additional sidebands appear. When this occurs, the amplitude of the nth sideband relative to the magnitude of the unmodulated carrier, can be calculated using the nth ordinary Bessel function, with the modulation index, β, as an argument.

$$A_n = J_n(\beta)$$
$$\beta = \pi \times UI \tag{8.14}$$

Figure 8.40 shows a series of measurements of a jittered clock signal in both the frequency and time domains, along with the demodulated jitter time function and spectrum. Fig. 8.40a shows a 2.48832 GHz (OC-48) clock signal in the time domain. The jitter function in this example is a sinusoid at a 10 kHz rate with an amplitude that corresponds to a phase deviation of 0.25 UI peak-to-peak. This display is similar to what would be seen on a high-speed oscilloscope. Figure 8.40b shows the clock spectrum and corresponding jitter sidebands. This display is similar to what would be seen on an RF-spectrum analyzer.

8.5.4 Jitter Measurement Categories

Both SONET and SDH standards specify the jitter requirements at the optical interfaces necessary to control jitter accumulation within the transmission system.[16,17] The transmission equipment specifications are organized into the following categories: *jitter tolerance, jitter transfer,* and *jitter generation.*

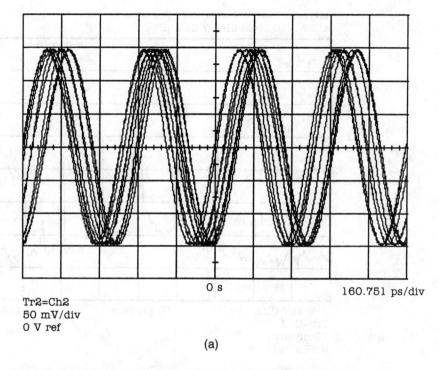

Tr2=Ch2
50 mV/div
0 V ref

0 s   160.751 ps/div

(a)

Figure 8.40   Jittered 2.48832 GHz clock signal. (a) Time domain representation. (Hewlett-Packard Journal. © Copyright 1995 Hewlett-Packard Company. Reproduced with permission.) *(continued)*

Jitter Tolerance.   This is a measure of how well the receiver can tolerate a jittered incoming signal. It is defined as the amplitude of applied sinusoidal jitter applied to an equipment input that causes a specified degradation in error performance. It is determined by measuring BER in the presence of the applied jitter signal. For telecom equipment, jitter tolerance is specified using jitter tolerance templates. Each template defines the region over which the equipment must operate while maintaining better-than-specified BER. The difference between the template level and actual equipment tolerance curve represents the operating jitter margin, and determines the pass/fail status. Each transmission rate typically has its own input jitter tolerance template. In some cases, there are two templates for a given transmission rate to accommodate different regenerator types.

To perform a jitter tolerance measurement, a method for supplying data with known levels of jitter is required. A common technique for jitter tolerance testing is to attenuate the nonjittered signal power until the onset of errors or a specific BER is obtained. The attenuation is reduced 1 dB and sinusoidal jitter is impressed on the data. SONET/SDH standards call for specific jitter magnitudes and frequency ranges over which a compliance measurement is to be made.

Figure 8.41 is an example of a jitter tolerance measurement. The horizontal axis, logarithmically scaled, indicates the jitter frequency. The vertical axis, also logarithmi-

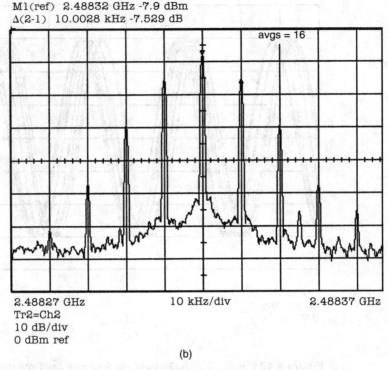

M1(ref) 2.48832 GHz -7.9 dBm
Δ(2-1) 10.0028 kHz -7.529 dB

avgs = 16

2.48827 GHz   10 kHz/div   2.48837 GHz
Tr2=Ch2
10 dB/div
0 dBm ref

(b)

Figure 8.40   (b) Frequency domain representation. (Hewlett-Packard Journal. © Copyright 1995 Hewlett-Packard Company. Reproduced with permission.) *(continued)*

cally scaled, is the magnitude of the jitter. The solid line beginning at 15 UI at 10 Hz and ending at 0.15 UI at 20 MHz is the jitter input template. It indicates the minimum level of jitter vs. jitter frequency that SONET equipment must tolerate at the OC-48 data rate. The test results are indicated at various jitter magnitudes and frequencies. At each frequency point where a box appears, the BER remains within acceptable limits in the presence of jitter. At points where an X appears, the jitter tolerance threshold of the device or system under test has been exceeded.

    Jitter Transfer.   This is a measure of the amount of jitter transmitted from input to output of a regenerator. It is the ratio of the amplitude of the equipment's sinusoidal output jitter to the applied sinusoidal input jitter. Jitter transfer specifications help ensure that once installed in a system, the equipment won't cause an unacceptable increase in jitter in any part of the spectrum. A cascade of similar units, each with just a small increase in jitter, could result in an unmanageable jitter level. SONET/SDH specifications define allowable jitter transfer functions for various transmission rates and regenerator types. Jitter transfer requirements on clock recovery circuits allow a small amount of jitter gain up to a given cutoff frequency, beyond which the jitter must be attenuated.

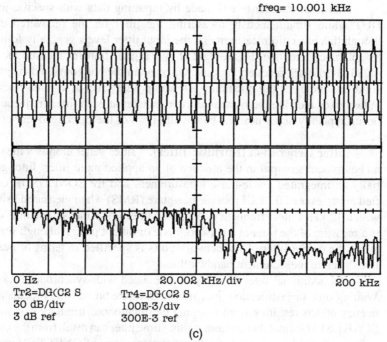

Figure 8.40 (c) Demodulated 10 kHz jitter waveform and spectrum. (Hewlett-Packard Journal. © Copyright 1995 Hewlett-Packard Company. Reproduced with permission.)

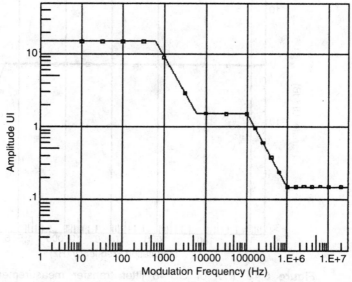

Figure 8.41 Plot of OC-48 jitter tolerance measurement. (Hewlett-Packard Journal. © Copyright 1995 Hewlett-Packard Company. Reproduced with permission.)

The jitter transfer test is made by inputting data with specific jitter levels into the DUT while simultaneously measuring the jitter on the recovered clock signal. Even though this is a ratio measurement, the input jitter levels generally follow those of the jitter tolerance template. Figure 8.42 shows a jitter transfer measurement made on a OC-48/STM-16 clock recovery module. Once again the horizontal axis is jitter frequency. The solid line indicates the maximum allowable jitter level at a given frequency. The rectangular boxes indicate the measured jitter transfer at each jitter frequency. Failures, if they occur, typically occur near the bandwidth limit of the clock recovery circuit.

Jitter Generation (Intrinsic Jitter).   Jitter generation is a measure of the jitter at an equipment's output in the absence of an applied input jitter. Jitter generation is essentially an integrated phase-noise measurement and for SONET/SDH equipment is specified not to exceed 0.01 UI root mean square (RMS) when measured using a highpass filter with a 12 kHz cutoff frequency. A related jitter noise measurement is output jitter, which is a measure of the jitter at a network node or output port. Although similar to jitter generation, the output jitter of the network ports is specified in terms of peak-to-peak UI over two different measurement bandwidths.

An additional jitter category is associated with synchronous transmission systems. Waiting time or justification jitter results from the bit stuffing process associated with frequency offsets resulting from mapping plesiochronous tributary data into the synchronous SONET/SDH format. In addition, waiting time jitter can result from the payload mapping and pointer adjustments associated with the construction of SONET/SDH transported payloads.

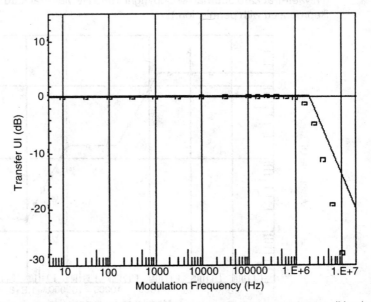

Figure 8.42 Plot of OC-48 jitter transfer measurement. (Hewlett-Packard Journal. © Copyright 1995 Hewlett-Packard Company. Reproduced with permission.)

8.5.5 Jitter Measurement Techniques

Although jitter measurements are made on digital waveforms, the tests themselves tend to be analog in nature. The most frequently encountered techniques to measure jitter usually employ either an oscilloscope, a phase detector, or instrumentation than can recover the phase modulation on a signal.

Intrinsic data jitter, intrinsic clock jitter, and jitter transfer can be directly measured with a high-speed digital sampling oscilloscope. As shown in Figure 8.43a, a jitter-free trigger signal for the oscilloscope is provided by clock source B, whose frequency reference is locked to that of clock source A. Clock source A, which is modulated by the jitter source, drives the pattern generator. The pattern generator supplies jittered data for the jitter transfer measurement to the DUT. The jittered input and output waveforms can be analyzed using the built-in oscilloscope histogram functions.

The oscilloscope measurement technique has several limitations. The maximum jitter amplitude that can be measured is limited to 1 UI peak-to-peak. Above this level, the eye diagram is totally closed. Secondly, this technique offers poor measurement sensitivity because of the inherently high noise level caused by the large measurement bandwidth involved. In addition, the technique does not provide any information about the jitter spectral characteristics or the jitter time function. Finally, the technique requires an extra clock source to provide the oscilloscope trigger signal.

Many of the limitations of the sampling oscilloscope technique can be addressed by using a phase detector. The phase detector in Figure 8.43b, compares the phase of the recovered clock from the device or equipment under test with a jitter-free clock source. The output of the phase detector is a voltage proportional to the jitter on the recovered clock signal. The range of the phase detector can be extended beyond 1 UI by using a frequency

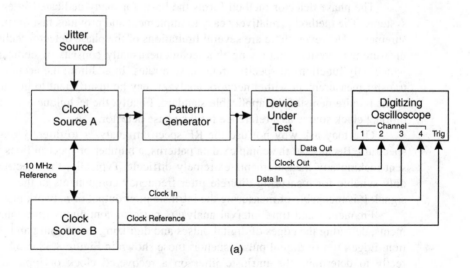

Figure 8.43 Jitter measurement techniques (a) Oscilloscope-based system. *(continued)*

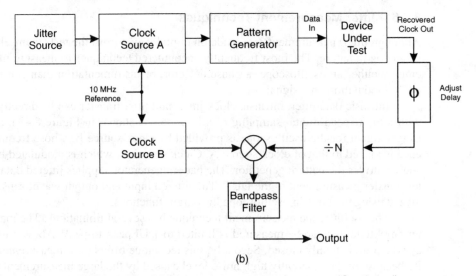

Figure 8.43   (b) Phase-detector-based system. *(Continued)*

divider. Intrinsic jitter is measured by connecting the output of the phase detector to a voltmeter with appropriate bandpass filters. An RF spectrum analyzer can also be connected to the output of the phase detector to observe the baseband jitter spectrum. A low-frequency network analyzer can be connected to the output of the phase detector to measure jitter transfer.

The phase detector method forms the basis for many dedicated jitter measurement systems. The method is relatively easy to implement and provides fast intrinsic jitter measurements. However, there are several limitations of the phase detector technique. A jitter measurement system employing this technique usually consists of dedicated hardware, which only functions at specific transmission rates. In addition, the accuracy of the jitter transfer measurement with a network analyzer may be insufficient to guarantee accuracy to the requirements of the applicable standard. Finally, the technique also requires an additional clock source as a reference for the phase detector.

One may ask why not use the RF spectrum analyzer to directly observe the data spectrum. Because of the complex data patterns, a number of spectral lines would be present making any measurements extremely difficult. Typically, the spectrum analyzer is only suitable for measuring discrete jitter frequency components in the recovered clock signal. It is incapable of measuring the intrinsic jitter output of network elements.

Frequency and time interval analyzers perform continuous time-interval measurements, detecting the edges of digital pulses and digitizing their timing and phase information. Edge jitter of digital pulses such as those shown in Figure 8.38 can be measured directly to determine the intrinsic jitter on a recovered clock or input data without a jitter-free clock as a reference. However, this jitter measurement tool offers limited jitter transfer measurement capability and no jitter tolerance measurement capability.[18]

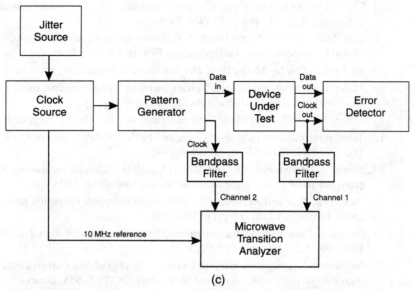

Figure 8.43 (c) Microwave-transition-analyzer-based system. (Hewlett-Packard Journal. © Copyright 1995 Hewlett-Packard Company. Reproduced with permission.)

Sampler-based instruments offer a general purpose jitter measurement solution.[19] These instruments typically operate by taking time samples of the data, then analyze the sampled data using digital signal processing techniques. Thus, they can offer built-in phase demodulation capability. A sampler-based instrument serves as the measurement receiver in the jitter system shown in Figure 8.43c, which also includes a pattern generator, error detector, and clock source. A synthesized signal generator serves as the jitter modulation source.

REFERENCES

1. *Synchronous Optical Network (SONET) Transport Systems: Common Generic Criteria,* GR-253 CORE, Bellcore, Piscataway, NJ, Dec. 1994.

2. American National Standards Institute. 1992. *Digital hierarchy—Optical interface rates and formats specifications (SONET),* ANSI T1.105–1991, New York.

3. International Telecommunications Union. 1990. *Optical interfaces for equipments and systems relating to the synchronous digital hierarchy.* ITU G.957, Geneva.

4. International Telecommunications Union. 1993. *Error performance measuring equipment operating at the primary rate and above.* ITU O.151, Geneva.

5. Bissell, C.C. and Chapman, D.A. 1992. *Digital signal transmission.* Cambridge MA: Cambridge University Press.

6. Ballo, D.J. and Wendler, J.A. 1992. The microwave transition analyzer: a new instrument architecture for component and signal analysis. *Hewlett-Packard Journal* 43(5):48–62.

7. The Institute of Electrical and Electronic Engineers, Inc. 1977. *IEEE standard pulse terms and definitions,* IEEE Std 194-1977, New York.

8. Hart, M.G., Duff, C.P., and Hinch, S.W. Firmware Measurement Algorithms for the HP 83480 Digital Communications Analyzer. 1996. *Hewlett-Packard Journal,* 47, No. 6:13–21.

9. HP Eyeline Display Mode. 1996. *Hewlett-Packard Journal,* 47, no. 6:18–19.

10. Miller, C.M. 1994. High-speed digital transmitter characterization using eye diagram analysis. *Hewlett-Packard Journal* 45 (4):29–37.

11. Green, P.E. 1993. *Fiber Optic Networks.* Englewood Cliffs, NJ: Prentice Hall.

12. Telecommunications Industry Association. 1995. *OFSTP-4, Optical Eye Pattern Measurement Procedure,* TIA/EIA-526-4. Arlington.

13. Hinch, S., Woodward, M., and Miller, C., 1995. Accurate measurement of laser extinction ratio, *Hewlett-Packard Lightwave Symposium,* Santa Rosa. 1995, 1–12.

14. Andersson, P.O. and Akermark, K. 1994. Accurate optical extinction ratio measurements. *Photonics Technology Letters* 6 (11):1356–1358.

15. "Apparatus and method for determining extinction ratio of digital laser transmitters," U.S. Patent No. 5,535,038, Jul. 9, 1996.

16. International Telecommunications Union. 1990. *Digital line systems based on the synchronous digital hierarchy for use on optical fibre cables,* CCITT G.958. Geneva.

17. *Synchronous Optical Network (SONET) Transport Systems: Common Generic Criteria,* GR-253-CORE, Bellcore, Piscataway, NJ, Dec. 1994.

18. Hewlett-Packard. 1988. *HP 5371A frequency and time interval analyzer—Jitter and wander analysis in digital communications.* Application Note 358 2. Palo Alto, CA.

19. Miller, C.M. and McQuate, D.J. 1995. Jitter analysis of high-speed digital systems. *Hewlett-Packard Journal* 46 (1):49–56.

CHAPTER
9

Insertion Loss Measurements

Christian Hentschel, Dennis Derickson

9.1 INTRODUCTION

The insertion loss of optical components used in a transmission link define the system's power budget and the system margin. Therefore, it is important to be able to measure insertion loss accurately. This chapter discusses insertion loss measurements of optical components such as fibers, connectors, attenuators, filters, couplers, and multiplexers. The same principles can also be used for optical components with gain, such as optical fiber amplifiers and semiconductor optical amplifiers. These measurements are more thoroughly discussed in Chapter 13. Most of the techniques used for measuring insertion loss can be applied to measurements of optical return loss as well.

9.2 HOW THE COMPONENT INFLUENCES THE MEASUREMENT TECHNIQUE

Insertion loss measurement is always a two-step procedure: in the first step (calibration), the reference power, in other words, the input power to the test device, is measured. In the second step, the test device is inserted and the power is measured again. The two power results are divided by each other, and the attenuation is usually expressed in dB.

It turns out that different types of components usually dictate different measurement principles. The most frequent optical component features two fiber pigtails with connectors on each end. However, there is a large variety of other types: components with bare-fiber pigtails, components with flange-type connectors (fixed to the housing), bulk-optical components with relatively large optical ports and integrated-optics components with port

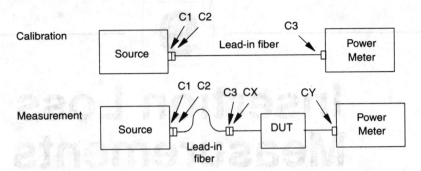

Figure 9.1 Insertion loss measurement of a pigtailed and connector-ized component.

dimensions similar to fiber cores. Only the most important types of components will be discussed here.

9.2.1   Measurement of Pigtailed and Connectorized Components

We want to start our discussion with a device under test (DUT) with fiber pigtails and connectors. The measurement setup and sequence are shown in Figure 9.1

In the calibration step, a lead-in fiber, preferably of the same type as the DUT's fiber pigtails, connects the source and the power meter directly. In the next step, the DUT's input connector CX is mated with the system's connector C3, and the DUT's output connector CY is brought to the power meter.

Ideally, exchanging connectors C3 and CY at the input port of the power meter should not influence the power measurement results. However, the second step adds another connector pair to the setup. Therefore, the measured insertion loss includes the loss of the DUT plus the insertion loss of the input connector pair C3/CX. Since the loss of this connector pair can only be estimated (in the order of 0.5 to 1 dB), there is a relatively large uncertainty of typically ±0.3 dB even if good connectors are used.

Figure 9.1 can also be viewed as a measurement method for connector insertion loss if the DUT is a short cable, because a short cable will have negligible fiber loss.

9.2.2   Measurement of Flange-Mount Components

The situation becomes more complicated when the DUT has connectors that form part of the housing. A typical example is a programmable optical attenuator. This scenario is shown in Figure 9.2

The calibration setup includes two lead-in fibers, preferably of the same type as the DUT-internal fibers, and one connector pair, C3/C4. After the calibration, the procedure continues with opening connector pair C3/C4 and inserting the DUT. Incidentally, the following changes occur:

1. Connector pair C3 / C4 is replaced by CY / C4.
2. Another connector pair, C3 / CX, is added to the setup.

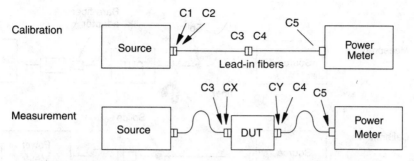

Figure 9.2 Insertion-loss measurement of component with flange-mount connectors.

Therefore, the measured insertion loss includes the loss of one connector pair again. Because there are two new connector matings, the uncertainty of the insertion loss is approximately twice that of the example in Figure 9.1. An typical uncertainty would be ±0.6 dB for these two new connector matings.

9.2.3 Measurement of Components with Bare-Fiber Pigtails

Optical components with bare-fiber ends exist because of two reasons: either to connectorize them with customer-specific connectors, or to splice them into a transmission or other system. In both cases, a performance verification includes the measurement of insertion loss. A possible measurement setup and procedure are illustrated in Figure 9.3.

This situation is similar to the one outlined in Figure 9.1. In the calibration step, the power from a bare lead-in fiber is measured using a bare-fiber adapter. The power meter should be capable of capturing the entire radiated power from the fiber end. Then, in the measurement step, the test device is connected with the help of a fusion splice or mechanical splice. The uncertainty of the measured insertion loss can be expected to be much

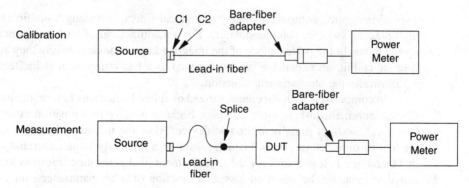

Figure 9.3 Insertion loss measurement of a component with bare-fiber pigtails.

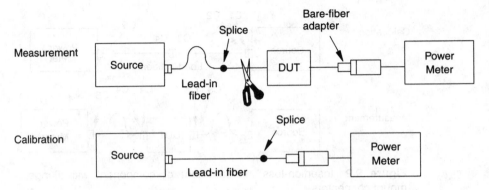

Figure 9.4   Cutback measurement.

lower than in the connectorized cases because a good fusion splice has very low insertion loss (typically less than 0.1 dB) and a correspondingly lower uncertainty. Again, the insertion loss includes the splice loss.

The highest precision of all insertion-loss methods can be expected from cutback measurements. This method is shown in Figure 9.4.

In comparison with all of the methods discussed above, the cutback measurement reverses the sequence of measurement steps. In the first step, the test component is spliced into the setup, and the power from the output pigtail is measured. Then the DUT's input pigtail is cleaved just after the splice, and the power exiting from the cleave is measured again. Very small uncertainties of better than ±0.01 dB can be expected if the cleave is of perfect optical quality, and if a good bare fiber adapter and a power meter with appropriate performance are used.

9.2.4 Insertion-Loss Measurement of Integrated Optics Components

Typical integrated components today are optical filters, wavelength multiplexers and demultiplexers, switches, and modulators. For a manufacturer of such components, it is essential to verify the performance of the integrated optics devices before they are pigtailed, because adding an expensive production step to a bad component is ineffective. Figure 9.5 illustrates the measurement situation.

In comparison with the connectorized or spliced situations before, the integrated optics measurement task is more complex because usually two positioners are needed to "connect" the fiber pigtails to the test device. Also, the measured insertion loss includes the loss of both "connections," which results in a relatively large uncertainty. Therefore, this technique is less useful for the measurement of the absolute insertion loss, but well-suited to measure the insertion loss as a function of other parameters, for example, the wavelength or the state of polarization.

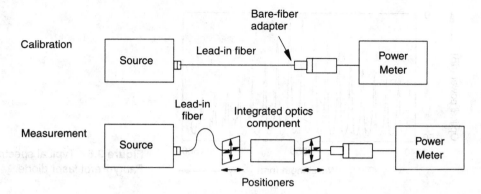

Figure 9.5   Insertion-loss measurement of an integrated optics component.

9.2.5   Imaging Techniques

Up to here, we have only discussed techniques in which fibers are used to connect the test component. Multimode or singlemode fibers may have to be used, depending on test component's internal fibers.

Alternatively, it is also possible to image light to the fiber input instead of connectorizing or splicing. The aim is to reduce the uncertainty introduced by the connector. This is important in the performance verification of low insertion-loss components that are being sold without connectors. Perfect imaging quality is necessary to accomplish this, particularly when singlemode fibers are involved. Tungsten lamp sources are often imaged onto the end of singlemode fibers. The tungsten lamp can completely fill the numerical aperture of the fiber and couple a well-known amount of light into the fiber.

Imaging techniques can also be used to create certain modal distributions in multimode fibers. The two preferred launch conditions in graded-index multimode fibers are full excitation and the 70% excitation. Full excitation means that the fiber is overfilled, both in terms of core diameter and numerical aperture, and that undesired cladding modes are absorbed. 70% excitation means that the launch optics produce a beam which, at the fiber input, has a diameter of 70% of the core diameter, and a numerical aperture of 70% of the fiber's numerical aperture. See Reference 1 for a more complete discussion.

9.3   SINGLE-WAVELENGTH LOSS MEASUREMENTS

Historically, most insertion-loss measurements were carried out using sources with fixed wavelengths, mostly Fabry-Perot (FP) lasers. These lasers are well-suited for measuring the loss of broadband optical components such as fibers, connectors, attenuators, etc., because they produce high power (1 mW and more) within a spectral window of several

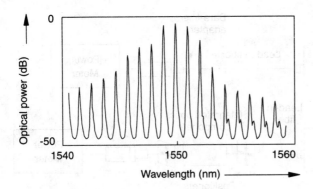

Figure 9.6 Typical spectrum of a Fabry-Perot laser diode.

nanometers. This bandwidth is wide enough to avoid optical interference problems (see below), and narrow enough to be considered as single wavelength.

Figure 9.6 shows a typical spectrum of an FP laser. It consists of several lines, which represent longitudinal modes of the laser's resonant cavity. For some applications, it is important to know that there is continuous competition between these longitudinal modes, which means that the powers of all modes fluctuate and only the total power is stable. For insertion-loss measurements, only the total power counts. In addition, some averaging is usually implemented in the optical power meters, which effectively reduces the remnant noise of the source.

Other possible optical sources for single-wavelength measurements are:

1. Surface-emitting LEDs; however, their power is usually 30 dB lower than the one from an FP laser when coupled into a fiber. In addition, their wide spectral width makes it difficult to attribute a center wavelength to the measurement result.
2. Distributed-feedback lasers (DFBs); they produce sufficient power. However, their typical spectral width of 20 to 50 MHz is so narrow that optical interference problems are likely to occur.

The measurement range is defined by both the source and the power meter. In compliance with the typical definition used in the field of microwaves, we define the measurement range for single-wavelength insertion loss measurements:

Measurement range: the difference between the source power and the sensitivity of the receiver (in this case the optical power meter).

Example

If the source power is 0 dBm (1 mW), and the noise level of the optical power meter, at a given averaging time, is −90 dBm, then the measurement range is 90 dB. Practical measurements require a signal-to-noise ratio (SNR) of more than 1:1. For example, if the SNR at the lowest power level needs to be 20 dB for reasons of accuracy, then the measurement range is 70 dB.

9.4 UNCERTAINTIES OF SINGLE-WAVELENGTH LOSS MEASUREMENTS

In most cases, the largest uncertainty in insertion-loss measurements must be attributed to the connectors. The loss of a connector pair is not perfectly repeatable. Even worse, many measurement procedures include mating different connector specimens. These problems and their solution are thoroughly discussed in Section 9.1. A number of other uncertainty contributions must be taken into account in single-wavelength insertion-loss measurements. This is the subject of the discussion below. As always, root-sum-squaring the partial uncertainties to calculate a total uncertainty is applicable, provided that the partial uncertainties are statistically independent.

The list below gives an overview of the uncertainties in insertion-loss measurements. Of course, not all of them will be applicable in the individual situation. On the contrary, it should be an objective to avoid most or all of them. The uncertainties are:

1. connector-related uncertainty (see the discussion in Section 9.1),
2. power-meter-related uncertainties,
3. polarization-dependent loss of the DUT,
4. optical interference,
5. wavelength characteristics of the source, and
6. incompatible fibers.

These uncertainties are analyzed in more detail below.

9.4.1 Power-Meter-Related Uncertainties

Insertion loss is always calculated as the ratio of two power levels. Therefore, absolute accuracy in power measurements is not required. Only the uncertainty of the power ratio is important.

The uncertainty of the power ratio has several components which can usually be added by root-sum-squaring:

1) The power meter's nonlinearity. This uncertainty is identical to the power meter's nonlinearity between the two relevant power levels and depends on the magnitude of the power ratio. Power ratios near one will usually be very accurate. This is the case when the DUT exhibits an insertion loss near 0 dB. On the contrary, a noticeable uncertainty may have to be taken into account when the power ratio is large. This subject "nonlinearity" is thoroughly discussed in Chapter 2. One of the results was that the correspondent loss measurement error is the difference between the nonlinearities at the two relevant power levels. This difference is independent from the reference level.

2) The power meter's polarization dependence. The two measurement steps usually mean two different states of polarization for the power meter. Therefore, the power meter's polarization-dependent responsivity (PDR) is another contribution to the uncertainty. This is particularly important when low-loss components are to be measured. This problem is further discussed in Section 9.4.2 below.

3) The power meter's spatial homogeneity. Inhomogeneous detector surfaces are also discussed in Chapter 2. This uncertainty can be substantial in conjunction with angled connectors, for example, when the reference measurement is done with a straight connector and the DUT measurement is done with an angled connector. In this case, the two beams strike two different areas of the detector surface.

4) The power meter's numerical aperture. The maximum angle of acceptance of most power meters is in the order of 0.2 or 0.3. If the power meter cannot accept all light from the fiber, an additional error can occur, because the ratio of coupled to total power is usually subject to strong variation. This can be the case when fibers with large numerical aperture are used, possibly in conjunction with angled connectors.

9.4.2 Uncertainty Caused by Polarization-Dependent Loss (PDL)

Another important uncertainty can be expected when a test device exhibits polarization-dependent loss (PDL). Laser sources produce nearly 100% polarized radiation. When the wave arrives at the input of the test device, the polarization state is completely unpredictable. If we assume that the power meter is perfect and that the DUT has a PDL of 0.2 dB p-p, then the measured insertion loss will have an additional uncertainty of ±0.2 dB. In many cases, it is sufficient to use an insertion-loss technique which reduces this uncertainty as much as possible, see below. In other cases, the PDL must be measured in addition to the insertion loss; see Section 9.5.

To make the situation more complicated, the PDL of the test device must be viewed in conjunction with the power meter's polarization-dependent responsivity (PDR). The possible error depends on the (unpredictable) relative geometrical alignment between the component's PDL axis and the power meter's PDR axis: one extreme is that the PDL and the PDR are added, in the other extreme they are subtracted. The two extremes of the uncertainty caused by polarization dependence can be described by:

$$U_{PDL} = \pm \ (PDL_{DUT} \pm PDR_{PM}) \quad [dB] \tag{9.1}$$

There are two possible ways to remove the PDL uncertainty from the test result: either use a broadband, unpolarized source, such as an LED or the amplified spontaneous emission (ASE) of an optical amplifier, or randomize the polarization state of a polarized source such as a laser. As an example, Figure 9.7 shows a measurement setup which uses a fiber-loop-type polarization controller as polarization randomizer. The measurement procedure involves generating all possible polarization states and averaging during the same time.

Notice that an isolator may have to be inserted after the source, in order to prevent the changing polarization state of any reflected waves from influencing the laser's output power.

Another way of removing the PDL effects is using the average of the displayed minimum and maximum values obtained during polarization randomizing.

Does polarization randomizing also work when both the DUT and the power meter exhibit polarization dependence? To analyze this problem, we recall that the minimum/

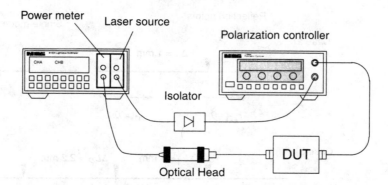

Figure 9.7   Polarization randomizing minimizes the uncertainty caused by polarization dependence.

maximum axes (principle axes) of the DUT and the power meter have a random orientation against each other. It can be shown that polarization randomizing reduces the partial uncertainty to the *product* of the PDL and the PDR (both in dB), instead of the sum as in Equation 9.1:

$$U_{\mathrm{PDL}} = \pm \mathrm{PDL}_{\mathrm{DUT}}\, \mathrm{PDR}_{\mathrm{PM}} \quad [\mathrm{dB}] \tag{9.2}$$

Example

If both PDL values are ±0.1 dB, then the correspondent insertion loss uncertainty is ±0.01 dB. This means that the uncertainty remains negligible as long as the polarization dependencies of both the DUT and the power meter are relatively small. This way, the test device's PDL can effectively be removed from the measurement result.

9.4.3 Uncertainty Caused by Optical Interference

Generally, reflections can cause measurement uncertainty because of two reasons. The first reason is that laser sources are sensitive to back-reflections, particularly since the two measurement steps usually represent different loads for the optical wave. The second reason is that two or more reflections can interfere with each other, which means that only a slight change of wavelength changes the insertion loss.

As a side aspect, it is interesting to know that optical interference can also create broadband noise. This happens when an optical wave interferes with a doubly reflected and delayed replica of itself. In this case, any photodetector, through its square-law characteristics, can mix the two partial waves so that a low-frequency beat spectrum is generated. In the context of insertion loss, however, it is assumed that these measurements are carried out with a relatively long averaging time which effectively averages out the photocurrent noise.

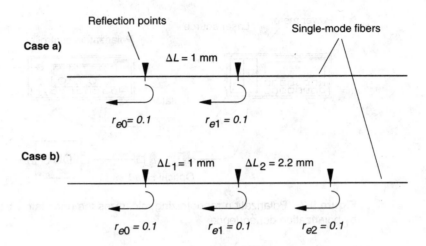

Figure 9.8   Fiber with internal reflections.

The laser's sensitivity to back-reflections can be cured using an optical isolator. More difficult is the solution to optical interference. To understand this problem, a fiber path with two and three internal reflections is analyzed; see Figure 9.8.

The power-reflectivities are all assumed to be 0.01, equivalent to −20 dB. For the mathematical analysis, it is better to think of reflected electric fields instead of optical power. Accordingly, the electric field reflectivities (symbol r_e) are the square roots of the power reflectivities, in other words, 0.1 as shown in Figure 9.8.

In the analysis, the fact that the optical power at the second reflection point is slightly reduced is ignored. Also neglected are all secondary reflections. Both assumptions are appropriate if the reflectances are relatively small. In contrast, the interference between large reflections must be viewed as optical resonators; this is not further discussed here. For case (a), the total power reflection, in linear notation, oscillates around a constant value:

$$r_{p,\,total} = r_{e0}^2 + r_{e1}^2 + 2r_{e0}\,r_{e1}\cos(\phi) \tag{9.3}$$

or:

$$r_{p,\,total} = r_{p0} + r_{p1} + 2\sqrt{r_{p0}\,r_{p1}}\cos(\phi) \tag{9.4}$$

where r_e are the electric field reflectivities, r_p are the optical power reflectivities, and ϕ is the phase difference between the two reflected fields:

$$\phi = \frac{4\pi\,n\Delta L}{\lambda} \equiv \frac{4\pi\,n\,\Delta L\nu}{c} \tag{9.5}$$

where ΔL is the distance between the two reflection points, n is the refractive index of the fiber, ν is the optical frequency, and c is the speed of light in vacuum. Equations 9.4 and 9.5 can be used to calculate both the average reflection and the oscillation amplitude around it. Note that the oscillation amplitude depends on the product of the two power reflectivities.

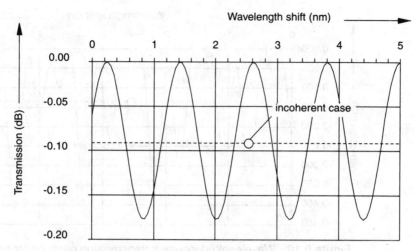

Figure 9.9 Wavelength-dependent transmission changes for case (a) caused by two electric field reflectivities of 0.1.

The power transmission is the complement to 1:

$$t_{p, \text{ total}} = 1 - r_{p, \text{ total}} \tag{9.6}$$

Figure 9.9 shows the calculated transmission for case (a) for the wavelength range of 1550 nm to 1555 nm. It is obvious that a strong fluctuation of the transmitted power is observed, which depends on the reflectances and the spacing between the reflection points.

It was assumed that the two interfering waves are coherent with each other, in other words, that they have a stable phase relation. This is the case when the spectral width of the source is relatively narrow. When the spectral width is wide, then the two reflected powers add, and the wavelength dependence vanishes. This is shown as the incoherent case in Figure 9.9. There is no interference uncertainty in this case. A more detailed discussion on the influence of the spectral width of the source is in Section 9.4.4.

Figure 9.10 shows the calculated transmission for case (b). The mathematics for this case are simply extensions of Equations 9.3 to 9.6. For example, a modified Equation 9.4 contains three quadratic terms and three mixed terms, the latter because three reflections correspond to three interaction possibilities.

Evidently, three reflection points can create an irregular transmission pattern. Strong attenuation peaks are also observed; this is because the maximum total reflection is proportional to the *square* of the sum of the electric field reflectances.

Notice that polarization alignment was assumed in the analyses above, in other words, that the reflected waves have the same polarization states. If this is not the case, then the interference will be reduced or eliminated.

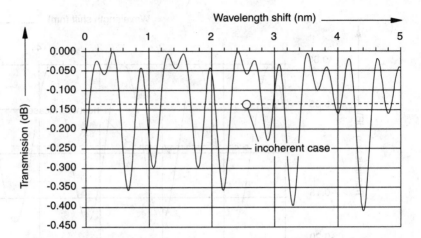

Figure 9.10   Wavelength-dependent transmission changes for case (b) caused by three electric field reflectivities of 0.1.

9.4.4   Uncertainty Caused by the Wavelength Characteristics of the Source

The wavelength characteristics of the source can be superficially described by its center wavelength and its spectral width, each of which has its own influence of the insertion loss uncertainty.

The center wavelength of the laser source contributes to the uncertainty because it is often not well known and temperature-dependent. The correspondent insertion-loss uncertainty is the product of the wavelength uncertainty and the DUT's sensitivity to wavelength changes. However, this uncertainty is often negligible because single-wavelength measurements are mostly carried out on test devices with relatively small wavelength dependence. Components with strong wavelength dependence must be tested with a tunable source, or with a broadband source and a wavelength-selective power meter, for example, an optical spectrum analyzer (OSA).

More important is the spectral width of the source. As indicated in the section above, the creation of optical interference depends on the spectral width of the source. To analyze this problem, let us consider case (a) of Figure 9.8 again. Two reflected waves of identical spectral content are recombined after one of them is delayed by a path length $2\Delta L$. It would be instructive to be able to vary the path length and observe the interference effects. This is indeed possible with the Michelson interferometer shown in Figure. 9.11 Corner-cube reflectors are used in this interferometer example to remove optical feedback to the source. Signals reflected back to a laser diode can influence its spectral characteristics.

If a laser beam is injected into the interferometer, and the two path lengths are approximately equal, then strong optical interference is produced. However, when the path lengths are sufficiently different, then there is no stable phase relation between the two

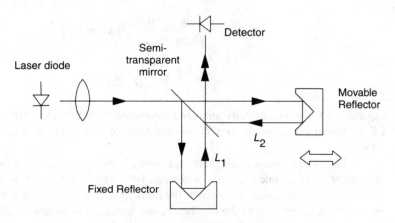

Figure 9.11   Michelson interferometer.

partial waves, and the interference vanishes. This problem is often analyzed under the assumption that the source has a Lorentzian spectral shape. Equation 9.7 represents the spectral power density of a Lorentzian source (in its single-sided form):

$$p(v) = \frac{p_0}{1 + k(v - v_m)^2} \tag{9.7}$$

where $k = 4 / \Delta v^2$, v is the optical frequency, v_m is the center frequency, p_0 is the power density at the center frequency, and Δv is the full-width, half-maximum spectral width (FWHM).

Under this spectral assumption, the dependence of the photodetector power on the path length difference is given by the following equation.[2]

$$P(\tau) = P_0\left[1 + \exp\left(-\frac{\tau}{\tau_c}\right)\cos(2\pi v_m\tau)\right] \tag{9.8}$$

where the coherence time τ_c is the time difference τ after which the optical interference begins to vanish, and τ is calculated from twice the length difference ΔL:

$$\tau = \frac{2\,\Delta L}{c/n} = \frac{2\,(L_2 - L_1)}{c/n} \tag{9.9}$$

where c is the speed of light in vacuum, n is the group refractive index of the medium. τ_c can alternatively be expressed as the coherence length l_c, which is the length difference (in vacuum) that corresponds to τ_c.

$$l_c = c\tau_c \tag{9.10}$$

Both l_c and τ_c are a measure of the spectral width of the source. For a Lorentzian spectrum, the coherence time is given by:

$$\tau_c = \sqrt{\frac{k}{2\pi}} = \frac{0.8}{\Delta\nu} \quad [\Delta\nu \text{ in Hz}] \tag{9.11}$$

or:

$$\tau_c = \frac{0.8\,\lambda^2}{c\,\Delta\lambda} \quad [\lambda,\ \Delta\lambda \text{ in } m]$$

Equation 9.8 was numerically analyzed for a spectral width $\Delta\lambda = 100$ nm (typical for an LED), which corresponds to a coherence time of 0.064 ps, or a coherence length of 19 μm.

The example shows that there is a strong interference between the two reflections, equivalent to large uncertainties, when the path length difference is small. This corresponds to short distances between reflections in a fiber path. When the distance is increased, then the powers of the two beams are combined without optical interference. Figure 9.12 shows that the interference is substantially reduced after a one-way length difference of approximately one coherence length, in this case after approximately 20 μm (notice that the effective path length is approximately three times larger).

With coherence lengths of meters and even kilometers, all laser sources are substantially more coherent than the LED example above. This means that strong optical interference may be observed even when the reflection points are separated by very long distances. Therefore, the spectral width should be as large as possible to avoid interference-related uncertainties. Table 9.1 lists the recommended spectral characteristics at the wavelength of 1550 nm. The values were calculated on the basis that the effective path length difference is longer than three coherence lengths, and by using a group refractive index of 1.5 (for silica). The wavelength of 1550 nm was chosen as a worst case, because narrower spectral widths are required at shorter wavelengths.

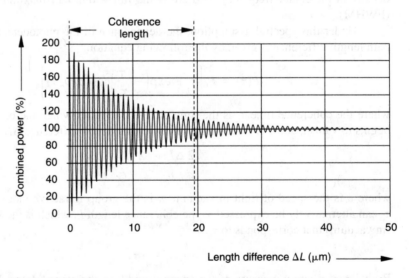

Figure 9.12 Dependence of the photodetector power, in relative units, on the length difference for a spectral width of 100 nm.

Table 9.1   Recommended spectral characteristics to avoid optical interference, as a function of separation between reflection points, ΔL, at 1550 nm.

	Recommended Values			
ΔL	Coherence length	Spectral width (m)	Spectral width (Hz)	Recommended source
0.1 mm*	<0.1 mm	>19 nm	>2400 GHz	LED
1 mm	<1 mm	>1.9 nm	>240 GHz	LED
10 mm	<10 mm	>190 pm	>24 GHz	LED + FP laser
100 mm	<100 mm	>19 pm	>2.4 GHz	LED + FP laser
1 m	<1 m	>1.9 pm	>240 MHz	LED + FP laser
10 m	<10 m	>190 fm	>24 MHz	LED + FP + DFB + modulated ECL
100 m	<100 m	>19 fm	>2.4 MHz	LED + FP + DFB + modulated ECL
1 km	<1 km	>1.9 fm	>240 kHz	LED + FP + DFB + modulated ECL
10 km	<10 km	>0.19 fm	>24 kHz	LED + FP + DFB + all ECL

*$\Delta L = 01$ mm means a path length difference of 0.3 mm.

Let us discuss some examples:

1. To avoid optical interference in a glass plate of 1 mm thickness, for example, in a photodetector window, a spectral width of 1.9 nm (FWHM) is required. Only LEDs and white light sources can reliably provide such a wide bandwidth.

2. For bulk-optical components of 100 mm length, a spectral width of 19 pm is sufficient, which is equivalent to 2.4 GHz. Such a spectral width (or more) is typical for FP laser sources. Notice that FP lasers oscillate at several wavelength (longitudinal modes) which are separated by typically 0.8 to 1 nm. This cannot be described by the Lorentzian model of Equation 9.7, and gives rise to a more complex interference pattern than the one shown in Figure 9.12. The effective spectral width for these lasers is somewhere in the middle between the width of each of the longitudinal modes (possibly less than 1 GHz) and the width of the spectral envelope (usually several nanometers, equivalent to several hundred GHz).

3. To eliminate the interference between connectors of a jumper cable, for example, of 2 m length, a spectral width of 120 MHz is recommended. Again, an FP type laser source is the right solution for this case. Alternatively, a modulated DFB laser or a modulated external-cavity laser (ECL) may be used. Intensity modulation can be used to increase the bandwidth. A practical example is discussed in Section 9.7.1.

4. In principle, much wider separations than 2 m should be used when the source is either an unmodulated DFB laser or unmodulated external-cavity tunable laser (ECL). An unmodulated DFB typically has a spectral width between 20 and 50 MHz, where the separation should be around 10 m. The spectral width of an ECL is only around 100 kHz. In this case, a separation of 2 km and more would be necessary. This shows that in insertion-loss measurements using these sources, opti-

cal interference can no longer be avoided by separating the reflection points. Instead, all possible reflections should be carefully minimized. One large reflection can usually be tolerated, whereas all other reflections should be reduced to at least −40 dB.

9.4.5 Uncertainty Caused by Incompatible Fibers

Ideally, the lead-in fibers to a test device should be of the same type as the device's own fibers. Differences in the refractive index profile (in multimode fibers) or in the mode field diameter (in singlemode fibers) will usually cause additional insertion losses. Typical examples for fibers with smaller-than-standard mode field diameters are those used for erbium-doped fiber amplifiers (EDFAs) and fibers used for pumping these amplifiers. Larger-than-standard mode field diameters in singlemode fibers are becoming more important because they are less sensitive to nonlinear effects at high power levels.

In all of these cases, adding jumper fibers to the measurement setup is recommended, so that the DUT is connected by fibers of the same type as the DUT-internal fibers. This way, the transition between different fiber types can be avoided. The critical transition will be part of the setup and therefore has no influence on the measurement result.

9.5 PDL MEASUREMENT

In many cases, it is not sufficient to eliminate the PDL influence from the insertion-loss result. Instead, the PDL is considered as a separate characteristic and must be measured in addition to the insertion loss. The two most important PDL-measurement techniques are polarization scanning and the Mueller method. A third method which uses the Jones calculus is outlined in Heffner.[3]

9.5.1 Polarization-Scanning Method

Figure 9.13 shows the principle measurement setup for polarization scanning. This measurement is usually based on the fact that laser sources produce nearly 100% polarized

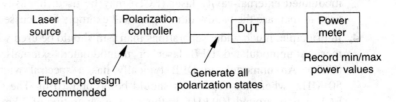

Figure 9.13 Principle setup for measuring the PDL of optical filters.

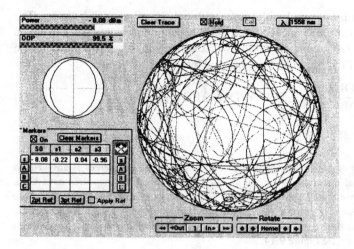

Figure 9.14 Poincaré plot of polarization states obtained from the HP 11896A polarization controller (measurement time 30 s, scan speed 5).

light. The polarization controller has to convert the fixed input polarization state to all possible output states, including all linear, elliptical, and circular states.

The typical polarization controller used for this measurement consists of several motorized fiber loops. To generate polarization scanning, the motors run at different speeds, so that completely random polarization states are generated. Figure 9.14 shows a Poincaré plot of the polarization states generated with a fiber-loop type polarization controller within 30 s. Displaying polarization states using the Poincaré sphere is covered in Chapter 6. The traces were measured with a real-time polarization analyzer. The points leading to minimum and maximum transmission are always opposite to each other, for example, the North Pole and the South Pole or the west and the east.

The measurement procedure is described in detail in Reference 4. The PDL value is finally calculated using:

$$\text{PDL}_{\text{dB}} = 10 \log \frac{T_{\max}}{T_{\min}} \equiv 10 \log \frac{P_{\max}}{P_{\min}} \tag{9.12}$$

where $T_{\max}$, $T_{\min}$ are the transmission coefficients and $P_{\max}$, $P_{\min}$ are the output powers.

To obtain good accuracy in this measurement, the following is important:

1. Constant input power to the DUT. Changing the polarization state should not change the input power to the DUT. A well-designed polarization controller is necessary to accomplish this. Quite often, it is overlooked that laser sources react with unstable power when light of varying polarization state is back-reflected like in this measurement. Therefore, an attenuator or isolator should be inserted between the source and the polarization controller.

2. Generation of all polarization states. The evolution of the polarization state is most easily understood using a real-time polarization analyzer: After some time, the

Poincaré sphere is fully covered with traces with only small gaps in between. The question is: How much time is needed to make an accurate measurement? It can be answered by asking for the maximum allowable error angle, δ_{max}, on the Poincaré sphere. The relative PDL error can be expressed by:

$$\frac{\Delta PDL}{PDL} \cong -\frac{\delta_{max}^2}{2} \quad [PDL, \Delta PDL \text{ in dB}] \qquad (9.13)$$

Example

If a partial uncertainty of ±5% is the goal, then an error angle of 18 degrees is allowed. In other words, substantial error angles can be allowed in accurate PDL measurements. On a statistical basis, the error angle is proportional to 1/square root of time. Therefore, the relative PDL error (Equation 9.13) is proportional to 1/time. The following formula describes the relative PDL error. It is based on a power-meter averaging time of 20 ms and the same scan speed that was used to create Figure 9.14.

$$\frac{\Delta PDL}{PDL} = \frac{50\%}{\text{scan time}} \qquad (9.14)$$

Accordingly, a 5% error margin is obtained after 10 s.

3. Polarization-independent power meter. Many optical power meters have unacceptable polarization dependence for this type of measurement. Therefore, the power meter should be characterized or specified for polarization-dependent responsivity (PDR). See the discussion of power-meter polarization dependence in Chapter 2.

As always, the total uncertainty is essentially the root-sum-square of the individual uncertainty contributions. Equations 9.13 and 9.14 suggest that the total uncertainty (in dB) depends on the absolute value of the PDL. For PDL values of 0.1 dB, total uncertainties in the order of ±0.003 to ±0.005 dB are attainable.

In summarizing the scanning method, it can be said that this is a robust, accurate technique that is easy to implement, provided that the appropriate measurement equipment is selected. No programming is needed if the polarization controller features automatic scanning and the power meter is sufficiently fast and capable of recording minimum and maximum values. The relatively long measurement time of 10 s is acceptable for PDL measurements at only one wavelength.

9.5.2 Mueller Method

This method was first mentioned in Nyman.[5] The method is based on applying four well-known polarization states to the test device. The optical power transmission is measured at these four states only. The PDL is calculated from the four transmission results. The theory of the Mueller method is explained in Schmidt and Hentschel.[6] Only a short abstract is presented here.

A setup for PDL measurements with the Mueller method is shown in Figure 9.15. It is based on a polarization controller using waveplates. The particular design shown syn-

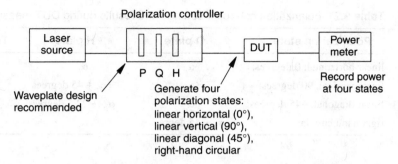

Figure 9.15   Setup for PDL measurements with the Mueller method.

thesizes the different polarization states with a quarter-waveplate (Q) and a half-waveplate (H).

A polarizer (P) is added in front of the waveplates to ensure a polarized signal of fixed orientation. The first step in using this setup is to rotate the polarizer (α_p) until a best match with the incoming signal occurs and maximum transmission is achieved. The Q and H plates are always to be aligned with respect to the reference angle α_p (see Table 9.2).

The measurement procedure starts without the DUT by measuring the optical power at the four well-defined polarization states. Then the DUT is inserted and the power is measured again at the same polarization states. Table 9.2 illustrates this process (power levels with capital subscripts indicate measurements with the DUT, and power levels with lower-case subscripts are without the DUT).

The Mueller matrix is a 4×4 matrix which describes the transmission and polarization characteristics of the DUT, similar to the Jones matrix. It can be shown[6] that only the elements of the first row, m_{11} to m_{14}, are needed to calculate the PDL. Equation 9.15 shows how these elements can be calculated from the transmission coefficients T:

$$
\begin{bmatrix} m_{11} \\[2ex] m_{12} \\[2ex] m_{13} \\[2ex] m_{14} \end{bmatrix}
=
\begin{bmatrix} \frac{1}{2}\left(\frac{P_A}{P_a} + \frac{P_B}{P_b}\right) \\[2ex] \frac{1}{2}\left(\frac{P_A}{P_a} - \frac{P_B}{P_b}\right) \\[2ex] \frac{P_C}{P_c} - m_{11} \\[2ex] \frac{P_D}{P_d} - m_{11} \end{bmatrix}
=
\begin{bmatrix} \frac{T_1 + T_2}{2} \\[2ex] \frac{T_1 - T_2}{2} \\[2ex] T_3 - m_{11} \\[2ex] T_4 - m_{11} \end{bmatrix}
\tag{9.15}
$$

Table 9.2 Polarization controller settings and results during DUT measurement.

Polarization state	Q plate	H plate	Transmission T
linear horizontal, 0 degrees:	α_p	α_p	$T_1 = P_A/P_a$
linear vertical, 90 degrees:	α_p	$\alpha_p + 45$ degrees	$T_2 = P_B/P_b$
linear diagonal, +45 degrees:	α_p	$\alpha_p + 22.5$ degrees	$T_3 = P_C/P_c$
right hand circular:	$\alpha_p + 45$ degrees	α_p	$T_4 = P_D/P_d$

The maximum and minimum transmissions are given by:

$$T_{max} = m_{11} + \sqrt{m_{12}^2 + m_{13}^2 + m_{14}^2} \tag{9.16}$$

$$T_{min} = m_{11} - \sqrt{m_{12}^2 + m_{13}^2 + m_{14}^2} \tag{9.17}$$

Finally, the PDL can be calculated using:

$$PDL_{dB} = 10 \log \frac{T_{max}}{T_{min}} \tag{9.18}$$

The critical points in this measurement are similar to those of the scanning method, for example, constant power from the laser source and polarization-independent power meter. The requirement for constant power from the polarization controller is relaxed because the polarization dependence of the power level is mathematically corrected. This is necessary because polarization controllers with waveplates often produce larger power variations than polarization controllers with fiber loops.

In comparison with the scanning method, this method is mathematically more complex, and it consists of two measurement steps, one without and one with the DUT inserted. However, the calibration step is only needed once in a series of measurements. Altogether, the Mueller method features at least the same accuracy as the polarization scanning method: For example, accuracies in the order of ±0.003 to ±0.005 dB are attainable for PDL values of 0.1 dB (this includes the repeatability of the waveplate positions).

The big advantage of the Mueller method is its speed: less than 2 s per PDL result can be achieved. This makes the Mueller method attractive for wavelength-dependent PDL measurements.

9.6 INTRODUCTION TO WAVELENGTH-DEPENDENT LOSS MEASUREMENTS

The commercial interest in wavelength-division multiplexing (WDM) and optically amplified systems created a large variety of optical filters and the necessity to measure their insertion loss as a function of the wavelength. Different requirements have to be met depending on the type of filter and its spectral bandwidth. A pump coupler at the input of an optical amplifier, for example, may have a spectral bandwidth of 50 nm or more. In contrast, a WDM demultiplexer may have a spectral bandwidth of only 0.2 nm for each of its channels. As a consequence, different types of sources and receivers may have to be used for such different components.

A large percentage of optical filters carry insertion-loss specifications for both the passband and the stopband. In a WDM demultiplexer, for example, the stopband loss defines the amount of crosstalk between the different channels. Depending on the allowable crosstalk, different test setups and associated measurement capabilities may have to be used to test these characteristics.

Most of the above-discussed uncertainties in single-wavelength measurements are also applicable to wavelength-dependent measurements. Of particular interest is the necessity to avoid optical interference, because (wavelength-dependent) optical interference will directly be visible in the measurement results. This will be further discussed below.

Another important uncertainty is PDL, because the state of polarization may change over the wavelength range of interest, thereby influencing the measurement results. Quite often, the PDL is considered a separate characteristic of the component.

In the following, several different source-receiver combinations are compared and their differences discussed.

9.7 WAVELENGTH-DEPENDENT LOSS MEASUREMENTS USING A TUNABLE LASER

The basic measurement configuration for wavelength-dependent loss measurements is similar to those discussed in conjunction with the single-wavelength measurements. There are two options: one in which the receiver is a simple optical power meter, a second, in which the receiver is an OSA. These two alternatives are discussed below.

9.7.1 Loss Measurements with a Tunable Laser and a Power Meter

Figure 9.16 shows the measurement setup for wavelength-dependent insertion-loss measurements involving a tunable laser and an optical power meter. The DUT is of the pigtailed and connectorized type.

In the calibration step, the lead-in fiber is directly connected with the power meter. The wavelength of the tunable laser (TLS) is tuned in predefined steps from the lowest to the highest wavelength, and the wavelength-dependent power levels are recorded. In the actual measurement of the DUT, the wavelength scan is repeated and the power levels recorded again. The logarithmic difference between the two sets of power levels is the wavelength-dependent insertion loss.

For the discussion of the limitations of this measurement method, it is important to understand some typical characteristics of tunable lasers, namely ECL. Figure 9.17 shows the schematic diagram of an ECL.

The gain medium is a conventional laser diode in which the laser resonator is disabled by antireflection coating on one side. Then the resonator is rebuilt by adding an external reflector, the diffraction grating. The grating simply acts like a wavelength-selective plane mirror. Depending on the resonator length, a comb of wavelengths can

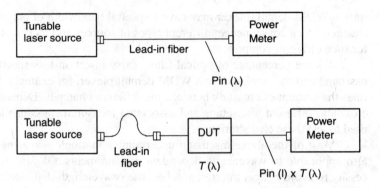

Figure 9.16 Measurement setup for wavelength-dependent insertion loss.

oscillate in principle. To select a single wavelength out of the comb, the grating angle must be set appropriately. Fine-tuning therefore involves changing the cavity length and rotating the diffraction grating simultaneously, all of this with sub-micrometer precision. See Duarte[7] for a more complete discussion.

Wavelength Accuracy. In wavelength-dependent insertion-loss measurements, the most obvious quality criterion of the tunable laser source is its wavelength accuracy. Both absolute wavelength uncertainty and tuning linearity are important and must be distinguished. Quite often, a typical absolute uncertainty of ±0.1 nm is of no concern. For the measurements of narrowband filters however, the tuning linearity is most important. Figure 9.18 shows a measurement result of the wavelength error of a modern tunable laser over a wide wavelength range. The measurement was carried out with a wavelength meter of much higher accuracy.

A ±25 pm peak-to-peak variation of the wavelength error (TLS wavelength minus measured wavelength) is observed. Also shown is the wavelength error when the tuning is

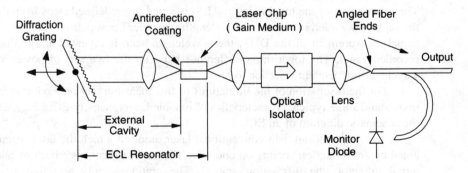

Figure 9.17 Schematic of a tunable external-cavity laser.

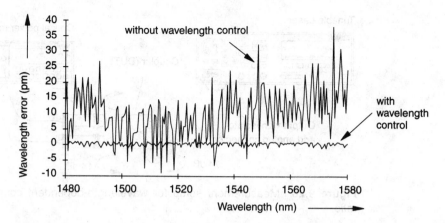

Figure 9.18 State-of-the-art wavelength accuracies of tunable lasers.

accomplished under control of the wavemeter, provided that the electronics of the specific TLS include such a possibility. In our case, a much-improved tuning linearity of ±1 pm was achieved.

To demonstrate the benefit from highly accurate tuning linearity, a narrow passband filter was measured without and with wavelength control. This is a critical application because the filter's 3 dB-bandwidth was only 0.09 nm wide. Accordingly, the tuning range in Figure 9.19 was only 1 nm (much narrower than in Figure 9.18).

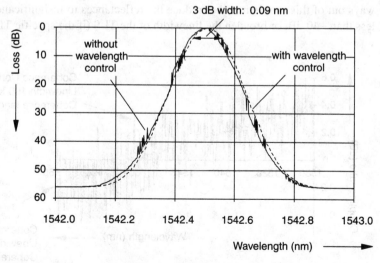

Figure 9.19 Measurement of a narrow filter with a power meter and a tunable laser, with and without wavelength control.

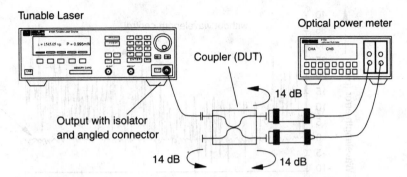

Figure 9.20 Measurement setup for wavelength-dependent coupling ratio.

Optical Interference Effects. Another relevant point in wavelength-dependent insertion-loss measurements with tunable lasers is to avoid optical interference effects. This subject was already discussed in Section 9.4.3. Figure 9.20 shows a setup for the measurement of the wavelength-dependent coupling ratio of a coupler, in which the reflections were not well controlled. One of the coupler arms was well terminated because it connects to the output of the TLS that has a return loss of 60 dB. The three unterminated ports represent three reflections of 14 dB (3.5%) which are separated by 2 m. The measurement result is shown in Figure 9.21.

Obviously, optical interference effects can be a substantial problem. This problem is due to the fact that a typical TLS linewidth of 100 kHz is far too narrow to avoid interference effects when the spacing between the reflection points is only 2 m. There are two ways out of this problem: either reduce the reflectances to insignificance, for example, to less than −40 dB, or broaden the linewidth of the TLS (if the specific TLS features this ca-

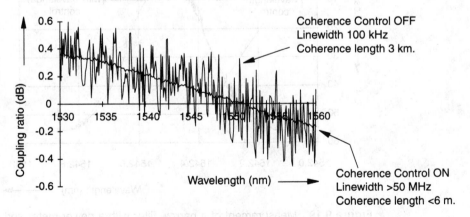

Figure 9.21 Wavelength-dependent coupling ratio of a coupler.

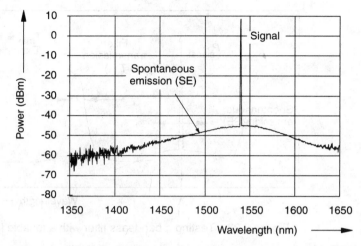

Figure 9.22   Spectral power density of an external cavity laser.

pability). A measurement result with a broad linewidth of >50 MHz (coherence control ON) is also shown in Figure 9.21.

Limitations Caused by Spontaneous Emission. Finally, another limitation comes from the fact that all lasers, including tunable lasers, produce spontaneous emission (SE) in addition to the spectrally narrow signal. Figure 9.22 shows the spectrum of a modern TLS. A resolution bandwidth of 1 nm was used in this measurement. The wavelength was set to 1540 nm, its signal output power was +8 dBm (6 mW), and the peak of the SE spectral power density was −50 dB/nm relative to the signal. The total integrated SE power was −32 dB relative to the signal power.

On this basis, what are the loss limits of this measurement technique, respectively, how high can the loss of the test component be before the measurement accuracy starts to degrade? To answer this question, it is useful to define for all swept-wavelength measurements:

Measurement range:
> **The largest measurable loss of a passive component. Alternatively, the difference between the source power and the receiver's sensitivity.**

Dynamic range:
> **The difference between the smallest and the largest measurable loss on a single component, for example, as a function of the wavelength. It depends on the component characteristics and on the characteristics of the source and the receiver.**

If we assume a source power of 8 dBm (as in the example above) and a receiver sensitivity (noise level) of −100 dBm, then the measurement range is 108 dB. If a SNR of 20 dB is required for accuracy reasons, then the measurement range is still 88 dB. The dynamic range can be substantially lower as shown below.

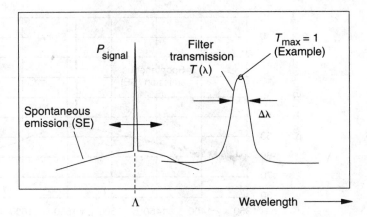

Figure 9.23   Testing a bandpass filter with a tunable laser.

Measurement of a Bandpass Filter.   To analyze the dynamic range of wavelength-dependent insertion-loss measurements involving a tunable laser and a power meter (see the measurement setup of Figure 9.16), let us discuss the test of a bandpass filter. Figure 9.23 illustrates the situation.

The power at the power meter can generally be expressed by:

$$P(\Lambda) = \int \rho_{in} (\lambda, \Lambda) \, T(\lambda) \, d\lambda \tag{9.19}$$

where Λ is the signal wavelength of the TLS, ρ_{in} is the spectral power density of the TLS which depends on Λ, and $T(\lambda)$ is the wavelength-dependent transmission through the component. The wavelength range for the integration should be the same as the wavelength range of the SE. If the filter bandwidth is small, then the largest power meter result is simply:

$$P_{max} \cong P_{signal} \, T_{max} \tag{9.20}$$

Example

If the signal power is 8 dBm, and the maximum transmission is 1 (0 dB), then the maximum power meter result is also 8 dBm. This simple result is caused by the fact that, when the TLS wavelength coincides with the filter's center wavelength, the SE is attenuated by the filter and does not add to the power meter result.

Conversely, when the TLS wavelength is set outside the filter's passband, a small part of the SE emission (the one which falls into the filter's passband) is not attenuated. Then the minimum power level is:

$$P_{min} \cong P_{signal} \, T_{min} + \rho_{se} \, \Delta\lambda_{filter} \, T_{max} \tag{9.21}$$

where ρ_{se} is the spectral power density of the SE. The first part of the equation represents the desired measurement result. The second part is unwanted and represents the limit of the dynamic range.

Example

Using logarithmic notation: If we ignore the first part of the equation and assume an SE of −50 dBm/nm (as in Figure 9.22) and a noise-equivalent filter bandwidth $\Delta\lambda = 0.1$ nm, then $P_{min} = -60$ dBm. Therefore, the dynamic range for this specific narrow-band filter is 60 dB. This is obviously less than the measurement range of 88 dB mentioned above. Notice that the dynamic range is not limited by the power meter's noise level here.

Measurement of a Stopband Filter.   An even smaller dynamic range is obtained in the measurement of a stopband filter. See Figure 9.24.

In this case, the maximum power is obtained when the TLS is set outside the stopband. For a filter with a narrow bandwidth, if we assume wavelength-independent signal power and passband transmission, the maximum power is given by:

$$P_{max} \cong P_{signal}\, T_{max} + \int_{SSE\ band} \rho_{se}(\lambda)\, T(\lambda)\, d\lambda \tag{9.22}$$

where ρ_{se} is the spectral density of the SE.

Example

Using the same values as before: With a signal power of 8 dBm and a maximum passband transmission of 0 dB, the first part of the equation is 8 dBm. The second part can be calculated using the TLS spectrum of Figure 9.22: The integrated SE power was 32 dB below the signal power. If we assume $T(\lambda) = 1$ for the entire wavelength range of interest, then the sec-

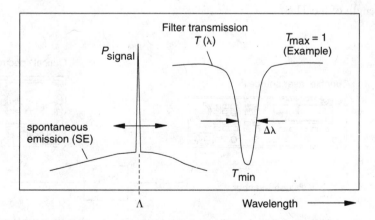

Figure 9.24   Testing a stopband filter with a tunable laser.

ond part of the equation represents a value of −26 dBm. This is considered sufficiently small to be ignored. Therefore, $P_{max} = 8$ dBm.

Conversely, when the TLS wavelength is set to the center of the stopband, most of the SE is still unattenuated. The minimum power level is:

$$P_{min} \cong P_{signal} \, T_{min} + \int_{SSE\ band} \rho_{se}(\lambda) \, T(\lambda) \, d\lambda \qquad (9.23)$$

The second part is again unwanted and represents the limit of the dynamic range. Using an integrated SE of −32 dB relative to the signal, and $T(\lambda) = 1$, then analysis of Equations 9.22 and 9.23 yields a dynamic range of only 32 dB. This is certainly insufficient for the characterization of some stopband filters, although it represents today's state-of-the-art performance of tunable lasers.

Summarizing Measurements With a TLS and a Power Meter. The dynamic range of wavelength-dependent insertion-loss measurements using a modern tunable laser and a sensitive power meter is between 55 and 32 dB, depending on the type of filter. An SE of −50 dB/nm relative to the signal power was assumed for this estimation.

The big advantages of tunable lasers are their high power and fine wavelength resolution, though. This way, even the narrowest filters can be measured. If using a power meter does not provide sufficient dynamic range, then an OSA can be used instead; see below. Alternatively, a tunable laser with much lower SE would be needed.

9.7.2 Loss Measurements with a Tunable Laser and an Optical Spectrum Analyzer

The dynamic range limitations of the TLS / power meter combination can be overcome by using an OSA as the receiver. Figure 9.25 shows the measurement setup. For this measurement, the OSA's center wavelength should be set to always follow the signal wavelength of the TLS.

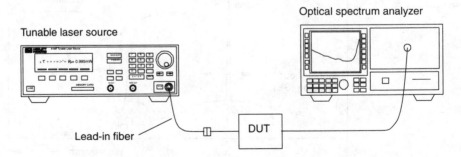

Figure 9.25 Insertion-loss measurement setup using a tunable laser and an optical spectrum analyzer.

Before discussing the dynamic range of this measurement, it is important to understand some of the properties of OSAs, such as the achievable measurement range and the finite filter rejection.

TLS/OSA Measurement Range.   Figure 9.26 illustrates the largest measurement range that is available for a narrow-linewidth laser used in conjunction with a sensitive OSA. Both the TLS spectrum and the OSA noise level (without signal) were measured with a resolution bandwidth of 1 nm.

The difference between the laser's signal level and the spectrum analyzer noise floor, the measurement range, is approximately 80 dB in this case. The dynamic range is usually smaller, even though the OSA acts as a wavelength-selective receiver. One of these limitations comes from the OSA itself.

Limited Filter Rejection in the Stopband.   An increase of the noise floor will result from high power bypassing the OSA filter. As an example, Figure 9.26 shows that there is a noise floor which is approximately 15 dB higher than the noise floor without any signal (at 1300 nm, the actual SE of this specific TLS is negligible). This noise floor rise is due to the main signal bypassing the OSA filter. A filter rejection of approximately 60 dB was observed in this case.

Dynamic Range Improvement Due to Using an OSA.   It is interesting to analyze the dynamic range improvement due to replacing the power meter of section 9.7.1 by an OSA. To do this, let us reconsider the measurement of the stopband filter, because this was the most critical case. The smallest signal was obtained when the TLS wavelength was set to the center of the stopband. In this situation, the OSA's photodetector detects the

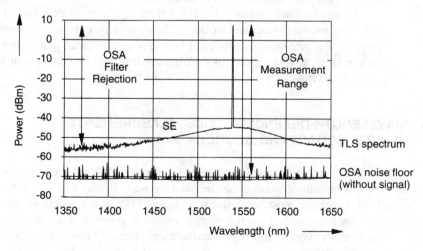

Figure 9.26 Measurement range for the combination of narrow-linewidth laser and OSA.

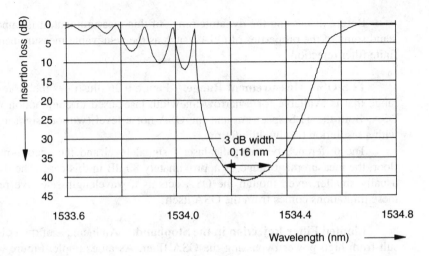

Figure 9.27 Measurement of a fiber grating with a tunable laser and an OSA, with wavelength control to obtain best resolution.

attenuated signal (to be ignored here) and the spontaneous emission of the TLS which is filtered by the OSA. If we assume that the OSA's resolution bandwidth is set to 1 nm, and that the OSA attenuates the SE spectrum by 60 dB (as discussed above), then the dynamic range is increased from 32 dB (with the power meter) to 92 dB with the OSA. Of course, the OSA's photodetector noise will effectively limit the dynamic range to approximately 70 dB as shown in Figure 9.26. However, such a dynamic range should meet even the most stringent requirements in filter measurement.

As an example, a narrow stopband filter on the basis of a fiber grating was measured using a tunable laser and an OSA. The measurement result is shown in Figure 9.27. This particular grating had a stopband loss of 40.7 dB. Clearly, no dynamic range limitations can be seen, which indicates the high dynamic range of the combination of a tunable laser with an OSA.

9.8 WAVELENGTH-DEPENDENT LOSS MEASUREMENTS USING A BROADBAND SOURCE

Instead of using a tunable laser source for wavelength-dependent insertion-loss measurements, a broadband source can be used as well. In this case, however, a wavelength-selective receiver must be used, for which an OSA is a convenient solution. The advantages for this method of wavelength-resolved insertion loss measurements are wide wavelength coverage and high measurement speed. Figure 9.28 shows a possible measurement setup.

As usual, the procedure includes two steps:

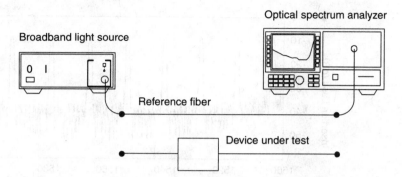

Figure 9.28   Experimental setup for wavelength-resolved insertion loss measurements using a broadband light source and an optical spectrum analyzer.

Step 1: The broadband light source is first connected to the OSA with a reference low-loss fiber. All future insertion loss measurements will be compared to the loss of the reference fiber. The OSA measures the spectral shape of the broadband source as shown in Figure 9.29a. This particular source is an edge-emitting LED (EELED) with a center wavelength of 1550 nm.

Step 2. The DUT is then inserted between the broadband source and the OSA. The resulting output spectrum from the DUT is shown in Figure 9.29b. The display shows a wavelength division multiplexer filter with a repeating passband.

Finally, the filter output spectrum (Figure 9.29b) is subtracted (in dB) from the input source spectrum (Figure 9.29a) resulting in the loss versus wavelength plot of Fig-

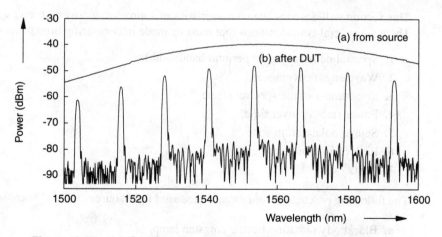

Figure 9.29   (a) Spectral density of EELED source measured with a 0.1 nm resolution bandwidth. (b) Spectral density at the output of the filter.

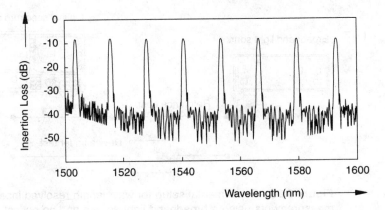

Figure 9.30   Insertion loss measurement from the data of Figure 9.29.

ure 9.30. The filter has a passband insertion loss of 8.5 dB. The noise floor is noticeably wavelength dependent. This occurs because the spectral density of the broadband source is wavelength dependent. The sensitivity of the spectrum analyzer is relatively wavelength independent. The result is a wavelength-dependent measurement range.

The following discussion will describe some of the measurement considerations for this method. Section 9.8.1 discusses in detail the types of broadband sources that are available for such measurements along with their characteristics. Section 9.8.2 discusses some of the OSA characteristics that must be considered when designing and interpreting the measurement.

9.8.1 Broadband Light Sources

This section will describe the characteristics of sources that cover a broad spectral width. There are several considerations that must be made in comparing broadband sources:

1. Spectral density (power per unit bandwidth),
2. Wavelength coverage,
3. Smoothness of the spectral shape,
4. Power stability over time,
5. Source polarization state,
6. Reflection sensitivity, and
7. Cost.

The following practical solutions for broadband light sources will be discussed:

a. Blackbody radiation from a tungsten lamp.
b. Amplified SE from an EELED.
c. Amplified SE from an optical amplifier.

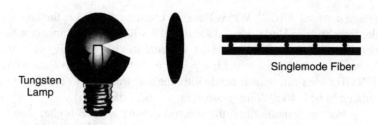

Figure 9.31   Tungsten lamp source imaged to a singlemode fiber.

Blackbody Radiation Sources: Tungsten Lamp Source.   A commonly used noise source for fiber-optic measurement applications is a tungsten lamp coupled to an optical fiber as shown in Figure 9.31.

The lens focuses a fraction of the lamp filament onto the end of the singlemode optical fiber. A key characteristic of a source is the amount of power available in a given bandwidth, the spectral power density. Figure 9.32 shows the spectral density of a light bulb source coupled into a singlemode fiber.

This lamp has a tungsten filament at a temperature of 2700 K resulting in a relatively low spectral density of −63 dBm/nm. It is essentially flat over the entire infrared wavelength range. The spectral density that can be coupled into a fiber depends only on the temperature of the tungsten filament.[8] It does not matter if a 1 W or 100 W bulb is used. In tungsten lamps, higher power is created by using a larger filament size. However, the larger filament cannot be imaged down to the 9 μm singlemode fiber core without much of the light arriving at too steep of an angle to be accepted by the fiber.

The maximum spectral power density per fiber mode from a blackbody radiator is given by Siegman's radiation formula:[9]

$$S_\nu = \frac{h\nu \, \epsilon}{e^{h\nu/kT} - 1} \quad [\text{W/Hz}] \tag{9.24}$$

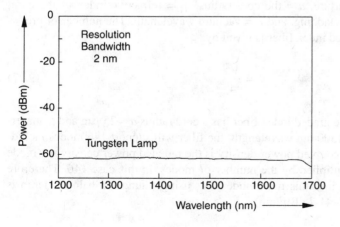

Figure 9.32   Spectral power density for a tungsten lamp coupled to a singlemode fiber.

where $h = 6.62 \times 10^{-34}$, Ws^2 is Planck's constant, $\epsilon \cong 0.3$ is the emissivity of tungsten, v is the frequency in Hertz, $k = 1.38 \times 10^{-23}$, Ws/K is Boltzmann's constant, and T is the temperature in Kelvin. The emissivity is equal to $(1 - \text{reflectivity})$ of the filament material. Using $T = 3000$ K, $\epsilon \cong 0.33$, $v = c/\lambda = 231$ THz (equivalent to 1300 nm), and $dv = 178$ GHz for 1 nm optical bandwidth, the spectral power density per mode, S_v, can be calculated to be 0.46 nW/nm (equivalent to -63.4 dBm/nm).

For singlemode fiber, the spectral density is 3 dB higher than stated, because there are two orthogonal modes that propagate in singlemode fiber. Multimode fiber can couple a significantly higher spectral density. The only parameters that are under the source designer's control are the temperature, the emissivity, and the core size of the optical fiber. Unfortunately, material limitations preclude any significant increase in the temperatures of resistively heated filament lamps. The temperature of the tungsten lamp can be increased from 2700 K to over 3300 K, which results in a 1.8 dB increase in spectral density. Simultaneously, the lifetime of the lamp is decreased by a factor of 10.

Coupling the Lamp to Fiber. Conceptually, the simplest and most efficient way to couple the blackbody source to the fiber is by butting the fiber directly against the much larger emitting filament.[8] However, as a practical matter, the light from the lamp must be imaged onto the fiber end as is shown in Figure 9.31. As long as the numerical aperture of the imaging optics is greater than the numerical aperture of the fiber, and the optical losses are negligible, then optimal coupling is accomplished.

Using Equation 9.24, it is possible to calculate the coupling into multimode fibers as well. First, the number of fiber modes must be calculated. Key fiber parameters are the numerical aperture, NA, and the normalized frequency, V, from Hentschel:[2]

$$NA = \sqrt{n_1^2 - n_2^2} \qquad (9.25)$$

$$V = \frac{2\pi a}{\lambda} NA \qquad (9.26)$$

where NA = numerical aperture, a = the core radius, n_1 = refractive index of the core, n_2 = refractive index of the cladding, and λ = vacuum wavelength. The number of propagating fiber modes for graded index fiber is given by:[2]

$$N = \frac{V^2}{4} \qquad (9.27)$$

A commonly available graded index fiber has a core radius $a = 25$ μm and a numerical aperture $NA = 0.2$. At 1300 nm wavelength, the fiber will support 146 modes including both polarizations. The coupled power density is the radiated power density per mode given by Equation 9.24 multiplied by the number of modes, in this case 146. Therefore the power density coupled into this multimode fiber using a tungsten halogen lamp is 67.2 nW/nm (equivalent to -41.7 dBm/nm).

Edge-Emitting LEDs.   Substantially higher spectral power density can be generated with light-emitting diodes (LEDs), particularly with edge-emitting LEDs (EELEDs). A second class of LEDs, namely surface-emitting LEDs, produces lower power density and will therefore not be discussed here.

The principle construction of an EELED is similar to the one of an FP laser, except that the two output facets are antireflection-coated to avoid any resonances or lasing in the device. A common implementation is the two-section EELED of Figure 9.33. In this case, the EELED consists of a semiconductor optical amplifier (SOA) and an absorber on the unused output side.[10]

The EELED has an optical waveguide fabricated along its entire length. The optical amplifier section is forward biased with a current source. To achieve gain, electrons are excited from the valence band to the conduction band with the electrical current. In the absence of an input signal, a certain fraction of the excited carriers will fall back to the low energy level and produce SE photons. These photons emit in random directions and will have a wavelength near the bandgap energy of the semiconductor. A small fraction of the spontaneously emitted photons (0.1%) will be captured by the optical waveguide and propagate towards the output of the amplifier. As the photons travel along, they will stimulate the generation of other photons. The stimulated photons will have the same direction, frequency, phase, and polarization as the photon that caused the stimulated emission.

It is desirable to have the noise output produced by a single pass through the amplifier. Multiple reflections in the LED, or between the coupling optics and the LED, can degrade the spectral smoothness of the source. In the worst case, lasing can occur. The EELED design in Figure 9.33 uses a reverse-biased absorber section to prevent any light from being reflected from that interface. The output side carries an antireflection coating to increase transmission to the fiber and to reduce the reflectivity of the transition from the waveguide to air of normally 30%. Angling the waveguide with respect to the output will also act to suppress multiple reflections.

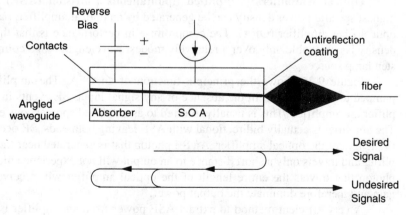

Figure 9.33 Two-section EELED.

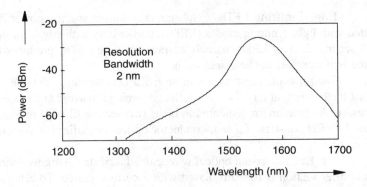

Figure 9.34   Spectral density of an edge-emitting LED.

Figure 9.34 shows the spectral density for an EELED with 15 dB of net gain in the optical amplifier coupled to a singlemode fiber. This SOA was designed to have a bandgap corresponding to a 1550 nm center wavelength. The spectral density in this case is −30 dBm/nm maximum with a 55 nm spectral width (full-width, half-maximum).

The relatively narrow width of optical amplifiers can be overcome by connecting the outputs of many optical amplifiers that are offset in wavelength. Semiconductor optical amplifiers are ideal for this application since the center wavelength of each amplifier can be engineered to obtain nearly any wavelength in the 500 nm to 2000 nm range.

The output of an EELED tends to have a substantial degree of polarization. The optical waveguide has a rectangular cross-section that tends to give more gain to one polarization than another. Researchers have been able to reduce the degree of polarization by waveguide symmetry and by the introduction of active regions with strained quantum wells.[11]

Optical Amplifiers: Amplified Spontaneous Emission (ASE) Sources.   The highest spectral power density can be generated by an optical amplifier, particularly by an optical fiber amplifier (OFA). The compromise in performance is that this high spectral density is achievable only over a relatively modest wavelength range compared to a tungsten lamp source.

Figure 9.35 shows the principle function of an OFA. The amplifier is optically pumped to produce gain. In the absence of any signal input, SE events in the optical amplifier are amplified. This is usually referred to as amplified spontaneous emission (ASE). The amplifier is actually bidirectional with ASE leaving both ends. SE occurs over the entire length of the optical amplifier. An SE photon that is generated near the end of the amplifier and travels only a short distance to an output will not experience much gain. An SE photon that travels the entire length of the optical amplifier will receive the maximum gain and therefore dominate the output power.

A very efficient method to extract ASE power from an amplifier is to add a 100% reflector at the input of the amplifier. In this case ASE that originates near the output of

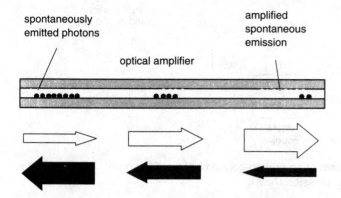

Figure 9.35   Amplified spontaneous emission.

the amplifier experiences two passes through the optical amplifier before leaving the output. The disadvantage for this method is spectral narrowing of the source and the possibility of lasing to occur if undesired reflections are present in the amplifier. Small-signal gains of up to 40 dB are achievable in single-pass optical amplifiers.

For an optical amplifier, the noise spectral density, S_v, per mode is:[12]

$$S_v = F(G - 1) hv \tag{9.28}$$

where G is the amplifier power gain and F is the noise factor, which for a two-level material is:

$$F = \frac{N_2 - N_1}{N_2} \tag{9.29}$$

N_2 and N_1 are the population densities of the upper and lower energy levels. The noise factor, F, has a minimum value of 1 for a complete population inversion. For erbium amplifiers,[12] the noise factor depends on the pump and emission wavelength but is typically larger than 2. Assuming a modest amplifier gain of 20 dB and a noise factor of 2, the power density at 1550 nm in a single polarization is −25 dBm/nm. This is a spectral density increase of 37 dB over the tungsten lamp. A useful result for future use is that the spectral density of an optical amplifier at 1550 nm is −48 dBm/nm (this is the value hv) + noise figure (dB) + gain (dB).

Figure 9.36 shows a possible diagram for a broadband source using an erbium-doped fiber amplifier (EDFA). The EDFA is optically pumped at either the 980 nm or 1480 nm absorption band of the amplifier. EDFAs are very convenient ASE sources because they connect directly to a singlemode fiber system with low loss.

Figure 9.37 shows an example spectral density plot for a 1480 nm pumped EDFA coupled into a singlemode fiber. The spectral density in this case is −10 dBm/nm over a spectral range of 1525 to 1565 nm with a total power of 4 mW. Using Equation 9.28 and assuming $F = 2$, one can infer the approximate gain for this EDFA. Expressed in logarithmic terms, the factor $(G - 1)$ is equal to the spectral power density per mode (−13 dBm/nm, because the fiber carries two polarization modes) minus noise figure

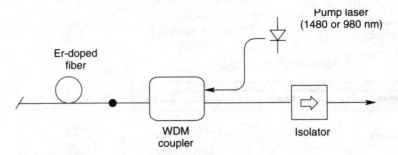

Figure 9.36　Diagram of an erbium-doped fiber amplifier producing amplified spontaneous emission (ASE).

(3 dB) minus $h\nu$ (−48 dBm/nm), which yields 32 dB. Neglecting the quantity −1 yields a gain of 32 dB for this EDFA.

　　　An optical isolator is often used at the output of the amplifier to reduce the reflection sensitivity and increase the output return loss. Large reflections back into an EDFA can be detrimental in several ways. The external reflection could combine with an internal reflection to cause the amplifier to turn into a laser. If the internal reflection level of the EDFA is very low, the external reflection will cause a reduction in the output ASE power because of gain saturation in the reverse amplification direction.

　　　EDFA ASE sources also have the advantage of being relatively unpolarized. The initial SE is randomly polarized. Since the fiber optical waveguide is circularly symmetric, all the polarizations receive essential equal amplification resulting in an unpolarized output.

　　　EDFAs cover the wavelength range near 1550 nm. This wavelength is of tremendous importance for WDM systems and associated insertion-loss measurements. There

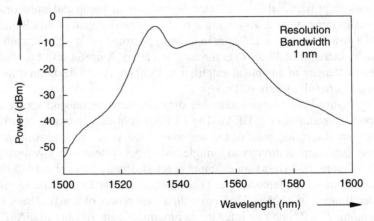

Figure 9.37　Spectral density of the ASE from an EDFA, at a resolution bandwidth of 1 nm.

Table 9.3 Comparison of broadband sources.

Parameter	Tungsten Lamp	Edge-emitting LEDs	Fiber Amplifiers
Center wavelength (nm)	Broadband	Adjustable by material	1550 Erbium 1300 Praseodymium
Spectral width (nm)	all of infrared	50–100 nm	30 to 40
Peak spectral density SM fiber (dBm/nm)	−63	−25	−10
Total power into a singlemode fiber	1 μW	100 μW	1 to 10 mW
Degree of polarization	0%	20–60%	< 5%
Reflection sensitivity	none	moderate	low when isolated
Cost	low	moderate	expensive

are fiber amplifiers available at other wavelengths as well: praseodymium-doped fluoride glass fibers[13] operate near 1310 nm.

Comparison of Broadband Sources. Table 9.3 compares the important characteristics of the broadband light sources described in this section. The EDFA is a good all-around source for most applications at 1550 nm. It has a high output power, moderate spectral width, and low degree of polarization. Its cost is its drawback. The EELED has somewhat lower spectral density, wider spectral width, and can be adjusted to a desired center wavelength. The light bulb source has a very wide spectral width but its usefulness is limited by its low spectral density.

9.8.2 Receiver Characteristics Relevant to Loss Measurement with Broadband Sources

This section discusses some of the characteristics of the OSA that are important for insertion-loss measurements with a broadband light source and an OSA. The type of receiver that will be discussed is the grating-based OSA as was presented in Chapter 2. This section will discuss the measurement range, measurement resolution, and dynamic range.

Measurement Range and Resolution. The measurement range for a broadband source/OSA measurement is determined by the following variables:

1. The spectral density of the source.
2. The filter bandwidth of the receiver.
3. The receiver sensitivity.

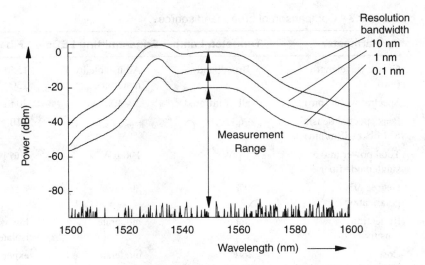

Figure 9.38 Measurement range for an erbium-ASE source measured at various resolution bandwidths.

Figure 9.38 illustrates the largest measurement range that is available for a broadband source used in conjunction with a sensitive OSA. The sensitivity setting was −90 dBm in this case. Accordingly, a relatively long sweep time of 25 s was necessary for the 100 nm span.

The maximum difference between the erbium-ASE source level and the spectrum analyzer noise floor, the measurement range, is over 90 dB in this case. It is important to realize that the displayed level of the source (and thereby the measurement range) depends on the filter width of the OSA, because the broadband source has continuous wavelength coverage. If the filter width of the spectrum analyzer is changed from 10 nm to 0.1 nm, the displayed level of the source drops by 10 log (0.1 nm / 10 nm) = 20 dB. A minimum resolution of 0.1 nm is typical for the spectrum analyzers used in the telecommunication industry.

The applicable filter width, and thereby the dynamic range, will depend on the wavelength resolution requirements of the specific test device. Obviously, a filter with a 1 nm width cannot be measured with a resolution bandwidth of 10 nm.

Notice that, in contrast to broadband sources, the signal of a tunable laser will be displayed with a constant amplitude, independent of the OSA resolution bandwidth. The TLS spectral width is much narrower than the resolution bandwidth of the OSA. Accordingly, the measurement range for a tunable laser will not depend on the OSA's resolution bandwidth.

Dynamic Range.   There are several factors that can limit the dynamic range of a measurement using a broadband source and an OSA. Some of these are related to receiver

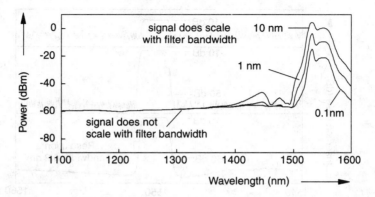

Figure 9.39 Dynamic range limitation of monochromator-bypass light.

characteristics. For low spectral density sources such as the tungsten lamp, the measurement range and the dynamic range are nearly identical.

For high spectral density sources such as the EDFA source, an increase of the noise floor can result from the finite filter rejection of the OSA filter. As an example, Figure 9.39 shows the spectral density of an EDFA source measured over the 1100 to 1600 nm wavelength range with the OSA filter bandwidth set to three different values.

Near the peak spectral density wavelength of 1550 nm, the displayed level scales properly with resolution bandwidth. At shorter wavelengths, there appears to be a signal that is independent of filter bandwidth. This is the same effect as the one observed in Figure 9.26: There is actually no signal present at these shorter wavelengths. The display level at short wavelengths is the result of the finite stopband rejection of the OSA filter. In our example, the noise floor is 70 dB below the *total* input power to the filter, independent of the filter bandwidth. The dynamic range will be equal to the filter stopband rejection, independent of the total power of the input signal.

9.8.3   Examples of Filter Measurements Using Broadband Sources

Two examples will be given to further illustrate measurements using broadband sources with OSAs. The first example shown in Figure 9.40 represents the measurement of a narrow bandpass filter using three different broadband sources: a tungsten lamp, an EELED, and an EDFA.

Because the filter has a passband width of only 2 nm, an OSA filter bandwidth of 0.2 nm was used to resolve the filter shape. Due to the low spectral density of the tungsten source, the passband insertion loss is measured but only 10 dB of the filter skirt is displayed. The EELED provides a significantly higher dynamic range. Only the erbium-ASE source produces adequate dynamic range to fully resolve the stopband of the filter.

Figure 9.41 shows a second example in which a notch filter on the basis of a fiber Bragg grating was measured using the same three sources. For the tungsten lamp source, the measurement of the filter notch is limited by the measurement range of the system. For the EELED and EDFA source, the notch depth measurement is limited to 40 dB for

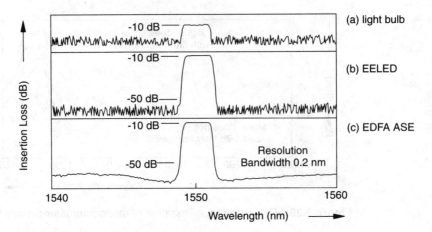

Figure 9.40   Loss measurement of narrow bandpass filter with different sources.

each case. The actual depth of the notch is larger than that measured in Figure 9.41. Obviously, filters with depths of more than 40 dB are difficult to measure with broadband sources. Most of the power from the broadband source is received by the input of the OSA since the transmission of the filter is high at most wavelengths. The finite level of filter rejection in the OSA causes a substantial amount of optical power to reach the detector after the optical filter and limits the dynamic range when measuring a notch filter. Only the combination of a tunable laser and OSA is capable of fully resolving deep notch filters. See Section 9.7.2.

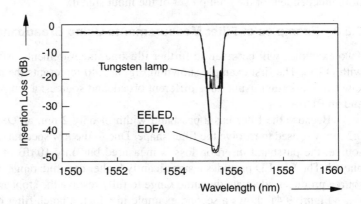

Figure 9.41   Comparison of a narrow notch filter measurement with various sources.

Table 9.4 Comparing the different methods of wavelength-dependent insertion-loss measurements (RBW = resolution bandwidth).

Method	Maximum power	Dynamic range	λ resolution	λ range	Speed
Narrowband source: TLS with power meter	0–8 dBm	30–60 dB depending on DUT	1–20 pm depending on TLS	>100 nm	200 ms per λ step
Narrowband source: TLS with OSA	0–8 dBm	70 dB	1–20 pm depending on TLS	> 100 nm	200 ms per λ step
Broadband source: Tungsten with OSA	−63 dBm/nm	7 dB @ 1 nm RBW	0.1 nm best case	all of infrared	1 s for a 100 nm sweep
Broadband source: EELED with OSA	−25 dBm/nm	45 dB @ 1 nm RBW	0.1 nm best case	> 100 nm	1 s for a 100 nm sweep
Broadband source: OFA with OSA	−10 dBm/nm	60 dB @ 1 nm RBW	0.1 nm best case	40 nm	1 s for a 100 nm sweep

9.9 SUMMARY

Table 9.4 summarizes the discussions above. The performance parameters given represent the current state of the art. Not all commercial equipment will be suitable to achieve these characteristics.

The dynamic range numbers for the methods involving broadband sources are given for an OSA sensitivity setting of −70 dBm and a resolution bandwidth (RBW) of 1 nm, where narrow filters may require narrower resolution bandwidths with less dynamic range. Also, the OSA sweep time of 1 s applies to an OSA sensitivity setting of −70 dBm.

LITERATURE

1. FOTP 50 Standard. *Light launch conditions for long length graded-index optical fiber spectral attenuation measurements.* Electronic Industries Association (EIA)

2. Hentschel, C. 1989. *Fiber Optics Handbook.* Böblingen, Germany: Hewlett Packard, PN 5952-9654.

3. Heffner, B. 1992. *Deterministic, analytically complete measurement of polarization-dependent transmission through optical devices. IEEE Photonics Technology Letters.* 4 (5): 451–454.

4. 1996. *Polarization dependent loss measurements using modular test system configurations.* Santa Rosa, CA: Hewlett Packard, product note 11896–1.

5. Nyman, B. 1994. *Automated system for measuring polarisation-dependent loss.* Optical Fiber Conference. San Jose, CA: Technical Digest, ThK6, p. 230.

6. Schmidt, S. and C. Hentschel. 1997. *PDL measurements using the HP 8169A polarization controller.* Böblingen, Germany: Hewlett Packard, PN 5964–9937E.

7. Duarte, F.J. 1995. *Tunable lasers handbook.* San Diego: Academic Press.

8. Stokes, L. F. 1994. *Coupling light from incoherent sources to optical waveguides. IEEE Circuits and Devices.* 10 (1): 46–47.

9. Siegman, A. E. 1971. *An introduction to lasers and masers.* New York: McGraw-Hill.

10. Derickson, D. et al. 1995. *High power, low-internal-reflection, edge-emitting LEDs. Hewlett Packard Journal* 46 (1): 43–48.

11. Durhus, P. et al. 1994. *1.55 µm polarisation independent semiconductor optical amplifier with 25 dB fiber to fiber gain. IEEE Photonics Technology Letters* 6:170–172.

12. Giles, C. R. and E. Desurvire. 1991. *Modeling erbium-doped fiber amplifiers. Journal of Lightwave Technology.* 9 (2): 271–283.

13. Potenza, M. 1996. *Optical fiber amplifiers for telecommunication systems. IEEE Communications Magazine.* 96–102.

10

Optical Reflectometry for Component Characterization

Wayne V. Sorin

10.1 INTRODUCTION

This chapter discusses several optical reflectometry techniques that are capable of characterizing and testing fiber connectorized optical components. The first part of this chapter discusses the measurement of total return loss, or in other words, the total integrated reflectivity from the device under test (DUT). In many instances, it is also important to know the location and magnitude of the individual reflectors that contribute to the total return loss. The second part of this chapter discusses methods of spatially resolving reflections with high resolution. High spatial resolution, in the context of this chapter, will refer to techniques capable of resolving reflective surfaces spaced by 1 cm or less. These high spatial resolution techniques are important in characterizing or trouble-shooting optical components such as fiber-pigtailed laser diodes, photodetectors, and optical isolators. Typically, resolutions of 1 mm or less are needed for the useful characterization of these devices. In contrast, reflectivity measurements performed on long fiber spans typically have spatial resolutions in excess of 1 m. A detailed discussion on this topic is given in Chapter 11.

The goal of spatially resolved optical reflectometry is to measure optical reflectivity as a function of distance. Figure 10.1 illustrates the basic concept behind reflectometry measurements. An optical probe signal is sent into the fiber or DUT. The probe signal can take a variety of forms, for example, the signal can be pulsed, chirped, sinusoidally modulated, or just continuous wave (cw). Reflected signals from the test device return after various time delays depending on the locations of the reflective sites. These signals are then processed to determine the magnitude of each reflection and its associated time delay. Knowing the speed of light within the test medium, the time delay can be converted into a

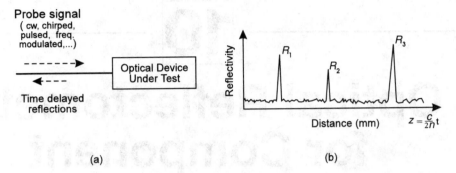

Figure 10.1 Basic concept for an optical reflectometry measurement. Many different types of probe signals can be used.

physical distance. From a signal processing point of view, the goal is to measure the impulse response for the optical power reflected from the test object. One important characteristic of a reflectometry measurement is that access to only one end of a fiber is required. Therefore, these measurements provide a very practical means for testing installed fibers or pigtailed components.

The layout for the chapter consists of first discussing a popular and practical technique that measures the total reflected power (total return loss) from a fiber-pigtailed component. Although this method does not provide any of the spatial information often needed for diagnostic purposes, it is usually adequate for testing whether a component passes a specified quality check based on total reflectivity. Next, Section 10.3 discusses many of the basic concepts that will be used throughout the remainder of the chapter. This section includes topics such as spatial resolution, effects of dispersion, the effect of spatial resolution on Rayleigh backscattered power, the concept of coherent speckle, and the differences between direct and coherent detection measurement techniques. For readers eager to get to specific reflectometry techniques, this section can be skipped and referred to later as these specific topics are encountered.

Section 10.4, a major topic in this chapter, deals with the technique known as optical low-coherence reflectometry (OLCR). Currently, this technique has demonstrated both the highest spatial resolution (< 2 μm) and the highest reflection sensitivity (< -160 dB). The section starts off with a general description of the technique followed by a detailed discussion of specific topics dealing with OLCR.

Section 10.5 discusses four alternative reflectometry techniques. These techniques are direct-detection optical time-domain reflectometry (OTDR), photon-counting OTDR, incoherent optical frequency-domain reflectometry (OFDR) and coherent OFDR. The first three techniques are included for completeness, but it is the belief of the author that these three techniques are limited when applied to high-resolution reflectometry applications. The final technique, described as coherent OFDR, has the promise of achieving results equal to OLCR. One specific implementation of coherent OFDR, called coherent

FMCW (frequency modulated continuous wave) reflectometry, will be analyzed and discussed in considerable detail.

The final section provides a comparison between the different techniques discussed in this chapter. Table 10.4 provides a compilation of experimentally achieved results using the various techniques. This table is instructive since it provides a useful tool for assessing the capabilities of each measurement technique.

10.1.1 Motivation for High Resolution Measurements

Standard long-haul OTDR is a popular and indispensable diagnostic technique for measuring long distance fiber-optic communication links (see Chapter 11). Characteristics such as fiber attenuation, connector loss, and the location of fiber breaks can all be determined by access to only one end of the test fiber. For these long-distance applications, spatial resolutions of 1 m or greater are usually adequate. In a similar manner, high-resolution reflectometry with spatial resolutions of about 1 mm or less will benefit manufacturers of components and subsystems in the area of optical component testing.

Unwanted optical reflections can degrade the performance of communication systems by affecting laser stability. Reflected light from fiber connectors or external components provide a source of unwanted feedback to the laser cavity. These reflective sites result in environmentally varying external cavities. External cavities can affect the frequency modulation response and the relative intensity noise (RIN) of the laser. To illustrate this effect, Figure 10.2 shows the intensity noise from a distributed feedback (DFB) laser diode with and without isolation from external reflections. Even with only small amounts of feedback, the RIN can be increased by more than 25 dB. Besides affecting the output intensity, external reflections can also cause variations in the optical frequency and linewidth of the laser. For example, external feedback from distant reflectors can cause the coherence length of a DFB laser to vary from its normal value of a few meters to lengths in excess of several kilometers. These long coherence lengths can then result in unpredictable intensity noise caused by coherent interference between other reflective sites within a fiberoptic link. These modulation and linewidth effects can cause considerable trouble in cable TV systems where analog transmission and high signal-to-noise ratios (SNR) are required.

Besides affecting the stability and noise characteristics of a laser, the presence of unwanted reflections can degrade system performance even when the laser is sufficiently isolated from external feedback. This occurs since weak etalons, caused by these unwanted reflections, can convert laser phase or frequency fluctuations into intensity noise. These etalon effects can occur from lens surfaces, windows on detector packages, optical elements within isolators and numerous other optical component configurations. For example, two 4% Fresnel reflections spaced by 10 cm can convert the phase noise of a DFB laser with a 50 MHz linewidth into a relative intensity noise of -117 dB/Hz (see example in Appendix A.4). This noise level can be over 30 dB larger than the typical intensity noise from a DFB laser. To predict the maximum conversion of phase noise to intensity noise requires knowing the laser linewidth, the magnitude of the reflections, and their optical path separation. These last two parameters are precisely the information that can be

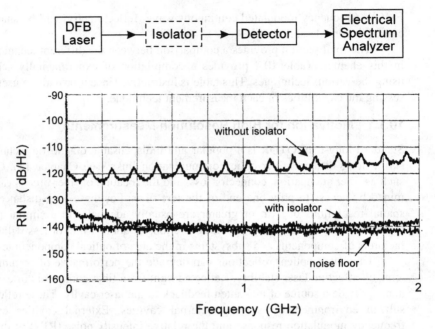

Figure 10.2 Experimental result showing the effect of optical reflections on the relative intensity noise (RIN) of a distributed feedback (DFB) laser.

obtained from a high-resolution reflectometry measurement. With this information, the system designer can then predict the worst-case phase noise conversion as a function of the optical sources that may be used. A more detailed discussion of optical phase noise conversion is given in Appendix A.4.

High-resolution reflectometry has been very useful in the design and development of optical components requiring small total reflectivities. For example, if a fiber-pigtailed isolator does not meet its required return-loss specification (typically −60 dB), a high-resolution reflectometry measurement can quickly identify which of its many internal surfaces is causing the unwanted reflection. This can be very useful in isolator design since these measurements can help determine the optimum location and angling of its many internal surfaces for minimizing isolator return loss. High-resolution reflectometry measurements are also very useful when trouble-shooting broken or malfunctioning pigtailed components. For example, during the assembly of pigtailed packages, the fiber can be broken or cracked due to the crimping process near the strain relief structure. Using a high-resolution reflectometer, this problem can often be quickly and easily identified, allowing modification of the assembly procedure to eliminate the manufacturing problem.

High-resolution reflectometry can also be used to accurately measure optical path delays in fiber and waveguide interferometers. New application areas such as fiber optic data and sensor networks in automobiles and airplanes will also benefit from these measurements. In the area of process monitoring, high-resolution reflectometry can provide a noncontact method for measuring the physical thickness and group index of manufactured film products.[1]

The above examples, although not comprehensive, illustrate that high-resolution reflectometry will play an important role in the trouble-shooting and characterization of future optical components and subsystems.

10.2   TOTAL RETURN-LOSS TECHNIQUE

A total return-loss measurement results in a single number that represents the total fractional power that is reflected from a test device. This technique is sometimes referred to as optical continuous wave reflectometry (OCWR) and is performed by sending a continuous wave (cw) optical signal to the test device and measuring the total reflected cw power.[2] This measurement is straightforward to perform and uses standard fiber-optic telecommunication parts and equipment. The term "total return loss" refers to the total reflectivity expressed in dB with the negative sign omitted.

The experimental arrangement for this type of measurement is shown in Figure 10.3. The cw source should have a stable output power and similar coherence properties to the source eventually used with the component. The power meter can be any cw optical detector with sufficient dynamic range. An absolute power calibration is not required. Calibration of the measurement system is achieved by replacing the DUT with a reflection of known value. A commonly used calibration reflector is the Fresnel reflection from the fiber-to-air interface found on a standard polished connector. This reflectivity is approximately -14.7 dB at a wavelength of 1.55 μm.

10.2.1   Reflection Sensitivity

In order to make an accurate total return-loss measurement, all unwanted powers reflected back into the power meter should be measured and taken into account. These unwanted reflections limit the reflection sensitivity of the total return loss measurement. Referring to Figure 10.3, these powers can come from discrete reflections such as those occurring at the terminated end of the fiber coupler or the connection to the test device. The reflection

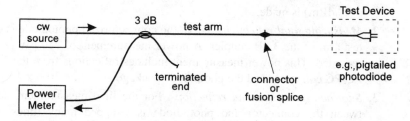

Figure 10.3   Experimental setup for measuring the total return loss from a pigtailed optical component.

from the unused coupler end can be decreased to less than -70 dB by cleaving or polishing the fiber end with a sufficient angle. In practice, reflected power from the connection to the test device often limits reflection sensitivity. If the test device is connectorized with a standard physical contact (PC) fiber connector, residual reflectivities can be in the -30 to -40 dB range. Using a precision-polished PC connector, this value can be reduced down to the -40 to -50 dB range. Reflectivities of less than -60 dB can be achieved using angle-polished connectors. Another alternative which effectively eliminates the connection reflectivity is to fusion splice the fiber connection.

With care, the effects of discrete reflections in the measurement setup can be made insignificant. However, there are certain reflected powers that cannot be eliminated, such as Rayleigh backscatter from the fiber pigtails and possible backscatter from the directional coupler. These reflections produce background signals which limit the ultimate reflection sensitivity for this technique. For example, since the directional coupler reflectivity can be on the order of -60 dB and Rayleigh backscatter from standard communications fiber is about -70 dB/meter, total return-loss measurements are limited to reflectivities greater than about -70 dB in practice.

Measurement Example. This example will give a relatively detailed description of a total return loss measurement. The arrangement in Figure 10.3 will be used to measure the total reflectivity from a connectorized fiber-pigtailed photodiode. The measurement can be divided into three main parts. The first step is to calibrate the power meter using a known reflectivity so that measured power can be related to optical reflectivity. The second step measures the total reflected power from the test device along with all the stray reflections from the measurement system. The final step is to eliminate the reflections from the test device so the background signal can be measured and then subtracted from the power reading taken in the second step.

1. *Calibrating the power meter.* For the measurement setup in Figure 10.3, it will be assumed that the source is a power-stabilized 1.55 μm EELED which delivers an output power of about 10 μW (-20 dBm) into a singlemode fiber. It is also assumed that the test arm of the 3 dB (50:50) coupler is terminated with a FC-PC connector. The reflection from this connector can be used to calibrate the power meter. With the connector end cleaned and exposed only to air, the glass-air interface results in a reflectivity of about -14.7 dB. For this arrangement, a power reading of 63 nW (-42 dBm) is made.

2. *Measuring total reflectivity.* Next, the connectorized photodiode is connected to the test arm of the 3 dB coupler. A power measurement of 250 pW (-66 dBm) is now recorded. This power measurement includes reflections from the 3 dB coupler, the FC-PC connector and the photodiode package.

3. *Subtracting background reflections.* For the final measurement step, the fiber between the connector and photodiode is wrapped many times around a small-diameter mandrel to attenuate all reflections from the photodiode package. The resulting measured power now becomes 187 pW (-67.3 dBm). This power reading

represents all the background reflections of the measurement setup. The reflected power caused by the photodiode package can now be calculated by subtracting the background reflections, in other words, 250 pW − 187 pW = 63 pW (−72 dBm). This power is 30 dB smaller than the −42 dBm measured for the connector-air reflection. The resulting photodiode reflectivity is therefore equal to −14.7 dB − 30 dB = −44.7 dB. Expressed as a total return loss we get 44.7 dB.

10.2.2 Multiple Reflections

When a test device contains multiple reflections, a total return-loss measurement can give inconsistent results. This occurs since the multiple reflections can add interferometrically resulting in the total return loss being wavelength dependent. A narrow-band source may produce a large or small reflectivity value depending on its exact wavelength. A broadband spectral-width source will return an average reflectivity value. One way of quantifying this effect is by using a frequency domain approach where the total reflectivity varies as a function of the probe wavelength. The following example illustrates this effect by considering the simplified case of only two reflections of approximately equal magnitude.

Example

Consider the case of an optical component that contains a 1 mm thick glass window. This could represent a hermetic window inside a photodiode package. Assume that the only two significant reflections come from the front and back surface of the window and each has a reflectivity of about −30 dB ($R = 10^{-3}$). For a narrow linewidth optical source, these two reflections can add either constructively or destructively depending on the precise wavelength of the source. This effect is illustrated in Figure 10.4. Since the two reflectivities are approxi-

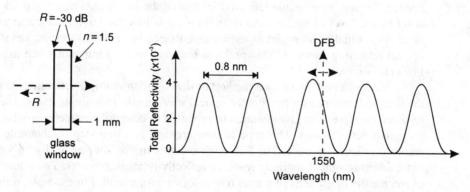

Figure 10.4 Multiple reflections produce a wavelength-dependent reflectivity which can result in an inconsistent total return-loss measurement.

mately equal, destructive interference causes an almost complete cancellation of any reflected power. Assuming a refractive index of $n = 1.5$ for the glass, and $L = 1$ mm for the window thickness, reflectivity nulls are produced with a period of about $c/2nL = 100$ GHz (0.8 nm). The wavelength dependence of the total return loss is shown in Figure 10.4. For a DFB laser, whose spectral width is typically less than 0.1 nm, the return loss can vary significantly depending on its exact wavelength. When the two reflections add constructively, the return loss can be as large as 24 dB (4×10^{-3}). When they add destructively, the total reflectivity can be more than 10 dB smaller depending on how closely the two reflections are matched. In fact, for a typical DFB laser with $\Delta\lambda/\Delta T \approx 0.1$ nm/C, a temperature change of only 4C is sufficient to convert a reflectivity maximum to a reflectivity minimum. On the other hand, if an EELED ($\Delta\lambda \approx 50$ nm) is used, its wide spectral width would average the wavelength variations resulting in a relatively stable return loss of about 27 dB (2×10^{-3}). This simple example illustrates that a single total return-loss measurement can be inadequate for devices containing multiple reflectors.

One way of dealing with multiple reflectors is to measure total reflectivity as a function of wavelength. This provides much more information than a single wavelength measurement, allowing system designers to predict reflectivity variations depending on the source used. There are several ways in which wavelength-resolved return loss measurements can be made. For example, if a wavelength-tunable narrow-linewidth source is used in Figure 10.3, the power meter can record reflectivity as a function of wavelength. Another approach is to use a broadband source such as an EELED and replace the power meter with an optical spectrum analyzer (OSA). In this case, the receiver scans across wavelength instead of the source. These procedures are very similar to the wavelength-resolved insertion loss measurements discussed in Chapter 9 except for the addition of a directional coupler to separate the incident and reflected signals.

In summary, there are a few general considerations to note when making total return-loss measurements. First, the measurement technique is relatively simple and inexpensive to implement since only commonly available components and instruments are needed. Second, measurable reflectivities need to be larger than about −70 dB. This limit is set by the level of background reflections caused by Rayleigh backscatter and coupler directivity. Finally, inconsistent measurements can occur when the test device contains several reflecting surfaces. This problem becomes more significant when using narrow-linewidth sources.

As mentioned above, one method of dealing with multiple reflections is to measure the total reflectivity as a function of optical wavelength. For simple cases, a wavelength-resolved measurement gives enough information to determine reflector spacings and their individual reflectivities. This type of measurement is a first step in obtaining a spatially resolved reflectivity trace. The following sections in this chapter deal with various techniques used to obtain spatially resolved reflectivity measurements. These techniques can overcome the limitations of a total return-loss measurement. For example, reflection sensitivity is no longer limited to about −70 dB but can be extended to better than −160 dB[3], and multiple reflections can be spatially resolved therefore eliminating the uncertainty caused by coherent interference effects.

10.3   BASIC CONCEPTS FOR SPATIALLY RESOLVED REFLECTOMETRY

Before discussing the individual spatially resolved reflectometry techniques, a brief description of some of the basic concepts will be presented. These common concepts allow the similarities between the different measurement techniques to become more apparent. Some readers may want to initially skip this section and refer back to it as the specific topics arise during the following sections.

10.3.1   Spatial Resolution

This section will describe the relationship between spatial resolution and pulse width (or frequency span) of the probing signal. These relationships will be independent of whether the measurement technique uses a cw broadband signal, a frequency-chirped optical carrier, or a short optical pulse.

It is important to first explain what spatial resolution refers to in the context of this chapter. The term "spatial resolution," without any additional qualifiers, can refer to either "single-point" or "two-point" spatial resolution. Two-point spatial resolution refers to the minimum distance between two reflectors that can still be resolved by the measurement system. This value is approximately equal to the FWHM (full-width half-maximum) of a single reflection on a reflectometry trace. The term single-point spatial resolution refers to the accuracy for determining the location of a single reflector. It is usually equal to the accuracy of locating the peak position on a reflectometry trace. Sometimes the term "distance resolution" will be used in place of single-point resolution. Single-point resolution is usually much smaller than two-point resolution. In this chapter, the term spatial resolution will refer to two-point resolution unless otherwise specified.

For a pulsed probe signal, as in the case of time-domain reflectometry, spatial resolution is limited by the pulse width of the source. Shorter pulse widths result in finer spatial resolution. The relationship between spatial resolution, Δz_r, and pulse width is given by

$$\Delta z_r \cong \frac{v_g}{2} \Delta t_s \qquad (10.1)$$

where Δt_s is the system response time, which is simply equal to the pulse width assuming a sufficiently fast response time for the receiver. The group velocity, v_g, is the speed at which the pulse travels along the fiber. The group velocity can be expressed as $v_g = c/n_g$ where c is the speed of light in a vacuum and n_g is the group index for the medium (for example, for standard telecommunications fiber $n_g = 1.47$).

Equation 10.1 states that the minimum resolvable separation between two reflectors is equal to one-half the spatial extent of the optical pulse. The reason for this factor of one-half can be seen by referring to Figure 10.5. To spatially resolve the two reflecting surfaces, the trailing edge of the input pulse must pass the first reflector before the leading edge travels twice the reflector spacing.

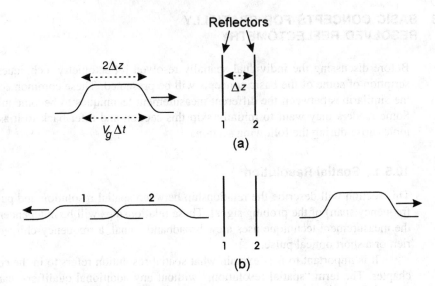

Figure 10.5 Two-point spatial resolution is approximately equal to one-half the length of the optical probe pulse.

If the receiver's response time is slower than the optical pulse duration, the spatial resolution degrades and the parameter Δt_s in Equation 10.1 should be modified using

$$\Delta t_s \cong \sqrt{\tau_p^2 + \tau_r^2} \qquad (10.2)$$

where τ_p is the optical pulse width and τ_r is the response time of the receiver. When trying to achieve the highest resolution using pulsed techniques, the response time of the detector usually limits the minimum spatial resolution. This occurs since very short optical pulses can be generated using mode-locked techniques, while detector speeds of less than about 10 psec are difficult to achieve. Using Equation 10.1, a high-speed receiver with a 10 psec response time is limited to a spatial resolution of about 1 mm. It will be shown later that other reflectometry techniques, which do not use optical pulses, can achieve spatial resolutions of much less than 1 mm.

Frequency-domain techniques rely on probing the test device (or fiber) with a signal whose frequency is scanned or stepped over some bandwidth interval. By measuring both the amplitude and phase of the reflected signal at each frequency, the frequency-domain transfer function can be obtained. This can then be Fourier transformed to obtain the desired time-domain response. In making frequency-domain measurements, the spatial resolution, Δz_r, can be related to the frequency span, Δf_s, of the probe signal by the relationship

$$\Delta z_r \cong \frac{v_g}{2} \frac{1}{\Delta f_s} \qquad (10.3)$$

where v_g is the group velocity as described in Equation 10.1. Comparing Equation 10.3 with Equation 10.1, one can see that the only difference is that Δt_s was replaced by $1/\Delta f_s$. It is interesting to note that Equation 10.3 is equivalent to the equation relating the free-spectral-range (*FSR*) of an etalon to its mirror spacing (*FSR* = *c*/2*nL*). See Chapter 4 for more information regarding the *FSR* of an etalon. This equivalence occurs since a complete sinusoidal period is required in the frequency domain to resolve two reflections in the Fourier (time) domain.

Equation 10.3 is valid for both electronic and optical frequency scan ranges. For example, consider a microwave network analyzer which can modulate an optical intensity over a frequency span of about 20 GHz. This large electronic frequency span would result in a spatial resolution of about 5 mm in fiber. On the other hand, an external cavity semiconductor laser can scan an optical frequency range of over 10 THz (80 nm at a wavelength of 1.55 μm). This frequency range would allow a spatial resolution of less than 10 μm in fiber. The above example shows that spatial resolutions can be orders of magnitude smaller when using optical as opposed to electronic frequency spans.

Equation 10.3 is quite general and is also valid for the case of optical low-coherence reflectometry. For this technique, the spectral width of the low-coherence source should be used for the frequency span, Δf_s. Since optical spectral widths can be greater than 100 nm, spatial resolutions of less than 10 μm are possible. Table 10.1 is provided to show the relationship between the spatial resolution in a fiber and the required pulse width or frequency span of the probing optical source.

10.3.2 Dispersion Limit

As spatial resolution is decreased, the effects of fiber dispersion must eventually be considered. This problem arises since the accompanying increase in spectral width causes spreading of the probe signal as it travels along the fiber. The effects of dispersion depend on the spectral content of the probe signal combined with the length and dispersion characteristics of the fiber. This resolution broadening caused by dispersion can be written as

$$\Delta z_r = \Delta z_i \sqrt{1 + \left(\frac{2L}{L_D}\right)^2} \tag{10.4}$$

Table 10.1 Spatial resolution and the required pulse width or frequency span.

Spatial resolution Δz, in fiber	Pulse width Δt_s	Frequency span Δf_s	Wavelength span $\Delta\lambda$ at 1.55 μm
10 μm	100 fsec	10 THz	80 nm
100 μm	1 psec	1 THz	8 nm
1 mm	10 psec	100 GHz	0.8 nm
10 mm	100 psec	10 GHz	0.08 nm

where Δz_i is the initial spatial resolution before any dispersion, L_D is the dispersion length in the fiber (which depends on the spectral width of the probe pulse), and L is the one-way length of fiber to the reflection site. The dispersion length can be approximated by

$$L_D \cong \frac{4\, n_g^2\, \Delta z_i^2}{\lambda^2\, c\, D} \qquad (10.5)$$

where n_g is the group index, c is the speed of light, and D is the fiber dispersion parameter at the signal wavelength λ (a detailed discussion of fiber dispersion can be found in Chapter 12). Equation 10.5 uses the result, $D = l/L \cdot \Delta\tau/\Delta\lambda$, combined with the relationships between pulse width, frequency span, and spatial resolution given in Equations 10.1 and 10.3.

Table 10.2 gives some representative values for the initial spatial resolution and the resulting dispersion length. The equivalent pulse widths, τ_p, required to achieve each spatial resolution is also included. A fiber dispersion of $D = 17$ ps/km · nm is assumed, which is a standard value for nondispersion-shifted fiber at a wavelength of 1.55 μm. For spatial resolutions of less than 100 μm, the dispersion from a few meters of fiber can become significant.

Table 10.2 implies that it should not be possible to achieve a 10 μm resolution at the end of a few-meter-long fiber pigtail. While this is true for certain reflectometry techniques, there is a class of techniques where this limitation can be overcome. Through the use of optical interference techniques, it is possible to cancel the effects of fiber dispersion by balancing the dispersion in the test arm with an equal dispersion in the reference arm. One technique that uses this method is optical low-coherence reflectometry. Using this technique, spatial resolutions as small as a few microns can be achieved even at the ends of fiber pigtails that are meters in length.

10.3.3 Rayleigh Backscatter and Spatial Resolution

The measurement of Rayleigh backscatter (RBS) is a very powerful method for characterizing fiber optic links. Conventional OTDRs use RBS to determine the location of fiber breaks, the fiber attenuation coefficient, splice loss, and various other link characteristics. The success of using RBS in long- and medium-resolution applications has spurred a

Table 10.2 Dispersion lengths in nondispersion shifted fiber at 1.55 μm.

Δz_i in fiber	τ_p	L_D
10 μm	0.1 psec	7 cm
100 μm	1 psec	7 m
1 mm	10 psec	700 m
10 mm	100 psec	70 km

strong desire that high-resolution reflectometry techniques be also capable of measuring RBS. For example, measuring the internal attenuation along an integrated optical wave-guide.[4]

Two difficulties arise when attempting to measure RBS with high spatial resolution. First, the strength of the RBS signal becomes very small as the spatial resolution is de-creased. Second, the measured RBS amplitude becomes very noisy because of coherent interference effects. This noisiness is sometimes referred to as *coherent speckle* and will be discussed in the next section. As the spatial resolution is reduced, less scatterers con-tribute to the backreflected signal causing the average RBS signal to decrease. The rela-tionship between the average Rayleigh backscattered reflectivity, R_{rbs}, and spatial resolu-tion, Δz_r, is given by

$$R_{rbs} = S\alpha_s\Delta z_r \tag{10.6}$$

where α_s is the scattering attenuation coefficient per unit length and S is the backscatter capture ratio. For telecommunication fiber at a wavelength of 1.55 μm, the attenuation coefficient (due only to Rayleigh scattering) is about $\alpha_s = 3.9 \times 10^{-5}$ m^{-1} (in other words 0.17 dB/km). Additional sources of loss contribute to a total attenuation coefficient of about 0.2 dB/km.

The backscatter capture ratio, S, is the fraction of the total Rayleigh-scattered power that gets captured in the fiber and propagates in the backwards direction. In singlemode fiber the capture ratio is inversely proportional to the square of the mode diameter which results in higher backscatter signals from certain specialty fibers. For example, the mode diameter in an EDFA is made small to improve the amplifier efficiency. This results in a larger fraction of the scattered power captured and guided in the backwards direction. Fig-ure 10.6 illustrates this effect by showing the measured RBS power near a fusion splice between a standard telecommunications fiber and an erbium-doped fiber.[5] There is almost a ten times increase in RBS power from the higher numerical aperture (NA) erbium fiber.

Approximate reflectivity strengths for RBS can be calculated using Equation 10.6. Using standard values for step-index fiber at 1.55 μm, we get $S = 1.2 \times 10^{-3}$ and $\alpha_s = 3.9 \times 10^{-5}$ m^{-1}. This gives a reflectivity of $R_{rbs} = 4.7 \times 10^{-8}$ for a spatial resolution of 1 m. Ex-pressed in decibels, this results in a RBS level of about -73 dB/m at a wavelength of 1.55 μm.

Figure 10.7 shows a plot of the mean RBS reflectivity as a function of spatial reso-lution. The equivalent pulse width and frequency span, as described in Section 10.3.1, is also given. The extreme differences in RBS power between long-haul and high-resolution applications are evident from Figure 10.7. In long-haul reflectometry, a resolution of 100 m results in a backscattered reflectivity of about -53 dB. In contrast, the RBS power for an OLCR measurement with a 10 μm resolution is 70 dB smaller.

Figure 10.7 helps illustrate the difficulty in making high-resolution backscatter measurements using pulsed OTDR techniques. As the pulse width is decreased to im-prove spatial resolution, the receiver noise increases due to the larger required bandwidth. At the same time the RBS signal also decreases. These two effects make it increasingly

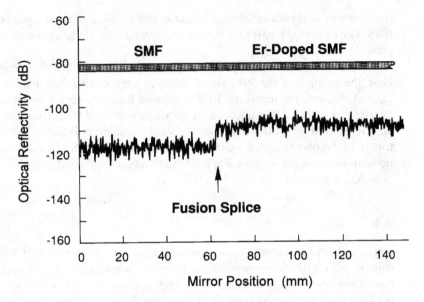

Figure 10.6   OLCR measurement of Rayleigh backscatter at a fusion splice between a standard singlemode fiber (SMF) and a higher numerical aperture erbium-doped fiber. Probe wavelength = 1.55 μm, spatial resolution = 32 μm.

difficult for commercial OTDRs to make useful backscatter measurements at resolutions of less than about 1 m.

10.3.4   Rayleigh Backscatter and Coherent Speckle

As the spatial resolution is made smaller, the uncertainty in the amplitude of the RBS signal increases and it becomes more noise-like in nature. This amplitude uncertainty occurs whenever the spatial resolution becomes comparable to the coherence length of the probe signal. Commercial long-haul OTDR measurements do not exhibit this noise-like effect since the probe pulse is typically many orders of magnitude larger than the coherence length of the source. For the case of high-resolution reflectometry, spatial resolution and coherence length tend to become similar since large spectral widths are required for the probe signal. It should be noted that this uncertainty in reflected power only occurs for randomly distributed reflections. This amplitude uncertainty does not occur for discrete reflections from a single interface or from periodically distributed reflections such as those from a fiber grating. This noise-like effect makes it difficult to use RBS signals for measuring parameters such as waveguide attenuation or joint losses in integrated waveguides.

    The cause of this amplitude uncertainty is due to the random nature of the distributed scatterers that contribute to the backscattered signal. When the coherence length of the probe signal is equal to the spatial resolution of the measurement, all the scattering

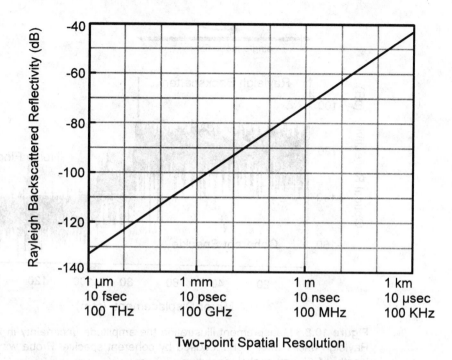

Figure 10.7   Average reflected power caused by Rayleigh backscatter in standard telecommunications fiber at 1.55 μm. Spatial resolution is within the fiber.

sites within the spatial resolution cell add in a coherent manner. This can result in constructive or destructive interference making the uncertainty in backreflected power very large. Although the variance of the RBS amplitude can become quite large, its average value is still determined by Equation 10.6. A similar effect, with the same basic origin, is the apparent graininess in light intensity when a laser beam is reflected off of a diffuse wall. For this reason we will refer to this effect as *coherent speckle* although it is also referred to as coherent fading.[6,7]

Figure 10.8 shows the measured RBS power near the end of an angle-cleaved singlemode fiber (SMF).[5] This OLCR measurement was performed at 1.55 μm and had a spatial resolution within the fiber of about 32 μm. Figure 10.8 shows that the peak-to-peak variation in the RBS power can be larger than 30 dB. Although the coherent speckle looks similar to the noise floor trace, it is actually a stable reflection signal whose structure does not average away with multiple measurements. To obtain a different structure either the location of the individual scatterers must change (for example, by heating the fiber) or the probe signal must be shifted to a different spectral band of frequencies.

Two methods for reducing the amplitude uncertainty have been demonstrated.[5,6,8] The first is to spatially average adjacent points. As more points are averaged, the uncertainty (or variance) decreases and the RBS approaches its average value.[5] The disadvan-

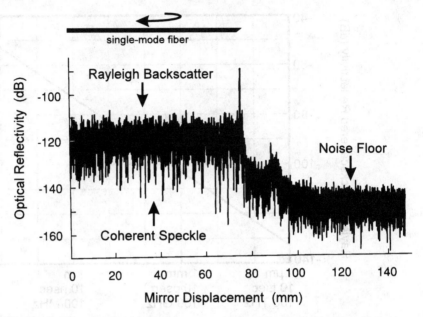

Figure 10.8 Measurement illustrating the amplitude uncertainty in the Rayleigh backscatter signal caused by coherent speckle. Probe wavelength = 1.55 μm, spatial resolution = 32 μm.

tage with this method is that the spatial resolution is degraded as a result of the averaging process. The second method is to perform repeated measurements, each using a different range of spectral frequencies.[6] Probing with different spectral frequencies causes the interference between scatters to change, resulting in different reflected power levels. By averaging these uncorrelated powers, the reflection uncertainty is reduced. One difficulty with this approach, for high-spatial-resolution applications, is that very large spectral widths are needed to obtain a sufficient number of averages.

For both the spatial and frequency averaging methods, the uncertainty is reduced as more independent, uncorrelated values are averaged together. The fractional uncertainty in the measured reflectivity, R, can be approximated as

$$\frac{\Delta R}{R} \cong \frac{1}{\sqrt{N}} \qquad (10.7)$$

where N is the number of uncorrelated averages. To obtain an accuracy of 0.1 dB (2.3%), the required number of independent averages is on the order of 1000 or more. The requirement for this large number of independent averages makes it difficult to use RBS for the precise measurement of loss in high-resolution reflectometry measurements.

10.3.5   Coherent vs. Direct Detection

Optical reflectometry techniques can be classified into two general categories. These two categories are direct-detection and coherent-detection techniques. Coherent techniques can be more complex to implement, but they offer advantages not possible using direct-detection techniques. These advantages include larger dynamic ranges, increased signal sensitivity, and the possibility for dispersion cancellation.

When using direct detection, only the reflected optical signal is incident on the detector. This is illustrated in Figure 10.9a. The generated detector photocurrent, I_d, is proportional to the optical power and can be written simply as

$$I_d = \mathcal{R} P_s \tag{10.8}$$

where P_s is the reflected optical power incident on the photodetector and $\mathcal{R}$ is the responsitivity of the detector. Direct detection is relatively insensitive to the spectral characteristics and polarization state of the optical source. Both pulsed and frequency modulated techniques[9] have been implemented using direct detection. Currently, direct detection is used by most commercial OTDRs.

Coherent detection differs from direct detection in that an extra optical signal is added to the backreflected light before it is sent to the photodetector (see Figure 10.9b). This extra signal is commonly referred to as the local oscillator (LO) or reference signal. The photocurrent using coherent detection can be written as

$$I_d = \mathcal{R}[P_s + P_{LO} + 2\sqrt{P_s P_{LO}} \cos\Delta\phi] \tag{10.9}$$

where $\Delta\phi$ is the optical phase difference between the backreflected signal, P_s, and the local oscillator, P_{LO}. This equation assumes that both the LO and signal have identical polarization states when they interfere on the detector. Information on the origin of this equation can be found in Chapter 5.2.2. The signal of interest in Equation 10.9 is the term containing the square root which describes the optical interference between the signal and local oscillator. This interference term is commonly separated from the first two direct-

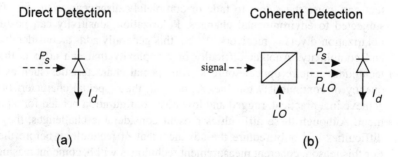

Direct Detection   Coherent Detection

(a)   (b)

Figure 10.9 Schematic representation showing the differences between direct and coherent detection. The local oscillator (LO) acts as a gain factor for weak signal powers.

detection terms by modulating the phase difference, $\Delta\phi$. This allows a bandpass filter to select just the interference term while filtering out the two direct-detection terms.

Two important differences can be seen when comparing direct and coherent detection. The first is that the dynamic range for coherent detection can be much larger. This occurs because the photocurrent signal is proportional to the square root of the reflected power. For direct detection, the photocurrent signal is linearly proportional to the reflected power. The square-root relationship allows a fixed photocurrent range to handle a much larger range of incident optical powers. For example, a receiver with a dynamic range for the photocurrent of 10^6 allows a dynamic range for the optical signal of 60 dB when using direct detection. If coherent detection is used, a dynamic range of 120 dB is obtained. This corresponds to a 10^6 times larger range for the reflected optical power when using coherent detection.

The second difference is related to minimum signal sensitivity. In coherent detection, the LO power, P_{LO}, acts as an amplifying factor which multiplies the signal power, P_s. This means that small signals, that would be below the receiver noise floor in a direct-detection technique, can be multiplied by a strong LO power to bring the interference term above the receiver noise floor. Increasing the LO power continues to improve the signal sensitivity until the shot-noise limit for the signal is reached. After this point, increasing the LO power does not improve signal sensitivity. In the shot-noise limit, the receiver sensitivity is approximately equal to one single photon within its integration time (inverse of receiver bandwidth). This single photon sensitivity can be orders of magnitude greater than direct-detections systems.

Another factor that differentiates the two detection schemes is the possibility of canceling the effects of dispersion using coherent detection. This is achieved by having the LO signal travel through a medium with the same dispersion characteristics as the test medium. Since the interfering signals have the same frequency-dependent delays, the broadening effects of dispersion can be eliminated. This effect is used in OLCR to achieve spatial resolutions below 10 μm.

Although coherent detection has several very important advantages over direct detection, it also has some added difficulties that have slowed its use in commercial instruments. The first difficulty is the polarization sensitivity of the interference signal. This effect can cause the signal to fade or completely disappear as the test fiber is moved or subjected to environmental changes. Polarization sensitivity can be removed by using polarization-diversity receivers,[10,11] but this generally adds a considerable increase in cost and complexity. Another difficulty or complexity that can arise is that some coherent techniques require optical sources with special characteristics such as being single frequency or continuously tunable. At present, these special characteristics are difficult to achieve in a practical, rugged and low-cost configuration needed for a commercial instrument. Although these difficulties present considerable challenges, they are not inherent difficulties and only require the advancement of technology before they are overcome. For this reason, coherent measurement techniques will become increasingly popular in future commercial instruments.

In summary, the advantages coherent detection offers over direct detection are increased sensitivity to reflected power, larger dynamic range, and the possibility of dispersion cancellation. These advantages are very important in achieving high-resolution optical reflectometry measurements. The disadvantages are that some form of polarization diversity must be used and that a more complex optical source may be required. As practical solutions to these problems are developed, coherent techniques will become more widespread in their use.

10.4   OPTICAL LOW COHERENCE REFLECTOMETRY

10.4.1   Introduction

This section is devoted to an in-depth discussion of the measurement technique commonly referred to as optical low-coherence reflectometry (OLCR). OLCR is based on a coherent detection scheme that is often referred to as white-light interferometry. This type of interferometry differs from other coherent techniques because it uses broad spectral-width optical sources with short coherence lengths. Compared to the other high-resolution reflectometry techniques, OLCR currently offers substantial advantages in both theoretical performance and practical implementation.

A simple description of OLCR can be given by referring to Figure 10.10. A cw low-coherence source with a coherence length on the order of 10 μm is coupled to a fiber-optic directional coupler. The low-coherence signal is then divided into the test and reference arms. The optical delay for light returning from the reference path can be varied by translating a movable reference mirror. When the optical delay for light reflected from the reference arm matches the delay for light reflected from the DUT, optical interference takes place at the photodiode. This interference signal only occurs if the optical distance to the reference mirror is within a coherence length compared to the optical distance to the DUT. When the two distances are separated by more than the coherence length, the photodiode only sees a constant cw power since the reflected signals do not interfere. Now, if the single DUT reflector is replaced by multiple reflectors, separate interference signals will occur whenever the position of the reference mirror matches that for one of the DUT

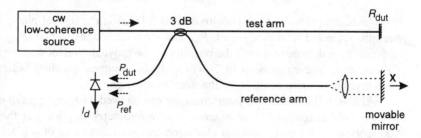

Figure 10.10   A variable-delay Michelson interferometer using single-mode optical fiber.

reflectors. By scanning the position of the reference mirror and recording the magnitude of the interference signals, a plot of reflectivity versus distance can be obtained for the DUT. A more detailed description will be given in Section 10.4.2.

Compared to direct-detection techniques, OLCR has several major advantages. First, the dynamic range for reflectivity measurements is much larger since the photocurrent signal is proportional to the square root of the optical reflectivity. Second, shot-noise-limited detection is possible which results in single-photon sensitivity. And finally, fiber dispersion can be canceled, allowing for extremely high spatial resolutions. Compared to other coherent techniques, such as those based on optical frequency scanning, OLCR currently has advantages in achieving higher spatial resolution and in ease of implementation.

Experimentally, researchers have demonstrated two-point spatial resolutions below 2 μm[12] and reflection sensitivities greater than −160 dB.[3] Instruments based on OLCR are commercially available[13] and currently are the only commercially practical solutions for achieving spatial resolutions of less than 1 mm.

10.4.2 Description of Operation

The basic concepts used in OLCR were demonstrated by Albert A. Michelson, more than 100 years ago, when he constructed an optical interferometer using the sun as his low-coherence source. Using his interferometer, Michelson was able to disprove the existence of an ether which some scientists thought was needed to explain the propagation of optical radiation. He later used these concepts to accurately define the length of a meter with respect to various optical wavelength standards.

The optical arrangement Michelson used is now referred to as the Michelson interferometer. A fiber optic version is shown in Figure 10.10. In this interferometer, light is split into two separate paths. One path connects to the DUT and the other to a movable mirror. Reflections from the DUT and movable mirror are then recombined at the optical splitter where a portion of the signals is sent to an optical detector. On the detector, optical interference can be observed depending on the differential time delay between the two reflected light signals. As described in Chapter 5.2.2., the photocurrent as a result of this coherent detection process is given by

$$I_d = \mathcal{R}[P_{\text{ref}} + P_{\text{dut}} + 2\sqrt{P_{\text{ref}}P_{\text{dut}}}\,\cos\Delta\phi(x, t)] \qquad (10.10)$$

where $\mathcal{R}$ is the photodiode responsivity and $\Delta\phi(x,t)$ is the optical phase difference between the reflected reference signal, P_{ref}, and the reflected test signal, P_{dut}. This differential phase term depends on both the position of the movable mirror and the time statistics of the source. The expression in Equation 10.10 assumes matched polarization states, which gives maximum contrast for the interference signal.

Although the Michelson interferometer can be used with any type of optical source, the term "white-light" or "low-coherence" interferometry implies that the source has a wide spectral width and statistical characteristics similar to that of a blackbody thermal radiator. This type of source is different than that of a laser where optical feedback results in a narrowing of the spectral width and a suppression of optical intensity fluctuations.

Typical sources used for low-coherence interferometry are Tungsten lamps, edge-emitting light-emitting diodes (EELED), superluminescent laser diodes (SLD), and EDFAs.

A typical photocurrent signal as a function of the position of the movable mirror is shown in Figure 10.11. The sinusoidal frequency burst occurs when the delay time along the reference arm matches that for a reflection found in the test arm. The sinusoidal nature of the interference signal is a result of the changing optical phase of the reflected reference signal. To conserve the average reflected power, a complementary version of this response is reflected back into the source. Each period of the interference signal occurs whenever the reference mirror moves one-half of the average optical wavelength. This factor of one-half is a result of the optical signal traveling to the mirror and back again causing a doubling of the optical delay. The peak height of the interference signal is proportional to the square root of the reflectivity from the test arm reflector.

Next, a brief description will explain the spatial extent of the interference packet shown in Figure 10.11. The envelope of the interference packet is caused by the time-averaged statistical properties of the differential phase term, $\Delta\phi(x,t)$, in Equation 10.10. When the reference mirror is far away (relative to the coherence length) from the zero-delay position, the optical phases from the returning reference and test signals are uncorrelated and randomly vary with respect to one another. This causes the differential phase, $\Delta\phi$, to randomly vary with a time constant equal to the coherence time of the source. For typical EELED sources, this variation time is on the order of 10^{-13} s. Since the response time of the photodetector is much slower, only the time-averaged quantity will be observed. For large time delays, the time-averaged interference signal is equal to zero.

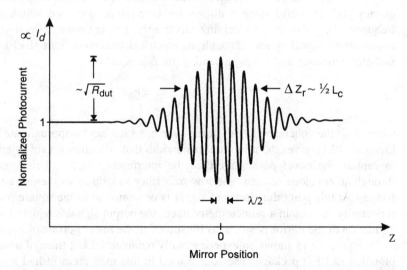

Figure 10.11 A typical photocurrent response from an OLCR measurement. This waveform represents the coherence function of the low-coherence source.

If the mirror position is set so the differential time delay is less than the coherence time, the time variations in $\Delta\phi(x,t)$ are not as large and the cosine term in Equation 10.10 will not average to zero. The shape of the interference envelope depends on the time average of the cosine function as it nonlinearly transforms the differential phase fluctuations. The width of the interference packet is approximately equal to one-half the coherence length of the source, which determines the spatial resolution in OLCR. A typical EELED with a 60 nm spectral width centered at a wavelength of 1.3 μm has a coherence length of about 28 μm which results in an interference packet with a width of about 14 μm.

Although the above description for the simple Michelson interferometer has been understood for more than 100 years, it was not until about 1987 that researchers first demonstrated how these concepts could be applied to high-resolution reflectometry in optical fibers.[14–16] It was shown that if the single reflector in the test arm was replaced by an optical component with multiple reflecting surfaces, the reflectivity and location of each surface could be determined as long as the surfaces were spaced by more than the coherence length of the source.

A practical implementation of an OLCR is shown in Figure 10.12. The DUT now consists of multiple reflections and a polarization controller (PC) is added to one of the arms to ensure matched polarization states at the photodiode. In practice, the mirror in the reference arm is often moved at a constant velocity, v_m, which causes the interference fringes to occur at a Doppler frequency, f_d, given by

$$f_d = \frac{2v_m}{\lambda} \tag{10.11}$$

where λ is the average wavelength of the low-coherence source. This offset Doppler frequency shift is useful since it allows for heterodyne detection which avoids the low-frequency $1/f$ noise in the receiving electronics. For optimum reflection sensitivity, the photocurrent signal is sent through an electrical bandpass filter (BPF) centered at the Doppler frequency and having a bandwidth, Δf_d, equal to

$$\Delta f_d \cong \frac{2v_m}{L_c} \tag{10.12}$$

where L_c is the coherence length of the source, which can be approximated by $L_c \cong \lambda^2/\Delta\lambda$. Equation 10.12 gives the minimum bandwidth that still allows a sufficient response time to capture the envelope variation on the interference signal. This signal is then sent through an envelope detector and low-pass filter to strip away the unwanted interference fringes. At this point the electrical signal is proportional to the square root of the optical reflectivity. To obtain a reflectometry trace, the output signal is squared and displayed as a function of the mirror position, as illustrated in the lower portion of Figure 10.12.

Figure 10.13 shows an experimentally obtained OLCR trace of a singlemode fiber-pigtailed EELED package. The source used in this measurement had a spectral width of 60 nm centered at 1.3 μm with a fiber-coupled output power of about 20 μW. The reference mirror was moved at a speed of 17.5 mm/s which generated an interference signal with a Doppler frequency shift of 27 kHz (see Equation 10.11). The signal then passed

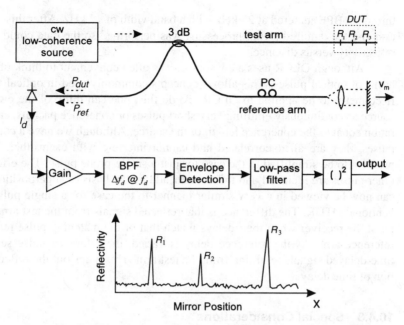

Figure 10.12 Practical implementation of an OLCR measurement.

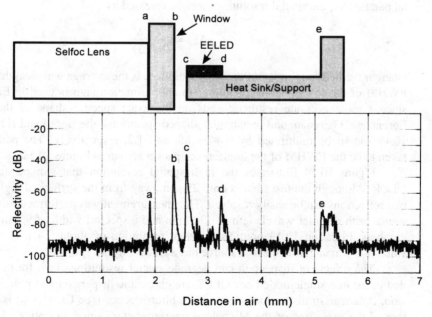

Figure 10.13 OLCR measurement of a pigtailed-EELED (edge-emitting light-emitting diode) package.

through a BPF centered at 27 kHz with a bandwidth of 12 kHz. After this, envelope detection and the required signal processing was performed so the data could be displayed as reflectivity versus distance.

Although OLCR uses a cw source, it is often convenient to think of its operation in terms of optical pulses. This allows concepts commonly used in optical time-domain reflectometry to be applied to OLCR. To do this, one can think of the cw low-coherence source as continuously emitting very short pulses or coherence packets, each having a duration equal to the coherence length of the source. Although we have a continuous train of pulses, they are all uncorrelated and cannot interfere with each other. This allows the problem to be simplified to the case for just a single-probe pulse. The effects from all the other pulses are identical and their results add on the basis of superposition. The situation can now be viewed in a very similar manner to the case for a single pulse found in conventional OTDR. The difference is that reflected signals from the test arm can only register at the receiver when their delays match that of the matching pulse returning from the reference arm. As the reference delay is varied, the reference pulse scans through the time-delayed signals returning from the test arm. This maps out the reflectivity as a function of time delay.

10.4.3 Special Considerations

Spatial Resolution.   The spectral width of the low-coherence source sets the spatial resolution in OLCR. By rewriting Equation 10.3 in more commonly used experimental parameters, the spatial resolution can be expressed as

$$\Delta z_r \cong \frac{1}{2n_g} \frac{\lambda^2}{\Delta\lambda} \tag{10.13}$$

where n_g is the group index of the test medium, λ is the average wavelength and $\Delta\lambda$ is the FWHM of the low-coherence source. An approximation sign is used in Equation 10.13 since a more accurate result depends on the exact spectral shape of the source. For Lorentzian, Gaussian, and rectangular-shaped spectrums, the right-hand side of Equation 10.13 should be multiplied by 0.44, 0.88, and 1.2, respectively. The parameter Δz_r is taken to be the FWHM of the interference packet shown in Figure 10.11.

Figure 10.14 illustrates the high spatial resolution that can be achieved using OLCR. Although the test fiber is only 25 μm away from the surface of a glass block, the two reflections can be easily resolved. This measurement was performed using an EELED source with a center wavelength of 1300 nm and a spectral width of 90 nm. Using these numbers, Equation 10.13 predicts a spatial resolution of about 9.4 μm. The measured value of 8 μm suggests that the source has a Gaussian-like spectrum.

The effects of dispersion can degrade spatial resolution. Two forms of dispersion that occur in a singlemode fiber OLCR are chromatic dispersion and polarization dispersion. Polarization dispersion occurs when birefringence (see Chapter 6) is present in either of the two arms of the Michelson interferometer. Since an optical signal traveling through a birefringent material can experience more than one delay, this results in a

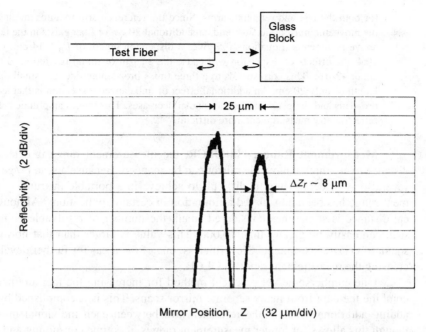

Figure 10.14   Experimental result illustrating the high spatial resolution obtainable using OLCR.

broadening of the detected interference packet. This effect is usually not significant unless the fiber or optical components are purposely chosen to be birefringent.

Chromatic dispersion (also called material dispersion) occurs when the different wavelengths from a broadband optical source propagate at different velocities through an optical medium. If the dispersion in both arms of the Michelson interferometer are identical, the effects of dispersion are canceled and there is no degradation in the spatial resolution. In practice, exact dispersion cancellation is often not possible. This is because the reference arm contains an air path for the movable mirror and the test arm usually consists of a continuous fiber pigtail. This difference generally does not present a problem for measurements made at 1.3 μm since the dispersion for standard step-index fiber is near zero and, therefore, matches that of the air path. However, at other wavelengths such as 1.55 μm, this unbalanced dispersion can have a significant effect on spatial resolution.[17] The effects of unbalanced dispersion can be estimated using the results given in Section 10.3.2.

Example

Assume an OLCR measurement is made at 1.55 μm with an EELED whose spectral width is 60 nm. The initial spatial resolution, without any dispersion broadening, is calculated to be about 14 μm (using $n_g = 1.47$). Assume standard step-index fiber ($D = 17$ ps/km · nm) is used for both the test and reference arms. Since the reference arm requires an air space between the moveable mirror and fiber end, an additional 20 cm of fiber exists in the test arm. For the above parameters, Equation 10.5 gives a dispersion length of $L_D \sim 14$ cm. Using Equation 10.4, the effects of dispersion will cause the 14 μm of initial resolution to be degraded to about 42 μm. This corresponds to a three times broadening due to a small differential fiber length of only 20 cm. An additional effect of unbalanced dispersion is that as the resolution broadens and the peak interference signal decreases. This makes amplitude calibration of the reflectometry measurement more difficult.[18]

Measurement Range. In OLCR, the measurement range is determined by the distance over which a movable mirror can be scanned. Because a scan range of more than several tens of centimeters is difficult to achieve in a portable instrument, the measurement range has been considered a limitation in certain applications. Although this physical distance is sometimes considered small, the number of resolvable points in a single scan can easily be greater than 10,000. This value is larger than that normally found in standard OTDR instruments. Nevertheless, one of the areas for further development is increasing the measurement range in OLCR.

One cumbersome yet practical method for increasing the measurement range is to combine the data from many separate mirror scans. This is accomplished by sequentially adding additional fiber to the reference arm between each individual mirror scan. Although this allows for longer measurement ranges, it is time consuming and a nuisance to the user and, therefore, not considered an adequate long-term solution. Another method for increasing the range is to fold the optical path between the fiber and the movable reference mirror.[19] This multiplies the physical distance scanned by the reference mirror. One difficulty with this approach is that differential dispersion is increased due to the effective increase of the air path for the reference signal.

Another more sophisticated method for extending the measurement range consists of adding a recirculating optical delay into the reference arm of a modified interferometer configuration.[20,21] This has the effect of generating a periodic comb of delays that all move together as the reference mirror is scanned. Using this concept, the measurement range can be increased by more than two orders of magnitude. Further development work is still needed before this method becomes a practical solution.

Reflection Sensitivity. The reflection sensitivity in OLCR is a function of both the optical probe power and the amount of noise that passes through the bandwidth of the receiver. For the basic OLCR configuration shown in Figure 10.12, the SNR for a single, weak reflector is given by[22]

$$\text{SNR} = \frac{2\mathcal{R}^2 P_{\text{ref}} P_{\text{dut}}}{\underbrace{4kT\Delta f/R}_{\text{(thermal)}} + \underbrace{2q\mathcal{R}P_{\text{ref}}\Delta f}_{\text{(shot)}} + \underbrace{\text{RIN}\,\mathcal{R}^2 P_{\text{ref}}^2 \Delta f}_{\text{(intensity)}}} \qquad (10.14)$$

where P_{ref} and P_{dut} are detected powers from the reference and test arms. The noise terms in the denominator are described in Appendix A. These three noise terms, from left to right, are thermal noise, shot noise, and intensity noise. The parameter, $\mathcal{R}$, is the responsitivity of the photodiode. The peak root-mean-square (rms) signal power in the numerator is obtained from Equation 10.10. The constraint of considering only a single, weak reflector allows the assumption $P_{dut} << P_{ref}$ to be made. This constraint makes the reference power responsible for the shot noise and intensity noise and simplifies the expression given in Equation 10.14. The above expression assumes matched polarization states, therefore, giving optimum signal contrast.

The dominant noise source in Equation 10.14 depends on the value of the reference power P_{ref}. For high impedance receivers (for example, $R = 1$ MΩ) and typical low-coherence sources (with spectral widths of several tens of nanometers) the intensity noise term starts to dominate for reference powers larger than a few microwatts. As discussed in Appendix A, this intensity noise is an inherent characteristic of low-coherence optical sources. Increasing the source power when the intensity noise is dominant does not increase the reflection sensitivity since both the signal and noise increase at the same rate. In the basic OLCR configuration shown in Figure 10.12, the source RIN sets a limit on the maximum reflection sensitivity.

Several methods have been demonstrated to overcome the sensitivity limitation due to intensity noise. One method consists of modifying the interferometer configuration shown in Figure 10.12 to make use of a balanced-detection scheme which cancels the intensity noise.[5,23,24] The basic concept behind this scheme is illustrated in Figure 10.15 where the reference power is redirected so that it can be combined with the test signal in a separate 3 dB coupler. Since the two detectors see in-phase intensity fluctuations and 180 degree out-of-phase interference signals, subtracting the photocurrents eliminates any intensity noise while enhancing the interference signal.

Another method for eliminating the effects of intensity noise requires only a slight modification to the basic OLCR configuration shown in Figure 10.12. This method is implemented by adding optical attenuation to the reference arm to decrease the value of

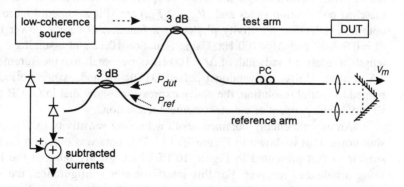

Figure 10.15 OLCR configuration using a balanced-detection scheme. This scheme eliminates intensity noise while enhancing the interference signal.

P_{ref}.[22] With sufficient attenuation, the intensity noise term in Equation 10.14 can be reduced to a value below the shot-noise level. For typical EELED sources, the attenuated reference power should be less than about 1 μW. If a high-impedance low-noise receiver is used, both the thermal noise and intensity noise can be made smaller than the shot noise. The following example uses the discussion on noise sources, given in Appendix A, to give typical noise values in an OLCR measurement.

Example

A 1.55 μm EELED with a spectral width of 60 nm is used to make an OLCR measurement. Assume that the reflected signal power is weak and much smaller than the reference power. If the reference power is attenuated to give a photodiode current of 1 μA, we get a shot-noise level equal to 0.57 $pA/\sqrt{Hz}$ (see Table A.2 in Appendix A.3). The intensity noise from the 1 μA of reference current is about 0.36 $pA/\sqrt{Hz}$ (see Equation A.7 in Appendix A.2). For a 1 MΩ transimpedance receiver, the minimum thermal noise level for the receiver is about 0.13 $pA/\sqrt{Hz}$ (see Table A.1 in Appendix A.1). This example shows that by attenuating the reflected reference power to give a 1 μA reference current, the shot noise can be made dominant over the other noise sources. For this condition, maximum reflection sensitivity is obtained.

By making the shot noise the dominant noise source, the maximum reflection sensitivity can be achieved for a given source power. This is a special condition since it means that the receiver sensitivity is approximately equal to a single photon within the integration time set by its bandpass filter. Assuming shot-noise-limited detection, the OLCR arrangement shown in Figure 10.12 has a maximum reflectivity sensitivity given by

$$R_{min} = \frac{4\,q\,\Delta f}{\mathcal{R}\,P_o} \tag{10.15}$$

where P_o is the source power and the relation $P_{dut} = P_o R_{min}/4$ is used in Equation 10.14 to account for losses through the 3 dB coupler. As mentioned earlier, the conditions of matched polarization states and ($P_{dut} \ll P_{ref}$) are still assumed. Figure 10.16 shows this optimum reflection sensitivity plotted as a function of source power for the conditions $\mathcal{R} = 0.8$ A/W and $\Delta f = 100$ Hz. Using Equation 10.12 and assuming a 30 μm coherence length, a receiver bandwidth of $\Delta f = 100$ Hz corresponds to a measurement speed of about 1.5 mm/sec. Unlike conventional pulsed OTDR methods, where reflection sensitivity is related to spatial resolution, the above expressions show that in OLCR the reflection sensitivity is independent of the measurement resolution.

An experimentally obtained result achieving sensitivity to within a few dB of the shot-noise limit is shown in Figure 10.17.[5] This data was obtained using a configuration similar to that presented in Figure 10.15 which greatly reduced the intensity noise by using a balanced receiver. For this interferometer configuration, the shot-noise-limited sensitivity is 3 dB worse than that given by Equation 10.15. This 3 dB reduction in sensitivity is due to the use of the balanced receiver which adds additional noise since it sums

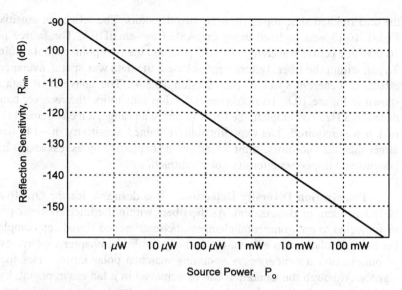

Figure 10.16 Theoretical minimum reflection sensitivity for OLCR assuming shot-noise-limited detection. The receiver bandwidth is 100 Hz and the photodiode responsivity is 0.8 A/W.

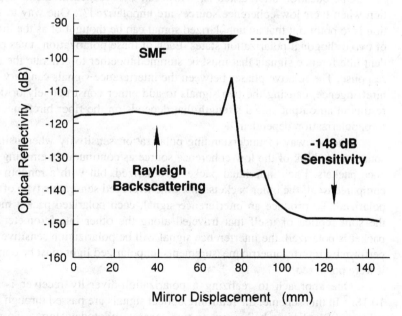

Figure 10.17 OLCR measurement of Rayleigh backscatter from a singlemode fiber (SMF). Probe wavelength = 1.55 μm, source power = 10 mW, spatial resolution = 32 μm and receiver bandwidth = 3 Hz.

the uncorrelated shot noises from the two detectors. The reflection sensitivity obtained in Figure 10.17 was realized using the ASE from an EDFA. The source power was $P_o =$ 10 mW at a center wavelength of 1.55 μm, and the two-point spatial resolution was about 32 μm within the fiber. Before being plotted, the data was spatial averaged to remove the effects of coherent speckle[5] (see Section 10.3.4). The unaveraged data for this plot is shown in Figure 10.8. To maximize reflection sensitivity, the receiver bandwidth was reduced to 3 Hz. To complete the full 150 mm scan range, a measurement time on the order of 1 h was required. The experimentally obtained sensitivity of −148 dB is about 5 dB larger than the shot-noise limit which can be explained by excess losses in the fiber components and imperfect intensity noise subtraction.

Polarization Diversity Detection.   It is desirable for the OLCR measurement to be independent of polarization. As the fibers within the interferometer paths are handled or subjected to environmental changes, the interference signal can completely fade away because of polarization effects. Up to this point in the chapter we have avoided the issue of polarization dependence by assuming matched polarization states for the interfering signals. Although this condition can be achieved in a lab environment, by use of manual polarization controllers, this solution is not acceptable in a general purpose commercial instrument. Polarization sensitivity can be one of the most difficult problems to solve when developing a commercial, fiber optic, coherent measurement system.

One question often asked is, "How can OLCR measurements depend on polarization when their low-coherence sources are unpolarized?". One way to answer this question is to point out that an unpolarized signal can be thought of as the linear superposition of two orthogonal polarization states. Each of these polarization states produces independent interference signals that must be summed together to generate the total photocurrent response. The relative phases between the interference signals can vary depending on the birefringence, causing the two signals to add either constructively or destructively. This results in an output signal strength that depends on the fiber birefringence and is, therefore, polarization dependent.

Another way of understanding polarization sensitivity when using an unpolarized source is to think of the low-coherence source as continuously emitting very short coherence packets. Each individual packet is polarized, but with a random polarization state compared to all the other packets. In a time-averaged sense, this type of signal will be unpolarized. To produce an interference signal, each polarized packet must combine with the split replica of itself that traveled along the other interferometer path. Since each packet is polarized, the interference signal will be polarization sensitive. Therefore, when performing interferometric measurements, unpolarized light must be considered as though it were polarized.

One approach to realizing a polarization-diversity receiver is shown in Figure 10.18.[17] In this scheme, the returning optical signals are passed through a polarizing beam splitter (PBS) before being sent to two separate photodetectors.[25] To make this scheme work properly, the light returning from the reference arm should be polarized and divided equally onto each detector. Polarizing the reference signal can be accomplished by insert-

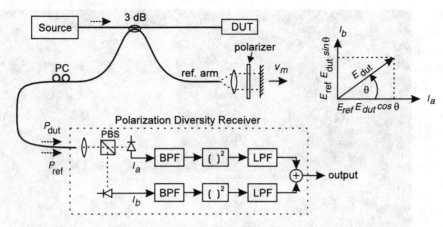

Figure 10.18 Implementation of a polarization-diversity receiver. The polarization controller (PC) is adjusted to set equal reference powers onto each detector.

ing a polarizer into the air path before the movable reference mirror or by using a polarized low-coherence source.

The incoming test signal will be divided by the PBS depending on its returning polarization state. This polarization state will vary since small movements in the test fiber can alter the returning polarization state. The insert in Figure 10.18 shows that for the case of a linearly polarized test signal, with an unknown angle, θ, with respect to the polarizing beam splitter, the optical field is split in accordance to $\cos\theta$ and $\sin\theta$. Since the reference signal is equally divided on the two detectors, squaring and summing the two photocurrents produces a signal independent of the unknown polarization angle θ (this result uses the relationship $\cos^2\theta + \sin^2\theta = 1$). If the returning test signal is elliptically polarized, the only difference will be a phase shift in the fringe pattern on one of the detectors. This phase shift is not important since the envelope detector eliminates any phase information before the summing process. Therefore, the output signal is completely independent of the polarization state returning from the test arm. Other polarization-diversity schemes are available,[11] but they are normally complex and costly.

This polarization-diversity scheme described above is used in the Hewlett-Packard HP 8504B optical low-coherence reflectometer.[17] A photo of this instrument is shown in Figure 10.19. The two knobs on the front panel are used to adjust the polarization state returning from the reference arm so that it divides equally onto the two detectors as illustrated in Figure 10.18. Once this calibration process is done, reflectometry measurements are insensitive to any polarization changes occurring along the fiber path of the test device. These polarization changes occur whenever a fiber is moved or handled. It is important that the reference fiber not be disturbed after performing this calibration process. The HP 8504B has a spatial resolution of about 25 μm with a total scan range of 40 cm (air distance).

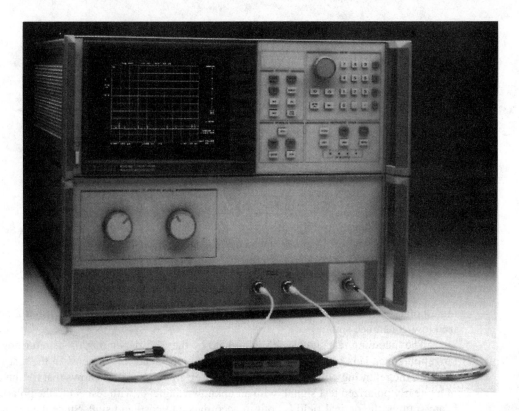

Figure 10.19 The Hewlett-Packard 8504B Precision Reflectometer is insensitive to polarization changes in its test arm due to its polarization-diversity receiver.

Spurious Sidelobes.   Fake signals or spurious sidelobes can occur in OLCR measurements if unwanted multiple reflections occur in the measurement instrument. These unwanted sidelobes can be confused with actual reflections from the test device. They are generated by weak etalons in the optical path which cause the primary signal to be followed by a weaker time-delayed signal due to the double reflection within the etalon. Depending on where in the interferometer these weak double reflections occur, spurious signals can show up as either symmetric or asymmetric sidelobes relative to a large primary reflection peak.

Figure 10.20 shows two configurations for illustrating the generation of unwanted sidelobes. In the upper interferometer there are two weak reflectors in the source arm. Thinking of the low-coherence source as emitting a continuous stream of short coherence packets, each transmitted packet is followed by a weaker duplicate after passing through the two reflectors. The resulting interference signal, as a function of mirror position, is obtained by a cross-correlation between the packets returning from the test and reference arms. This results in a large reflection peak with two symmetrically placed sidelobes, as

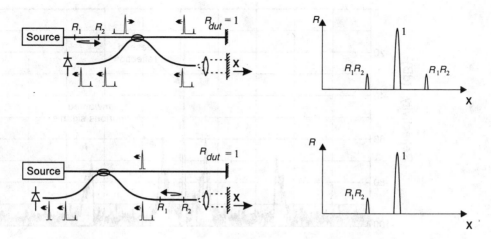

Figure 10.20 Unwanted sidelobes are caused by weak etalons in the optical paths. The location of the etalons determine the symmetry of the sidelobes.

shown on the right side of the figure. The magnitude of these sidelobes are smaller than the primary reflection by the product $R_1 R_2$ and their spacing from the primary reflection is equal to the optical path length between the two reflectors.

If the weak etalon is instead located in the detector arm of the interferometer, the same response is obtained. In practice, these weak etalons can occur within the EELED semiconductor chip or the photodiode package, therefore, care is needed in selecting suitable components. An example of the effects of weak etalons within a packaged EELED source is shown in Figure 10.21. The main peak in the center of the plot is due to the Fresnel reflection from a fiber-air interface. Ideally, only this single reflection should show up on an OLCR measurement. The two sidelobes closest to the main peak are due to the residual facet reflectivities from the semiconductor chip. The outer two sidelobes are a result of reflections within the structure of the packaged EELED. To minimize these spurious sidelobes, researchers have designed special EELED chips with ultra-low facet reflectivities.[26]

The bottom section of Figure 10.20 shows the effect of a weak etalon located in the reference arm. By cross-correlating the impulse responses returning from the two arms, only a single sidelobe occurs to the left of the primary reflection peak. This can cause confusion since it appears that a weak reflecting surface exists before the actual reflector is encountered. If the etalon is placed in the test arm, the single sidelobe occurs on the other side of the primary peak. The above results can be very useful when troubleshooting the sources of unwanted sidelobes in an OLCR interferometer.

Source Considerations. This section will briefly discuss some of the low-coherence sources near the wavelength region of 1.5 μm that can be used to make OLCR measurements. The output from these broadband sources usually consists of amplified

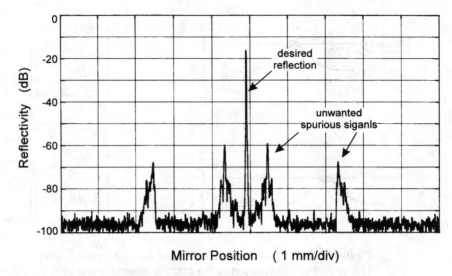

Figure 10.21 Impulse response showing spurious sidelobes centered about a single reflection peak. The symmetrical structure occurs since these stray reflections occur within the EELED source module.

spontaneous emission (ASE). Lasers are usually not suitable since the use of feedback tends to narrow their spectral content. One possible exception to this is a mode-locked fiber laser which can generate an effectively broadband signal.

The two most commonly used sources are EELEDs and EDFAs. These sources can couple sufficient power into a singlemode fiber so that reflection sensitivities in excess of −100 dB can be obtained. Spatial resolutions from these sources can vary from about 10 to 50 μm. Table 10.3 gives some typical parameters for these two sources as well as for a white-light Tungsten lamp. Although a white-light Tungsten source has an optical spectrum capable of achieving a spatial resolution as small as a couple of microns, the amount of power that can be coupled into a singlemode fiber is usually not adequate. Spectral plots showing the relative powers and spectral widths of these three sources can be found in Chapter 9.

EDFA sources can achieve very high reflection sensitivities (greater than −160 dB) because of their high output powers. Possible drawbacks of using EDFAs are cost and poorer spatial resolution. EELEDs are more economical and can have adequate sensitivity and resolution. One possible drawback is that spurious sidelobes can occur if the EELED is not designed to suppress facet reflectivities. This sidelobe problem usually prevents superluminescent diodes (SLDs) from being adequate low-coherence sources.

Measurement Examples. This section will show four examples to illustrate some of the types of measurements that can be made using OLCR. The first example illustrates the ability of OLCR to spatially resolve the many reflections from a multi-element optical

Table 10.3   Characteristics of some low-coherence sources near the 1.55 μm wavelength region.

Source	Typical power	Typical spectral width	Resolution (in fiber)	Comments
EDFA	> 10 mW	20 nm	40 μm	- high sensitivity - low sidelobes - more costly
EELED	100 μW	50 nm	16 μm	- good sensitivity - economical - possible sidelobes
Tungsten Lamp	0.25 nW/nm	> 600 nm	< 2μm	- poor sensitivity - high resolution - dispersion concerns

component. The next two examples show that interesting results can be obtained by measuring the Rayleigh backscatter in fiber and integrated waveguides. The final example shows that active components such as semiconductor lasers can be probed. This example also shows the effects of measuring a distributed Bragg grating.

Figure 10.22 shows the internal reflections from a two-stage polarization-independent optical isolator. The measurement was made at its operating wavelength of 1.55 μm using an EDFA low-coherence source. This isolator contains many optical elements, each being antireflection coated and angled to avoid any backcoupled reflections. This type of reflectometry measurement was very useful in the initial design of this isolator since it provided quick feedback on the best arrangement and angular orientation for its many optical elements. OLCR is also very useful in trouble-shooting components whose total return loss (see Section 10.2) does not fall within specifications. It quickly identifies which elements are causing the reflections and allows corrective action if necessary.

The effects of a tight bend on the Rayleigh backscatter from a singlemode fiber is shown in Figure 10.23. The measurement was made on a standard step-index fiber at 1.55 μm. The EDFA probe signal resulted in a spatial resolution of about 32 μm within the optical fiber. The fiber bend had a diameter of about 5 mm which attenuated the forward-traveling signal by about 20 dB. The rise in the backscatter at the fiber bend, from about -117 dB to -97 dB, was initially not expected. It is now believed to be caused by the light that escaped from the core of the singlemode fiber and entered the cladding and plastic jacket of the fiber. The larger scattering levels from these regions are then coupled back into the singlemode fiber through the same fiber bend that allowed the light to escape in the first place. The backscatter from these external regions can be measured for more than 30 mm along the fiber before it drops below the noise floor of the measurement.

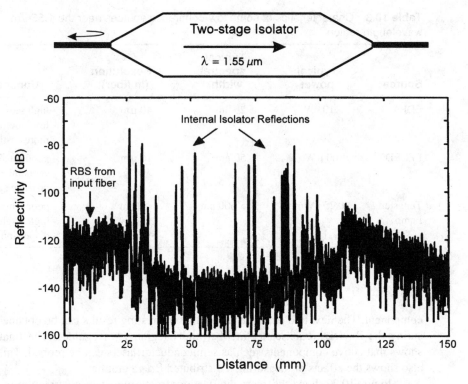

Figure 10.22 OLCR measurement showing multiple reflections from within a two-stage polarization-independent isolator.

Figure 10.24 shows the backscatter from an integrated optical waveguide made in a LiNbO$_3$ substrate. The measurement was made to test the operation of an in-line polarizer constructed by placing a metal region on top of the titanium-doped waveguide. By adjusting the polarization state of the input probe signal, the drop in the backscatter level after the polarizer could be varied from 0 to 15 dB The 15 dB change indicates that the extinction ratio for the polarizer was about 7 dB. This type of measurement could be very useful for more complex waveguide arrangements such as multiple polarizers, metal electrodes for phase modulation and Y-junctions in the waveguide. In both Figure 10.23 and Figure 10.24, the plotted data was spatially averaged to reduce the effects of coherent speckle (see Section 10.3.4).

The final example illustrates that OLCR can be used to probe active devices such as semiconductor lasers. Because semiconductor lasers can act as absorbers for the probe signal, they typically require a bias current to make the waveguide region more transparent. With high levels of bias current, the reflected signals can be increased because the probe signal can experience optical gain. This effect works until the laser signal starts to become the dominant noise source in the receiver of the OLCR. Since the spectral width

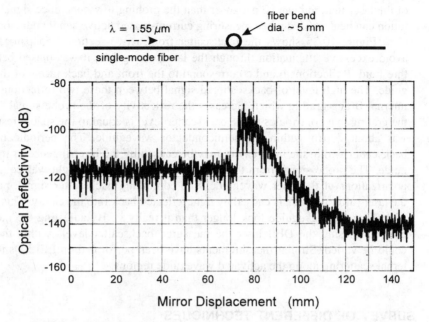

Figure 10.23 Backscatter signal from a singlemode fiber near the location of a tight bend. Spatial resolution is 32 μm within the fiber.

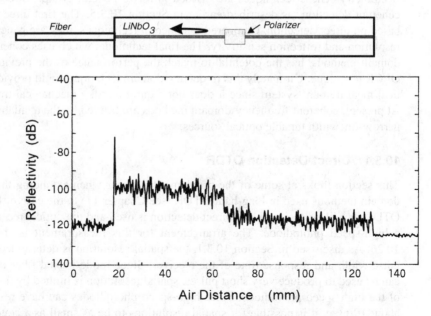

Figure 10.24 Backscatter from a lithium niobate (LiNbO₃) waveguide incorporating an evanescent-field polarizer. The probe signal was polarized orthogonal to the polarizer.

of the laser material can be narrower than the probing low-coherence signal, spatial resolution can be degraded when measuring current-biased active semiconductor devices.

Figure 10.25 shows the backscatter from a two-section 1.55 μm DFB laser. To avoid excessive attenuation through the laser waveguide, it was biased below its lasing threshold. Reflections b and c correspond to the front and back facets of the laser waveguide. The high level of backscattered signal between these two reflections is due to the internal Bragg grating which increases the reflectivity of the probe signal. The region of the grating that contributes to the backscatter level is equal to the spatial resolution which was about 20μm within the semiconductor waveguide. The periodic nature for the backscatter within the laser cavity may be caused by the birefringence of the laser waveguide. The probe signal from the EDFA was unpolarized so light was coupled into both polarizations of the laser waveguide. The continuing backscatter signal, past the end of the laser (after peak c), is caused by light reflected back into the cavity. This has the effect of making the waveguide look longer than it really is. By adjusting the ratio of the two currents biasing the DFB laser the measured backscatter levels within the laser cavity could be varied. Similar measurements have been performed on EELEDs to measure internal scattering along the activated waveguide regions.

10.5 SURVEY OF DIFFERENT TECHNIQUES

This section discusses several techniques used to demonstrate high-resolution optical reflectometry. These techniques are divided into two general groups, direct detection and coherent detection, as described earlier in Section 10.3.5. The first three techniques are based on direct detection. Compared to coherent techniques, they are generally limited in resolution and reflection sensitivity. The final technique, which uses coherent frequency-domain methods, has the potential to match the performance of the previously discussed OLCR technique. Ultimately this frequency-domain technique could provide a more practical measurement system since it does not require a bulky mechanical translation stage. At present, coherent frequency-domain methods are limited by the availability of suitable narrow-linewidth tunable optical sources.

10.5.1 Direct-Detection OTDR

This section looks at some of the problems that occur when extending the pulsed, time-domain methods used in long-haul OTDR (see Chapter 11) to the case of high-resolution OTDR. It will be assumed that direct detection is used and the optical receiver consists of a high-speed photodiode. The arrangement for this measurement is shown in Figure 10.26. As discussed in Section 10.3.1, the spatial resolution is determined by the optical pulse width and response time of the receiver. Since mode-locked fiber laser techniques can be used to produce very short pulses, spatial resolution is limited by the response time of the high-speed photodiode. Since high-speed photodiodes can have response times of about 10 psec, it is possible for spatial resolutions to be as small as a couple of millime-

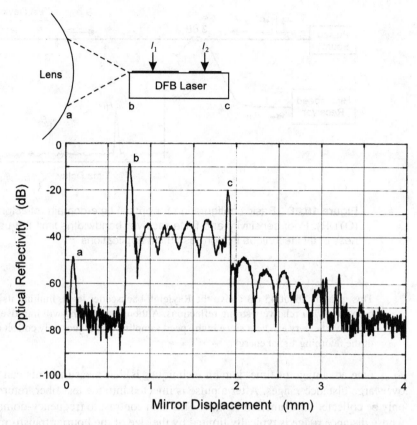

Figure 10.25   Backscatter from a two-section DFB laser. The high level of scattering within the laser chip is due to its internal Bragg grating.

ters. The main problems associated with pulsed techniques using direct detection are poor receiver sensitivity and the difficulties associated with efficient data collection and averaging of the reflected signals at gigabit data speeds.

Example

Consider a 1.55 µm OTDR designed to achieve a spatial resolution of 10 mm by using a high-speed InGaAs photodiode and receiver. Assuming a 50 ohm amplifier, with a 3.5 GHz bandwidth and a 10 dB noise figure, the receiver sensitivity will be on the order of 5 µW. This sensitivity is approximately 5 orders of magnitude poorer than that obtained using a commercial OTDR with a 100 m spatial resolution. If a mode-locked laser is used that generates pulses with 1 W of peak power, the single-shot reflection sensitivity is about −47 dB.

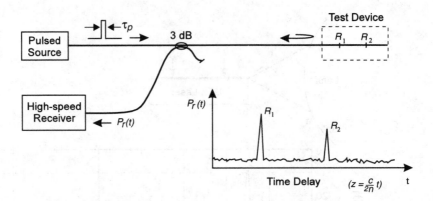

Figure 10.26 Basic configuration for optical time-domain reflectometry (OTDR). Poor sensitivity and finite receiver bandwidths limit the usefulness of this technique in high-resolution applications.

This value is about 45 dB above the Rayleigh backscatter level, limiting its use to applications involving relatively strong reflections. Although averaging will improve the sensitivity, it will not be very efficient since high-speed sampling techniques only collect a small fraction of the returning signal energy.

One advantage of time-domain techniques is that measurements can be performed over large distance ranges. After a pulse is injected into the test fiber, return signals need only be collected as a function of time. This is in contrast to frequency-domain techniques whose distance range is typically limited by the size of the Fourier transform. But as illustrated in the above example, due to poor sensitivity, high-resolution direct-detection OTDR techniques will be limited to applications involving high reflectivity signals.

10.5.2 Photon-Counting OTDR

The poor sensitivity associated with the use of high-speed photodiodes can be greatly improved by using a photon-counting detection scheme. Photon counting can be performed using photo-multiplier tubes or gain-quenched avalanche photodiodes (APD) operating in the Geiger mode.[27,28] Currently, practical methods for photon-counting at wavelengths longer than about 1 μm are difficult to achieve.

A simplified photon-counting scheme is shown in Figure 10.27. A mode-locked fiber laser is used to generate the short optical probe pulses. As the pulse is injected into the test fiber, an electrical trigger pulse is sent to start a time counter. The counter increases linearly with time until a returning photon is detected by the photon-counting receiver and stops the timer. The output from the counter is used to build a histogram of the photon arrival times. A maximum of one photon is detected for each probe pulse. The histogram eventually represents the probability of a reflected photon as a function of time delay which is proportional to optical reflectivity versus distance.

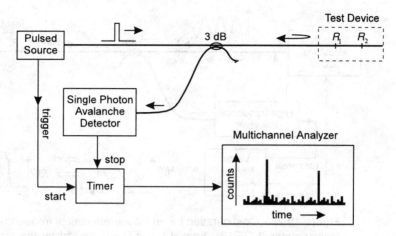

Figure 10.27   Basic configuration for a photon-counting OTDR. Single-photon detection allows for a high reflection sensitivity.

By using photon-counting, the receiver noise is drastically reduced when compared to the previous case for the high-speed photodiode. One fundamental difference is that the noise floor is independent of the receiver bandwidth. The commonly found trade-off between spatial resolution and receiver sensitivity no longer exists. The noise floor is now set by the shot noise from the background dark counts and the quantum efficiency of the photon counting detector. Another important difference is that the spatial resolution is no longer inversely proportional to the bandwidth of the receiver as described by Equation 10.3. Instead, the important factor becomes the timing resolution for detecting the leading edge of the avalanche process which stops the time counter. This allows for a relatively high spatial resolution using a modest bandwidth for the photon counting receiver.

Photon-counting OTDRs have been demonstrated with spatial resolutions as small as 1.5 cm.[27] In this experiment, a room-temperature silicon APD operating in a Geiger mode was used to detect single photons at a wavelength of 0.85 μm. The receiver had a 5% quantum efficiency and a noise equivalent power of 3×10^{-15} W. This sensitivity was enough to measure the Rayleigh backscatter from a multimode optical fiber with a spatial resolution of 1.5 cm. Although the sensitivity of a photon-counting OTDR is very good, it is difficult to achieve spatial resolutions below about 1 cm since the measurement technique is limited by the performance of high-speed electronics. This technique should be useful for characterizing short fiber links, but is limited to applications where high spatial resolution is not required.

10.5.3   Incoherent Frequency-Domain Techniques

Incoherent or direct detection optical frequency-domain reflectometry (OFDR) is the frequency domain equivalent of a standard pulsed OTDR measurement. This can be under-

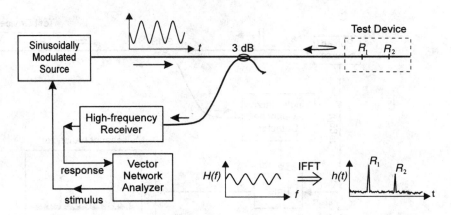

Figure 10.28   Configuration for an incoherent optical frequency-domain reflectometer (I-OFDR). Spatial resolution is limited by the modulation bandwidth.

stood by noting that the reflected power from a typical test device acts like a linear time-invariant system. This means that a reflectometry trace can be obtained by either measuring the impulse response directly or equivalently, by measuring the frequency-domain transfer function (magnitude and phase of the reflected signal at each modulation frequency) and performing an inverse Fourier transform.

The basic concepts used in incoherent OFDR are illustrated in Figure 10.28. An electrical vector network analyzer performs a stimulus-response measurement by probing the test fiber with sinusoidally modulated optical power. The frequency-domain transfer function is obtained by measuring the amplitude and phase of the reflected signal at each probe frequency. Optical reflectivity versus distance is obtained by taking the inverse Fourier transform of the frequency response and scaling the time axis to represent distance. The minimum spatial resolution is inversely proportional to the range over which the frequency is scanned. This relationship was given earlier in Equation 10.3, where Δf_s is the range of the frequency scan.

Using this technique, researchers have demonstrated spatial resolutions as small as several millimeters.[9,29] These results were obtained using a microwave network analyzer and a high-speed LiNbO$_3$ amplitude modulator to intensity modulate the output of a semiconductor laser. The receiver consisted of a high-speed InGaAs photodiode and microwave amplifier. This arrangement provided a frequency scan range of 25.6 GHz which corresponds to a spatial resolution of about 4 mm within an optical fiber.[9]

For high-resolution reflectometry, OFDR offers several advantages when compared to conventional pulsed OTDR. One advantage occurs in reflection sensitivity since signal averaging can be done more efficiently. This is because the high frequency sinusoidal signals can be measured with a narrow bandpass filter. Whereas in the pulsed case, data collection must be done over the full electrical bandwidth. Another advantage is that higher spatial resolution is easier to implement using OFDR. This is because the frequency response of the measurement electronics can be easily deconvolved from the frequency-domain measurement, allowing the full system bandwidth to be used in determining spatial resolution.

An equivalent method, referred to as incoherent FMCW (frequency-modulated continuous wave) reflectometry performs the same frequency-domain measurement but without the need of a vector network analyzer.[30,31] In this method the probe frequency is linearly chirped instead of being stepped and the returning signals are mixed with the nondelayed input frequency. The resulting output contains low-frequency beat signals whose frequencies are proportional to the delay times experienced by the reflected test signal. Reflectivity versus distance is obtained by either observing the mixer output on a spectrum analyzer or by time-sampling the output and taking its Fourier transform. Because of the difficulty of generating large continuous frequency sweeps, spatial resolution using incoherent FMCW is typically not as good as in the above network analyzer approach.

Achieving spatial resolution of less than a few millimeters is very difficult using any of the above direct-detection techniques since electrical bandwidths in excess of 30 GHz are required. Although frequency-domain techniques offer advantages over pulsed techniques, suitability for optical component testing is still limited. The next section deals with a coherent detection technique whose frequency span is not limited by an electronic bandwidth. By utilizing the much larger bandwidth of an optical signal, spatial resolutions can be orders of magnitude smaller.

10.5.4   Coherent Frequency-Domain Techniques

Coherent optical frequency-domain reflectometry (C-OFDR) has the potential for becoming a dominant high-resolution reflectometry technique. It can provide spatial resolutions below 100 μm[32,33] and can have single photon reflection sensitivity. Although similar in performance to OLCR, it does not require a bulky translation stage for scanning an optical delay. C-OFDR is similar to incoherent OFDR except that instead of modulating the optical intensity, the actual optical field is used as the sinusoidally modulated probe signal. In C-OFDR, both the amplitude and optical phase of the reflected signal are measured as a function of the laser frequency.

Researchers have demonstrated high-resolution C-OFDR by step-tuning an external cavity laser over a wavelength span of 100 nm.[33] This resulted in a spatial resolution (in air) of about 12 μm. To allow for accurate optical phase measurements, the experiment was performed in a stable free-space environment. The requirement of a stable environment points out a limitation for making C-OFDR measurements using a relatively slow, step-tuned frequency scan. If the DUT were located at the end of a fiber-optic pigtail, environmental variations would distort the measurement of the returning optical phase.

To solve this problem, a variation of C-OFDR called coherent FMCW reflectometry can be used. This technique uses a frequency chirped laser, allowing for a short measurement time, so that the relatively slow environmental phase drifts become less significant. The remaining part of this section will focus on some of the details of this measurement technique.

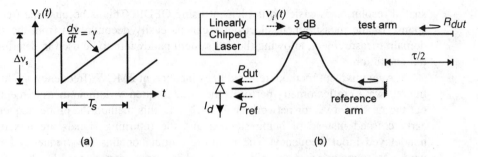

Figure 10.29 Implementation of a coherent FMCW reflectometer. This technique offers the potential for high spatial resolution combined with good reflection sensitivity.

Coherent FMCW Reflectometry. The concepts used in coherent FMCW (frequency-modulated continuous wave) reflectometry were initially developed for early RF radar systems.[34] A singlemode fiber-optic implementation of a coherent FMCW reflectometer is shown in Figure 10.29. The key element in this type of measurement is a single-frequency narrow-linewidth laser whose optical frequency can be chirped, preferably in a phase-continuous manner (without mode hops). Mode hops limit the frequency resolution which in turn limits the maximum distance range that can be measured. The output from this source is sent into a 3 dB coupler which divides the power into a test and reference arm. The optical signal returning from the reference mirror coherently mixes with time delayed reflections from the test arm. Because of the linear chirp on the laser frequency, any optical interference will have a beat frequency that is proportional to the time delay between the returning reference and test arm signals. A plot of reflectivity as a function of distance is obtained from the Fourier transform of the photocurrent. A more detailed description and analysis is given below.

As mentioned above, the key element in coherent FMCW reflectometry is a narrow linewidth source whose optical frequency can be linearly chirped as a function of time. Ideally, the laser should provide constant optical power while its frequency is periodically ramped, as shown in Figure 10.29a. During a ramp cycle the instantaneous frequency of the optical carrier can be written as

$$v_i(t) = v_o + \gamma t \tag{10.16}$$

where v_o is a large constant optical frequency and γ is the rate of change of frequency. The interference is monitored over the scan time T_s and the total optical frequency deviation is given by Δv_s. For simplicity of analysis, it will be assumed that there is only a single reflection (R_{dut}) in the test arm whose returning signal has an optical delay of τ relative to that from the reference mirror. The detected photocurrent for the coherent detection process is given by Equation 10.9 as described earlier in the chapter. The differential optical phase, $\Delta\phi(t)$, provides the information to determine the distance to the test reflection.

This term will be calculated from the instantaneous frequency given above in Equation 10.16. The returning optical phase for the test and reference signals can be determined using

$$\phi(t) = 2\pi \int v_i(t)dt = 2\pi v_o t + \pi \gamma t^2 \qquad (10.17)$$

where the only difference is that the signal from the test arm is delayed by an additional optical delay of τ. The differential phase term needed for Equation 10.9 can now be calculated as

$$\Delta\phi(t) = \phi(t) - \phi(t - \tau) = 2\pi(\gamma\tau)t + 2\pi v_o \tau - \pi\gamma\tau^2 \qquad (10.18)$$

where the last two constant phase terms are not important for understanding the operation of the reflectometer. The interesting term is the time dependent one which generates a photocurrent beat frequency which is proportional to the time delay between the test and reference signals. From Equation 10.18 this beat frequency can be identified as

$$f_b = \gamma\tau \qquad (10.19)$$

A pictorial understanding of this beat frequency term is illustrated in Figure 10.30a, where the instantaneous frequencies for the two reflected signals are displayed as a function of time. Although the two reflected optical frequencies are constantly changing, the difference or beat frequency between them remains constant for a given optical delay.

A plot of reflectivity as a function of distance is obtained by taking the Fourier transform of the sampled photocurrent and displaying its magnitude squared in the frequency domain as shown in Figure 10.30b. Since the frequency axis is proportional to optical time delay, which is in turn proportional to distance, the horizontal axis can be converted to distance. The proportionality constant between the frequency axis and distance is given by

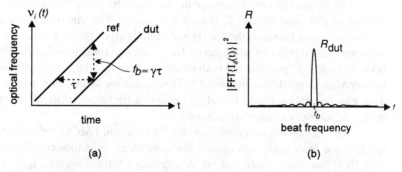

(a)   (b)

Figure 10.30 Graphical illustration describing the generation of beat frequencies in FMCW reflectometry.

$$z = \left(\frac{c}{2n_g \gamma}\right) f \tag{10.20}$$

where c is the speed of light and n_g is the group index seen by the optical test signal. The above result can be used to determine the spatial resolution for the measurement. The spatial resolution depends on the frequency resolution of the Fourier transform. Assuming data is collected over the total scan time T_s, the frequency resolution will be approximately $1/T_s$. By inserting this frequency resolution into Equation 10.20 and using the result $\gamma = \Delta v_s / T_s$, we get the general result for spatial resolution given earlier in Equation 10.3.

$$\Delta z_r = \frac{c}{2n_g \Delta v_s} \tag{10.21}$$

where Δf_s is replaced by the total optical frequency span Δv_s.

Example

To get a feeling for the magnitude of the parameters, consider the case where we wish to obtain a spatial resolution of 1 mm (in fiber) using a measurement time of $T_s = 100$ msec. Using Equation 10.21, the required optical frequency span, Δv_s, is about 100 GHz. To make this measurement in 100 msec requires a frequency chirp, γ, of 1 GHz/msec. The difference in beat frequencies for two reflectors spaced by the minimum distance of 1 mm is 10 Hz which is equal to the frequency resolution of the measurement.

When the test arm contains many reflectors, a separate beat frequency occurs for each reflection that mixes with the reference signal. In addition to these desired mixing terms, undesired mixing terms exist because of the interference between the multiple reflections from the test arm. This can be seen by the following expression for the photocurrent when there are many test arm reflectors.

$$I_d(t) \propto \left| E_{\text{ref}} e^{j\omega t} + e^{j\omega(t-\tau_1)} + E_2 e^{j\omega(t-\tau_2)} + E_3 e^{j\omega(t-\tau_3)} + ... \right|^2 \tag{10.22}$$

The desired mixing terms are the products between the reference field E_{ref} and the individual reflected signals E_i from the test arm. The undesired signals are the mixing terms that occur between the individual signals from the test arm. The beat frequencies between these terms can be mistaken for additional reflectors that do not exist. Two solutions to avoid this problem are both based on creating a large enough frequency difference between the desired and undesired beat frequencies so that they can be easily separated. This can be accomplished by adding either a frequency shifter[35] or an additional fiber delay to either of the interferometer arms.

The signal-to-noise considerations for coherent FMCW measurements are very similar to those associated with optical low-coherence reflectometry discussed earlier in Section 10.4. But since coherent FMCW requires a long coherence length source, it has the additional effect of phase noise. Phase noise can be caused by either the finite laser linewidth or by environmentally induced variations in the optical delay between the reference and test signals. The effect of phase noise is to reduce dynamic range; this prevents the measurement of small reflectivities when large reflections are present. Phase noise is minimized by choosing a laser with a narrow linewidth and by scanning the optical frequency over a short time interval relative to any environmental changes in the optical paths. A detailed analysis of phase noise is beyond the scope of this text but can be found elsewhere.[36]

Although we have assumed a linearly chirped laser for the above analysis, measurements using a nonlinear chirp are also possible.[32,37,38] Nonlinear tuning can be compensated for by coupling a portion of the signal into a stable reference cavity to generate trigger signals at equally spaced frequency intervals. Sampling the returning interference signal at these equal frequency intervals is equivalent to having a linearly chirp and sampling at equal time intervals. Another issue to deal with is the polarization sensitivity of the measurement. As in the case for OLCR, a polarization-diversity receiver (see Figure 10.19) can be used to solve this problem.

Coherent FMCW reflectometry has the potential to become the measurement technique of choice for high-resolution reflectometry. For this to occur, more progress is still needed in developing suitable frequency-tunable optical sources. For high-resolution applications, a continuously tunable, grating based, external cavity laser may offer a good solution. A possibly better choice might be an external cavity laser with an internal acousto-optic tunable filter and frequency shifter. This would allow fast linear-tuning rates without the worry of mechanical motors and bearings which are prone to wear.

10.6 COMPARISON OF TECHNIQUES

In this section, a comparison of the different reflectometry techniques discussed in this chapter will be made. The experimentally demonstrated results, given in Table 10.4, are not comprehensive but are meant to give a general feeling for the abilities of each measurement technique.

The spatial-resolution column in Table 10.4 illustrates an important differentiating feature. The upper four techniques are limited to resolutions on the order of millimeters or greater. This lower resolution is the result of having to measure the reflected optical power using direct detection. Direct detection limits the measurement bandwidth to that achievable with electronic circuitry, which in turn limits the spatial resolution to the millimeter range. The last two techniques, C-OFDR and OLCR, are based on coherent detection where the reflected signals optically interfere with a reference signal during the detection process. This allows the full optical spectrum of the probe signal to be utilized, resulting in an effective measurement bandwidth that is about three orders of magnitude larger than direct-detection techniques. The spatial resolution, which is inversely proportional to this bandwidth (see Section 10.3.1), is therefore reduced by about three orders of magnitude. Although OLCR has currently demonstrated resolutions that are better than C-OFDR, this difference is not inherent but only set by ease of practical implementation.

Table 10.4 Experimentally demonstrated results for various high-resolution reflectometry techniques.

	Spatial Resolution	Reflection Sensitivity	Dynamic Range	Measurement Range	Measurement of RBS
OCWR	none	−70 dB	70 dB	> 100 m	No
DD-OTDR	5 mm	−50 dB	50 dB	> 100 m	No
I-OFDR	4 mm[29]	−47 dB[29]	50 dB[31]	> 100 m	No
PC-OTDR	10 mm[39]	< −120 dB[39]	~ 60 dB[28]	> 100 m	Yes[27]
C-OFDR	12 μm[33]	−152 dB[40]	106 dB[40]	> 10 m	Yes[40]
OLCR	< 2 μm[12]	−162 dB[3]	> 120 dB	> 1 m[20]	Yes[1]

OCWR: Optical Continuous Wave Reflectometry—Section 10.2
DD-OTDR: Direct Detection Optical Time Domain Reflectometry—Section 10.5.1
I-OFDR: Incoherent Optical Frequency Domain Reflectometry—Section 10.5.3
PC-OTDR: Photon Counting Optical Time Domain Reflectometry—Section 10.5.2
C-OFDR: Coherent Optical Frequency Domain Reflectometry—Section 10.5.4
OLCR: Optical Low-Coherence Reflectometry—Section 10.4

Currently it is easier to generate the broadband ASE spectrum used in OLCR than the frequency tunable source (over the same spectral extent) needed in C-OFDR.

The next area of discussion is the reflection sensitivity for the different techniques. Here again, the lower two coherent techniques have demonstrated much better results than the upper direct-detection methods. The reason for this is that in an optimized coherent detection scheme, where the shot noise of the reference signal is the dominant noise source, the receiver is sensitive to single photons reflected back from the test device. With practical levels of probe power, this can result in reflection sensitivities greater than −150 dB. Although not a coherent technique, a photon-counting OTDR can also offer good sensitivities since the receiver is also sensitive to single photon events. Photon counting does have some additional limitations when compared to the coherent techniques. For example, averaging is slower since the probability of detecting a single photon should be less than unity for each probe pulse sent into the test fiber. Also, it is presently very difficult and inefficient to perform photon counting at the longer wavelengths used for optical communications.

The relatively poor sensitivity of DD-OTDR and I-OFDR is the result of a much higher receiver noise. The combination of requiring a large detection bandwidth and a small load impedance for the photodiode results in noise levels many orders of magnitude higher than required for single photon sensitivity. As for OCWR (total return-loss technique), its sensitivity is not limited by the receiver but by the background signal level due to Rayleigh backscatter from the fiber leads. Since OCWR does not spatially resolve individual reflections, this technique does not have to deal with high-speed detection or large measurement bandwidths.

Dynamic range is also an important parameter in making a reflectometry measurement. Dynamic range is the ability to measure a small reflection in the presence of a large reflection. It will be assumed that the receiver gain can be dynamically varied to account for the large changes in signal strength. Once again, the direct-detection techniques (top four columns in Table 10.4) show limited dynamic range when compared to the bottom two coherent techniques. One inherent difference between direct and coherent detection is that for direct detection the measured photocurrent is proportional to the reflected optical power, whereas in coherent detection, the photocurrent is proportional to the square root of the reflected power (see Section 10.3.5). This proportionality difference gives coherent techniques a big advantage when measuring signals with large amplitude differences. The reason why the dynamic ranges for C-OFDR and OLCR do not match the reflection sensitivities is that the intensity noise associated with large reflections can limit the detection sensitivity. For the case of photon counting, the dynamic range is limited by the number averages one is willing to perform.

Measurement range is one area where direct-detection techniques have advantages over coherent techniques. For DD-OTDR and PC-OTDR, the measurement range depends only on tracking the time delay for the reflected signals. This can greatly exceed 100 m. In I-OFDR, the measurement range depends on the smallest frequency step that can be generated. For these direct-detection schemes, the use of standard electronics can provide large measurement ranges. The coherent techniques have potential problems in dealing with large measurement ranges. For C-OFDR, optical phase noise can become quite large as the time delay between the reference and reflected signals become large. This causes both a loss in signal strength and spatial resolution. For OLCR, the measurement range is limited by the maximum delay that can be achieved using a mechanical translation stage. Distances over 1 m can be difficult to implement. This limitation may be somewhat relaxed by making use of recirculating delay structures.[19–21]

In conclusion, high-resolution reflectometry with submillimeter spatial resolution requires the use of coherent detection techniques. These techniques also provide excellent reflection sensitivity and dynamic range. The additional difficulties associated with coherent measurement techniques are polarization sensitivity and the requirement of more sophisticated optical sources. Both C-OFDR (for example, coherent FMCW) and OLCR have the same fundamental limits on achieving ultimate performance. In practice, OLCR is currently the simpler technique to implement. Its main limitation is that the measurement range is currently limited to about 1 m or less. C-OFDR may ultimately provide a more practical solution since the measurement is performed by tuning a laser frequency as opposed to translating a mechanical stage. As the technology related to tunable optical sources becomes more mature, C-OFDR techniques should increase in popularity.

REFERENCES

1. Sorin, W.V., and D.F. Gray. 1992. Simultaneous thickness and group index measurement using optical low-coherence reflectometry. *Photonics Tech. Lett.* 4:374–376.

2. Kapron, F.P., B.P. Adams, E.A. Thomas, and J.W. Peters. 1989. Fiber-optic reflection measurement using OCWR and OTDR techniques. *J. of Lightwave Tech.* 7:1234–1241.

3. Takada, K., T. Kitagawa, M. Shimizu, and M. Horiguchi. 1993. High-sensitivity low coherence reflectometer using erbium-doped superfluorescent fibre source and erbium-doped power amplifier. *Electron. Lett.* 29:365–367.

4. Takada, K., H. Yamada, and M. Horiguchi. 1994. Loss distribution measurement of silica-based waveguides by using a jaggedness-free OLCR. *Electron Lett.* 30:1441–1443.

5. Sorin, W.V., and D.M. Baney. 1992. Measurement of Rayleigh backscattering at 1.55 μm with 32 μm spatial resolution. *Photonics Tech. Lett.* 4:374–376.

6. Izumita, H., S. Furukawa, Y. Koyamada, and I. Sankawa. 1992. Fading noise reduction in coherent OTDR. *Photonics Tech. Lett.* 4:201–203.

7. Takada, K., A. Himeno, and K. Yukimatsu. 1991. Jagged appearance of Rayleigh-backscatter signal in ultrahigh-resolution optical time-domain reflectometry based on low-coherence interference. *Optics Letters,* 16:1433–1435.

8. Mark, J. 1987. Coherent measuring technique and Rayleigh backscatter from singlemode fibers. Thesis dissertation, Technical University of Denmark: 80.

9. Dolfi, D.W., M. Nazarathy, and S.A. Newton. 1988. 5 mm-resolution optical-frequency-domain reflectometry using a coded phase-reversal modulator. *Optics Lett.* 13:678–680.

10. Kobayashi, M., H. Hanafusa, K. Takada, and J. Noda. 1991. Polarization independent interferometric optical time domain reflectometer. *J. of Lightwave Tech.* 9:623–628.

11. Kazovsky, L.G. 1989. Phase- and polarization-diversity coherent optical techniques. *J. of Lightwave Tech.* 7:279–292.

12. Clivaz, X., F. Marquis-Weible, and R.P. Salathe. 1992. Optical low-coherence reflectometry with 1.9 μm spatial resolution. *Electron Lett.* 28:1553–1555.

13. The "HP 8504B Precision Reflectometer" sold by Hewlett-Packard Company and the "WIN-R Optical Coherence Domain Reflectometer" sold by Photonetics, Inc., in France.

14. Youngquist, R.C., S. Carr, and D.E.N. Davies. 1987. Optical coherence-domain reflectometry: a new optical evaluation technique. *Optics Lett.* 12:158–160.

15. Danielson, B.L., and C.D. Whittenberg. 1987. Guided-wave reflectometry with micrometer resolution. *Applied Opt.* 26:2836–2842.

16. Takada, K., I. Yokohama, K. Chida, and J. Noda. 1987. New measurement system for fault location in optical waveguide devices based on an interferometric technique. *Applied Opt.* 26:1603–1606.

17. Chou, H. and W.V. Sorin. 1993. High-resolution and high-sensitivity optical reflection measurements using white-light interferometry. *Hewlett-Packard J.*: 52–59.

18. Chou, H. 1992. Optical low coherence reflectometry: Improving reflectivity accuracy in the presence of chromatic dispersion. *Symposium on Opt. Fiber Meas.,* Boulder, CO: NIST Special Pub. 839, 167–170.

19. Takada, K., H. Yamada, Y. Hibino, and S. Mitachi. 1995. Range extension in optical low coherence reflectometry achieved by using a pair of retroreflectors. *Electron. Lett.* 31:1565–1567.

20. Baney, D.M., and W.V. Sorin. 1993. Extended-range optical low-coherence reflectometry using a recirculating delay technique. *Photonics Tech. Lett.* 5:1109–1112.

21. Baney, D.M., and W.V. Sorin. 1995. Optical low coherence reflectometry with range extension > 150 m. *Electronics Lett.* 31:1775–1776.

22. Sorin, W.V., and D.M. Baney. 1992. A simple intensity noise reduction technique for optical low-coherence reflectometry. *Photonics Tech. Lett.* 4:1404–1406.

23. Kobayashi, M., H.F. Taylor, K. Takada and J. Noda. 1991. Optical fiber component characterization by high-intensity and high-spatial-resolution interferometric optical-time-domain reflectometer. *Photonics Tech. Lett.* 3:564–566.

24. Takada, K., M. Shimizu, M. Yamada, M. Horiguchi, A. Himeno, and K. Yukimatsu. 1992. Ultrahigh-sensitivity low coherence OTDR using Er^{3+}-doped high-power superfluorescent fiber source. *Electron Lett.* 28:29–31.

25. Kreit, D. and R.C. Youngquist. 1987. Polarisation-insensitive optical heterodyne receiver for coherent FSK communications. *Electron. Lett.* 23:168–169.

26. Fouquet, J.E., G.R. Trott, W.V. Sorin, M.J. Ludowise, and D.M. Braun. 1995. High-power semiconductor edge-emitting light-emitting diodes for optical low coherence reflectometry. *IEEE J. Quantum Electron.* 31:1494–1503.

27. Bethea, C.G., B.F. Levine, S. Cova, and G. Ripamonti. 1988. High-resolution and high-sensitivity optical-time-domain reflectometer. *Optics Lett.* 13:233–235.

28. Lacaita, A.L., P.A. Francese, S.D. Cova, and G. Riparmonti. 1993. Single-photon optical-time-domain reflectometer at 1.3 μm with 5 cm resolution and high sensitivity. *Optics Lett.* 18:1110–1112.

29. Dolfi, D.W., and M. Nazarathy. 1989. Optical frequency domain reflectometry with high sensitivity and resolution using optical synchronous detection with coded modulators. *Electron Lett.* 25:160–161.

30. MacDonald, R.I. 1981. Frequency domain optical reflectometer. *Applied Opt.* 20:1840–1844.

31. Venkatesh, S. and D.W. Dolfi. 1990. Incoherent frequency modulated cw optical reflectometry with centimeter resolution. *Applied Opt.* 29:1323–1326.

32. Brinkmeyer, E. and U. Glombitza. 1991. High-resolution coherent frequency-domain reflectometry using continuously tuned laser diodes. *Optical Fiber Communication Conference,* paper WN2, San Diego, California, Feb. 18–22, p. 129.

33. Iizuka, K. and S. Fujii. 1992. A fault locator for integrated optics. *8th Optical Fiber Sensors Conference,* paper TH3.2, Monterey, California, Jan. 29–31, p. 297.

34. Hymans, A.J., and J. Lait. 1960. Analysis of a frequency-modulated continuous-wave ranging system. *Proc. IEE,* 107, pt. B:365–372.

35. von der Weid, J.P., R. Passy, G. Mussi, and N. Gisin. 1995. Self-heterodyne coherent optical frequency domain reflectometry. *Electron. Lett.* 31:2037–2038.

36. Venkatesh, S., and W.V. Sorin. 1993. Phase noise considerations in coherent optical FMCW reflectometry. *J. Lightwave Tech.* 11:1694–1700.

37. Takada, K. 1992. High-resolution OFDR with incorporated fiber-optic frequency encoder. *Photonics Tech. Lett.* 4:1069–1072.

38. Passy, R., N. Gisin, J.P. von der Weid, and H.H. Gilgen. 1994. Experimental and theoretical investigations of coherent OFDR with semiconductor sources. *J. Lightwave Technol.* 12:1622–1630.

39. Ripamonti, G. and S. Cova. 1986. Optical time domain reflectometry with centimetre resolution and 10–15 fw sensitivity. *Electron Lett.* 22:818–819.

40. Mussi, G., N. Gisin, R. Passy, and J.P. von der Weid. 1996. 152.5 dB sensitivity high dynamic-range optical frequency-domain reflectometry. *Electron. Lett.* 32:926–927.

OTDRs
and Backscatter
Measurements

Josef Beller

OVERVIEW

The preceding chapter, 10, covered spatially resolved reflectometry that enables a view inside of optical waveguides and other components with sub-cm spatial resolution. This chapter covers direct-detection optical time-domain reflectometers (OTDRs) which are best-suited for medium and long-haul distance ranges with submeter resolution. These OTDRs are designed to not only provide information about reflections, but also about attenuation properties and loss of the fiber under test. This is accomplished by exploiting backscattered and backreflected light returning from the fiber when probing it with a short laser pulse. Backscattered light provides a signature from which information about signal strength is deduced, sometimes in an ambiguous manner. Interpretation of the OTDR measurement features is aided by knowledge of the fiber backscatter and reflection mechanisms together with familiarity of the measurement process.

This chapter begins with a close look at the single-pulse OTDR. This is the most successful commercial method due to both its inherent simplicity and excellent results. After describing the operational principle, its limitations and performance parameters, a short treatment of fiber-attenuation and loss-mechanisms follows. The succeeding section analyzes the backscattering signal as a function of pulsewidth, fiber-length, and fiber parameters. Then two sections focus on insertion-loss measurements with an analysis of the uncertainty on the measurement data due to superimposed noise, and on return-loss and reflectance measurements. Section 11.6 provides an overview on remote-fiber-testing and some of its applications.

11.1 INTRODUCTION

Most optical fiber parameters like core and cladding diameter, refractive index profile, NA (numerical aperture), or cutoff wavelength are unlikely to change after fiber manufacture and installation. Yet, experience shows that properties like fiber loss and attenuation are prone to alter through environmental influences like humidity, temperature, or physical stress. Fiber loss is of great importance for installed fiber links since fiber attenuation directly determines the total loss and thus the quality of a transmission system.

A direct method to measure insertion loss is the cut-back technique.[1,2] The power coming from an unconnectorized fiber end is measured. While keeping the launching condition fixed, a known length of fiber is then removed and the fiber power is remeasured. The difference in power measurements is the fiber loss. The cut-back technique provides accurate results but is a destructive measurement that requires access to both ends of the fiber. In general, installed fiber-optic cables can not be measured by this method. Another technique is the OTDR backscatter measurement that indirectly allows for determination of the fiber attenuation. It is a nondestructrive technique capable of doing *in situ* measurements and requires access to only one end of the fiber. Furthermore it provides extra information about the fiber under test, like length dependence of the fiber attenuation, and insertion loss of defects, splices, bends, or connectors. The location and type of faults occurring in fiber fabrication, the homogeneity of fiber characteristics, and fiber length can all be tested.

In 1976 Barnsen and Jensen authored the first publications about backscatter in optical waveguides and provided proof for its suitability to fiber characterization.[3,4,5,6] OTDR analysis has evolved to a standard technique for testing of optical fibers links. The OTDR has become the primary instrument for single-ended characterization of optical fibers based on backscatter measurements.

OTDRs can be classified by several measurement techniques. Most instruments are based on time-domain techniques, such as single-pulse OTDRs and correlation OTDRs.[7,8,9] Frequency-domain OFDRs (Chapter 10), on the other hand, promise some advantages, especially in terms of highest resolution (sub-cm). OFDRs have yet to emerge from the laboratory state.[10,11,12] OTDRs can be further classified by with respect to the optical-detection technique. These are analogue direct detection using a photodetector, digital direct detection using photon counting,[13,14] and coherent detection with either homodyne or heterodyne methods.[15,16] The physical principle of backscatter measurements is the same independent of OTDR technique. For simplicity, the following text refers primarily to the standard single-pulse OTDR with many of the results applicable to other OTDRs.

11.2 PRINCIPLE OF OTDR OPERATION

OTDRs launch short duration light pulses into a fiber and then measure, as a function of time after the launch, the optical signal returned to the instrument. As the optical pulses propagate along the fiber, they encounter reflecting and scattering sites resulting in a fraction of the signal being reflected back in the opposite direction. Rayleigh scattering and

Fresnel reflections are physical causes for this behavior. By measuring the arrival time of the returning light, the locations and magnitudes of faults can be determined and the fiber link can be characterized.

A block diagram of a generic OTDR is shown in Figure 11.1. A pulse generator triggered by the signal-processing unit is used to modulate the intensity of a laser. While there are a few special designs using codes and correlation techniques,[9] the probe signal in conventional OTDRs is a single square-pulse. Pulse widths between 5 ns and 10 μs are used depending on the spatial resolution and sensitivity requirements of the measurement. Dual-wavelength OTDRs equipped with two laser diodes (typically 1310 nm and 1550 nm) combine the light sources via a wavelength-division-multiplexing (WDM) coupler. In order to prevent the laser signal from saturating the receiver, the source is coupled into the fiber under test by a directional coupler with sufficient isolation between ports A and B.

The most common coupler type is a 3 dB-fusion type fiber coupler with low polarization sensitivity and a split ratio near 50:50 at the measurement wavelengths. This keeps the round-trip loss to the receiver close to the minimum of 6 dB. In principle, beam splitters, circulators, polarizing prisms, or lithium-niobate acousto-optical switches can perform the same function as the fused directional coupler with potentially higher performance at a higher cost.

The directional-coupler guides the returning signal to the photodetector that is either a p-i-n diode or an avalanche photo diode (APD) acting as current source for a low-noise transimpedance amplifier with high linearity. Signals covering several orders of magnitude are incident on the photodetector. This requires the receiver to have a high dynamic range together with high sensitivity. A flash type analog-to-digital-converter (ADC)

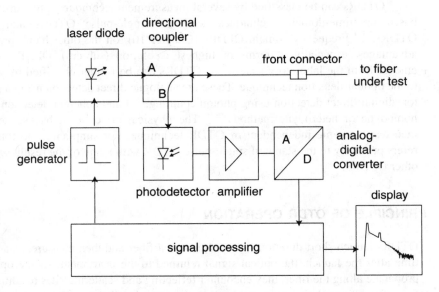

Figure 11.1 Block diagram of a single-pulse OTDR.

forms the interface to the digital data world where the measurement data is processed and the fiber signature is computed. The ADC sampling rate determines the spatial separation of adjacent data samples. Short-distance fiber links require spatial resolutions far smaller than those achievable by direct data acquisition. A sampling rate of 50 MHz corresponds to a data spacing of approximately 2 m.[17] It is not practical to increase the resolution by increasing the data sampling rate. OTDRs commonly use an interleaved processing scheme to improve spatial resolution down to the centimeter region.

The result of an interleaved measurement is the composition of individual measurement shots delayed by a varying fraction of the ADCs sampling time, as depicted in Figure 11.2. The delay is always referenced to the launched laser pulse. Depending on the de-

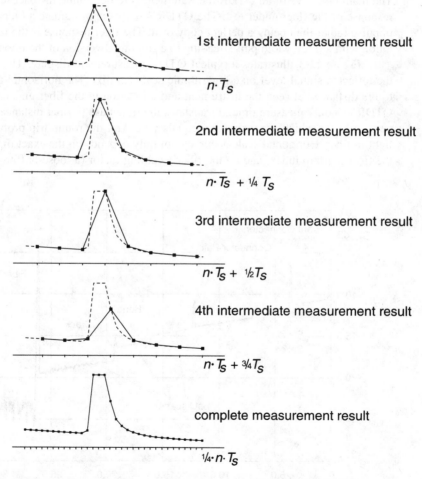

Figure 11.2 Improving spatial resolution by interleaved data acquisition.

sired resolution, the number of individual measurement shots can double, quadruple, or in general multiply as a power of two, compared to a noninterleaved measurement. The example above shows a four-fold improvement in sample spacing, at the expense of four times the measurement time. Taking full advantage of even higher interleaving rates requires a sufficiently high bandwidth in the signal-processing path to avoid smoothing of the final result.

Because the backscattered signal typically is very weak, it is often covered by noise. To overcome this problem, the process of sending a pulse and receiving the echo is repeated many times to improve the signal-to-noise ratio (SNR) by averaging.[17]

11.2.1 OTDR Fiber Signature

The main objective of an OTDR measurement is to determine the backscattering impulse response of the fiber under test. The OTDR's pulses approximate an ideal delta-function impulse rather than being a perfect copy of it. The fiber response is the result of a convolution with a finite width pulse, leading to a smoothed version of the impulse response.

Figure 11.3 illustrates a typical OTDR measurement display. The vertical scale is the reflected signal level on a logarithmic scale (in dB). The horizontal axis corresponds to the distance between the instrument and a location in the fiber under test. Because an OTDR can only measure time, it translates the time base to fiber distance by using a conversion factor which approximately equals $10\mu s/km$, the round-trip propagation delay of light in fiber. Horizontal scale accuracy not only depends on the exact timing, but also on the fiber's group index, and on the fiber's cabling factor (a loose-buffered fiber is longer

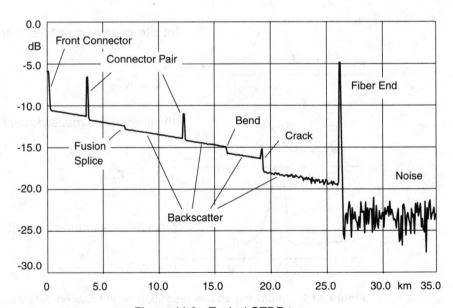

Figure 11.3 Typical OTDR trace.

than a tightly buffered one). With a typical time-base accuracy better than 0.01%, in practice, both the index of refraction and the cabling factor always are the limiting factors to overall distance accuracy.

An OTDR measures round-trip loss. Light captured by the OTDR moves in the opposite direction to the launched pulse. By traveling forward and backward to the OTDR it experiences a two-fold fiber attenuation and it needs twice the time it takes to reach a distinct location on the fiber. This is why traces are scaled by a factor of two on the display in both the vertical (5 · log instead of 10 · log) and horizontal direction, the so called one-way representation.

The measured response typically exhibits three types of features: straight lines caused by distributed Rayleigh backscattering, positive spikes caused by discrete reflections, and finally steps that can either be positive or negative depending on physical fiber properties.

The first event that can be seen on an OTDR trace is the reflection of the front connector that mates the OTDR to the fiber under test. As this reflection covers the near-end measurement zone, it is an undesirable event hiding information about the fiber to be tested. A clean high-quality connector with low reflectance is mandatory to achieve best results. A bad connector not only decreases launch power by its insertion loss, but also causes the returning light to be re-reflected back into the fiber under test again generating multiple echoes or ghost patterns. Connector care is an important part of OTDR operation (see Appendix C).

As the trace is plotted as power vs. distance, the slope of the straight lines gives the fiber attenuation in dB/km. Fusion splices cause a sudden drop in the backscatter level. The step size corresponds to the insertion loss only if the two mated fibers are identical. Otherwise the true insertion loss can only be determined by the average of two measurements from both fiber ends.[18] A similar signature on an OTDR display is caused by a stressed curve in the fiber. This is called a bend, and results in light reflecting out through the cladding, instead of continuing on down the fiber. Fusion splices and bends show only insertion loss without a reflection. They are called nonreflective events. A mismatch in the refractive index causes Fresnel reflections leading to spikes superimposed on the backscatter signal (see Equation 11.26 for details). Mechanical splices, connectors, and cracks in general show a tiny air gap which reflects light rather than scatters it. They belong to reflective events. An open nonterminated fiber end can cause a strong reflection depending on the condition of the fiber end surface. At a glass-to-air transition, up to 4% of the optical power can be reflected back to the OTDR. Behind the fiber end no optical signal can be detected and the curve drops to the receiver noise which ultimately limits the detectable power level.

11.2.2 Level Diagram

The level diagram in Figure 11.4 illustrates the range of powers that an OTDR has to work with. At a glance, both the highest and the lowest signal levels that return from the fiber, either because of reflections or because of backscatter, can be seen. This is useful information for an OTDR design with a certain required dynamic range. For example, the

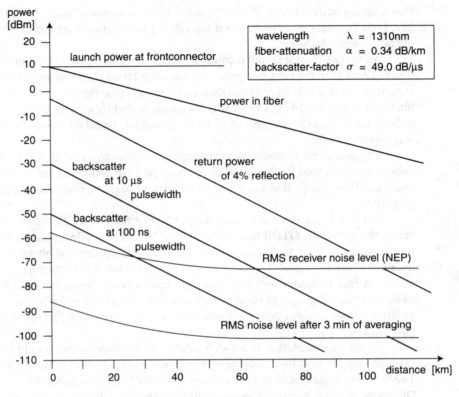

Figure 11.4   OTDR level diagram.

necessary noise equivalent power (NEP) that the receiver must not exceed to achieve the desired dynamic range performance can easily be determined.

To calculate the maximum power levels of reflected and backscattered light we assume an OTDR laser-source with a peak pulse power of +13 dBm. Then the launch power at the front-connector is about +10 dBm because of the 3 dB-coupler loss.

In case of a noncontact front panel connector, the resulting 4% reflection of a glass-to-air transition at the near end corresponds to a reflected light pulse approximately 14 dB below the incident pulse, in other words, −4 dBm. For a singlemode fiber at $\lambda = 1310$ nm, the backscatter power level is roughly 49 dB/μs below peak power (see Equation 11.23). This yields about −30 dBm backscatter power level for 10 μs pulsewidth, and −50 dBm for 100 ns, respectively. Light scattered and reflected back from a remote location in the fiber under test is exposed to twice the fiber loss as is indicated in Figure 11.4 by the slopes of the corresponding lines. Taking the fiber attenuation into account, the backscatter signal weakens with increasing distance, and eventually dips into the receiver noise. Often the receiver's bandwidth is adapted to the chosen measurement span to provide optimal resolution and dynamic range as well. This leads to a higher NEP at short-distance ranges. A wider pulse with its higher energy enables the OTDR to look deeper into the fiber. Also, receiver sensitivity can be improved with signal averaging. Depending on fiber length, laser pulse repetition rate, and processing speed, up to 25 to 30 dB of noise reduction within a 3 min measurement time can be achieved.[17] As the fiber's attenuation is considerably lower at $\lambda = 1550$ nm, the level diagram at this wavelength would reveal a much longer distance before the backscatter curves hit the noise level.

Example

A 20 km fiber link is being tested with an OTDR. Determine the noise reduction that can be achieved by signal averaging compared to a single measurement shot:

a. within the first second,
b. within a 3 min measurement time,
    under the assumption that 10% of the time is needed for processing overhead.

Solution

With a time-distance conversion factor of approximately 10 μs/km, the round-trip time T_{RT} for a 20 km link length is:

$$T_{RT} = 10 \frac{\mu s}{km} \times 20 \, km = 200 \, \mu s$$

This corresponds to N_{1s} and $N_{3\min}$ measurement shots in 1 s and 3 min measurement time, respectively, where:

$$N_{1s} = \frac{1}{200 \cdot 10^{-6}} \times 90\% = 5000 \times 0.9 = 4500$$

$$N_{3\min} = N_{1s} \times 180 = 810{,}000$$

The noise reduction is proportional to the square-root of N. Thus, the (two-way) SNR improvement is:

$$\Delta SNR_{1s} = 10 \times \log(\sqrt{N_{1s}}) = 5 \times \log(4500) = 18.3 \, dB$$

$$\Delta SNR_{3\min} = 10 \times \log(\sqrt{N_{3\min}}) = 5 \times \log(810{,}000) = 29.5 \, dB$$

Obviously the major part of the noise reduction is achieved within the very first second of the measurement. Keep in mind that OTDRs display one-way dBs, in other words, after 3 min the noise level on the screen decreases by $(29.5 - 18.3)/2$ dB = 5.6 dB compared to the first display update after a 1 s measurement time.

11.2.3 Performance Parameters

The performance of an OTDR is specified by a set of parameters that describe the quality of the measurement and allow the user to understand how much the instrument fits the application needs. Generic requirements for OTDRs are proposed in Reference 19. Figures 11.5 to 11.7 depict the key parameters: dynamic range, measurement range, attenuation deadzone, event deadzone, and resolution.

Dynamic Range and Measurement Range.   Dynamic range is an important parameter, as it is often used to rank an OTDR among a particular performance class. It provides information not only on the maximum fiber loss that can be measured, but also on

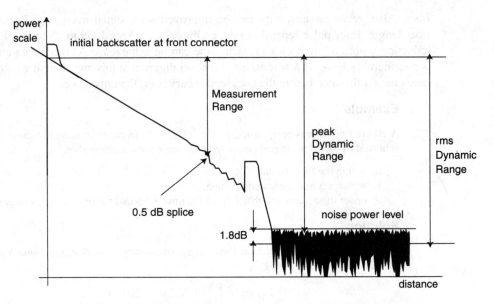

Figure 11.5 Dynamic range and measurement range.

the measurement time required for a given fiber loss. Dynamic range is defined as the difference between the initial backscatter level and the noise level after 3 min of measurement time, expressed in decibels of one-way fiber loss. The noise level either can be defined as rms (SNR = 1), or 98% peak noise level. Assuming purely Gaussian noise distribution, 98% of all noise samples are covered within approximately 2.4 times the standard deviation. This corresponds to a 1.8 dB increase if referring to the rms value rather than to the 98% peak level. As the reference condition SNR = 1 is the one giving the maximum value for the dynamic range, it is often used in the literature.

Measurement range deals with identification of events. The measurement range is defined as the maximum attenuation that can be inserted between the OTDR and an event for which the OTDR is still able to accurately measure the event.[19] Commonly a 0.5 dB splice is chosen as the event to be identified.

As the accuracy of a loss measurement made with an OTDR primarily depends on the SNR at that point, instruments with a high dynamic range are highly valued. Software algorithms to identify patterns in a noisy environment have great impact on measurement range performance.[20]

Dead Zones. Dead zones are always related to the presence of reflections. Dead zones occur when the reflected signal saturates the OTDR receiver. The receiver is slow to recover its sensitivity after the saturation resulting in the loss of information. If the receiver saturates because of strong signals, it will take some time to recover from this overload condition. As a consequence, the measured fiber response is superimposed by the receiver's overload behavior, yielding a distinct fiber segment covered by an exponentially

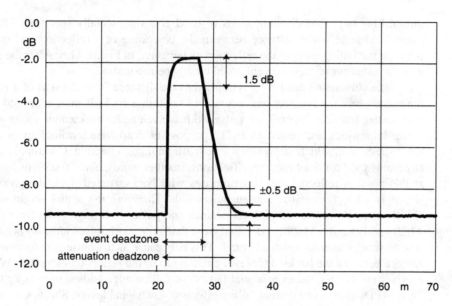

Figure 11.6 Attenuation and event dead zone.

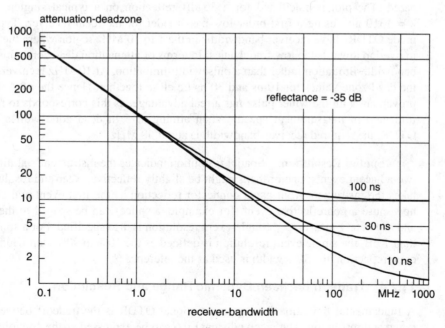

Figure 11.7 Minimum attenuation dead zone as function of OTDR receiver-bandwidth.

diminishing tail. Two different kinds of dead zones are usually specified. The OTDR's event dead zone is the distance between the beginning of a reflection and the −1.5 dB point on the falling edge of the reflection as indicated in Figure 11.6. After the event dead zone, an adjacent reflective event could clearly be recognized.

The attenuation dead zone is defined as the distance from the start of a reflection to the point where the receiver has recovered to within a ±0.5 dB margin around the settled backscatter trace. It depends on pulsewidth, wavelength (backscatter factor σ), the receiver bandwidth, and reflectance. The purpose of dead-zone terminology is to quantify the distance over which information is lost after a large reflection. Commonly a 35 dB reflection is used for dead zone specifications, in other words, about 0.03% of optical power at this point is reflected and superimposes with backscattered light, the power level of which is a function of the chosen pulsewidth. Therefore the actual height seen on the OTDR display depends on both reflectance and pulsewidth for a given fiber (see Figure 11.20 and Equation 11.29 for details). Note that very short pulsewidths do not necessarily lead to shorter attenuation dead zones. This is because as the pulse gets narrower, the difference between the backscatter level and the top of the reflection increases. With a finite receiver bandwidth, the exponential tail of the falling edge adds significantly to the dead zone. APDs also have inherent tailing effects at low signal levels. Short pulse widths also lower SNR. Therefore, OTDR specifications for attenuation dead zone are often given at a pulsewidth next to the shortest one. In Figure 11.7 the minimum-achievable attenuation dead zones for three different pulsewidths as a function of receiver-bandwidth are compared. The plot is calculated for a 35 dB reflection on a typical singlemode fiber at λ = 1310 nm, using a first-order low-pass model for the ideal receiver. Typical singlemode OTDRs have receiver bandwidths in the 1 to 10 MHz region where the curves start to overlap towards the low frequencies. In terms of attenuation dead zone, this indicates a bandwidth-limitation rather than a pulsewidth-limitation. At 10 MHz receiver-bandwidth, the dead-zone values for 10 ns and 30 ns lie close together. Hence the +2.4 dB SNR improvement with the wider pulse has a real advantage, as this corresponds to nearly a tenfold gain in measurement speed. When aiming at 5 m dead zone with a singlemode OTDR, the required receiver-bandwidth is about 50 MHz.

Spatial Resolution.   Spatial resolution indicates the instrument's ability to resolve two adjacent events; one of them might be slightly reflective. Near-end resolution simply takes the instrument front-panel connector reflection as the first event and characterizes how close a nonreflective event (for example, a splice) can be spaced to the instrument and accurately measured.[19] Single-event resolution is also specified. For a splice with less than 1 dB, the single event resolution is defined as the 10% to 90% step width. For a discrete reflection, the 50% width is used as the reference.[21]

11.2.4   Tradeoff between Dynamic Range and Resolution

A fundamental limitation for any conventional OTDR is the tradeoff between dynamic range and resolution. The received signal $s(t)$ can be expressed as the convolution ($\otimes$) of $p(t)$, the probing pulse, $f(t)$, the backscattering impulse response of the fiber, and $r(t)$, the impulse response of the receiver.

$$s(t) = p(t) \otimes f(t) \otimes r(t) \tag{11.1}$$

The achievable resolution is therefore limited by the receiver response and the geometrical width of the probe signal. For high-spatial resolution, the probe pulsewidth has to be as small as possible with a correspondingly wide receiver bandwidth. This leads to a reduced SNR. Increasing the strength of the received signal by using longer probe pulses and low noise (low bandwidth) receivers leads to improved sensitivity with correspondingly less resolution. This tradeoff of pulse width and sensitivity is shown in Figure 11.8. Two reflective events spaced about 100 m apart have been measured with a pulsewidth of 1 µs as well as 100 ns. Though the upper trace shows a smoother backscatter than the lower one, the drawback of insufficient spatial resolution is evident.

Increasing the laser output power also maximizes the backscatter level at a given pulsewidth. Unfortunately, the use of very high-power sources is normally precluded in a practical system because of reliability, cost, safety regulations, availability, or nonlinear scattering phenomena. Spread-spectrum techniques such as correlation[22] overcome this limitation and offer the possibility of improving the SNR without sacrificing resolution. Such techniques are commonly used in radar[23,24] and other peak-power limited systems. In OTDRs, however, this principle has only limited success, mainly because even small nonlinearities in the analog hardware lead to spurious signals that disturb the measurement results.

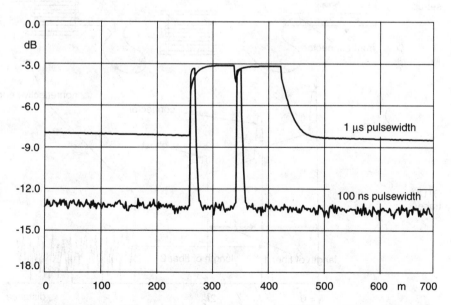

Figure 11.8   Differences in resolution and SNR caused by different pulsewidths.

11.2.5 Ghost Features Caused by Multiple Reflections

Sometimes the reflectometry trace includes "ghost" features. If the fiber under test contains connectors that reflect strongly, echoes generated by multiple reflections can produce spikes at false locations. The only solution for (real) ghosts is to avoid high reflectivity connections. Figure 11.9 shows a measurement setup with an OTDR and two connected fiber spools. Even in case of physical contact, ordinary connectors show discrete spikes on the OTDR display. Any echo pulse returning from the DUT to the OTDR is partially reflected back at the front-panel connector, acting like an additional delayed probe pulse. For well-maintained connectors, the echo amplitude is too small to generate a visible (and delayed) ghost picture.

Dirty or scratched connectors, however, can reflect a big part of a pulse's energy. In this case, a visible ghost spike can appear at a location l_{ghost} which can be calculated as

$$l_{ghost} = 2 \cdot l_2 - l_1 \tag{11.2}$$

with l_1 and l_2 being the location of the two reflections involved. In the example above $l_1 = 0$, yielding $l_{ghost} = 2\, l_2$.

In contrast to real ghosts, an OTDR can create ghost patterns if the repetition rate of the laser pulses is too fast and not adapted to the fiber under test. If the OTDR emits a sec-

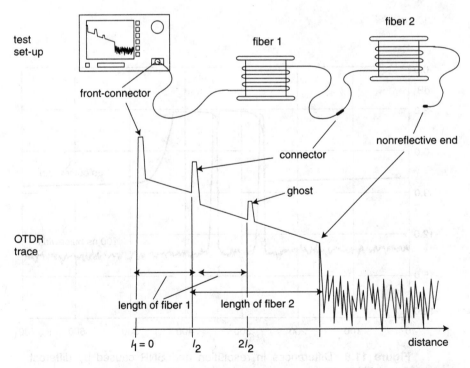

Figure 11.9 Example of a ghost generated by multiple reflections.

ond light pulse and starts data acquisition before the response of the previous one has been completely received, echoes from both pulses will overlap, leading to erroneous measurement results.

11.3 FIBER LOSS, SCATTER, AND BACKSCATTER

11.3.1 Loss in Fiber

Fiber loss is one of the most important properties of an optical fiber. It largely determines the maximum repeaterless span for an optical communication link. Absorption, scattering, and bending are the three major loss mechanisms in fused silica glass (SiO_2) fibers. Intrinsic absorption is extraordinarily low in the commercially used wavelength-window between the ultraviolet and the infrared region and is at short wavelengths negligible compared to scattering loss. Above 1700 nm, glass starts absorbing light energy because of vibrational transitions of the Si–O bond.[25] Figure 11.10 shows the typical tub-shaped total loss in optical silica fibers with its minimum attenuation around 1600 nm.

The dominant extrinsic absorption factor is the presence of impurities in the fiber material, for example, metal ions or water (OH^-) ions.[26,27] OH^- absorption can be used to monitor undersea links for the tiniest water penetration. Overtones of the fundamental absorption wavelength of water at 2.7 µm reside near the wavelength regions of commercial interest, for example, 1383 nm, 1250 nm, or 950 nm, respectively. The absorption lines

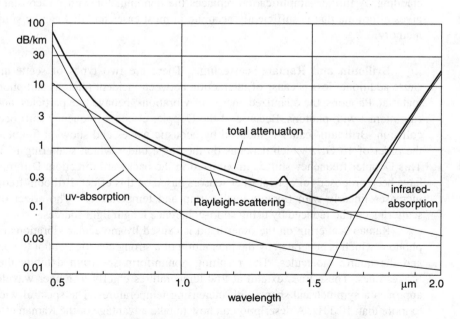

Figure 11.10 Wavelength dependence of silica-glass fiber attenuation α.

cover very narrow spectral widths. The locations and magnitudes of these absorption peaks in the earlier fiber manufacturing are the prime reason for the three major transmission windows at 0.85 μm, 1.3 μm, and 1.55 μm. In typical GeO_2-doped fibers, the zero-point for chromatic dispersion is near 1310 nm. The third transmission window at 1.55 μm shows the minimum fiber attenuation of approximately 0.2 dB/km, or even lower.

Rayleigh-Scattering Phenoma Description.   Scattering losses in glass arise from light interacting with density fluctuations in the fiber. Variations in material density and compositional fluctuations occurring during fiber manufacture create random inhomogeneities that give rise to refractive index variations. This isotropic phenomenon is termed Rayleigh scattering if the size of the defect is less than one-tenth of the wavelength of the incident light.[28,29] The mechanism does not cause elimination or conversion of optical energy but simply forces a part of the optical wave to escape from the waveguide. The scattered intensity is proportional to $1/\lambda$,[4] so that longer wavelengths in the standard transmission windows exhibit lower attenuation losses than shorter wavelengths. Equation 11.3 shows a simple empirical relation for the Rayleigh scattering loss α_s and the wavelength λ in germanium-doped silica fibers:[30]

$$\alpha_s = \frac{(0.76 + 0.51 \cdot \Delta n)}{(\lambda/\mu m)^4} \left[\frac{dB}{km}\right] \tag{11.3}$$

where Δn is the difference between the fiber core's refractive index and that of the fiber cladding. A further simplification replaces the denominator with the constant 0.8. This gives a formula that is sufficiently accurate in most cases and that can be simply held in memory.

Brillouin and Raman Scattering.   There are two types of scattering that can cause additional loss because of interaction between incident photons and phonons of the material. Phonons, the quantized energy of vibrations, behave like particles, and therefore can interact with photons. Because of the Doppler effect, a frequency shift occurs during collision. Brillouin scatter is induced by acoustic waves and shows a frequency shift in the order of 10 GHz which depends on the angle under which scattering is measured.[31] This Doppler frequency shift is maximized in the backward direction. Different methods have been proposed for optical fiber sensors based on this effect. Brillouin frequency shift increases linearly with (longitudinal) strain and temperature.[32] The effect of Brillouin scattering can be reduced by using short-coherence length light sources.

Raman scattering on the other hand is caused by molecular vibrations of "optical" phonons which is generated by the interaction of a strong electric field of the optical wave and the quartz molecules. The resulting nonuniform spectrum exhibits the so-called Stokes-lines. These lines extend at low temperatures chiefly to longer wavelengths and approach a symmetrical shape with increasing temperatures. The spectral width extends to more than 10 THz. A description on how to take advantage of the Raman effect for distributed temperature sensing can be found in Dakin and co-workers.[33] Both the Brillouin and the Raman scattering phenomena are nonlinear effects that occur only at higher power levels.[34,35] They can be a limiting factor in high power systems[36] (also see the treatment in Appendix B).

Micro- and Macrobending.   Bending is the third effect leading to increased fiber loss. Microbending is due to tiny imperfections in the geometry of the fiber caused either by the manufacturing process or by mechanical stress, such as pressure, tension, and twist. To overcome this problem, fibers are protected by either a loose or a tight buffer. Macrobending, on the other hand, occurs when the fiber is formed to a curvature with diameters on the order of centimeters. In this case, less-than-total internal reflection at the core-to-cladding boundary forces some of the light to leave the core.

11.3.2 Backscatter Signal Analysis

As a result of these attenuation effects, light traveling along a fiber exhibits an exponentially decreasing power level with the distance. The power transmission relation between incident light P_0 and transmitted power $P(z)$ at a distance z is

$$P(z) = P_0 \cdot e^{-\alpha z} \tag{11.4}$$

with the attenuation coefficient α measured in km^{-1} units. Since system losses in general are calculated in dB units, it is often more convenient to use the attenuation coefficient α_{dB} in dB/km units for loss calculations. Then

$$P(z) = P_0 \cdot 10^{-\frac{\alpha_{dB}}{10} z} \tag{11.5}$$

with

$$\alpha_{dB} = \frac{10}{\ln 10} \cdot \alpha \approx 4.34\, \alpha \tag{11.6}$$

To simplify matters, the total attenuation coefficient, α, is commonly composed of an absorption coefficient α_a and a scattering coefficient α_s

$$\alpha = \alpha_a + \alpha_s \tag{11.7}$$

It is obvious that the lower attenuation limit is bounded by α_s since $0 \leq \alpha_a \ll \alpha_s$.[37] Let's now focus on the effect of scattering and consider a laser pulse in a fiber with the temporal pulse duration τ. As the pulse propagates, light is scattered within a fiber element of length W, as depicted in Figure 11.11, with

$$W = \tau \cdot v_{gr} = \tau\, \frac{c}{n_{gr}} \tag{11.8}$$

where v_{gr} is the group velocity, c the speed of light in vacuum, and n_{gr} the group index of glass.

The scattered power dp_s at the position z within an infinitesimal small interval dz is proportional to the pulse power $P(z)$.[38]

$$dp_s = k \cdot P(z)\, dz \tag{11.9}$$

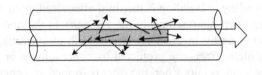

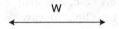

W

Figure 11.11 Scattering within a
fiber element of width *W*.

with
$$k = S \cdot \alpha_s \tag{11.10}$$

and
$$\alpha_s = \text{scattering coefficient} \sim \frac{1}{\lambda^4}$$

S is the fraction of the light scattered in all directions that is captured by the fiber core and guided back to the OTDR. The backscattering capture coefficient *S* is given in Equation 11.11.

$$S = \left(\frac{NA}{n_0}\right)^2 \cdot \frac{1}{m} \tag{11.11}$$

where *NA* is the fiber's numerical aperture, n_o is the refractive index of the fiber core center and *m* depends on the refractive index profile. A detailed derivation of the context can be found in References 38, 39, and 40. For singlemode fibers a typical value for *m* is 4.55.[41,42]

To derive the backscatter response caused by a rectangular pulse, we first assume the leading edge of the pulse to be at a location $L = T \cdot v_{gr}$. Light scattered back from exactly this distance will arrive at $t = 2T$, the round-trip time, at the OTDR port again. This situation is depicted in Figure 11.12a. After a time, $t = T + \tau/2$, the trailing edge hits the distance $L - W/2$. The light scattered from this position needs $t = T - \tau/2$ to travel back to the beginning, summing up to the same round-trip time $t = 2T$ as can be seen in Figure 11.12b. We can generalize the condition for light that returns after $t = 2T$. Let's look at a short interval ΔW that is $2\Delta z$ behind the pulse's leading edge at $t = T + \Delta t$. Backtraveling light from this part arrives after $t = T - \Delta t$ yielding again the round-trip time $t = 2T$. This close inspection reveals the important fact that the backscatter power seen by the OTDR at a time $t = 2T$ is actually the integrated sum of backscatter from the locations $z = L - W/2$ to $z = L$ when probing the fiber with a pulse of geometrical width *W*.

Summing up the light power backscattered from infinitesimal short intervals *dz* from the whole pulse and taking the fiber attenuation into account, yields

$$P_s(L) = \int_0^W S \cdot \alpha_s \cdot P_0 \cdot \exp\left(-2\alpha\left(L + \frac{z}{2}\right)\right) dz \tag{11.12}$$

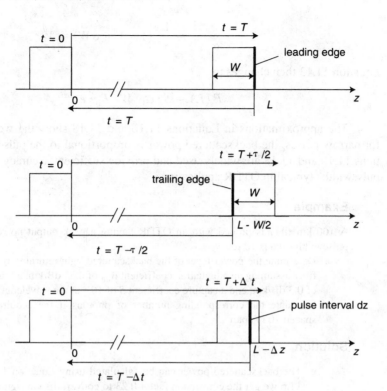

Figure 11.12   Backscatter round trip time $t = 2T$.

$$= S \cdot \frac{\alpha_s}{\alpha} \cdot P_0 \cdot e^{-2\alpha L} \left(1 - e^{-\alpha W}\right) \quad L \geq \frac{W}{2} \tag{11.13}$$

Equation 11.13 is valid only for $t \geq \tau/2$. For distances less than $W/2$, the lower integral limit has to be substituted by $W - 2L$. Then we get

$$P_s(L) = \int_{W-2L}^{W} S \cdot \alpha_s \cdot P_0 \cdot \exp\left(-2\alpha\left(L + \frac{z}{2}\right)\right) dz \tag{11.14}$$

$$= S \cdot \frac{\alpha_s}{\alpha} \cdot P_0 \cdot e^{-\alpha W}\left(1 - e^{-2\alpha L}\right) \quad 0 \leq L \leq \frac{W}{2} \tag{11.15}$$

An approximation for the initial value of the backscattered power P_{init} can be derived for $L = W/2$ from Equation 11.15 when developing the exponential functions into polynomials.

$$P_{init} = P_s\left(\frac{W}{2}\right) \approx S \cdot \alpha_s \cdot P_0 \cdot W \tag{11.16}$$

Likewise for short pulsewidth, $\alpha W \ll 1$, the last expression in parentheses in Equation 11.13 can be simplified to

$$\frac{1 - e^{-\alpha W}}{\alpha} = \frac{1}{\alpha} \cdot (1 - 1 + \alpha W) = W \tag{11.17}$$

Equation 11.13 then changes to

$$P_s(L) = S \cdot \alpha_s \cdot W \cdot P_0 \cdot e^{-2\alpha L} \tag{11.18}$$

The approximations in Equations 11.16 and 11.18 show the well-known fact that for narrow pulses, the backscattered power is proportional to the pulse duration τ. Equations 11.16 and 11.18 are widely used and provide sufficient accuracy when dealing with pulsewidths typical in OTDR applications.

Example

A 100 km fiber is probed with an OTDR having a peak output power of +13 dBm. The pulsewidth used is 10 μs.
    a. Determine the power level of the backscattered light returning from the far end of the fiber assuming an attenuation coefficient α_{dB} of 0.33 dB/km, a scattering coefficient α_s of 0.3 dB/km, and a capture coefficient S of 10^{-3} at the wavelength $\lambda = 1310$ nm.
    b. Calculate the corresponding number of photons if the acquired data samples are spaced 10 m apart.

Solution

    a. The backscattered power can be calculated using Equation 11.18. From Equation 11.6 we get the conversion factor 0.23 to convert dB/km figures in 1/km. Equation 11.8 gives the length W as 2 km.

$$P_s(100 \text{ km}) = 0.001 \times (0.3 \times 0.23) \times 2 \times 20 \text{ mW} \times e^{-2 \times (0.33 \times 0.23) \times 100}$$

$$= 35.3 \times 10^{-12} \times 20 \text{ mW} = 0.75 \text{ pW} \cong -91.5 \text{ dBm}$$

    b. With a time-distance conversion factor of 10 μs/km, a data spacing of 10 m is equivalent to the time period T

$$T = 10 \cdot 10^{-6} \times 0.01 = 10^{-7} s$$

As power is energy per time, the following relation between optical power and number of photons Z can be derived:

$$P = \frac{1}{T} \cdot h \cdot v \cdot Z = \frac{1}{T} \cdot \frac{h \cdot c}{\lambda} \cdot Z$$

Hence the average number of photons within a 100 ns timeframe that corresponds to an optical power of 0.7 pW is

$$Z = \frac{10^{-7} \times 1.3 \cdot 10^{-6}}{6.626 \cdot 10^{-34} \times 2.998 \cdot 10^{8}} \times 0.705 \cdot 10^{-12} = 0.46$$

After 3 min averaging with approximately 160,000 repetitive measurement shots, Z sums up to a total of 74,000 photons.

Substituting distance L in Equation 11.18 by the corresponding term $t \cdot c/n$ yields a time-dependent relation for the backscatter signal which shows a first-order lowpass behavior such as given in Equation 11.19. This can be used to derive the backscatter bandwidth BW_{Bsc} of optical fibers which is helpful in calculating a fiber's backscatter response to an arbitrary data signal.

$$P_s(t) \sim e^{-2\pi BW_{Bsc} \cdot t} \tag{11.19}$$

$$BW_{Bsc} = \frac{\alpha \cdot c}{\pi \cdot n} \tag{11.20}$$

Solving Equation 11.20 gives 5 kHz and 3 kHz figures for typical fibers at 1310 nm or 1550 nm wavelength, respectively.

An interesting aspect in regard to return-loss calculations is the total backscattered power $P_{s,total}(L)$ of a piece of fiber with length L, if excited by a cw (continuous-wave) signal. This can be readily derived from Equation 11.18 when integrating $P_s(L)$ for small pulsewidths, dW, over the total length of the fiber. With the substituted integration variable z we get

$$P_{s, total}(L) = \int_0^L P s(z) \, dz \tag{11.21}$$

$$= S \cdot \alpha_s \cdot P_0 \int_0^L e^{-2\alpha z} \, dz$$

The result is

$$P_{s, total}(L) = \frac{1}{2 \cdot \alpha} \cdot S \cdot \alpha_s \cdot P_0 \cdot [1 - e^{-2\alpha L}] \tag{11.22}$$

The exponential term $e^{-2\alpha L}$ indicates that very long fibers return a power value that approaches a maximum limit which mainly depends on the capture coefficient S.

Optical fibers are characterized by a backscatter factor σ in dB for a given pulsewidth. It is an indicator for the backscattered power that returns to the OTDR receiver. For rectangular pulses with peak power P_0, the near-end backscatter level is σ dB below peak power.

$$\sigma = -10 \cdot \log (S \cdot \alpha_s \cdot W) \tag{11.23}$$

This model is valid as long as the fiber's attenuation during the geometrical width W is negligible. Typical values for singlemode and multimode fibers are listed in the Table 11.1 below. The backscatter parameter η, defined in equation (11.24) is also included in the table. This parameter is independent of the pulsewidth and is sometimes used instead of σ.

$$\eta = \frac{c}{2 \cdot N} \cdot \alpha_s \cdot S \tag{11.24}$$

Table 11.1 Backscatter parameters for different kinds of optical fibers.

λ [nm]	Fibertype	α_s [km^{-1}]	S	σ [dB/1 μs]	η [W/J]
850	MM-SI 50μ	$3.5 \cdot 10^{-1}$	$1.1 \cdot 10^{-2}$	31	385
1300	MM-GI 62.5μ	$6.5 \cdot 10^{-2}$	$1.0 \cdot 10^{-2}$	38	65
1300	MM-GI 50μ	$6.5 \cdot 10^{-2}$	$5.0 \cdot 10^{-3}$	41	32
1310	SM 9μ	$6.3 \cdot 10^{-2}$	$1.0 \cdot 10^{-3}$	49	6.3
1550	SM 9μ	$3.2 \cdot 10^{-2}$	$1.0 \cdot 10^{-3}$	52	3.2

11.4 MEASURING SPLICE LOSS AND CONNECTOR LOSS

Splices, (macro-) bending, as well as physical and geometrical variations of fibers and connectors cause insertion losses that add to the total loss of fiber links. OTDRs can be used to measure loss from these events. Differences between Rayleigh-backscatter coefficients before and after the event affect insertion-loss accuracy of the OTDR measurement.

11.4.1 Fusion Splice Loss

Local insertion loss can be caused by reflective and by nonreflective events. For both cases, it is basically the same procedure to determine the loss from the step-size of the displayed data. Thus we can focus on an event like that depicted in Figure 11.13. Ideally insertion loss is represented as a sharp step, however, the convolution of the fiber's impulse response with the probing pulse leads to a smoothed transition depending on the pulse width.

As illustrated in Figure 11.13, the splice loss is defined as the vertical distance between the two lines L_1 and L_2 representing the fiber's backscatter signal before and after the event. The width of the step is as a first approximation equal to half of the width W of the probing pulse, but can be broadened by a low receiver bandwidth. It is common practice to apply the LSA (least-square-approximation)[43] method to determine the slope and position of the two auxiliary lines L_1 and L_2. More sophisticated algorithms are based on pattern matching schemes, providing improved accuracy with noisy trace data.[20] If the lines L_1 and L_2 are not quite parallel, the loss depends on the evaluated splice position.[44] In this case, the accuracy of finding the splice location is also important for splice loss measurements. Modeling the splice loss with the receiver bandwidth and pulsewidth as parameters and deriving the position from the turning point is one possibility to determine the splice position. A much simpler approach is to determine the position where the curve dips at the right-sided end of L_1, by checking the deviation of the data samples to line L_1 until a given threshold is exceeded.

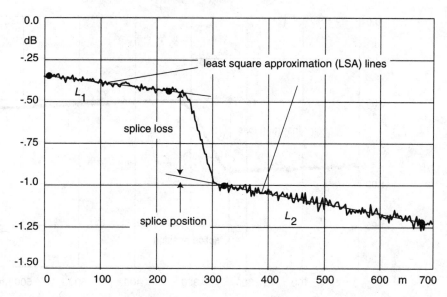

Figure 11.13 Nonreflective event (for example, fusion splice)—two different fibers spliced together.

11.4.2 Different Fibers

Different slopes of L_1 and L_2 occur if fibers with different attenuation coefficients are spliced together. If the backscatter after the loss event is higher than before the event, a "gainer" will occur. Figure 11.14 shows a gainer measurement. This example points out that backscatter information does not always precisely indicate what happens to a forward traveling signal as the light in the fiber certainly doesn't experience the gain. The OTDR calculates loss from differences in backscatter signals and not from the actual power in the fiber. Variations in fiber scatter coefficients α_s and backscatter capture coefficients S (differences in mode-field diameter on either side of a splice) affect the OTDR signal and can lead to ambiguity in the interpretation of data. If α_s and S of both fiber 1 and fiber 2 are known, then the true splice loss α_{splice} can be calculated according to Equation 11.25.[45]

$$\alpha_{\text{splice}} = 5 \cdot \log \frac{P_1}{P_2} - \alpha_2 \cdot \frac{W}{2} - 5 \cdot \log \frac{S_1 \alpha_1}{S_2 \alpha_2} \qquad (11.25)$$

In this equation P_1 and P_2 are the left and right-sided backscatter signal levels, α_2 is the attenuation coefficient of fiber 2, and W the geometrical pulse-width. The term on the right corrects for the different backscatter properties.

Different backscattering behavior of spliced fibers cause OTDR splice-loss measurements to be directional. Hence, in general, splice-loss measurements depend on the direction of the OTDR measurement. Measuring the loss from each end of a link and av-

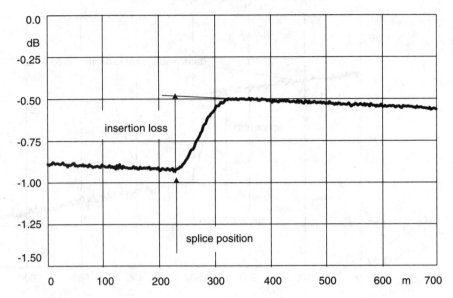

Figure 11.14 A gainer can occur when splicing two different fibers to-gether.

eraging the results will remove the directionality effect.[18] Bidirectional OTDR measure-ments have shown good correlation with measurements made using the cutback tech-nique.[46]

11.4.3 Insertion Loss of Reflective Events

An optical fiber can be perturbed by small changes of the refractive index. The air gap of a tiny crack, a mechanical splice, or connector are examples of such reflective events. Misalignment of connectors, mismatch in core diameter or NA, or nonconcentric fiber cores also induce additional loss. The OTDR display of a measured connector with low reflectance and an insertion loss of about 0.4dB is pictured in Figure 11.15. The slightly tilted roof of the pulse indicates that the shape is the sum of both the reflected rectangular signal and the transition of the backscattered signal (due to the insertion loss).

11.4.4 Bending Loss

Good fusion splices show very low variation of insertion loss with wavelength. Bending loss induced by a curvature however shows a strong dependence on wavelength. As the wavelength is increased, the mode becomes less confined to the higher doped core. The lower confinement can lead to higher radiation loss for bends induced in the fiber. In practice, loss induced by (macro-) bending is seen at $\lambda = 1550$ nm, and especially at $\lambda = 1625$ nm. An important application for the fiber's sensitivity to bending is to create a nonreflective termination. Wrapping a singlemode fiber several times around a mandrel (a

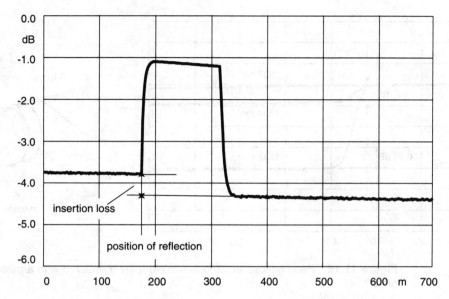

Figure 11.15   Reflective event (for example, a connector) with insertion loss.

small-diameter rod such as a pencil or a ballpoint pen) effectively attenuates the light in the order of 20 to 40 dB.

An example for induced bending-loss at different wavelengths by wrapping two turns of singlemode fiber around a mandrel with a diameter of 23 mm can be seen in Figure 11.16. For better comparison, the individual traces were aligned both horizontally and vertically. It is obvious that longer measurement wavelengths allow for easier detection of small fiber bends which can advantageously be used for security supervision of fiber optic networks.

11.4.5   Uncertainty of Loss Measurements

An important question is the accuracy of loss measurements made with OTDRs. The accuracy depends on the nonlinearity of electronic hardware and primarily on the SNR at a given point. Figure 11.17 shows measurement uncertainty plotted vs. the one-way rms SNR. The vertical axis gives the peak value of the uncertainty in terms of a 95% confidence interval. As an example, backscatter of an OTDR trace with 8 dB SNR shows a noise-ripple with a peak-amplitude that lies within ±0.1dB for 95 out of 100 trace samples.

In practice OTDR loss measurements surpass these results by applying curve fit algorithms to estimate the slope of the backscatter signal and improve the effective SNR shown in Figure 11.17 by several dBs.

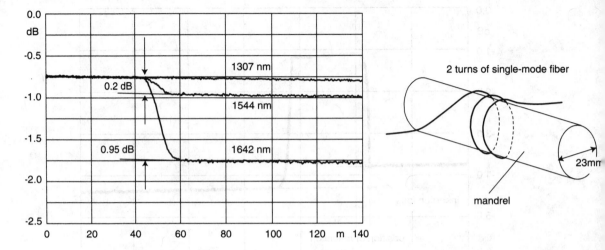

Figure 11.16 Bending-loss at various wavelengths measured with a pulsewidth of 100 ns.

Example

A 0.5 dB splice is located near the far end of a 50 km fiber. Which dynamic range is necessary to measure the event's insertion loss with ±0.05 dB accuracy assuming a fiber attenuation of $\alpha_{dB} = 0.33$ dB/km?

Solution

Figure 11.13 gives the necessary one-way SNR of 9.5 dB for ±0.05 dB measurement accuracy.

The total loss of the 50 km fiber is

$$\alpha_{total} = 0.33\frac{dB}{km} \times 50 \text{ km} = 16.5 \text{ dB}$$

Hence, the necessary OTDR dynamic range D is

$$D = 9.5 \text{ dB} + 16.5 \text{ dB} + 0.5 \text{ dB} = 26.5 \text{ dB}$$

The 0.5 dB insertion loss is added to the dynamic range because the noise that ultimately limits the achievable accuracy shows up after the event.

11.4.6 A Variable Splice-Loss Test Setup

So far, only simple nonbranched fibers were considered with OTDR measurements. An informative and very useful example of a branched fiber network will be discussed now. The ability to reliably produce a splice event with accurately adjustable insertion loss is of highest importance not only for the test of splice-loss uncertainty, but also for the development of measurement-range algorithms. A basic arrangement of a 3 dB coupler, an optical attenuator, and two fibers with different lengths L_1 and L_2, with $L_2 > L_1$, as shown in

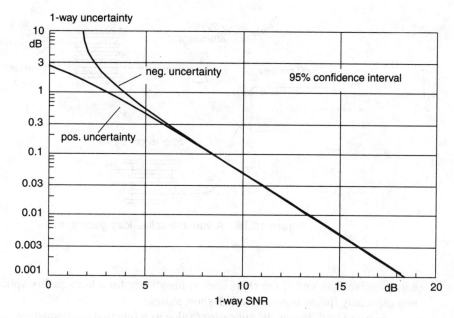

Figure 11.17   Uncertainty of OTDR loss measurements as a function of signal-to-noise ratio.

Figure 11.18, yields a variable step in the backscatter signal at a distance L_1 away from the position of the coupler.

This setup is a typical example where returning light from different paths overlap in the OTDR measurement result. When deriving the behavior of this setup, one should take into account that the OTDR actually measures the superimposed backscattered signals from both fibers. Assuming an attenuator with zero insertion loss and a perfect 50:50 coupling ratio, a 1.5 dB step could be measured both at the coupler itself and at a distance L_1. A simple train of thought reveals this number. At a distance L, with $L_1 < L < L_2$, the backscatter from the longest fiber only contributes to the signal seen by the OTDR. This signal is dampened twice by the 3 dB-coupler loss with the effect of a 3 dB total loss on the OTDR display (when omitting the fiber attenuation). At shorter distances $L < L_1$ the signal seen by the OTDR is twice as high as in the case before, since now both fibers contribute, leading to a one-way step of 1.5 dB at $L = L_1$. In reality, a finite insertion-loss of the attenuator decreases this step considerably. Bellcore recommends a 0.5 dB splice for measurement-range tests, therefore careful selection of attenuator and coupler is necessary to maintain the desired minimum 0.5 dB step for all measurement wavelengths applied. A slight deviation from the ideal 50:50 coupling ratio can be used to compensate partly for the attenuator's insertion loss if the low-loss path is connected to the attenuator. In principle, a more asymmetrical coupler, for example, with 80:20 coupling ratio, ensures even higher values for the achievable maximum splice-loss, however at the cost of an increasing total loss between OTDR and event. Last not least, a carefully terminated

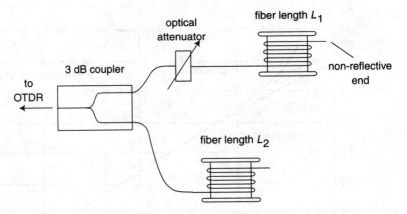

Figure 11.18 A variable splice-loss generator.

nonreflective fiber end at the short fiber is mandatory for a high-quality splice reproduction especially for measurements with short pulses.

Figure 11.19 depicts the splice-loss value as a function of attenuation achieved with a low-insertion loss attenuator and a 3 dB coupler with approximately 44:56% coupling ratio.

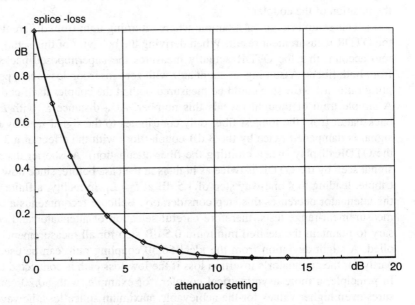

Figure 11.19 Splice loss vs. attenuator setting.

11.5 RETURN LOSS AND REFLECTANCE

At the boundary of two media with different refractive indices, reflection of light occurs. The reflectivity that results from an abrupt change in media can easily be calculated. The power reflection factor, r^2, at normal incidence is

$$r^2 = \frac{P_{refl}}{P_{inc}} = \left(\frac{n_1 - n_2}{n_1 + n_2}\right)^2 \leq 1 \qquad (11.26)$$

At a boundary between glass ($n_1 = 1.5$) and air ($n_2 = 1$), for example, $r^2 = 0.04$, so that 4% of the light is reflected. This is commonly referred to as a Fresnel reflection.

Common sources of reflections are glass-air interfaces at open fiber ends, mechanical splices, cracks, poorly mated connectors because of mud and dirt, and sometimes overpolished connectors. Reflections not only cause light returning back to the source, but also cause loss of optical power.

Reflections occurring in a fiber optic link can lead to performance degradation in high-speed systems.[47] Since any reflection causes a weak image of the transmitted pulse string, these ghost patterns return to the transmitter. In case of multiple reflections these echoes add to the traffic signal as a noise signal affecting bit-error-rate. With steadily increasing digital data rates and complex analog modulation schemes, this subject is going to be of more and more importance. Therefore it is common practice to test a fiber link for reflections at installation before system switch-on.

11.5.1 Return-Loss Measurements

There are two basic methods to measure the presence of reflections: return-loss-testing and reflectance-testing. (See Chapter 10 for further aspects on total return-loss methods.) Optical return loss (ORL) gives the ratio of incident light to total reflected light from an entire fiber optic span.

$$\text{ORL } (dB) = 10 \log \left(\frac{P_{inc}}{P_{refl}}\right) \geq 0 \qquad (11.27)$$

It applies to a series of components including fiber along a link and combines the effects of all reflective events in a transmission system. It is often used as part of an acceptance test. Either an optical-continuous-wave reflectometer (OCWR) or an OTDR with accumulative return-loss measurement capabilities is appropriate. Usually expressed as a positive dB value, ORL describes an optical system as a whole.

11.5.2 Reflectance Measurements

When knowledge of individual reflection magnitude and location is required as with system evaluation, then reflectance measurement is adequate. Reflectance is the term preferred for single components and is the ratio of reflected light at discrete point to light incident on that point. It can only be measured with OTDRs if the event is far away.

$$\text{refl} = 10 \log \left(\frac{P_{refl}}{P_{inc}}\right) \leq 0 \qquad (11.28)$$

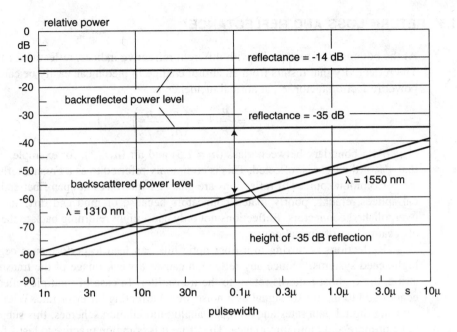

Figure 11.20 Reflected and backscattered power relative to the incident power in fiber.

The fact that the height of a reflection on an OTDR display does not exactly represent its reflectance, but rather is a relative indication for it, dependent on the selected pulsewidth, is depicted in Figure 11.20. There the power levels for both reflected and backscattered light relative to the incident power in the fiber are plotted vs. pulsewidth. It can clearly be seen that the reflected power is a function of reflectance only and independent of pulsewidth. The backscattered power is approximately proportional to the pulsewidth according to Equation 11.18. The proportionality is true only for short pulses where the fiber loss over the spatial pulse length is negligible. Also, backscatter power is wavelength dependent. This means that reflectance measurements with OTDRs is not quite straightforward, but in general the OTDR-firmware does the necessary calculations to correct for the pulsewidth and wavelength dependence.

The double-arrow gives the difference between the reflected power of a −35 dB reflection and the backscatter power measured with a pulsewidth $\tau = 100$ ns at $\lambda = 1310$ nm. As this graph is based on a 10 · log scale, an OTDR screen with its 5 · log scaling (see Section 11.2.1) would reveal such a reflection as a spike only half as high, in other words, approximately 12 dB. Another fact catches the eye when considering wide pulsewidths. There the backscattered power can reach or even exceed the power level caused by average Fresnel reflections, making it hard to discern the signature of standard connectors with about −40 dB reflectance.

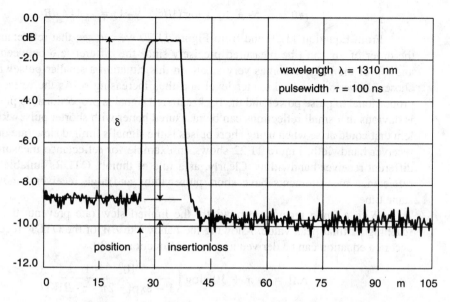

Figure 11.21   Example of reflectance measurement.

An example of a reflectance measurement of an isolated event can be seen in Figure 11.21. This plot shows a reflection with a noisy backscatter and auxiliary LSA (least-square-approximation) lines for backscatter level determination.

As an OTDR cannot directly measure incident power, this information is derived indirectly with the known (or estimated) backscatter factor σ. The relation between the measured pulse-height δ and the reflectance of an event is given in Equation 11.29 which can readily be derived from Equations 11.16, 11.23, and 11.28.

$$\text{refl } (dB) = -\sigma + 10 \cdot \log (10^{0.2 \cdot \delta} - 1) \qquad (11.29)$$

σ is the backscatter factor for the pulsewidth used. The exponent 0.2 comes from the OTDR's one-way data representation.

Example 11.4

Calculate the reflectance of the event shown in Figure 11.21 which has been measured at a pulsewidth $\tau = 100$ ns and a wavelength of $\lambda = 1310$ nm.

Solution

Initially for using Equation 11.29 we need to determine the appropriate backscatter factor σ for the given pulsewidth. This may be derived from Equation 11.23 or Table 11.1.

$$\sigma = 49 \frac{\text{dB}}{\mu s} + 10 \times \log\left(\frac{1 \ \mu s}{100 \ \text{ns}}\right) = 49 + 10 \ \text{dB} = 59 \ \text{dB}$$

The reflection height δ can directly be obtained from the drawing, yielding $\delta = 7.4$ dB. Hence, the reflectance is

$$refl = -59 + 10 \times \log(10^{0.2 \times 7.4} - 1) = -44.4 \, dB$$

From Equation 11.29 and from Figure 11.20 we can see that reflectance values in the order of $-\sigma$ can't be measured precisely since the difference δ between backscatter and top of the pulse becomes very small. In this situation a smaller pulsewidth must be chosen, lowering the backscatter level and thus increasing δ. As the reflected signal is proportional to pulse power and the backscattered signal is proportional to pulse energy, it is obvious that small reflections can be measured better with shorter pulsewidths. A problem that could arise when using short pulses is the signal's limited slew rate due to a finite receiver bandwidth. Figure 11.22 shows an example for reflectance measurements with different receiver bandwidths. Clearly, as a rule of thumb, OTDRs suitable for accurate reflectance measurements have short pulsewidths and high receiver bandwidth at the same time.

With a finite receiver bandwidth, the limited slew rate prevents the signal from reaching its final peak value. Knowing the bandwidth BW of the OTDR receiver, a more accurate equation can be derived as given in Equation 11.30.

$$refl = -\sigma + 10 \cdot \log\left(\frac{10^{0.2 \cdot \delta} - 1}{1 - \exp(-2\pi \cdot \tau \cdot BW)}\right) \tag{11.30}$$

This correction is not quite perfect since a simple first-order lowpass model was chosen for the approximation of the receiver transfer function. However, in regard to reflectance accuracy in OTDRs, this is a sufficiently precise formulation.

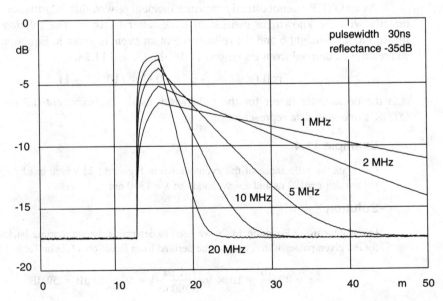

Figure 11.22 Influence of OTDR receiver bandwidth on reflection height ($\tau = 30$ ns, refl . = −35 dB).

If one happens to have a device with known reflectance, it is possible to determine the backscatter factor σ of a fiber by simply measuring the pulse-height δ according to Figure 11.21 and solving Equation 11.29 for σ.

Example

An FC/PC connector pair gives a reading of 33.5 dB total return loss when measured with a cw reflectometer setup. The same connector pair shows a reflectance of −36 dB with an uncompensated OTDR measurement according to Equation 11.29 using a pulsewidth of 30 ns. Calculate the OTDR receiver bandwidth BW.

Solution

Subtracting Equation 11.29 and Equation 11.30 gives the error ϵ caused by a finite receiver bandwidth

$$\epsilon = 10 \cdot \log\left[1 - \exp(-2\pi \cdot \tau \cdot BW)\right] = 33.5 \text{ dB} - 36 \text{ dB} = -2.5 \text{ dB}$$

Solving for BW yields

$$BW = -\frac{1}{2\pi \cdot \tau} \cdot \ln(1 - 10^{0.1\epsilon})$$

$$= -\frac{1}{2\pi \times 30 \cdot 10^{-9}} \times \ln(1 - 10^{-0.1 \times 2.5}) = 4.38 \text{ MHz}$$

11.5.3 Accumulative Return Loss

Some OTDRs can calculate the accumulative return loss vs. distance from a backscatter measurement. Omitting the backscatter information and displaying just the return loss of the fiber plus reflectances gives a clear and crisp overview of a fiber link. The total return loss of a link according to Equation 11.26 is given as the maximum value on the curve in Figure 11.23. This example of a 5 km fiber link measured at $\lambda = 1310$ nm demonstrates how the return loss increases monotonous with distance, showing distinct steps at positions of reflections. These are a −45 dB front-reflection, and two reflections with −40 dB and −36 dB reflectance at a distance of 2.0 km and 3.5 km, respectively. The total return loss (33dB) and the quality of reflective network components can be checked at a glance. For example, reliable OC_48 (2.5 Gb/s) data transmission requires a system return-loss of no more than 24 decibels with no individual reflectance exceeding −27 dB. A return-loss chart exactly reveals the information to check these two limits in a quick acceptance test.

From the total backscattered power $P_{s,\text{total}}(L)$ in Equation (11.22), the total return loss of a pure fiber with length L can be calculated according to Equation 11.27.

$$ORL(L) = -10 \log\left[S \frac{\alpha_s}{2\alpha} (1 - e^{-2\alpha L})\right] \tag{11.31}$$

For return loss measurements it is a valid approximation to equate α and α_s. On this basis the return loss can be simplified to

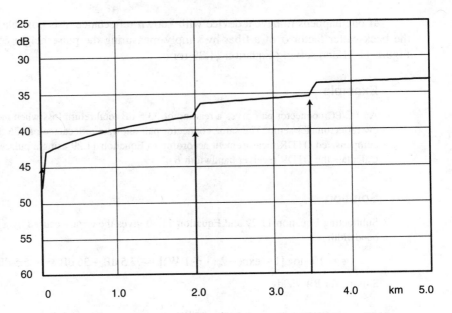

Figure 11.23   OTDR return-loss chart for a 5 km fiber at λ = 1310 nm.

$$\text{ORL}(L) = -10 \log \left[\frac{S}{2} (1 - e^{-2\alpha L}) \right] \tag{11.32}$$

This formula is particularly interesting because it suggests that S can be determined from Figure 11.23 if α is known. Or even both S and α can be determined either from measuring the return losses of different lengths of the same fiber, or from two different points on an accumulative return-loss chart.

A detailed report on return loss and reflectance measurements with OTDRs and with cw excitation can be found in Kapron and co-workers.[48] In Blanchard and co-workers,[49] further information about the calibration of OTDR return loss is given.

Example

Determine the total return loss for spans of singlemode fiber with lengths of 1 m, 1 km, 10 km, and 100 km and an attenuation coefficient of 0.19 dB/km, in other words, at a wavelength around λ = 1550 nm.

Solution

Using Equation 11.32, the total return loss as a function of fiber length L (in km) is

$$\text{ORL}(L) = -10 \times \log \left[\frac{0.001}{2} \times (1 - e^{-2 \times (0.19 \times 0.23) \times L}) \right]$$

Hence,   ORL($1\ m$) = 73.6 dB
ORL($1\ km$) = 43.8 dB
ORL($10\ km$) = 35.4 dB
ORL($100\ km$) = 33.0 dB

With increasing length L the total return loss approaches a limit which depends mainly on the fiber's capture coefficient S.

11.6   AUTOMATED REMOTE FIBER TESTING

OTDRs are currently used in installation and maintenance of optical fiber networks. To ensure the integrity of existing links, they are checked by a dispatched crew equipped with an OTDR. A typical fiber break takes more than 10 h to detect, locate, and repair.[50] With an increasing number of deployed fiber-optic cables, both practical and economical limitations arise with this kind of preventive maintenance. Since high transmission rates tie tens of thousands of subscribers to one fiber, network downtime is expensive in terms of lost revenue, customer satisfaction, reputation, and future business. Minimizing revenue loss from system downtime by quickly detecting, locating and then repairing breaks in operational cables, demands automated testing of the fiber plant. The significance of this issue is addressed by the Bellcore Generic Requirements for RFTS (Remote Fiber Test System).[51]

11.6.1   Link Loss Comparison

The typical test equipment for an RFTS consists of a remote test unit (RTU) together with its control unit, located at a central office. An optical test access unit (OTAU) (basically an optical switch) accomplishes direct access to the fiber optic network to share one test equipment with numerous fibers. The RFTS system is controlled by a computer with an operations system that encompasses databases and computing algorithms.

As the complexity of fiber optic network increases, maintenance cost becomes a significant issue. During network commissioning and installation, a reference file for each individual fiber of a cable is created and saved in the data base. A baseline OTDR measurement is made of the fiber link. As the test equipment remains permanently connected to the fibers, periodic testing with automatic analysis and trace comparison is performed as part of a proactive and preventive maintenance strategy.

A good indication for the onset of link degradation is a deviation in the end-to-end attenuation or total link loss. Link loss is defined as the difference between the initial backscatter level at the near end and the backscatter level at the far end. Calculating the link loss from an OTDR trace and comparing it to a reference value (recorded during commissioning of the fiber plant) as depicted in Figure 11.24 gives a first indication for a variation in the link characteristics.

If the deviation exceeds a given threshold level, the system responds with a detailed analysis providing a problem description, exact location, the nearest access point, the technical staff to be called and the task force's phone numbers. The repair process can start without delay.

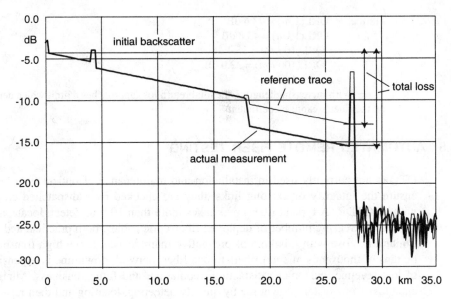

Figure 11.24 Comparison of the total link loss between reference trace and actual measurement.

11.6.2 Dark Fiber Testing

Many installed fiber cables contain spare fibers. These inactive fibers can be used for dark fiber testing. This allows OTDR measurements to be performed while traffic is carried on neighboring active fibers. Dark fiber testing has the advantage that no link modification is necessary and that the test wavelength can be identical to the transmission wavelength. A simple example of how this works is shown in Figure 11.25.

Justification for this practice is based on the assumption that if a critical failure occurs, most likely all fibers in the cable are affected by the breakdown. The percentage of detected failures with dark fiber testing is about 80%.[52]

11.6.3 Active Fiber Testing

If reliable fault coverage is mandatory, in-service or active fiber test is the measurement of choice. In contrast to dark fiber testing, this method uses the same fiber for test signal and data signal transmission.[53,54] WDM techniques are used to preserve the integrity of both live traffic and test signals. It is common practice to choose the test wavelength longer than the traffic wavelength. This ensures detection of all critical faults because longer wavelengths are more sensitive to physical abuse in the cable. With 1550 nm traffic, 1625 nm and 1650 nm have been established as test wavelengths. On 1310 nm data transmission links a 1550 nm test signal is often used. Commonly, measurement results at different wavelengths are not quite comparable, but especially at 1550 nm and 1625 nm, numerous fiber loss and splice loss measurements show similar loss results at both wave-

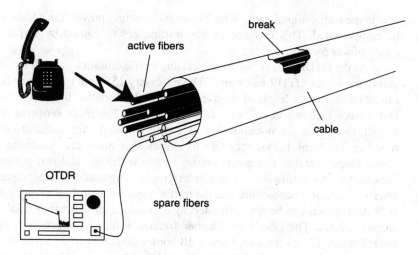

Figure 11.25   Dark fiber testing.

lengths.[55] Figure 11.26 depicts the way a 1625 nm OTDR can be connected with low loss via a WDM to a 1550 nm data link. There are two directions that the OTDR signal can be injected, toward the transmitter or toward the receiver. This example shows transmitter-side coupling. The WDM provides a low-loss test-signal path and isolates the transmitter from the OTDR test signal because of its directivity. Both the transmitter's signal and the OTDR test pulse together propagate towards the receiver. To ensure nonintrusive fiber testing, the OTDR test signal is removed from the traffic signal prior to reaching the receiver. This is accomplished by inserting a 1550 nm bandpass (filter F2) at the far end before the receiver, acting as a blocking filter at 1625 nm. The bandpass filter adds extra

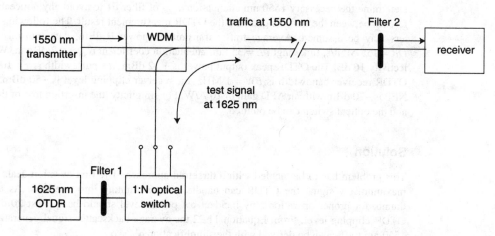

Figure 11.26   Out-of-band active fiber measurements at $\lambda = 1625$ nm.

loss to the traffic signal path. With higher transmitting power, the additional link loss can be compensated. This increase in transmitting power caused by active fiber testing is called power penalty.

At the OTDR side, filter F1 rejects any traffic crosstalk at 1550 nm, which could severely impair the OTDR test signal. The necessary 1550 nm isolation of filter F1 is determined by the power levels of the average traffic backscatter, the shot noise induced by it, and the residual rms backscatter at $\lambda = 1550$ nm. The more sensitive an OTDR is designed, the higher the susceptibility to interfering light. Any incident cw light superimposed on the weak backscatter signals causes shot noise that limits the achievable dynamic range. Another undesirable noise source is the residual rms portion of the traffic backscatter. An estimation of this interference magnitude can be done by taking the fiber's backscatter bandwidth, and the traffic's spectral distribution into account. The traffic's data stream can be approximated by a pseudorandom binary signal with a sin(x)/x-shaped spectra. The fiber's backscatter frequency transfer function is basically a first-order lowpass filter with a very low 3 dB corner frequency in the order of 3 kHz at a 1550 nm wavelength as is derived in Equation 11.20. A simple equation to estimate the suppression γ of the (random) traffic signal with data transfer rate DR when filtered through a first-order lowpass with bandwidth BW is given in Equation 11.33.

$$\gamma = 10 \cdot \log\left(2\,\frac{BW}{DR}\right) \tag{11.33}$$

The results are accurate within a couple of dBs if $DR > 10 \cdot BW$. With an improved out-of-band rejection of filter F1, the specified OTDR dynamic range could easily be maintained. Attention must be paid to any reflectivity that occurs on the fiber link. The interfering traffic noise level can increase drastically from these discrete reflections.

Example

Determine the necessary 1550 nm attenuation a_{1550} of filter F1 to avoid any noticeable interference between the traffic signal and the OTDR measurement result. The following parameters may be assumed: Average traffic transmitting power 0 dBm, data transmission rate $DR = 155$ Mbit/s, link length $L = 35$ km, attenuation coefficient $\alpha_{dB} = 0.2$ dB/km, WDM selectivity 16 dB. The OTDR peak output power is + 12 dBm, the pulsewidth is $\tau = 100$ ns, the OTDR receiver bandwidth is $BW = 1$ MHz, the receiver clipping level is -50 dBm, and the NEP is -70 dBm with an APD gain of 10 A/W. For simplicity, the insertion loss of the WDM and the optical switch can be omitted.

Solution

This problem has to be tackled with a threefold approach. The first question deals with the maximum cw signal the OTDR can handle without running into overload. As a rule of thumb, for proper operation, any incident cw power level should be at least 20 dB below OTDR clipping level. From Equation 11.22 the average backscatter signal generated by the 1550 nm traffic can be derived with the simplification $\alpha_{dB} = \alpha_s$:

$$P_{s,cw} = \frac{0.001}{2} \times 1 \text{ mW} \times [1 - e^{-2 \times (0.2 \times 0.23) \times 35}] = 480 \text{ nW} \cong -33 \text{ dBm}$$

The WDM's selectivity attenuates this signal typically by 16 dB, yielding

$$P_{s,cw} = -33 - 16 \; dBm = -49 \text{ dBm}$$

The requirement for the cw power level at the OTDR input to be lower than −70 dBm leads to the filter attenuation value:

$$a_{1550} = -49 - 70 \; dB = 21 \text{ dB}$$

The second issue is related to the shot-noise which is generated by the −70 dBm (= 0.1 nW) cw signal. The corresponding photocurrent i_{ph} in the APD is

$$i_{ph} = 0.1 \; nW \times 10 \frac{A}{W} = 1.0 \; nA$$

which leads to a rms shot noise of

$$\bar{i}_s = \sqrt{2 \times 1.6 \cdot 10^{-19} \times 1.5 \cdot 10^6 \times 1 \cdot 10^{-9}} = 21.9 \; pA$$

assuming an equivalent noise bandwidth of 1.5 times the 3 dB receiver bandwidth. This is equivalent to an optical input power of

$$P_{equiv} = 10 \times \log \left(\frac{21.9 \cdot 10^{-12} \times 10^3}{10} \right) = -86.6 \text{ dBm}$$

For the sake of simplicity, the calculation of the shot noise neglects the APD's excessive noise behavior which actually leads to a somewhat higher noise level. However, even with a 5 dB penalty increase the resulting shot noise is sufficiently below the receiver's NEP. Obviously the additional shot noise does not interfere noticeably with the measurement.

Finally the direct traffic crosstalk must be estimated. From Equation 11.33 we get the suppression γ of the (random) traffic signal when filtered through a lowpass

$$\gamma = 10 \times \log \left(2 \times \frac{3 \cdot 10^3}{155 \cdot 10^6} \right) = -44 \text{ dB}$$

Hence the rms noise level P_x due to traffic crosstalk is

$$P_x = 0 - 44 - 33 - 16 - 11 \text{ dBm} = -104 \text{ dBm}$$

This number takes the additional attenuation through backscattering, WDM, and filter F1 into account and is far below the receiver noise level. In the end, the most dominant factor for the filter F1 specification a_{1550} is the resulting cw power level at the OTDR input.

11.7 OUTLOOK

Though OTDR development has evolved over nearly two decades, the emerging changes in optical transmission technology and network topology, constantly give rise to new challenges. OTDRs primarily are suitable for ordinary point-to-point links. Point-to-multi-point links will soon gain importance. In addition to the ongoing and expected improvements in the standard performance parameters like dynamic- and measurement range, resolution, and deadzones, there remain at least three growing areas which need to be addressed with dedicated OTDRs and appropriate measurement methods. These are

1. Links equipped with optical fiber amplifiers;
2. Network structures with multiple branches, for example PONs (passive optical network), subscriber loop links, or LANs;
3. WDM-equipped multiple wavelengths links.

Convenient solutions will most likely be found in a combination of appropriately designed networks plus adapted test equipment, rather than in a single OTDR alone.

Another interesting area for derivatives of standard OTDRs develops in distributed fiber sensor applications which is, however, beyond the scope of this book. Not only temperature and strain, but even magnetic field and electrical current have been measured by use of Brillouin optical time-domain reflectometer (BOTDR) and polarization optical time-domain reflectometer (POTDR), respectively.

REFERENCES

1. CCITT Comité Consultatif International Télégraphique et Téléphonique-Recommendation G.652 "Characteristics of a Single-Mode Optical Fibre Cable".
2. TIA/EIA, Electronic Industries Association, Washington, DC, Fiber Optic Test Procedure FOTP-78 for singlemode fiber.
3. Barnoski, M.K. and S.M. Jensen. 1976. Fiber waveguides: a novel technique for investigating attenuation characteristics. *Appl. Optics* 15: 2112–2115.
4. Personik, S.D. 1977. Photon probe—an optical time domain reflectometer. *Bell Syst. Techn. J.* 56: 355–366.
5. Costa, B. and B. Sordo. 1977. Backscattering technique for investigating attenuation characteristics of optical fibers: a new experimental approach. *CSELT Rapporti Tecnici.* V.
6. Barnoski, M.K. et al. 1977. Optical time domain reflectometer. *Appl. Optics* 16: 2375–2379.
7. Zoboli, M. and Bassi, P., 1983. High spatial resolution OTDR attenuation measurements by a correlation technique. *Appl. Optics,* 22 (23), 3680–3681.
8. Newton, S.A. 1987. A new technique in optical time domain reflectometry. RF & Microwave Symposium and Exhibition. Hewlett-Packard.
9. Nazarathy, M. et al. 1989. Real-time long range complementary correlation optical time domain reflectometer. *J. Lightwave Tech.* 7 (1): 24–37.
10. Dolfi, D.W., and M. Nazarathy. 1989. Optical frequency domain reflectometer with high sensitivity and resolution using optical synchronous detection with coded modulators. *Electronics Lett.* 25 (2), 160–161.
11. Tsuji, K., K. Shimizu, T. Horiguchi, and Y. Koyamada. 1995. Coherent optical frequency domain reflectometry for a long singlemode optical fiber using a coherent lightwave source and an external phase modulator. *IEEE Photon. Technol. Lett.* 7.

12. Eickoff, W., R. Ulrich. 1981. Optical frequency domain reflectometry in singlemode fibre. *Appl. Phys. Lett.* 39, 693–695.

13. Bethea, C.G., B.F. Levine, L. Marchut, V.D. Mattera, and L.J. Peticolas. 1986. Photon counting optical time domain reflectometer using a planar InGaAsP avalanche detector. *Electron. Letters.* 22(6): 302–303.

14. Lacaita, A.L., P.A. Francese, and S. D. Cova. 1993. Single-photon optical-time-domain reflectometry at 1.3 μm with 5 cm resolution and high sensitivity. *Opt. Lett.* 18,

15. King, J.P., D.F. Smith, K. Richards, P. Timson, R.E. Epworth, S. Wright. 1987. Development of a Coherent OTDR Instrument. *Journ. Lightwave Techn.* LT-5(4), 616–623.

16. Horiuchi, Y. et al. 1990. Novel coherent heterodyne optical time domain reflectometry for fault localization of optical amplifier submarine cable systems. *Photon Technol. Lett.* 2(4):

17. Beller, J. 1993. A high performance digital signal processing system for the HP8146A OTDR. *Hewlett-Packard J.,* 63–68.

18. DiVita, P., and U. Rossi. 1980. The backscattering technique: its field of applicability in fibre diagnostic and attenuation measurements. *Opt. Quant. Elec.* 11(17): 17–22.

19. Generic Requirements for Optical Time Domain Reflectometer (OTDR) Type Equipment. Bellcore (Bell Communications Research) GR-196-CORE, Issue 1, 1995.

20. Anderson, D., and D. Judge. 1994. A pattern matching algorithm for remote systems that measures the distance and loss of fusion splices with high precision, NFOEC '94, San Diego, 139–154.

21. Caviglia, F. and P.G. Ricaldone. 1995. *OTDR performance specifications.* 3rd Optical Fiber Measurement Conference, OFMC '95, Liège 1995, paper VIII.1.

22. Healy, P. 1986. Instrumentation principles for optical time domain reflectometry. *J. Phys. E: Sci. Instrum.* 19, 334–341.

23. Gold, M.P. 1985. Design of long-range singlemode OTDR. *J. Lightwave Tech.* 3, 39–46.

24. Rihaczek, A.W. 1967. *Principles of high resolution radar.* NY: McGraw Hill Book Company.

25. Keiser, G. 1985. *Optical fiber communications.* NY:McGraw-Hill.

26. Stone, J., and C.A. Burrus. 1980. Reduction of the 1.38μm water peak in optical fibers by deuterium-hydrogen exchange. *Bell Syst. Techn. J.* 59.

27. Noguchi, N., N. Shibata, N. Uesugi, and Y. Negishi. 1985. Loss increase in optical fibers exposed to hydrogen atmosphere. *J. Lightwave Techn.* LT-3, 236–243.

28. Kerker, M. 1969. *The scattering of light and other electromagnetic radiation.* Academic Press.

29. Born, M., and E. Wolf. 1964. *Principles of optics.* Oxford: Pergamon Press.

30. Izawa, T. and S. Sudo. 1987. Optical fibers: material and fabrication, KTK Scient. Publ., Tokyo, p. 35.

31. Cotter, D. 1983. Stimulated Brillouin scattering in monomode optical fiber. *J. Opt. Comm.* 4(1), 10–19.

32. Horiguchi, T., K. Shimizu, T. Kurashima, M. Tateda, and Y. Koyamada. 1995. Development of a Distributed Sensing Technique Using Brillouin Scattering. *J. Lightwave Techn.,* 13(7), 1296–1302.

33. Dakin, J.P., D.J. Pratt, G.W. Bibby, and J.N. Ross. 1985. Distributed optical fiber Raman temperature sensor using a semiconductor light source and detector. *Electron. Lett.* 21(13), 569–570.

34. Koyamada, Y., H. Nakamoto, and N. Ohta. 1992. High performance coherent OTDR enhanced with erbium doped amplifiers. *Journ. Opt. Commun.* 13, 127–133.

35. Izumita, H., Y. Koyamada, S. Furakawa, and I. Sankawa. 1994. The performance limit of coherent OTDR enhanced with optical amplifiers due to optical nonlinear phenomena. *IEEE J. Lightwave Technol.* 12, p. 1230.

36. Stolen, R.H. 1980. Nonlinearity in fiber transmission. *Proc. IEEE* 68(10), 1232–1236.

37. Li, T. 1980. Structures, parameters, and transmission properties of optical fibers. *Proc. IEEE* 68(10), 1175–1180.

38. Neumann, E.G. 1980. Analysis of the backscattering method for testing optical fiber cables. AEÜ, Band 34, Heft 4, 157–160.

39. Brinkmeyer, E. 1980. Backscattering in singlemode fibers. Electronics Lett. 16(9), 329–330.

40. Nakazawa, M. 1983. Rayleigh backscattering theory for singlemode optical fibers. *Journ. Opt. Soc. Am.* 73.

41. Hentschel, C. 1989. *Fiber optic handbook.* Hewlett-Packard.

42. Danielson, D.L. 1981. *Backscatter measurements on optical fibers.* NBD Technical Note 1034, U.S. Department of Commerce.

43. Meyer, P.L. 1970. *Introductory probability and statistical applications.* 2nd ed. Addison-Wesley Publishing Company, Inc.

44. Hentschel, C. and J. Beller. 1995. *Taking the Guesswork out of OTDR Measurements.* Metrology for the Americas, Miami, November 1995.

45. Schickedanz, D. 1980. Theorie der Rückstreumessung bei Glasfasern. Siemens Forschungs-und Entwicklungsberichte, Springer-Verlag, Bd.9, 1980, Nr. 4, 242–248.

46. Costa, B. 1979. *Comparison between various fibre characterization techniques.* Opt. Commun. Conf. Proc., Amsterdam, II-1 - II-5.

47. Sasaki, S., H. Nakano, and M. Maeda. 1986. *Bit-error-rate characteristics with optical feedback in 1.5 μm DFB semiconductor laser.* Proc. Euro. Conf. Opt. Commun. ECOC '86, Barcelona, 1986.

48. Kapron, F.P., B.P. Adams, E.A. Thomas, and J.W. Peters. 1989. Fiber-optic reflection measurements using OCWR and OTDR techniques. *J. Lightwave Techn.* 7(9), 1234–1241.

49. Blanchard, P., P.H. Zongo, and P. Facq. 1990. Accurate reflectance and optical backscatter parameter measurement using an OTDR. *Electr. Lett.* 26(25).

50. Hou, V.T. 1991. Update on interim results of fiber optic system filed failure analysis. NFOEC '91, 539–548.

51. Bellcore: Generic Requirements for Remote Fiber Testing Systems (RFTS). TA-NWT-001295, Issue 2, 1993.

52. Davé, R. 1994. Next Generation FITL Remote Fiber Test Systems. NFOEC '94, 357–371.

53. Lee, N., D. Suarez. 1993. Active fiber monitoring using remote OTDR technology. NFOEC '93, 369–376.

54. Davé, R. 1993. Comparison of automated fiber maintenance technologies. NFOEC '93, 385–403.

55. Furukawa, S., Y. Koyamada, T. Horiguchi, and I. Sankawa. 1994. 1.65 μm optical surveillance and test system for subscriber lines and ultra-long span trunk. OFC '94, San Jose.

12

Dispersion Measurements

Paul Hernday

12.1 INTRODUCTION

All forms of dispersion degrade the modulation-phase relationships of lightwave signals, reducing information-carrying capacity through pulse-broadening in digital networks and distortion in analog systems. Designers of optical fiber transmission systems must cope with three basic forms of fiber dispersion, illustrated schematically in Figure 12.1. Intermodal dispersion, which limits data rates in systems using multimode fiber, results from the splitting of the signal into multiple modes that travel slightly different distances. Chromatic dispersion, which occurs in both singlemode and multimode fiber, results from a variation in propagation delay with wavelength that is shaped, in turn, by the interplay of fiber materials and dimensions. Polarization-mode dispersion, caused by the splitting of a polarized signal into orthogonal polarization modes with different speeds of propagation, becomes a limiting factor in singlemode fiber when chromatic dispersion is sufficiently reduced. This chapter deals with the dispersion characteristics of the physical layer of a telecommunications system—the fiber, components, and installed transmission paths—providing a brief introduction to the causes of dispersion and a sampling of measurement methods. Measurement approaches to these critical attributes take many forms; some of the methods described here are widely used commercial methods, and others are presented simply because they provide good insight into the dispersive phenomena. The detail with which the methods are discussed reflects, to some extent, the author's experience with the methods. Measurement methods for many fiber, component, and system attributes are developed by standards groups such as Telecommunications Industry Association and its international counterparts. Guidance in the measurement of dispersive effects is provided by formal fiber optic test procedures (FOTPs), referenced herein.

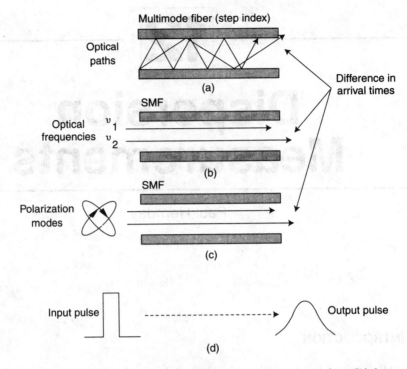

Figure 12.1 Schematic representations of fiber dispersion. (a) Intermodal dispersion. (b) Chromatic dispersion. (c) Polarization-mode dispersion. (d) Typical effect of dispersion on a transmitted pulse.

12.2 MEASUREMENT OF INTERMODAL DISPERSION

12.2.1 Introduction

Singlemode fiber supports the propagation of a single guided wave, or mode. In contrast, multimode fiber supports hundreds of modes, and this characteristic produces the phenomenon called intermodal dispersion. Because the modes travel slightly different distances, as indicated in Figure 12.1a, the modulation envelopes of light in the various modes become increasingly out of phase as the signal travels along the fiber. In the case of a digital system, each mode delivers a pulse, but the distribution of arrival times caused by the different propagation times causes the transmitted pulse to broaden in time and to drop in peak amplitude. Intermodal dispersion typically limits multimode fiber to transmission distances of less than 1 km and bit rates less than 1 Gb/s. For measurement purposes, intermodal dispersion is typically expressed as the lowest modulation frequency at which the baseband AM response of the fiber has rolled off by 3 dB. The following section discuss two common methods for measuring multimode fiber bandwidth.

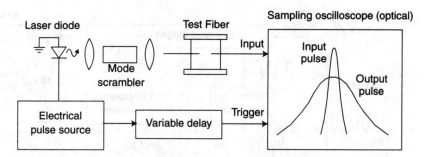

Figure 12.2 Measurement of multimode fiber bandwidth by the pulse distortion method.

12.2.2 The Pulse Distortion Method

The most common method for measuring multimode fiber bandwidth, illustrated in Figure 12.2, is based on measurement of the impulse response.[1] A pulsed laser source is coupled through a mode scrambler to the input of the test fiber. The source spectrum must be narrow enough that the results are not significantly influenced by chromatic dispersion. The output of the fiber is measured by a sampling oscilloscope with built-in optical receiver. A horizontal trigger is supplied through a variable delay to offset the difference in delay between a short reference fiber and the fiber to be tested. The reference fiber is typically a few meters in length. In operation, a fast pulse is launched into the test fiber and the output pulse is digitized, including leading and trailing edges down to 1% of the peak amplitude. Next, the input pulse is measured in the same way, using a short reference path in place of the test fiber. The reference fiber can be a short length cut from the input of the test fiber, or a short length of fiber having similar optical characteristics. The transfer function of the test fiber is given by

$$H(f) = \frac{B(f)}{A(f)} \tag{12.1}$$

where $B(f)$ and $A(f)$ are the Fourier transforms of the output and input pulses, respectively. Fiber bandwidth is defined as the lowest frequency for which $|H(f)| = 0.5$.

Multimode fiber bandwidth measurements are sensitive to optical launch conditions and the deployment of the test sample. For stable, repeatable measurements, a mode-scrambling device should be inserted ahead of the test device to assure excitation of a large number of modes. In addition, cladding light should be removed. In some cases, the fiber coating dissipates cladding modes. Alternatively, cladding mode strippers can be used at both ends of the test fiber.

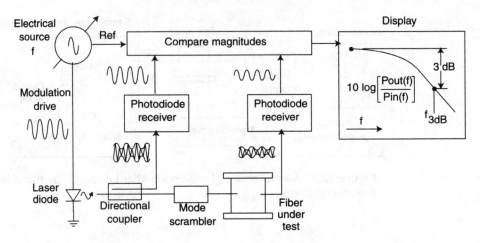

Figure 12.3 Measurement of multimode fiber bandwidth by the frequency domain method.

12.2.3   The Frequency Domain Method

An alternative method for the measurement of multimode fiber bandwidth is the baseband AM response method[2] shown in Figure 12.3. A narrowband cw optical carrier is sinusoidally intensity modulated by a swept-frequency RF or microwave signal source and coupled through a mode scrambler to the test fiber. Cladding mode strippers may also be needed. The modulation is detected by a photodiode receiver. The instrumentation records the received optical power as a function of modulation frequency. In operation, the test fiber is generally measured first, resulting in the measurement $P_{out}(f.)$ The input signal is then determined by measuring again with a short reference fiber. As in the pulse method, the reference fiber can be a short length cut from the input of the test fiber, or a short length of fiber having similar optical characteristics. The reference measurement yields the function $P_{in}(f)$. The frequency response in optical power is given by

$$H(f) = \log_{10}\left[\frac{P_{out}(f)}{P_{in}(f)}\right].$$

(12.2)

Fiber bandwidth is defined as the lowest frequency at which $H(f)$ has decreased by 3 dB from its zero-frequency value.

For convenience, source and receiver functions may be performed using a lightwave component analyzer or a network analyzer with appropriate transducers. In either case, sweeping of the modulation frequency, subtraction of the calibration values and display of the resulting transmission response are performed automatically. When an electrical network analyzer is used with external laser source and photodiode receiver, care must be taken to properly interpret the frequency response. Because an electrical network analyzer measures RF power and the photodetector produces a current that is proportional to optical power, the network analyzer will indicate a 6 dB change for a 3 dB change in optical level.

12.3   MEASUREMENT OF CHROMATIC DISPERSION

12.3.1   Introduction

Chromatic dispersion is simply a variation in the speed of propagation of a lightwave signal with wavelength. The optical source in a high-speed communication system is typically a single-line diode laser with nonzero spectral width. Pulse modulation increases the spectral width. Each wavelength component of the signal travels at a slightly different speed, resulting in the pulse broadening illustrated in Figure 12.1d. This section provides an introduction to chromatic dispersion, the primary dispersive mechanism in singlemode fiber, and describes several chromatic dispersion measurement methods appropriate to singlemode fiber. Chromatic dispersion is also an important parameter of multimode fiber.[3,4]

12.3.2   Causes of Chromatic Dispersion

In singlemode fiber, chromatic dispersion results from the interplay of two underlying effects—material dispersion and waveguide dispersion.[5] Material dispersion results from the nonlinear dependence upon wavelength of the refractive index, and the corresponding group velocity, of doped silica. Waveguide dispersion is rooted in the wavelength-dependent relationships of the group velocity to the core diameter and the difference in index between the core and the cladding. A third component, called second-order PMD or differential group delay dispersion, arises from the details of fiber PMD and produces an effect which is identical to chromatic dispersion.[5-7] Second-order PMD sets the ultimate limit to which a transmission path can be compensated for chromatic dispersion.

   The information-carrying capacity of a fiber optic system is highest when the group delay is flat with wavelength. In dispersion-unshifted singlemode fiber, the zero-dispersion wavelength λ_0 is near 1300 nm, where the two main underlying mechanisms, material dispersion and waveguide dispersion, naturally cancel one another. Control of the refractive index profile can place λ_0 anywhere in the 1300/1550 nm wavelength range. In dispersion shifted fiber, λ_0 is set to approximately 1550 nm.

12.3.3   Definitions and Relationships

The discussion will make use of several specialized terms, defined in Table 12.1

   An example of the chromatic dispersion of a disperson-shifted singlemode fiber in the region of 1550 nm is shown in Figure 12.4. The upper curve shows the typical wavelength dependence of the relative group delay. The chromatic dispersion coefficient D, shown in the lower curve, is the attribute used to estimate pulse broadening. Its value at a particular wavelength is determined by differentiating the relative group delay curve with respect to wavelength and dividing by the length of the transmission path

$$D_\lambda = \frac{1}{L}\frac{d\tau_g}{d\lambda} \tag{12.3}$$

where τ_g is the group delay in ps, L is the path length in km, and λ is the wavelength in nm. To a first approximation, the broadening T of a pulse is given by

Table 12.1 Definitions for Chromatic Dispersion.

Attribute	Units	Definition
V_g	m/s	The propagation speed of the modulation envelope of a lighwave signal.
τ_g or τ—Group delay	ps	The time required for a modulated signal to travel along the transmission path; also called the envelope delay.
$\frac{d\tau}{d\lambda}$—Chromatic dispersion	ps/nm	The slope of the group delay wavelength versus curve.
D—Dispersion coefficient	ps/nm-km	The chromatic dispersion of a path divided by the path length.
λ_0—Lambda-zero	nm	The wavelength at which chromatic dispersion is zero.
S_0—Slope at λ_0	ps/nm²km	The slope of the dispersion curve at λ_0.

$$T = D \, \Delta\lambda \, L, \tag{12.4}$$

where T is in ps, D is the dispersion in ps/nm-km, and $\Delta\lambda$ is the spectral width of the modulated lightwave signal in nm. For a given value of chromatic dispersion, the pulse becomes broader with increase in the path length or the spectral width of the signal.

A narrowband signal centered at λ_0 can propagate for great distances with minimal degradation from chromatic dispersion. Away from λ_0, dispersion increases according to

$$D(\lambda) = S_0(\lambda - \lambda_0) \tag{12.5}$$

where S_0 in ps/nm²km is the slope at λ_0. Knowledge of the slope of the dispersion curve at λ_0 allows system designers to estimate the dispersion at nearby wavelengths.

The zero dispersion wavelength of singlemode fiber is a function of temperature. The temperature dependence of λ_0 is approximately +0.03 nm/C for dispersion-shifted fiber and +0.025 nm/C for dispersion-unshifted fiber. Depending on the application, it may be appropriate to record the temperature of the fiber under test. The dispersion slope does not change appreciably with temperature.

12.3.4 Control of Chromatic Dispersion

Control of the total chromatic dispersion of transmission paths is critical to the design and construction of long-haul, high-speed telecommunications systems. The first objective is to reduce the total dispersion to the point where its contribution to the error rate of the system is acceptable. A system consists of many spans of fiber from different manufacturing runs, each with its own zero-dispersion wavelength and dispersion slope at lambda-zero. The dispersion of a single-channel system can be controlled by concatenating fibers of differing dispersion such that the total dispersion is near zero. Alternately, dispersion

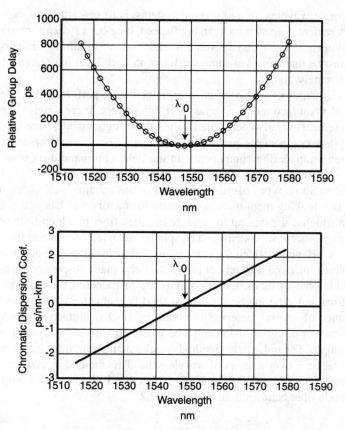

Figure 12.4   Chromatic dispersion measurement of dispersion-shifted fiber. (a) Relative group delay. (b) Dispersion coefficient.

may be allowed to accumulate along the path and then compensated at the output of the system. For example, if an installed dispersion-shifted fiber system exhibits negative dispersion at the chosen operating wavelength, it can be compensated with a relatively short length of dispersion-unshifted fiber, which has a large, positive dispersion coefficient at 1550 nm. Dispersion can also be compensated with chirped fiber Bragg gratings, made by exposing specially doped fiber to an interference pattern of intense ultraviolet light. Chirping refers to an increase in the period of the index variation as a function of distance along the fiber. Dispersion compensation is more complex in dense WDM systems.

The output power of erbium-doped fiber amplifiers (EDFAs) is sufficient to produce nonlinear effects in singlemode fiber. The second objective of chromatic dispersion compensation is to limit the impairments caused by these nonlinearities. For example, four-wave mixing can be controlled by maintaining a small negative dispersion in those portions of a span that are exposed to high power levels. (Another approach to this problem, in wavelength multiplexed systems, is to space the channels in unequal wavelength increments). In summary, in high-speed, long-haul systems, chromatic dispersion must be compensated on a system end-to-end basis to minimize ordinary pulse broadening, but must be maintained at some small but nonzero level in regions where the optical power is extremely high.

Chromatic dispersion measurements are performed by fiber and cable researchers and manufacturers, and by system integrators. Increasingly, chromatic dispersion measurements are also performed during the design, manufacture, and incoming inspection of system components, particularly chromatic dispersion compensators and wavelength division multiplexing components. In the field, chromatic dispersion is measured in connection with the installation of new systems or the upgrade of existing routes to higher bit rates. Some types of measurement systems require local access to both ends of the test fiber, making them more appropriate for factory and laboratory testing and special cases in which it is practical to measure installed fiber in a loop-back arrangement. Some types of measurement systems can be split between remote ends of the fiber.

Having defined the basic terms and measurement applications, we next examine the phase shift and differential phase shift chromatic dispersion measurement methods. Two additional methods, the interferometric method and the pulse delay method, will not be discussed. The interferometric method is useful for measurement of fibers in the length range of several meters. Unfortunately, fiber manufacturing processes are such that the behavior of a long span cannot be projected with accuracy from measurements of a short sample. The pulse delay method is a classic measurement based on the change in pulse arrival time over a series of wavelengths. This method is used for measurement of single-mode and multimode fibers and is discussed in connection with measurement of multimode fiber bandwidth in Section 12.2.2.

12.3.5   The Modulation Phase-Shift Method

A chromatic dispersion measurement apparatus based on modulation phase shift[8,9] is shown in Figure 12.5. In operation, the output of a narrowband, tunable optical source is intensity modulated and applied to the fiber under test. The transmitted signal is detected and the phase of its modulation is measured relative to the electrical modulation source. The phase measurement is repeated at intervals across the wavelength range of interest. From measurements at any two adjacent wavelengths, the change in group delay, in ps, corresponding to the wavelength interval $\Delta\lambda$ is

$$\Delta\tau_\lambda = -\frac{\phi_{\lambda+\frac{\Delta\lambda}{2}} - \phi_{\lambda-\frac{\Delta\lambda}{2}}}{360\,f_m} \times 10^{12} \tag{12.6}$$

where λ is the center of the wavelength interval, f_m is the modulation frequency in Hz, and ϕ is the phase of the recovered modulation relative to the electrical modulation source, in degrees. Wavelengths are in nm. The curve of relative group delay shown in Figure 12.4 is constructed by accumulating these changes in group delay across the measurement

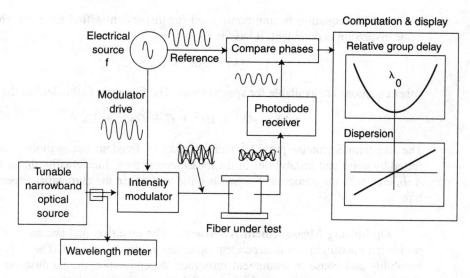

Figure 12.5 Measurement of chromatic dispersion by the modulation phase-shift method.

wavelength range. The vertical axis is labeled "relative" because measurement of dispersion does not require determination of absolute delay, although this same set of instrumentation can measure absolute delay with a change in measurement flow and computation. For convenience of interpretation, the relative group delay curve is typically offset vertically to bring the minimum value to zero.

As mentioned earlier, chromatic dispersion is defined as the rate of change of the group delay with wavelength. The result is also valid if change in relative group delay is used. Equation 12.3 becomes

$$D_\lambda = \frac{1}{L} \frac{d\Delta\tau}{d\lambda} \qquad (12.7)$$

where D is the dispersion coefficient in ps/nm/km, $\Delta\tau$ is the relative group delay in ps, and L is the path length in km. If the physical characteristics of the fiber are consistent along its length, the total chromatic dispersion, in ps/nm, scales linearly with fiber length.

Curve Fitting to Improve Measurement Precision. The precision of chromatic dispersion measurements is greatly improved by fitting an appropriate model equation to the measured relative group delay data. Because direct differentiation of the raw group delay data tends to amplify the effect of noise, the values of λ_0, S_0, and $D(\lambda)$, the value of dispersion at any particular wavelength, are calculated from the fitted curve. The model equation is chosen according to the type of fiber being measured. The three-term Sellmeier is commonly used for dispersion-unshifted fiber, in which material dispersion plays the major role. It has the form

$$\tau(\lambda) = A\lambda^2 + B + C\lambda^{-2}. \qquad (12.8)$$

The quadratic equation is commonly used for dispersion-shifted fiber, in which waveguide dispersion is dominant. It has the form

$$\tau(\lambda) = A\lambda^2 + B\lambda + C. \tag{12.9}$$

Other equations are available for special cases. The five-term Sellmeier has the form

$$\tau(\lambda) = A\lambda^4 + B\lambda^2 + C + D\lambda^{-2} + E\lambda^{-4}. \tag{12.10}$$

The five-term Sellmeier provides more degrees of freedom but is more sensitively affected by noise and instabilities in the measurement path. Incidentally, the association of A, B, etc., with the particular λ terms is quite variable in the chromatic dispersion literature.

Optimizing Measurement Accuracy.   The precision and accuracy of chromatic dispersion measurements is dependent upon test equipment design. The impact of phase instability and phase measurement resolution depends upon the modulation frequency. For a given wavelength step, a higher modulation frequency will produce a phase change that is larger in comparison with a given phase measurement uncertainty.

Phase-detection equipment can generally measure phase unambiguously only over a ±180 degree range. If the combination of wavelength step, modulation frequency, and device dispersion produces a phase change that exceeds the phase-detector range, the apparent phase will simply fold back within the measurable range and produce a $N \times 360$ degree error, where N represents the number of full cycles of phase that escaped notice of the instrumentation. For unambiguous phase measurements, the wavelength step must be limited so that the phase change produced by a wavelength step falls within the ±180 degree range of the phase detector. Some chromatic dispersion measurement systems provide automatic, real-time optimization of the modulation frequency, selecting as high a modulation frequency as possible without exceeding the range of the phase detector.

Wavelength accuracy is important because the actual phase shift is proportional to the wavelength step. Depending upon measurement objectives, the inherent wavelength accuracy of a tunable laser or filtered broadband optical source, for example, 0.1 nm, may be sufficient. When high precision is called for, as in high-speed, long-haul applications, a tunable external cavity laser and wavelength meter are commonly employed. Measurement of narrowband devices such as fiber Bragg gratings and wavelength division multiplexers demands both the high wavelength resolution of a tunable laser and the accuracy of a wavelength meter to faithfully reveal the fine structure of the relative group delay curve.

Thermal transients in the measurement setup and the fiber under test can contribute significant measurement error. Spools of fiber are very effective thermometers. A change in temperature produces a change in length and a corresponding change in group delay $\Delta\tau$ according to

$$\Delta\tau = \frac{\Delta L}{L}\tau \tag{12.11}$$

where τ is the absolute group delay through the path under test, L is the path length, and ΔL is the temperature-induced change in path length. Because the modulation phase-shift method does not instantly step across the measurement wavelength range, temperature-induced changes in group delay can accumulate throughout a measurement and distort the final results. To minimize temperature effects, the device under test (DUT) can be allowed to stabilize in the ambient air before starting the measurement. The test device can be covered with a physical baffle such as a spool cover or a cloth blanket to block the flow of air currents across the spool. Baffling should also be used when the spool being measured is inside of a circulating air type of thermal chamber.

The effects of thermal drift in both instrumentation and test device can be reduced by alternating the phase measurement between the test wavelengths and a fixed reference wavelength. Changes in the reference wavelength phase are caused only by temperature effects, and this information can be used to correct the phase measurements performed at the test wavelengths. Use of the reference wavelength trades off measurement speed for increased measurement accuracy.

12.3.6   The Differential Phase-Shift Method

The differential phase-shift method[10] determines the value of chromatic dispersion at a selected wavelength directly from measurement of the change in group delay across a small wavelength interval. The resulting chromatic dispersion can be assumed to be the average dispersion over the wavelength interval if the interval is sufficiently small. A reference fiber and test fiber are measured using the same wavelength increment. The chromatic dispersion in ps/nm·hm at each wavelength λ_i is given by

$$D_{\lambda_i} = \frac{\Delta\phi_{\lambda_i} - \Delta\phi'_{\lambda_i}}{360 \, f_m \, L \, \Delta\lambda} \times 10^{12} \qquad (12.12)$$

where $\Delta\lambda$ is the wavelength interval centered about λ_i, $\Delta\phi_{\lambda_i}$ is the measured phase change with the test fiber in the path, $\Delta\phi'_{\lambda_i}$ is the measured phase change with only a calibration fiber in the path, and f_m is the modulating frequency in Hz. If the test-fiber measurement is performed without the calibration fiber in the path, L is the length of the test fiber minus the length of the calibration fiber, in km. Wavelength is in nm and phase is in degrees. Dispersion is measured as a function of wavelength by repeating the process at a sequence of wavelengths. A model equation can be fitted to the curve to improve the precision with which λ_0, D, and S_0 can be determined.

The differential phase-shift measurement can be performed using several different equipment setups.[8] All of the methods require modulation of the optical test signal, but the approaches differ in other ways. Test wavelength may be controlled by switching among different laser diodes, filtering light from an EELED, or modulating the wavelength of an external cavity laser. The reference required for phase measurements is derived from the same electrical signal source used to modulate the optical source. The reference path may be a direct connection to the phase meter in the case of laboratory measurements, or a fiber link in the case of field measurements. In addition, the various implementations of the differential phase-shift method differ in the details of signal pro-

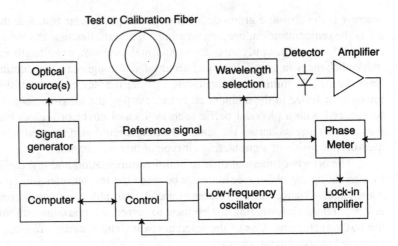

Figure 12.6 Setup for measurement of chromatic dispersion by the differential phase-shift method.

cessing. One method, referred to as the double demodulation method, is shown in Figure 12.6. Wavelength is changed at a high enough rate to avoid the $1/f$ noise of the electronics and to reduce the impact of drift of the instrumentation and the test device.

12.3.7 The Baseband AM Response Method

Chromatic dispersion changes the relative phase of the sidebands of modulated signals. In the case of a simple intensity modulated signal, chromatic dispersion converts amplitude modulation (AM) to frequency modulation (FM). The effect gives the AM response a characteristic shape which can be analyzed to determine the dispersion coefficient at the operating wavelength.[11,12] The setup shown in Figure 12.7 is based on a lightwave component analyzer (LCA). A tunable laser with narrow spectral width is tuned to the wavelength at which dispersion is to be determined. The light is intensity modulated by a lithium niobate Mach-Zehnder modulator. As the modulation frequency is swept, the baseband AM response exhibits a series of nulls. The null frequencies in GHz are predicted by

$$f_m = \sqrt{\frac{500\, c\, (1 + 2N)}{D\, L\, \lambda_0^2}} \qquad (12.13)$$

where $N = 0,1,2 \ldots$ is the index of the null, D is the chromatic dispersion in ps/nm-km, L is the length of the fiber in km, and λ_0 is the wavelength in nm. At 1550 nm, for a 20 km fiber with a dispersion coefficient of 17 ps/nm-km dispersion and a wavelength of 1550 nm, the first null occurs at 13.55 GHz. The chromatic dispersion of the test fiber at the first null is given by

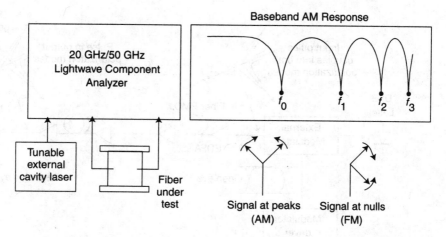

Figure 12.7 Measurement of chromatic dispersion by the baseband AM response method.

$$D_0 = \frac{500\,c}{f_m^2\,L\,\lambda_0^2} \tag{12.14}$$

As will be apparent from inspection of the example following Equation 12.13, this method is most applicable to measurement of relatively large values of dispersion, that is, well away from the zero dispersion wavelength.

12.4 POLARIZATION-MODE DISPERSION

12.4.1 Introduction

Polarization-mode dispersion, or PMD, is a fundamental property of singlemode optical fiber and components in which signal energy at a given wavelength is resolved into two orthogonal polarization modes of slightly different propagation velocity.[13] The resulting difference in propagation time between polarization modes is called the differential group delay, commonly symbolized as $\Delta\tau_g$, or simply $\Delta\tau$. PMD causes a number of serious capacity impairments,[14-16] including pulse broadening. In this respect, its effects resemble those of chromatic dispersion, but there is an important difference. Chromatic dispersion is a relatively stable phenomenon. The total chromatic dispersion of a telecommunications system can be calculated from the sum of its parts, and the location and value of dispersion compensators can be planned in advance. In contrast, the PMD of singlemode optical fiber at any given signal wavelength is not stable, forcing system designers to make statistical predictions of the effects of PMD and making passive compensation impossible. The effect of PMD on a digital transmission system is shown in Figure 12.8.

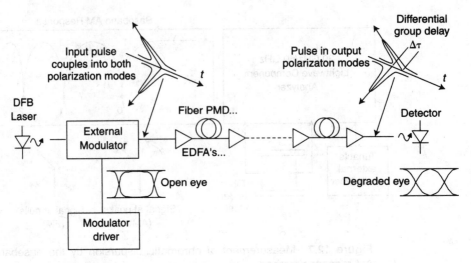

Figure 12.8 The effect of polarization mode dispersion in a digital communication system.

PMD can also affect the performance of high channel-capacity fiber optic CATV systems as shown in Figure 12.9. Direct intensity modulation of the DFB laser in the transmitter produces a signal of fixed state of polarization and chirped wavelength. In other words, the optical frequency is modulated to some extent by the changing laser current. Fiber PMD transforms each signal frequency component differently, adding a polarization modulation component to the signal. The final damage is done by the polarization-

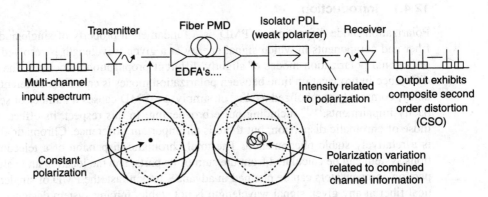

Figure 12.9 The effect of polarization-mode dispersion in a CATV distribution system.

dependent loss (PDL) of components. By attenuating polarization states selectively, PDL discriminates the changing polarization into an intensity component that is detected along with the desired signal, producing composite second-order distortion (CSO).[14]

System designers minimize the impact of PMD by specifying low PMD in fiber and components and by minimizing the PDL of components. PMD of some older installed fibers may be much higher than recently manufactured fibers, and system operators often measure these installed fibers as they plan to upgrade their systems to higher bit rates.

12.4.2 Causes of PMD

Optical fiber and fiber-optic components typically exhibit a small difference in refractive index for a particular pair of orthogonal polarization states, a property called birefringence. The index difference results in a difference in the propagation time—the differential group delay—for waves traveling in these polarization modes, as shown for the case of pulse transmission in Figure 12.10a. In singlemode fiber, birefringence originates from non-circularity, or ovality, of the fiber core in two ways: an oval waveguide is inherently birefringent, and the mechanical stress field set up by the oval core induces additional birefringence. The waveguide effect generally dominates in low PMD fiber. The phenomenon of mode coupling, to be discussed in the next section, complicates the picture, making the differential group delay a function of wavelength and of environmental conditions.

The birefringence of crystalline materials such as quartz is produced by the regular structure of the crystal. PMD in optical components may be caused by the birefringence of its sub-components, which may include quartz elements, for example, or segments of asymmetric fiber. Propagation of a signal along parallel paths of slightly different optical length, where the splitting is a function of polarization, also produces differential group delay. Certain types of optical isolators owe their PMD to this parallel-path mechanism. Components for state-of-the-art communication systems are typically specified at less than 0.05 ps of differential group delay.

Polarization maintaining fiber (PM or hi-bi fiber) is an extreme example of stress-induced birefringence. Elements of dissimilar glass embedded in the cladding assure that the core is exposed to an asymmetric stress field that is uniform along the fiber's length. When polarized light is coupled into a section of PM fiber, as illustrated in Figure 12.10a, the electric field of the input light resolves into two orthogonal polarization modes with slightly different speeds of propagation. The polarization modes are maintained along the fiber and energy does not couple between the modes. As indicated earlier, the difference in propagation time between these orthogonal polarization modes is called the differential group delay. The polarization modes and the differential group delay of a simple, linearly birefringent device are independent of wavelength. The degree of birefringence that PM fiber exhibits is often expressed as beat length, defined as the length of fiber over which the fast and slow waves change relative phase by 360 degrees. The beat length is typically a few millimeters in PM fiber.

Fiber birefringence is also induced by bending. Through the photoelastic effect, bending produces an asymmetry of the index of refraction. This is the basis of the common fiber polarization adjuster, discussed in Chapter 6. Bend-induced birefringence is not

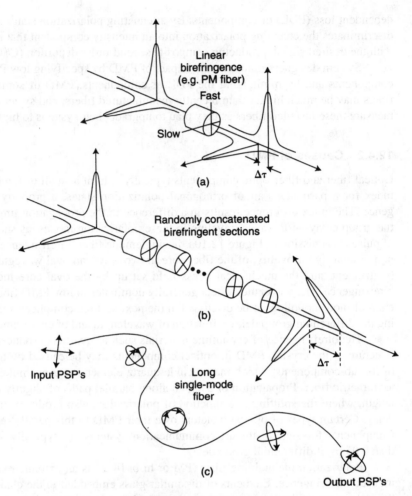

Figure 12.10 Conceptual model of PMD. (a) A simple birefringent device. (b) Randomly concatenated birefringences. (c) Input and output principal states of polarization.

in itself a significant cause of PMD, although by modifying the extent of mode coupling along a fiber, it can cause a difference in measured PMD between spooled, cabled, and installed fiber.[17]

12.4.3 Mode Coupling and the Principal States of Polarization

The birefringence of singlemode fiber varies along its length, an artifact of variation in the drawing and cabling processes. A detailed statistical model of PMD has been developed from the basis of accumulated local birefringences, and has been experimentally confirmed.[18–20] A long fiber is represented in Figure 12.10b as a series of random length, birefringent segments of random rotations. Each segment exhibits fast and slow polarization modes, generally with some of the signal light propagating in each. The electric field emerging from each segment is projected onto the polarization modes of the following segment, a process called mode coupling. It is repeated many times along a fiber span or communication link, and it is this mechanism that causes the differential group delay of most installed fiber systems to show a dependence upon wavelength and upon environmental conditions.

Fastest and slowest propagation through a fiber span can be related to orthogonal input polarization states particular to the fiber span and the environmental conditions. These relationships are most easily described using the Poincaré sphere. In general, the output polarization of an optical component or fiber will trace out an irregular path on the Poincaré sphere as wavelength is changed.[21] Over a small wavelength interval, any portion of this path can be represented as an arc of a circle. The center of the circle, projected normal to the plane of the circle to the surfaces of the sphere, locates two states that are diametrically opposed and therefore orthogonal. These states are called the principal states of polarization.[13] The defining characteristic of the principal states is that over a narrow wavelength interval, they locate the axis of the sphere about which the output state of polarization rotates as wavelength is changed. Figure 12.11 shows two examples. In the

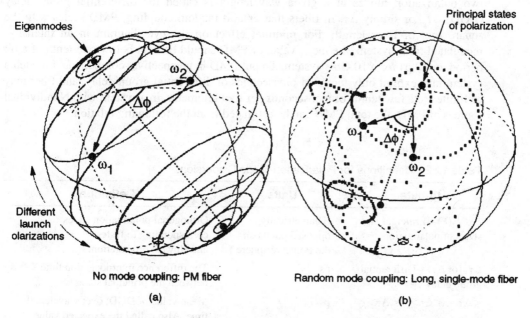

No mode coupling: PM fiber   Random mode coupling: Long, single-mode fiber

(a)   (b)

Figure 12.11   Variation of output polarization with wavelength. (a) Simple birefringent device; no mode coupling. (b) Long singlemode fiber; random mode coupling.

absence of mode coupling, the typical case for optical components, the principal states are fixed; as wavelength changes, the output state of polarization orbits regularly around the principal states axis of the sphere as shown in Figure 12.11a. In the general case of long, singlemode fiber paths, the principal states move randomly about the sphere as a function of wavelength, as shown in Figure 12.11b, and can be considered fixed only over narrow wavelength intervals.

A rigorous description of the PMD of a particular fiber at a given time requires specification of both the differential group delay and the principal states as functions of wavelength. Although only the mean or rms differential group delay is actually specified for commercial applications, the behavior of the principal states is of great interest to researchers and cable manufacturers.

12.4.4 Definitions and Relationships

The term PMD is used to denote the physical phenomenon in general and the mean, or expected, value of the differential group delay in particular. The attributes that define PMD are the differential group delay $\Delta\tau(\lambda)$ and the principal states of polarization $PSP_{1,2}(\lambda)$. Both are generally functions of wavelength in singlemode fiber systems. The principal states exist in orthogonal pairs. Some of the more common PMD terminology is defined in Table 12.2.

For a given transmission path, the difference in propagation time experienced by the two polarization modes at a given wavelength is called the differential group delay (DGD) $\Delta\tau_g$, or simply $\Delta\tau$. In fibers that exhibit random coupling, PMD scales with the square root of fiber length. For minimal effect on the eye diagram in an intensity-modulated NRZ system, the mean value of PMD should be kept below one-tenth of a bit period, or 10 ps for a 10 Gb/s system. To put PMD in perspective, components for such a system may exhibit only tenths of picoseconds of differential group delay, and fiber may be limited to a few tenths of picoseconds per root kilometer. In a system with N individual sources of PMD, the total PMD can be estimated from the following relation:

Table 12.2 Definitions for Polarization-Mode Dispersion.

Attribute	Units	Definition
$PSP_{1,2}(\lambda)$—Principal states of polarization	(Diametrically opposed points on the Poincaré sphere.)	Orthogonal polarization states corresponding to fastest and slowest propagation.
$\Delta\tau_g$ (or $\Delta\tau$)—Differential group delay	ps	The difference in propagation time between the principal states.
$<\Delta\tau>$, $<\Delta\tau>_\lambda$, $<\Delta\tau>_t$	ps	Mean value of DGD; over wavelength, time. Also called the expected value.
$[\Delta\tau]/\sqrt{L}$	ps/$\sqrt{\text{km}}$	PMD coefficient; the "PMD" of randomly mode-coupled fiber

$$\text{Mean total DGD} \approx \sqrt{<\Delta\tau_1>^2 + <\Delta\tau_2>^2 + \dots + <\Delta\tau_N>^2}. \quad (12.15)$$

A powerful and intuitively meaningful representation of PMD, defined in the same real, three-dimensional space as the Poincaré sphere, is provided by the polarization dispersion vector[22] $\boldsymbol{\Omega}$, illustrated in Figure 12.12. This vector originates at the center of the Poincaré sphere and points toward the principal state of polarization about which the output states of polarization rotate in a counterclockwise sense with increasing optical frequency ω.

The length of $\boldsymbol{\Omega}$, or $|\boldsymbol{\Omega}|$, is the differential group delay. When the output state of polarization is expressed as a three-dimensional vector $\mathbf{s}$ composed of normalized Stokes parameters locating the state of polarization on the sphere, the rotation about the principal states of polarization axis can be written as a cross product relation:

$$\frac{d\mathbf{s}}{d\omega} = \boldsymbol{\Omega} \times \mathbf{s}. \quad (12.16)$$

In the most general case, typical for spans of singlemode fiber, $\boldsymbol{\Omega}$ is a function of optical frequency.

The rate of rotation of the output state of polarization about the principal states axis is a measure of the differential group delay of the device. This is true regardless of the degree of mode coupling, although in highly mode-coupled devices the relationship must be evaluated over wavelength increments small enough that the principal state remains fixed. Differential group delay is given by

$$\Delta\tau = \frac{\Delta\theta}{\Delta\omega} \quad (12.17)$$

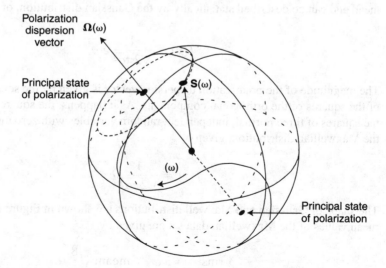

Polarization
dispersion $\Omega(\omega)$
vector

Principal state
of polarization

$S(\omega)$

(ω)

Principal state
of polarization

Figure 12.12   Dispersion vector representation of PMD.

where $\Delta\tau$ is the differential group delay in seconds, $\Delta\theta$ is the rotation about the principal states axis in radians, and $\Delta\omega$ is the optical frequency change that produced the arc, in radians/seconds. This relationship is a key element of the principal states model and the basis of a number of PMD measurement methods, most directly the Poincaré arc method described in Section 12.4.10 and shown in Figure 12.22.

12.4.5 Statistical Characterization of PMD in Mode-Coupled Fiber

The designers of undersea telecommunications systems predict the impact of PMD from the distribution of $\Delta\tau$, the differential group delay; it is this parameter which determines pulse broadening. We have discussed the variability of $\Delta\tau$ as a function of wavelength and the environmental conditions. It has been shown that in the random mode-coupled or "long fiber" regime, $\Delta\tau$ follows a Maxwell distribution.[14] That is, the distribution of values of $\Delta\tau$ measured over a wide wavelength range will be Maxwellian. The same distribution will result over time at a fixed wavelength, if the path is acted upon by a changing environment. As a result of this variability, the PMD of a path is expressed statistically, as either the mean or the rms differential group delay, related by

$$<\Delta\tau^2>^{1/2} = \sqrt{\frac{3\pi}{8}} <\Delta\tau> . \tag{12.18}$$

It is interesting to note the relationship of the Maxwell distribution to the dispersion vector $\mathbf{\Omega}$ discussed earlier. At a given wavelength, the direction of $\mathbf{\Omega}$ locates the principal states of polarization on the sphere's surface, and its length gives the magnitude of the differential group delay. The dispersion vector can be decomposed into three orthogonal vectors along the axes of the sphere. Because each of these components stems from cascaded, random birefringences, each is a normal independent random variable with zero mean and can be described statistically by the Gaussian distribution, of the form

$$f(x) = \frac{1}{\sqrt{2\pi\sigma^2}} e^{\frac{-x}{2\sigma^2}} \tag{12.19}$$

The magnitude of the polarization dispersion vector is given by the square root of the sum of the squares of the orthogonal components. As it happens, the square root of the sum of the squares of three normal, independent random variables with zero mean is described by the Maxwellian distribution, given by

$$f(x) = \frac{1}{\sqrt{2\pi\alpha^2}} e^{\frac{-x}{2\alpha^2}} . \tag{12.20}$$

The shape of Gaussian and Maxwell distributions are shown in Figure 12.13. The rms and mean values of the Maxwellian data set are given by

$$\text{rms:} \quad \sqrt{3}\,\alpha \quad \text{mean:} \quad \sqrt{\frac{8}{\pi}}\,\alpha. \tag{12.21}$$

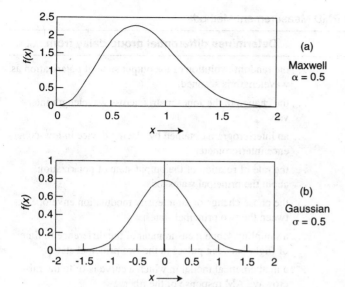

Figure 12.13 Gaussian and Maxwell distributions.

12.4.6   A Brief Summary of PMD Measurement Methods

The PMD problem has inspired the development of many measurement methods, most of which are summarized in Table 12.3.

The following discussion will focus on the first six methods. The first three are in wide commercial use. The Poincaré arc, or SOP (for state of polarization) method will be of interest to owners of polarimeters and has been submitted for standardization. The modulation phase-shift method is a useful application of a lightwave component analyzer or a network analyzer with electrical-to-optical and optical-to-electrical transducers. The pulse delay method is discussed for the insight that it offers into the phenomenon of PMD.

12.4.7   The Fixed-Analyzer Method

In the fixed analyzer PMD measurement method,[23] the mean differential group delay is determined statistically from the number of peaks and valleys in the optical power transmission through a polarizer as wavelength is scanned.[24,25] A polarizer placed directly before a detector is referred to as an "analyzer," hence the name of the method (it is also referred to as the wavelength-scanning method).

The output state of polarization of fibers and fiber optic devices generally trace out paths on the Poincaré sphere as wavelength is changed. In the case of components that exhibit simple birefringence, the path is a circle described by a ray sweeping about the principal states axis, as shown in Figure 12.11a. The diameter of the circle depends upon the balance of energy in the principal states of polarization. The circle collapses to a point at the principal states. In contrast, the output polarization of randomly mode-coupled, singlemode fibers moves irregularly about the sphere, reflecting the wavelength dependence of both the principal states and the differential group delay. This case is shown in Figure 12.11b.

Table 12.3   A Sampling of PMD Measurement Methods.

Method	Determines differential group delay from . . .
Fixed analyzer[23] (also called wavelength scanning)	. . . the random evolution of the output state of polarization as wavelength is scanned.
Jones matrix eigenanalysis[28]	. . . the change in the Jones matrix across wavelength intervals.
Interferometry[34]	. . . an interferogram obtained by placing device in low-coherence interferometer.
Poincaré arc (also called SOP)	. . . the rate of rotation of the output state of polarization about the principal states axis.
Modulation phase shift	. . . the phase change of an intensity modulation envelope between the two principal states.
Pulse delay	. . . a sampling scope measurement of the difference in time of flight between pulses in the two principal states.
Baseband curve fit	. . . a mathematical model in which a curve is fit to the microwave AM response of the fiber.

Although the variation of output polarization in the case of randomly mode-coupled fibers is erratic, at any particular wavelength, a pair of principle states of polarization exist that describe a sphere diameter about which the output polarization will rotate over a sufficiently small wavelength scan. As pointed out in the discussion of the dispersion vector, differential group delay $\Delta\tau$ can be calculated from the relation

$$\Delta\tau = \frac{\Delta\theta}{\Delta\omega}$$

(12.22)

where $\Delta\theta$ is the radian angle of the arc and $\Delta\omega$ is the incremental change in radian optical frequency that produced the arc. The radius of the arc does not figure into the calculation. Repeating this measurement at a series of wavelengths yields $\Delta\tau(\lambda)$, the differential group delay as a function of wavelength (this technique, a valid way of measuring PMD, will be discussed later in the chapter). Intuitively, the fixed analyzer method employs this same principle. It determines rate of rotation from the number of peaks and valleys in the transmission through a polarizer over a relatively large wavelength scan. It accounts for the movement of the principal states with a well-understood correction factor.

Measurement Setup and Process.   Several alternative hardware configurations of the fixed analyzer method are shown in Figure 12.14 (polarimetric detection is shown in Figure 12.16). The setups differ in the type of light source and the means of defining spectral width and tuning the wavelength.

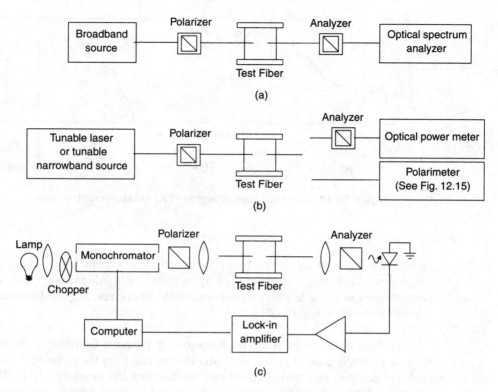

Figure 12.14 Alternative setups for the fixed analyzer PMD measurement.

A typical fixed-analyzer response is shown in Figure 12.15. To allow for adequate resolution of the features of the response, the spectral resolution of the setup should satisfy the requirement[23]

$$\frac{\Delta\lambda}{\lambda} < \frac{1}{8\,\upsilon\,\Delta\tau} \tag{12.23}$$

where $\Delta\lambda$ is the spectral width of the source or the resolution bandwidth of the receiver, λ is the nominal measurement wavelength, υ is the optical frequency in Hz and $\Delta\tau$ is the differential group delay of the test device in seconds. Wavelengths are in meters. In the region of 1550 nm, the requirement is approximated by

$$\Delta\lambda(\text{nm}) < \frac{1}{\Delta\tau(\text{ps})} \,. \tag{12.24}$$

Measurement data is collected while sweeping or stepping the wavelength of the source (or receiver, depending upon the setup). Typically, a reference measurement is taken with

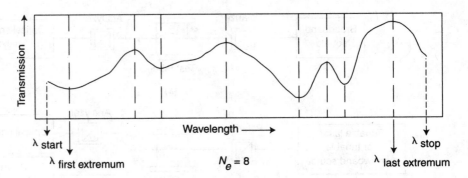

Figure 12.15   A typical fixed analyzer PMD measurement response.

the analyzer removed for correction of the actual measurement for the dependence of source power and device insertion loss upon wavelength. Alternatively, the reference measurement can be made with the analyzer rotated 90 degrees. A typical corrected measurement is shown in Figure 12.15.

Analysis of the Fixed Analyzer Response by Extrema Counting.   An automatic means is generally used to extract the PMD information from the resulting measurement trace. The methods generally used are extrema counting and Fourier analysis. The mean differential group delay $<\Delta\tau>$ of the test device can be determined by extrema counting from[23]

$$<\Delta\tau>_\lambda = \frac{k \, N_e \, \lambda_{\text{start}} \, \lambda_{\text{stop}}}{2(\lambda_{\text{stop}} - \lambda_{\text{start}})c} \tag{12.25}$$

where λ_{start} and λ_{stop} are the ends of the wavelength sweep in meters, N_e represents the number of transmission extrema (peaks and valleys) that occur across the scan, and c is the speed of light. The unitless factor k statistically accounts for the effects of the wavelength dependence of the principal states of polarization. It is called the mode-coupling factor and its value is 0.824 for randomly mode-coupled fibers and 1.0 for nonmode-coupled fibers and devices.[25] The λ subscript of the expected value $<\Delta\tau>$ indicates that the quantity is determined over a wavelength span.

An alternative form of Equation 12.25 replaces the start and stop wavelengths of the span with the wavelengths of the first and last extremum, respectively. Note that the number of extrema is reduced by one count to account for the change in wavelength endpoints.

$$<\Delta\tau>_\lambda = \frac{(N_e - 1) \, k \, \lambda_{\text{first extremum}} \, \lambda_{\text{last extremum}}}{2(\lambda_{\text{last extremum}} - \lambda_{\text{first extremum}})c} \tag{12.26}$$

When the number of extrema is large, the difference in mean differential group delay resulting from the two equations is small. However, many users prefer to let the test device determine the analysis wavelengths. This point is easier to see when we realize that the start and stop wavelengths can generally be changed up or down to some degree without changing the number of extrema.

If the test device shows no mode coupling, the analysis is based upon the first and last extremum and the mode coupling factor is set to unity.

$$\Delta\tau = \frac{(N_e - 1)\,\lambda_{\text{first extremum}}\,\lambda_{\text{last extremum}}}{2\,(\lambda_{\text{last extremum}} - \lambda_{\text{first extremum}})c} \tag{12.27}$$

This formula should be used for measurements of polarization maintaining fiber and optical components such as isolators and waveplates.

False peaks and valleys can be caused by wavelength dependence of the optical source output power or the test path insertion loss. A reference measurement is generally needed. If the optical power is repeatable over wavelength, false peaks can be avoided by ratioing the measurement to a second scan performed with the output analyzer removed. Alternatively, the second scan may be taken with the output analyzer rotated 90 degrees. Note that these precautions do not correct for short-term fluctuations of optical source level or optical coupling. Polarimetric detection, discussed in a following section, is insensitive to these power variations.

The wavelength range over which the fixed analyzer PMD measurement is performed should be great enough to produce a statistically significant number of extrema. For optical components without mode coupling, a basic measurement can be based on a single cycle of amplitude change, or even a single peak and valley pair. Near 1550 nm, the wavelength change required to span two peaks in the fixed analyzer response for a component, such as an optical isolator, is approximated by

$$\Delta\lambda_{\text{between peaks}}\,(\text{nm}) = \frac{7.8\,(\text{ps-nm})}{\Delta\tau\,(\text{ps})} \tag{12.28}$$

A component with 1 ps average differential group-delay produces peaks which are spaced an average of 7.8 nm apart. For random mode-coupled devices, the relationship is

$$\Delta\lambda_{\text{between peaks, average}}\,(\text{nm}) = \frac{6.5\,(\text{ps-nm})}{\Delta\tau\,(\text{ps})} \tag{12.29}$$

A 70 km fiber with 0.08 ps per root kilometer has an average differential group delay of 0.669 ps, requiring a scan width of 97 nm to produce approximately 10 cycles of wavelength scanning response. In singlemode fiber measurements, accuracy improves as the cycle count increases, although it may not be cost effective to extend the measurement beyond 10 or 20 cycles.

In the case of singlemode fiber, some variation in the measured PMD will occur over different launch polarizations. This is unavoidable, but a more central value can be obtained by repeating the measurement with different orientations of the input and output polarizers, or with a different lay of the input and output pigtails of the test device. Polarimetric detection, discussed later, reduces the effect of launch state and eliminates the effect of output polarizer orientation. In the case of optical components, the number of extrema is generally stable with polarizer orientation but it may be necessary to adjust the polarizers to obtain good interference between light in the two polarization modes.

Polarimetric Detection of the Output Polarization. In the fixed analyzer method, variation of the output state of polarization with wavelength is determined from the transmission through a fixed polarizer, called an analyzer in this application. Viewed in terms of the Poincaré sphere, the analyzer can be represented as a single point on the sphere. Each closest approach of the output state of polarization to this point corresponds to a peak in transmission through the analyzer.

Detection of the output polarization with a fast polarimeter, as shown in Figure 12.16, provides several advantages over the use of a single analyzer. (See Chapter 6 for a discussion of polarimeters.) The three normalized Stokes parameter traces provide a full description of output polarization over wavelength. Each trace is analyzed by extrema counting or Fourier analysis, and the resulting three mean differential group delays are av-

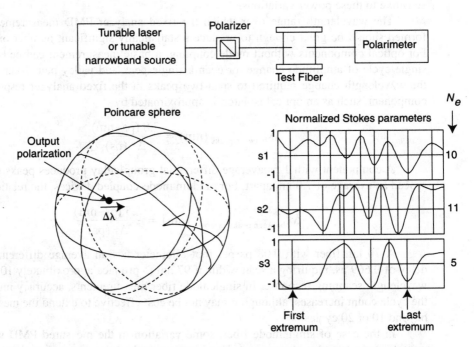

Figure 12.16 Fixed analyzer measurement of PMD using a polarimeter.

eraged. As a result of this completeness, fixed analyzer measurements taken with the polarimeter are less dependent upon launch polarization and less affected by the positions of the pigtails. A second advantage of polarimetric detection is that the normalized Stokes parameters are immune to changes of optical power. A reference or calibration measurement scan is not required and the absolute power level may vary during the measurement without affecting accuracy. The polarimeter also allows the user to view the output state of polarization on the Poincaré sphere for a sensitive indication of the stability of the test device. This is useful because the fixed analyzer response of a long fiber is sensitive to mechanical movement and temperature.

Fourier Analysis of the Fixed Analyzer Response.   Analysis of the fixed analyzer response can be shifted to the time domain by taking the Fourier transform of the frequency domain data.[23] The resulting spectrum, which is functionally equivalent to the interferometric measurement response shown in Figure 12.21d, is of Gaussian shape in the case of randomly mode-coupled fiber. The value of PMD is determined from the response by fitting a Gaussian to the data, or by the second moment calculation described in Section 12.4.9 in connection with the interferometric method. Evaluation of the fixed analyzer response by Fourier analysis has the advantage of giving a graphical indication of the mode coupling characteristics of the sample. In addition, Fourier analysis allows filtering out high-frequency features, induced by noise or vibration, that would be detected as peaks and valleys by the extrema-counting method of analysis.

Measurement of Short, Low-PMD Fibers.   Although fiber is drawn in continuous lengths of tens of kilometers, finished cable products for terrestrial applications are often only a few kilometers in length. Short lengths of fiber exhibit correspondingly small values of differential group delay and require a large wavelength range to produce a significant number of peaks. Short fibers also have low insertion loss, making the white light source or edge-emitting light-emitting diode (EELED) array and optical spectrum analyzer (OSA) or monochromator a reasonable PMD test solution. See Figure 12.14a, c. The broadband light source provides sufficient power and the tunable filter provides adequate wavelength selectivity. The measurement dynamic range achievable with the OSA or monochromator configuration is a function of the bandwidth of the tunable filter. Low-PMD devices typically exhibit very mild wavelength-dependence, allowing the use of larger filter bandwidth for higher signal level.

Measurement of Amplified Systems.   The fixed analyzer method is not commonly used to characterize EDFAs for undersea applications because the differential group delay is so low that extremely large wavelength-tuning range is required. However, the method is quite applicable to the measurement of long, multiamplifier systems. The most extreme requirements are posed by transoceanic paths which may contain over 100 amplifiers. The cascading effect reduces system bandwidth to a few nanometers, limiting the span over which the fixed analyzer response can be measured and typically forcing the use of a spectrally narrow optical source or narrow receiver bandwidth. In one approach, an amplified spontaneous emission (ASE) source is polarized and applied to the first amplifier of the chain, driving it and subsequent amplifiers into saturation. The output of the system is measured through a fixed analyzer using a spectrum analyzer. Although the system bandwidth is narrow, there are typically enough peaks to allow an estimate of PMD. The combination of tunable laser source and polarimeter is another effective approach.

The Impact of Movement in the Test Path.   Mechanical perturbation of the test path can affect the output polarization, adding spurious peaks and valleys to the fixed analyzer response. The test path includes the DUT and all fiber connecting the device to the instrument. In practice, the fiber path to and from the test device is always moving, if only microscopically. Measurement accuracy suffers when polarization changes caused by the environment are larger than the detection threshold of the instrumentation. It is good practice to constrain the motion of the pigtails that connect the measurement system and the DUT, and to avoid coupling severe mechanical vibrations into the spool of fiber being measured. The stability of the path can be checked by measuring the output polarization for a period of time at a fixed wavelength.

12.4.8   The Jones-Matrix-Eigenanalysis Method

The Jones matrix eigenanalysis (JME) method directly determines the difference in group delay between the principal states of polarization as a function of wavelength.[26] The analysis is based upon measurement of the transmission matrix of the test device at a series of wavelengths. The method can be applied to short and long fibers, regardless of the degree of mode coupling. It is applicable to the measurement of linear, time-invariant devices. The restriction of linearity precludes optical devices that generate new optical frequencies. The restriction of time invariance applies only to the polarization transformation caused by the device, and does not include the absolute optical-phase delay.

In the Jones calculus,[27] a polarized signal is expressed as a Jones vector, a complex, two-element column matrix. The Jones vector completely describes the amplitude and polarization state of the signal. The transmission path is represented by the Jones matrix, a complex two-by-two matrix. Input and output Jones vectors are related by the Jones matrix.

Measurement of the Jones matrix (discussed in Chapter 6) requires the application of three known states of linearly polarized light to the device or fiber under test. In the process described below, linear states oriented at 0, 45, and 90 degrees are used, as described by Jones, although the mathematics may be generalized to other input states. The Jones matrix is computed from the relationship of the measured output states to the known, applied input states. The resulting matrix describes the polarization transforming characteristic of the two-port device to within a complex constant that represents the absolute propagation delay. Absolute delay is not involved in the determination of differential group delay.

Measurement Setup and Process.   The measurement setup for the JME method is shown in Figure 12.17. Key elements are a tunable narrowband optical source, a

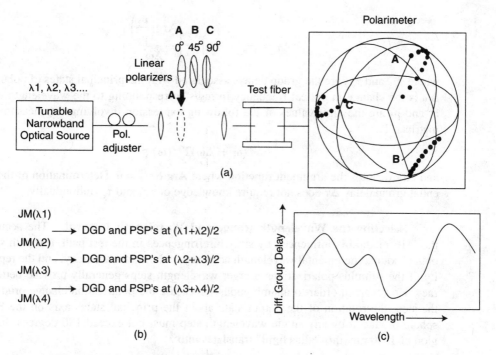

Figure 12.17 PMD measurement by the JME method. (a) Setup. (b) Measured attributes. (c) Differential group delay as a function of wavelength.

switchable polarizer for generating three linear polarization states, a fast polarimeter, and a computer for control of the hardware and processing of measurement results. Source polarization is adjusted to an approximately circular state to allow transmission through each polarizer. The Jones matrix of the path from the polarizers to the polarimeter is measured at a series of discrete wavelengths. A value of differential group delay $\Delta\tau$ at any wavelength λ_i is computed from a pair of Jones matrices measured at two wavelengths equally spaced about λ_i. The resulting series of $\Delta\tau$ values are displayed graphically to reveal the wavelength dependence, and are averaged to find the mean differential group delay $<\Delta\tau>_\lambda$. Fiber pigtails from the polarizer to the device and from the device to the polarimeter typically contribute less than 0.005 ps each to the measurement result.

The optical source used in the JME method must be tunable and sufficiently narrow-band to avoid depolarization by the test device. A tunable external cavity diode laser is commonly used.

Computing the Differential Group Delay. Differential group delay $\Delta\tau$ is computed from a pair of Jones matrices separated by a small wavelength step as follows[26,28]

$$\Delta\tau = \left|\tau_{g,1} - \tau_{g,2}\right| = \left|\frac{\text{Arg}\left(\frac{\rho_1}{\rho_2}\right)}{\Delta\omega}\right|, \tag{12.30}$$

where $\tau_{g,1}$ and $\tau_{g,2}$ are the group delays associated with the principal states of polarization, $\Delta\omega$ is the change in optical frequency in rad/s corresponding to the wavelength interval, ρ_1 and ρ_2 are the eigenvalues of the following expression involving the measured Jones matrices $\mathbf{T}$:

$$\mathbf{T}(\omega + \Delta\omega)\mathbf{T}^{-1}(\omega), \tag{12.31}$$

and Arg denotes the argument function where Arg $\alpha\epsilon^{j\theta} = \theta$. Determination of the differential group delay $\Delta\tau$ does not require knowledge of $\tau_{g,1}$ and $\tau_{g,2}$ individually.

Selecting the Wavelength Range and Wavelength Interval. The accuracy of the JME method is influenced by stray birefringences in the test path, test path stability, optical source incremental wavelength accuracy, polarimeter accuracy, and the repeatability of the stimulus polarizations. Larger wavelength steps generally provide better accuracy. However, in order to unambiguously measure the polarization change produced by the step, the rotation of the output state about the principal states axis on the Poincaré sphere produced by any single wavelength step must not exceed 180 degrees. In the region of 1550 nm, this "alias limit" translates into[28]

$$\Delta\tau(\text{ps})\, \Delta\lambda(\text{nm}) \leq 4.0\,(\text{ps-nm}) \tag{12.32}$$

For example, the maximum differential group delay that is measurable with a 0.1 nm wavelength change at 1550 nm is 40 ps. Very small wavelength steps, less than 0.1 nm or even 0.01 nm, may be required for measuring narrowband devices and devices that have extremely high PMD and/or strong variation of differential group delay with wavelength. In these cases, it may be necessary to improve the wavelength accuracy by adding a wavelength meter (see Chapter 4) to the system.

The effect of wavelength step size on the measurement of a 44 km spool of single-mode fiber is shown in Figure 12.18. Thirteen consecutive measurements were performed across the 1470 to 1570 nm wavelength range using different wavelength intervals. The average differential group delay is nearly independent of wavelength interval over a wide range. All of the curves roll off at the high end as aliasing occurs. Taking the maximum differential group delay to be 1.25 ps, aliasing is expected to occur above 3.2 nm according to Equation 12.32. The increase in the maximum values at low step size is related to laser tuning linearity and the repeatability of the insertable polarizers.

The wavelength range over which the measurement is performed can be selected according to the type of test device. The differential group delay of a broadband component such as an optical isolator is often independent of wavelength and a short series of 1 to 5 nm-steps will produce an accurate measurement. A single step may be sufficient, but a series of steps provides the benefit of averaging. Long, highly mode-coupled fibers, because of the statistical nature of PMD, require a broader wavelength range—typically the

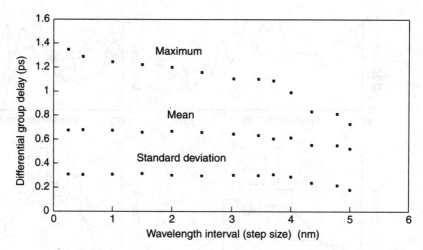

Figure 12.18 JME measurements of the DGD of a randomly mode-coupled fiber, as a function of wavelength step size.

full available tunable laser range—and a step size small enough to reveal the variation of differential group delay with wavelength. Measurement of the PMD of submeter length fibers[29] can be performed using a single step between the wavelengths of 1300 and 1550 nm. For fibers shorter than the mode-coupling length, PMD is not expected to vary significantly over the wavelength range in which the fiber remains singlemode.[17]

Statistical Characterization of Fiber PMD. The JME method is well-suited to measuring the distribution of differential group delay over wavelength, temperature, or time. Differential group delay values from individual wavelength steps of a single measurement can be displayed in a histogram as shown in Figure 12.19a. Comparison of the distribution to an ideal Maxwell distribution indicates the extent to which the variability of differential group delay has been sampled. The Maxwell curve is specified by the single parameter α, which can be determined for a particular data set using the maximum likelihood estimation, defined for an expected Maxwell distribution by[28]

$$\alpha^2 = \frac{1}{3N} \Sigma \, \Delta\tau_i^2 \tag{12.33}$$

where $\Delta\tau_i$ are the differential group delay values measured across N wavelenth intervals. A reasonably complete Maxwellian distribution assures that the fiber is adequately characterized and that the mean or rms values can be computed with good confidence. A poor Maxwell fit may result from measuring the fiber across too narrow a wavelength range, or from measuring a fiber sample that is not randomly mode-coupled. If the measurement

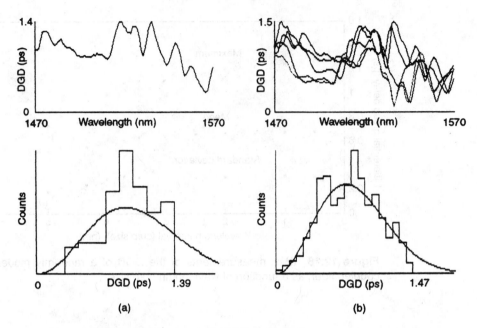

Figure 12.19 Distributions of DGD values from measurement of a 44 km fiber. (a) Single measurement. (b) Multiple measurements at different temperatures.

wavelength range is not sufficient to obtain a good distribution, it may be repeated at a series of temperatures, as shown in Figure 12.19b. For large-scale devices such as large coils of undersea cable, measurements can be made several times during the daily temperature cycle. It is sometimes possible to heat the fibers in a cable by running a current through a central metal wire.

Measurement of Amplifiers and Amplified Systems.   The degree of polarization of the output of an EDFA is influenced by the power and degree of polarization of the input signal and the unpolarized SE output of the amplifier. When the signal wavelength does not coincide with the gain peak, the output degree of polarization decreases. The JME method is quite effective in these applications by virtue of the polarimeter's ability to measure the polarized component of a signal. A degree of polarization of 50% or higher produces good results and useable measurements have been performed with as low as 25% degree of polarization. In long telecommunication systems, EDFAs are operated in saturation, generally providing a large enough output signal to maintain better than 90% degree of polarization within the system passband.

The Impact of Movement in the Test Path.   The JME method is valid when the PMD-related characteristics of the path under test are time invariant. The test path includes the DUT and all fiber connecting the device to the instrument. In principle, the test path is always moving, if only microscopically. Measurement accuracy suffers only when the movements are large enough to produce an output polarization change that is significant compared to the change produced by the wavelength step. The JME measurement method is particularly vulnerable to movement for test devices with less than 0.1 ps of differential group delay or when measuring with a small wavelength step. It is good practice to constrain the motion of the pigtails that connect the measurement system and the DUT, and to avoid coupling severe mechanical vibrations into the spool of fiber being measured.

The most demanding application of the method, from the standpoint of path movement, is the measurement of suspended or undersea cables. The varying output polarization can be displayed on the Poincaré sphere. If the variation of output polarization observed in the measurement time required for a pair of Jones matrix measurements is small relative to the changes produced by the wavelength interval, the method will yield useful measurements. The curve of differential group delay versus wavelength will be considerably less smooth because of the movement, but the mean value of differential group delay across a wavelength range will be statistically meaningful.

A simple, qualitative experiment was performed in our laboratory to see whether useful measurements could be obtained over a fiber path that is in motion. The DUT was connected to the instrumentation through a 13 m suspended span of unjacketed fiber. The span drooped about 1 m at the center and was caused to sway in a ½ m arc at the midpoint. This movement caused the output polarization state to move in a pattern 3 to 6 degrees peak-to-peak on the Poincaré sphere. Tests were performed over 1500 to 1550 nm with 1 nm step. The measurements shown in Figure 12.20a and b were taken on a component with 0.218 ps differential group delay. Although movement introduced noise, the average differential group delay values were reasonably consistent. In addition, it was found that doubling the wavelength step nearly halved the amplitude of the movement-induced peaks. The component was replaced with a 44 km spool of singlemode fiber to obtain the measurements shown in Figure 12.20c and d. Measurement results showed better agreement in this case because the test fiber's high PMD allowed the wavelength-induced polarization changes to dominate the movement induced changes.

12.4.9   The Interferometric Method

The interferometric PMD measurement method[30,31] is based upon measurement of the electric-field autocorrelation, or mutual coherence, of two signals derived from the same wideband source. Like the pulse delay and differential phase-shift methods, it is based on direct measurement of a time delay. The measurement is illustrated in Figure 12.21. Consider first the basic Michelson interferometer with no test device inserted in the detector path, shown in Figure 12.21a. Light from the broadband LED or white light source is coupled into both arms of the interferometer and light from the moving mirror and fixed mirror arms is superimposed at the detector. Interference occurs when the lengths of the two arms differ by less than the coherence length of the source. Maximum visibility occurs when the path lengths are perfectly matched. The width of the response is inversely proportional to the source spectral width. The amplitude of the photocurrent envelope is dis-

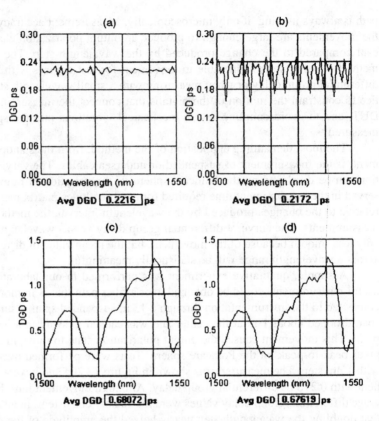

Figure 12.20 The effect of fiber movement on JME PMD measurement results. (a) Isolator, stationary fiber. (b) Isolator, moving fiber. (c) Fiber spool, stationary fiber. (d) Fiber spool, moving fiber.

played as a function of the time delay introduced by the moving mirror. The time delay Δt is given by

$$\Delta t = \frac{2\Delta x}{c} \tag{12.34}$$

where Δx is the distance of the mirror from balance, in other words, from the point where both paths are of equal length. Refer to Chapter 10 for a detailed discussion of this type of interferometry.

A basic interferometric PMD measurement setup requires the addition of polarizers at the optical source and photodetector. The setup shown in Figure 12.21b, although slightly more complex, allows adjustment for maximum visibility of a particular response peak. In this setup, the arms of the interferometer are orthogonally polarized. Mirror movement creates a delay between the orthogonally polarized incident waves. The polarizer before the detector, called an "analyzer" in this application, allows interference by

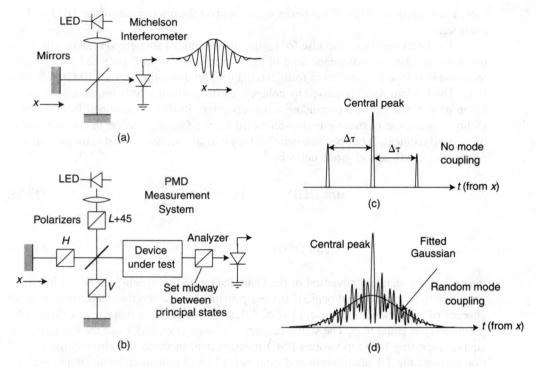

Figure 12.21   Interferometric PMD measurement. (a) Basic interferometer. (b) Example of a setup for testing PMD. (c) Response for simple birefringent device. (d) Response for random mode-coupled device.

coupling light from each of the output eigenmodes of the test device onto the photodetector with common polarization.

The interferometric method is applicable to optical components with well-defined eigenmodes, and to optical fiber in the "long-fiber" regime, where the principal states are strong functions of wavelength. The resulting interferograms, however, are quite different.

Consider first the case of a non-mode-coupled device. Referring to Figure 12.21b, the test device is placed in the optical beam and the mirror position is scanned. The response is as shown in Figure 12.21c. Both interferometer polarizations decompose into both eigenmodes of the test device. The central peak occurs when the interferometer arms are of equal length. The outlying peaks occur when the movable mirror introduces a delay equal to the differential group delay of the test device. One outlying peak is produced by interference between slow-mode light originating in the fixed mirror path and fast-mode light delayed by the movable mirror. The other outlying peak is produced by interference between fast-mode light originating in the fixed-mirror path and slow-mode light advanced by an identical movement of the mirror in the opposite direction. The time separation of the central and either of the outer peaks is the differential group delay of the device. The relative intensity of the peaks depends upon the orientations of the DUT and the analyzer.

The interferometric response for highly mode-coupled singlemode fibers with PMD much greater than the coherence time of the source is shown in Figure 12.21d. The photocurrent envelope is Gaussian in form, having a width determined by the PMD of the test fiber. The random detail is caused by coherence effects and has little impact on the calculation of PMD in the case of random mode coupling. PMD is determined from the photocurrent using one of two methods—direct fitting of a Gaussian curve, or computation of the second moment. It has been shown that for an ideal data set, a fitted Gaussian curve is related to the differential group delay by[32]

$$\text{rms DGD:} \quad <\Delta\tau^2>^{1/2} = \sqrt{\frac{3}{4}}\,\sigma \qquad (12.35)$$

$$\text{mean DGD:} \quad <\Delta\tau> = \sqrt{\frac{2}{\pi}}\,\sigma \qquad (12.36)$$

where σ is the standard deviation of the Gaussian curve that would best fit the interferogram, excluding the central peak, if it were possible to measure the interferogram in the absence of instrumentation noise. (The traditional view has been that the rms differential group delay is equal to σ. The statistical nature of long-fiber PMD and the differences in signal processing between various PMD measurement methods has allowed the distinction between the 1:1 relationship and Equation 12.35 to remain undemonstrated until recently[32,33]).

More commonly, PMD is determined using an algorithm based upon the square root of the second moment of the photocurrent response.[34] The second moment is given by

$$\sqrt{\frac{\int I(t)t^2\,dt}{\int I(t)dt}} \qquad (12.37)$$

where $I(t)$ represents the photocurrent. In practice, the algorithm involves amplitude-shifting of the response to reduce the effect of noise, removal of the central autocorrelation peak, truncation of the interferogram, computation of the second moment of the truncated interferogram, and determination of a Gaussian which, when substituted for the photocurrent and integrated over the same limits, produces the same value for the second moment. The rms differential group delay is determined from the value σ produced by this process, according to[34]

$$<\Delta\tau^2>^{1/2} = \sqrt{\frac{3}{4}}\,\sigma. \qquad (12.38)$$

Alternately, PMD can be found from the second moment σ_ϵ without substituting the Gaussian. Assuming all other steps of the algorithm are followed, σ_ϵ relates to σ as

$$\sigma_{\epsilon} \approx \sqrt{\frac{3}{4}} \, \sigma \tag{12.39}$$

The rms differential group delay is given, to within a few percent, by:

$$<\Delta \tau^2> \, \approx \sigma_{\epsilon}. \tag{12.40}$$

The value of σ_{ϵ} is assigned to the rms differential group delay by some interferometric PMD measurement instruments.

The above discussion has assumed test devices with large values of PMD in relation to the coherence time of the source. For lower values of PMD, or specifically low-PMD source-width products, the PMD measured by interferometry depends upon the exact spectral shape of the optical source.[32]

In some of the PMD literature, it has been suggested measurements of small values of PMD can be corrected for the effect of the source autocorrelation peak according to

$$\sigma_{\text{device}}^2 = \sigma_{\text{Meas}}^2 - \sigma_{\text{Source}}^2 . \tag{12.41}$$

However, it has been pointed out that this relation is not valid and that we do not yet have a rigorous method of correcting for optical source spectral width.[35] This reference provides several other excellent techniques for more accurately extracting the PMD value from the measured response.

The interferometric method is tolerant of movement along the fiber path during the measurement. Movement changes the details of the interferogram, but not the overall shape. Because interferometry measures large values of PMD quickly and interferometric setup is easily split into source and receiver units, the method also lends itself to measurement of high-PMD, installed fiber. Using modulation techniques, it is also possible to measure EDFAs using interferometry.[36]

12.4.10   The Poincaré Arc Method

The fixed analyzer, JME, and Poincaré arc (or SOP) methods are considered frequency-domain methods because they derive information from the change in output polarization state of the test device as wavelength is changed. Poincaré arc measurement results for a non-mode-coupled and a randomly mode-coupled test device are shown in Figure 12.22. The traces are generated by coupling polarized light from a tunable narrowband optical source partly into each of the input principal states of polarization of the test device. As the wavelength is adjusted incrementally, an arc is traced out on the Poincaré sphere by a ray perpendicular to the principal states axis. The differential group delay over the wavelength increment is computed from

$$\Delta \tau = \frac{\Delta \theta}{\Delta \omega} \tag{12.42}$$

where $\Delta \tau$ is the differential group delay in seconds, $\Delta \theta$ is the rotation about the principal states axis in radians, and ω is the radian optical frequency change that produced the arc.

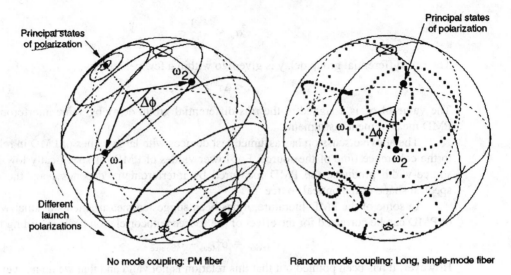

Figure 12.22 PMD measurement by the Poincaré arc (SOP) method.
(a) Simple birefringent device. (b) Spool of fiber.

Although the Poincaré arc method is applicable to both component and randomly mode-coupled fiber measurements, the strong wavelength dependence of the principal states in randomly mode-coupled fiber may require readjustment of the launch polarization to maintain a reasonable amount of light in each of the two principal states of polarization.

12.4.11 The Modulation Phase-Shift Method

The modulation phase-shift method determines differential group delay from the difference in modulation phase between the principal states of polarization. The setup is shown in Figure 12.23. An intensity-modulated lightwave is coupled into the test fiber. The output signal is detected and the modulation phase is measured relative to either the electrical modulation source or a detected sample of the light incident on the test device. A network analyzer or lightwave component analyzer is typically used in this setup because it provides the microwave source, phase detection, phase normalization, and display functions. The phase response for an arbitrary input polarization is captured in memory and excess phase is removed electronically. The resulting trace is a flat line. The input polarization is then varied to produce maximum and minimum excursions of phase. Differential group delay is computed from the phase difference at a particular modulation frequency according to

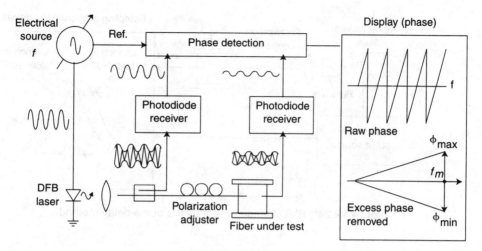

Figure 12.23 PMD measurement by the modulation phase-shift method.

$$\Delta \tau = \frac{\phi_{max} - \phi_{min}}{360 \times f_m}. \tag{12.43}$$

This technique is simple and intuitive but requires experimentally selecting the principal states of polarization. It also requires a stable operating temperature and isolation from mechanical vibrations.

12.4.12   The Pulse-Delay Method

One of the earliest and conceptually simplest methods of measuring differential group delay at a given wavelength involves sequentially launching very short pulses of light into the fast and slow polarization modes (input principal states) of the fiber and measuring the difference in arrival time of the pulses emerging from the corresponding output principal states. The setup is shown in Figure 12.24.

The differential group delay exhibited by spans of singlemode fiber, including fibers of newer and older technologies, range from less than 0.1 ps to several tens of picoseconds. An ultra-short-pulse source and high-speed detector are required. Pulses of 0.1 ps width have been produced by a demonstration EDFA pulse source at Hewlett-Packard Laboratories. Nonlinear crystal-autocorrelation techniques could be used to work around photodetector response limitations. In addition, since the principal states are measured sequentially, the instrumentation must be extremely stable.

The pulse-delay setup allows direct observation of the impact of PMD upon a fast pulse. Maximum broadening is observed when the polarized light couples equally to the input principal states of polarization.

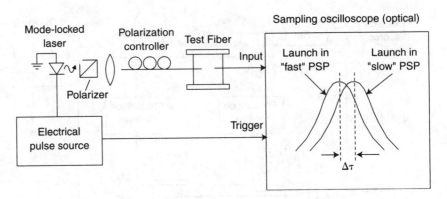

Figure 12.24   PMD measurement by the pulse-delay method.

12.4.13   Agreement between PMD Measurement Methods

As a result of the probabilistic character of PMD and the diversity of models, definitions, and measurement methods, the topic of numerical agreement between the methods has received a great deal of attention. Interest centers on the "long-fiber" regime, where mode coupling is random and the PMD coefficient is defined and has meaning. In this class of fiber, measurements of PMD coefficient commonly differ by 15 to 20%, with better agreement at higher values of PMD, worse at lower values. In contrast, for devices with simple birefringence, where the device is characterized by a single value of differential group delay, the methods agree to within a few percent.

The less impressive agreement observed for highly mode-coupled fibers is caused by a number of factors. First among these is that PMD is generally a function of temperature and mechanical stresses and thus is a less than ideal test device for inter-method comparisons. A second factor is the wavelength range, which is limited by instrumentation or by the bandwidth of the test device (for example, the narrowed passband of a long, amplified fiber link). This factor affects all of the measurement methods. For example, the fixed analyzer response of a fiber contains fewer peaks as the wavelength range is reduced, allowing a change of one count to more heavily influence the result. Closely linked is a third factor, the orientation of the launch polarization and output analyzer relative to the principal states of polarization. This is a factor in the interferometric and fixed analyzer techniques, particularly when measuring small values of PMD or using a narrow optical spectrum for the measurement. Taking the fixed analyzer as an example, rearrangement of the pigtails of the test fiber transforms the input and output polarizations and can change the extrema count.

A fourth factor bearing on measurement agreement involves the details of the instrument design and algorithms. These include filtering, thresholding, and the mathematical steps of extracting PMD from the conditioned data. A final aspect of measurement agreement is the conceptual model of PMD within which the measurement results are evaluated. The principal-states model describes PMD in terms of the wavelength-dependent differential group delay and principal states of polarization. It assumes narrowband light and considers the interference effects. An alternative model views PMD in terms of the repeated, random splitting of input pulses that are much shorter than the differential group delay. Splitting produces a distribution of pulse arrival times.[37] The PMD material presented in this chapter is based upon the principal-states model for three reasons. The high-speed telecommunications systems that are affected by PMD use narrowband light. Extremely short pulses, much shorter than the differential group delay, have wide spectra which suffer more severely from chromatic dispersion. Secondly, the pulse-splitting or incoherent light model can produce unexpected results in the intermediate mode-coupling regime because of the unaccounted-for coherent effects of the modeled, time-distributed output pulses. Finally, the principal-states model is directly verifiable by physical measurements.

12.5 SUMMARY

All forms of dispersion limit the capacity of systems to carry information. Degradation of the modulation phase relationships of lightwave signals causes pulse broadening in digital systems and equally serious distortion in analog systems. Intermodal dispersion, which limits data rates in systems using multimode fiber, is caused by the existence of multiple spatial modes of differing path lengths. Chromatic dispersion results from a variation in propagation delay with wavelength caused by the interplay of fiber materials and dimensions. PMD, caused by the division of signal energy into orthogonal polarization modes with different speeds of propagation, becomes a limiting factor after chromatic dispersion has been sufficiently reduced.

Chromatic dispersion is a relatively stable phenomenon. The total chromatic dispersion of a telecommunications system can be calculated from the sum of its parts. The location and value of dispersion compensators can be planned in advance. In contrast, the PMD of singlemode optical fiber at any given signal wavelength is not stable, forcing system designers to make statistical predictions of the effects of PMD and making passive compensation impossible.

The main chromatic dispersion measurement methods discussed in this chapter are based upon measurement of the variation of modulation phase with signal wavelength. The modulation phase-shift method measures the change in group delay over a series of wavelength intervals and determines the attributes of chromatic dispersion directly from a curve fit to the data. The differential phase-shift method measures the dispersion coefficient directly by stepping or sweeping between closely spaced wavelengths. Repeating the measurement at a series of wavelengths allows calculation of the remaining attributes. A method based on the conversion of AM to FM by the dispersion of the test device is also discussed.

The problem of PMD has inspired the development of many measurement methods, several of which were discussed. The fixed analyzer method determines a single value of

PMD from measurement of the transmission through a polarizer over wavelength. The fixed analyzer response may be Fourier transformed to yield a spectrum that gives insight into the degree of mode coupling and allows calculation of PMD from a Gaussian fit or from the second-moment algorithm. The JME method determines the differential group delay and principal states as functions of wavelength from measurements of the transmission matrix at a series of wavelengths. The interferometric method determines PMD from the electric field autocorrelation function using a broadband source. The value of PMD is computed with an algorithm based on the second moment. The Poincaré arc, or SOP (state of polarization) method uses a polarimeter to capture the arc traced out on the Poincaré sphere by the output polarization of the test device over a series of wavelength increments. The modulation-phase and pulse-delay methods determine PMD from measurements of the change in modulation phase and the change in pulse arrival time, respectively, between the principal states of polarization.

The reader is reminded that detailed measurement methods for many fiber, component and system parameters, or attributes, in the field of fiber optic communications are developed and published by standards committees under the auspices of the Telecommunications Industry Association and its international counterparts. Dispersion measurement technique is described in detail in EIA/TIA fiber optic test procedures (FOTPs), some of which are referenced herein.

REFERENCES

Intermodal Dispersion

1. TIA/EIA FOTP-51. 1991. *Pulse distortion measurement of multimode glass optical fiber information transmission capacity.* Washington, DC: Telecommunications Industry Association.

2. TIA/EIA FOTP-30. 1991. *Frequency domain measurement of multimode optical fiber information transmission capacity.* Washington, DC: Telecommunications Industry Association.

Chromatic Dispersion

3. Hackert, M.J. 1992. Development of chromatic dispersion measurement on multimode fiber using the relative time of flight measurement technique. *IEEE Photonics Technology Letters,* 4(2): 198–200.

4. EIA/TIA FOTP-168. 1992. *Chromatic dispersion measurement of multimode graded-index and singlemode optical fibers by spectral group delay measurement in the time domain.* Washington, DC: Telecommunications Industry Association.

5. Cohen, Leonard G. 1985. Comparison of singlemode fiber dispersion measurement techniques. *Journal of Lightwave Technology,* LT-3 (5): 958–966.

6. Heffner, B.L. 1993. Accurate, automated measurement of differential group delay dispersion and principal state variation using Jones matrix eigenanalysis. *IEEE Photonics Technology Letters,* 5(7), 814–817.

7. Poole, C.D. and C.R. Giles. 1988. Polarization-dependent pulse compression and broadening due to polarization dispersion in dispersion-shifted fiber. *Optics Letters,* 13: 155–157.

8. EIA/TIA FOTP-169. 1992. *Chromatic dispersion measurement of singlemode optical fibers by the phase shift method.* Washington, DC: Telecommunications Industry Association.

9. Costa, B., D. Mazzoni, M. Puleo, and E. Vezzoni. 1982. Phase shift technique for the measurement of chromatic dispersion in optical fibers using LED's. *IEEE Journal of Quantum Electronics,* QE-18 (10): 1509–1515.

10. TIA/EIA FOTP-175. 1992. *Chromatic dispersion measurement of singlemode optical fibers by the differential phase shift method.* Washington, DC: Telecommunications Industry Association.

11. Christensen, B., J. Mark, G. Jacobsen, and E. Bodtker. 1993. Simple dispersion measurement technique with high resolution. *Electronics Letters,* 29(1): 132–134.

12. Rosher, P.A., M.K. Compton, and A.D. Georgiou. Dispersive considerations in microwave optical system.

Polarization-Mode Dispersion

13. Poole, C.D. and R.E. Wagner. 1986. Phenomenological approach to polarization dispersion in long singlemode fibers. *Electronics Letters,* 22: 1029–1030.

14. Darcie, T.E. and C.D. Poole. 1992. Polarization-induced performance variables. *Communications Engineering and Design.*

15. Bahsoun, S., J. Nagel, and C. Poole, 1990. *Measurements of temporal variations in fiber transfer characteristics to 20 GHz due to polarization-mode dispersion.* Proc. 16th European Conf. Opt. Comm., Amsterdam: 1003–1006.

16. Poole, C.D., R.W. Tkach, A.R. Chraplyvy, and D.A. Fishman. 1991. Fading in lightwave systems due to polarization-mode dispersion. *IEEE Photonics Technology Letters,* 3: 68–70.

17. Poole, C.D. 1988. *Optics Letters.* 13: 687.

18. Poole, C.D., J.H. Winters, and N.A. Nagel. 1991. Dynamical equation for polarization mode dispersion. *Optics Letters,* 16: 372–374.

19. Foschini, G.J. and C.D. Poole. 1991. Statistical theory of polarization dispersion in singlemode fibers. *Journal of Lightwave Technology,* LT-9: 1439–1456.

20. Gisin, N. and J.P. Pellaux. 1992. Polarization mode dispersion: time versus frequency domains. *Optics Communications,* 89: 316–323.

21. Born, M. and E. Wolf. 1980. *Principles of optics. 6th ed.* Pergamon.

22. Bergano, N.S., C.D. Poole, and R.E. Wagner. 1987. Investigation of polarization dispersion in long lengths of singlemode fiber using multi-longitudinal mode lasers. *Journal of Lightwave Technology,* LT-5:1618–1622.

The Fixed Analyzer PMD Measurement Method

23. TIA/EIA FOTP-113. 1997. *Polarization-mode dispersion measurement for singlemode optical fibers by the fixed analyzer method.* Washington, DC: Telecommunications Industry Association.

24. Poole, C.D. 1989. Mesurement of polarization-mode dispersion in singlemode fibers with random mode coupling. *Optics Letters,* 14: 523.

25. Poole, C.D. and D.L. Favin. 1994. Polarization-mode dispersion measurements based on transmission spectra through a polarizer. *Journal of Lightwave Technology,* LT-12: 917–929.

The Jones Matrix Eigenanalysis PMD Measurement Method

26. Heffner, B.L. 1992. Automated measurement of polarization mode dispersion using Jones matrix eigenanalysis. *IEEE Photonics Technology Letters* 4:1066.

27. Jones, R.C. 1947. A new calculus for the treatment of optical systems. VI: Experimental determination of the matrix. *Journal of the Optical Society of America,* 37:110–112.

28. TIA/EIA FOTP-122, 1996. *Polarization-mode dispersion measurement for singlemode optical fibers by Jones matrix eigenanalysis,* Washington, DC: Telecommunications Industry Association.

29. Heffner, B.L. 1993. Attosecond-resolution measurement of polarization mode dispersion in short sections of optical fiber. *Optics Letters,* 18 (24):2102–2104.

The Interferometric PMD Measurement Method

30. Mochizuki, K., Y. Namihira, and H. Wakabayashi. 1981. Polarization mode dispersion measurements in long singlemode fibers. *Electronics Letters,* 17:153–154.

31. Gisin, N., J-P Von der Weid, and J-P Pellaux. 1991. Polarization mode dispersion of short and long singlemode fibers. *Journal of Lightwave Technology* LT-9:821–827 and references therein.

32. Heffner, B.L. 1985. Analysis of interferometric PMD measurements: Relation to principal states model for highly mode-coupled fibers. *Technical Digest—Optical Fibre Measurement Conference.* Liege, Belgium, September.

33. Williams, P.A. and P.R. Hernday. 1995. Anomalous relation between time and frequency domain PMD measurements. *Technical Digest—Optical Fibre Measurement Conference.* Liege, Belgium, September.

34. TIA/EIA FOTP-124. *Polarization-mode dispersion measurement for singlemode optical fibers by the interferometric method,* Telecommunications Industry Association.

35. Williams, P.A. 1996. Accuracy issues in comparisons of time- and frequency-domain polarization mode dispersion measurements. *Technical Digest—Symposium on Optical Fiber measurements.* NIST Special Publication 905: 125–129.

36. Namihira, Y., K. Nakajima, and T. Kawazawa. 1993. *Electronics Letters,* 29:1649.

37. Gisin, N., R. Passy, and J.P. Von der Weid. 1994. *IEEE Photonics Technology Letters* 6: 730–732.

CHAPTER

13

Characterization of Erbium-Doped Fiber Amplifiers

Douglas M. Baney

INTRODUCTION

This chapter presents several commonly used methods for optical amplifier gain and noise figure measurement. The methods are often modified in various ways depending on the instrumentation selected for the testing as well as the specifics of the measurement results obtained. With this in mind, the principle goal of this chapter is to convey enough information about the basic measurement methods so that the reader can make informed decisions on measurement procedures, experimental apparatus, and possible further improvements to the measurement techniques. As characterization methods are optimized, often the physics of the optical amplifier influences the measurement technique. For this reason, a working knowledge of the amplifier is useful to understand the basis for the measurement techniques.

Some of the amplifier properties that affect the communications systems are the optical gain, gain flatness, noise, temporal response, and polarization dependence.[1-5] These characteristics have a bearing on the methods employed for testing optical amplifiers. This chapter will be focused primarily on the erbium-doped fiber amplifier (EDFA), which has had a significant impact on lightwave telecommunications and the field of optics in general. A discussion of the various noise processes caused by optical amplification and interference effects[6,7] is also presented. This will be useful in understanding the origin of the noise figure of an optical amplifier.

The organization of the chapter is as follows. A discussion of basic fiber amplifier characteristics is presented in Section 13.1. Section 13.2 discusses the concept of optical gain. The noise processes that contribute to signal-to-noise ratio (SNR) degradation are introduced in Section 13.3. Noise figure definition is discussed in Section 13.4. Section

13.5 deals with measurement of gain and noise figure using both optical and electrical methods. Section 13.6 discusses other types of optical amplifiers which may impact future telecommunications systems. Section 13.7 highlights the most significant sources of measurement uncertainty in gain and noise figure measurements. Section 13.8 provides useful constants for noise figure calculations, and Section 13.9 provides a brief summary of this chapter.

13.1 FIBER AMPLIFIERS

The history of the rare-earth doped fiber amplifier dates back to the early 1960s with the demonstration of optical gain in neodymium-doped glass fiber at a wavelength of 1.06 μm.[8,9] Years later, the convergence of singlemode glass fiber drawing and semiconductor laser technologies set the stage for the 1.55 μm fiber optic amplifier. The optical propagation loss in silica glass fiber is lowest at 1.55 μm (~0.2 dB/km) making this wavelength region important for long-haul telecommunications. The demonstration, in 1986, of an erbium-doped silica fiber laser, and an EDFA in 1987[10,11] showed the great potential of fiber optic amplifiers. Soon after these initial results, telecommunications laboratories around the world began research and development efforts aimed at applying the EDFA towards optical communications systems.

The EDFA has a number of characteristics which make it an excellent amplifier for optical communications including: polarization-independent gain, low interchannel crosstalk, wide optical bandwidth and low-noise generation. In brief, the EDFA offers a nearly ideal way to compensate for signal propagation losses along high-speed singlemode fiber-optic links.

13.1.1 Basic Concepts

The basic *black-box* characteristics of an optical amplifier are shown in Figure 13.1. The incident optical signal is amplified after traversing the optical amplifier. In addition to providing for optical gain, the amplifier also adds other optical powers to the input and output optical fiber. These added optical powers include:

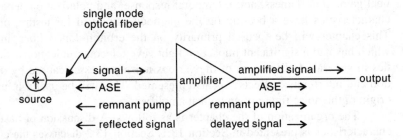

Figure 13.1 In addition to signal amplification in the forward direction, the amplifier adds other powers to the optical network.

- Amplified spontaneous emission (ASE);
- Remnant power from the pump laser;
- Time-delayed scaled-replicas of the signal power.

The degree to which these added powers are significant depends on the design of the EDFA.

The essential components of the EDFA are shown in Figure 13.2. These components are the laser pump, the wavelength division multiplexer (WDM), the optical isolators, and the erbium-doped fiber. With these basic components, many different amplifier topologies are possible.[12] To obtain gain, optical energy must be provided to the erbium-doped fiber. The energy source is called the pump. It delivers optical power at a wavelength of 980 nm or 1480 nm. The pump power is typically in the range of ~10 mW to ~400 mW. The WDM serves to efficiently couple signal and pump light into, or away from, the rare-earth doped fiber. Isolators reduce any light reflected from the system back to the amplifier to an acceptable level. Without an isolator, optical reflections may degrade the amplifier gain performance and increase noise generation. In Figure 13.2, the pump light is traveling along the fiber in a direction opposite that of the signal. This type of pumping is referred to as counter-directionally pumped, or counter-propagating, or simply counter-pumped. Codirectionally or copropagating-pumped amplifiers have the laser pump on the input end of the amplifying fiber. Sometimes multistage amplifiers are used with an isolator separating two erbium-doped fiber (EDF) gain sections. This design allows for improved amplifier noise and output power performance. Fiber Bragg gratings[13] are also used for flattening the EDFA gain variation with wavelength. This improves the amplifier performance in WDM applications as well as reducing optical noise.

The erbium ions are located in the central core region of the EDF as shown in Figure 13.3[14] The central core region (diameter ~5 μm) of the EDF is where the pump and signal wave intensities are the highest. Placement of the erbium ions in this region provides maximum overlap of pump and signal energy to the ions, resulting in better amplification. A lower index glass cladding layer surrounds the core region to complete the waveguide structure and provides for increased mechanical strength. A protective coating is added to the fiber bringing the total diameter to 250 μm. This coating, with its increased refractive index with respect to the cladding also serves to remove any nondesired light (higher order spatial modes) propagating within the cladding. Apart from the erbium dopant, this fiber construction is the same as standard singlemode telecommunications

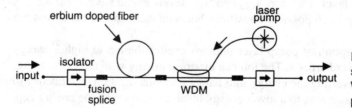

Figure 13.2   An EDFA design showing the essential components. WDM: wavelength division multiplexier.

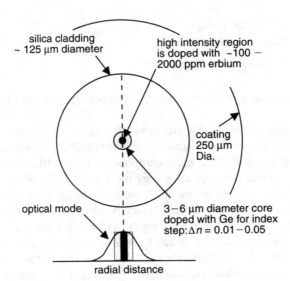

silica cladding
~ 125 μm diameter

high intensity region
is doped with ~100 –
2000 ppm erbium

coating
250 μm
Dia.

optical mode

3 – 6 μm diameter core
doped with Ge for index
step: $\Delta n = 0.01 - 0.05$

radial distance

Figure 13.3 Erbium-doped fiber core geometry.

fiber. The important characteristics for the EDF are its loss/gain per unit length at the pump and signal wavelengths. This information is often given in terms of emission and absorption cross-sections, and the confinement factors for the signal and pump light.

Energy Levels.   The trivalent erbium atoms (Er^{3+}) are the active elements in the amplifier responsible for optical gain. The relevant optical transitions are shown in Figure 13.4.[15] The approximate wavelengths of the transitions are indicated with respect to the ground state. The designations on the right side are the commonly used quantum numbers assigned to each transition. These numbers are of the form $^{2S+1}L_j$ where S is the spin quantum number, L is the orbital angular momentum, and J is the total ($L + S$) angular momentum. L is equal to one of: 0, 1, 2, 3, 4, 5, 6 . . . which is designated by the letters S, P, D, F, G, H, I. This *LSJ* scheme is used in the literature to indicate the ion energy levels.[16,17] The "local crystalline" fields perturb the ion energy structure resulting in splitting of each LSJ energy level into multiple levels. This splitting is referred to as Stark splitting (Stark effect). As a result of randomness in the glass molecular structure, each ion experiences a different field strength and orientation, resulting in different Stark-splitting. This splitting is responsible for the large gain bandwidth of rare-earth doped amplifiers. Within the *LSJ* description for the ion energy structure, the number of Stark-split lines is ($2J + 1$)/2 for each level. Thus the $^4I_{13/2}$ and $^4I_{15/2}$ levels would have 7 and 8 Stark lines respectively, resulting in 56 possible transitions between them spread out across the 1.55 μm band.

In Figure 13.4, absorption of pump laser photons excites the ion to higher energy states as shown by the upward arrows. The ion can dissipate energy with either the radiation of a photon, or by converting the energy into lattice phonons (heat). The tendency to radiate a photon when transitioning to a lower energy level increases with the energy gap.

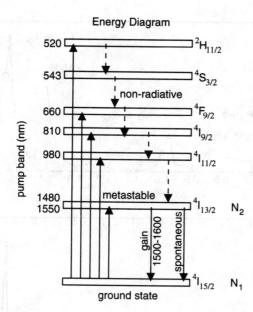

Figure 13.4 Partial energy diagram for the trivalent erbium ion.

Fortunately the $^4I_{13/2} - {}^4I_{15/2}$ transition is predominantly radiative in silica-based glasses resulting in excellent amplification characteristics in the 1.5 μm to 1.6 μm wavelength region.

The different transitions between the various energy levels result in detailed absorption and fluorescence spectrums from visible wavelengths to the infrared as seen in Figure 13.5.[18,19] The precise shape of the absorption characteristic and the magnitude of the fluorescence depends on the codopants added to the glass structure which modifies the ion energy structure. The choice of glass hosts with low phonon (vibration) energies, such as ZBLAN fluoride glass, will allow rare-earth ions to have strong fluorescence (light emission) between energy levels which normally undergo non-radiative decay in a silica glass host. In Figure 13.5a, a strong fluorescence in the 1.55 μm region is evident for erbium ions in silica glass.

The absorption characteristic, shown in Figure 13.5b is also useful to investigate potential wavelengths for pumping the erbium ion. From the figure, the wavelengths of 1480 nm, 980 nm, 800 nm, 670 nm, and 521 nm should permit excitation of the erbium ion. All of these wavelengths have been successfully used to pump EDFAs.[20,21] The addition of other codopants such as ytterbium allows for pumping at other wavelengths.

The 1480 nm pump wavelength is used in EDFAs for a number of reasons including: (1) the availability of high pump power from semiconductor laser diodes operating at this wavelength; (2) good power efficiency since there is a small energy difference between 1480 nm and 1550 nm; (3) lower attenuation in optical fiber for remotely pumped EDFAs; (4) the broad absorption spectrum places less stringent demands on pump laser wavelength accuracy.

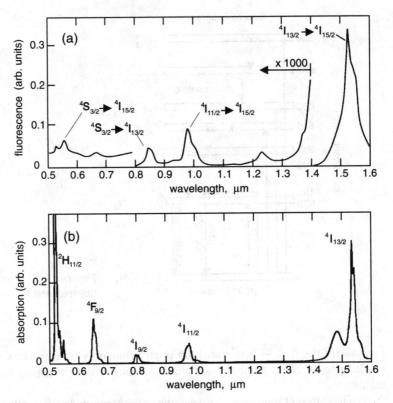

Figure 13.5 (a) Fluorescence intensity and (b) absorption characteristic for two different erbium doped silica fibers. [With permission after ref. [18], [19] ©1988 IEEE, ©1991 IEEE.]

The 980 nm pump band offers the best EDFA noise performance but also requires tighter pump wavelength accuracy to align to the narrow absorption band about 976 nm. The advent of the fiber Bragg-grating (FBG) has alleviated this problem by providing wavelength selective feedback to the pump laser to insure operation at the proper wavelength. FBGs are constructed by radiating a Ge-doped silica fiber core laterally with UV light to create periodic refractive index perturbations along a short length (~1 cm) of fiber.[13] This forms a wavelength-selectable narrow band (0.1 nm ~ 10 nm) reflective grating. A small reflection between approximately 1 to 10% provides feedback to lock the pump wavelength to the peak of the erbium absorption characteristic.

Two-, Three-, and Four-Level Systems. Optical amplifiers are classified as two-, three-, or four-level laser systems. An EDFA pumped into the $^4I_{13/2}$ band as shown in Figure 13.4 ($\lambda_p = 1480$ nm) is often approximated as a two-level system, since the pump and signal transitions are between the same energy bands. Pumping at 980 nm constitutes a three-level system where the ion energy quickly decays (~2 μs) nonradiatively from the $^4I_{11/2}$ level to the long-lived $^4I_{13/2}$ metastable state. Amplifiers based on two- or three-level systems must be designed properly to limit reabsorption of the signal due to the presence of ground-state absorption at the signal wavelength. A four-level system is an extension of the three-level where there is an additional energy level below the lower level of the gain transition. Four-level systems do not have the ground-state signal reabsorption which can degrade the performance of amplifiers based on three-level systems.

Stimulated Emission, Spontaneous Emission.   When the erbium ion (Er^{3+}) is excited from the ground state through absorption of pump light, it will decay nonradiatively from the higher lying energy levels until it reaches the metastable state ($^4I_{13/2}$ state). The incident signal light (see Figure 13.2) arrives at the excited erbium atoms distributed along the optical fiber core. Stimulated emission occurs creating additional photons with the same optical phase and direction as the incident signal, thus amplification is achieved. Excited ions that don't interact with the incident light spontaneously decay to the ground state with a time constant of approximately 10 ms. The captured spontaneous emission (SE) has a random phase and direction. Typically less than 1% of the SE is captured by the optical fiber mode and becomes a source of optical noise. This noise gets amplified resulting in amplified spontaneous emission (ASE). Once in the ground state, absorption of a pump photon activates the erbium ion again and the process repeats itself. The presence of ASE causes degradation of the SNR of signals passing through the amplifier. Proper design of the amplifier will minimize the SNR degradation.

Table 13.1 gives a quick overview of the capabilities of EDFAs in terms of ranges of values for key performance characteristics.

13.2   GAIN

Gain is the most fundamental parameter of an optical amplifier. In addition to optical gain, the amplifier produces ASE. The optical amplifier gain, G, is defined as

$$G = (P_{out} - P_{ASE})/P_s \tag{13.1}$$

Table 13.1   EDFA Characteristics

Specification	Value	Units
gain	0 – ~ 50	dB
power output	1 –> 4000	mW
noise figure	3.5 – 12	dB
wavelength range	1520 ~ 1570	nm

where P_s and P_{out} are the amplifier input and output signal powers and P_{ASE} is the noise power generated by the amplifier which lies within the optical bandwidth of the measurement.

Predicting the gain is complicated by the distributed bidirectional nature of the amplifier, this often requires a numerical solution. An understanding of the net amplifier gain, G, can be derived from an analysis of the gain from individual "slices" along the fiber. A simplified analysis is presented here. Once the concepts are understood, the equations can be readily generalized to create a more realistic amplifier model.[17,22] An ASE-free two-level approximation is assumed. An EDFA is actually a concatenation of many amplifiers of incremental length Δz as illustrated in Figure 13.6. The net gain, G, is composed of the contributions of all the gain elements, $g(z)$ along the amplifier fiber:

$$G = \lim_{\Delta Z \to 0} \{e^{g(z_1)\Delta z} \times e^{g(z_2)\Delta z} \times ... e^{g(z_n = L)\Delta z}\} = \exp\left(\int_0^L g(z)\,dz\right) \qquad (13.2)$$

The incremental signal gain, $g(z)$ for a photon propagating down the fiber is dependent on the metastable state population density, N_2, (see Figure 13.4), the ground-state population, N_1, the stimulated emission cross-section, σ_e, the absorption cross-section, σ_a, and the confinement (overlap) factor, Γ_s, between the signal field and the erbium-ion population. Γ_s can vary from zero to unity. A typical value is 0.3. The emission and absorption cross-sections represent the strength of the transition in other words, the ability to produce gain or absorption respectively. The gain coefficient is the difference between the upper and lower ion populations with a weighting taking into account their transition strengths:

$$g(z) = \Gamma_s[\sigma_{e,s} N_2(z) - \sigma_{a,s} N_1(z)] \qquad (13.3)$$

In the discussion, subscripts s and p refer to signal and pump respectively. Similarly, the pump loss in a slice of fiber is given as

$$\alpha_p(z) = \Gamma_p[\sigma_{e,p} N_2(z) - \sigma_{a,p} N_1(z)] \qquad (13.4)$$

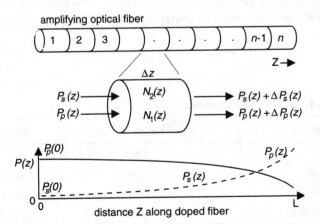

Figure 13.6 Fiber amplifier represented as a series of individual gain slices causing signal and pump power gain and absorption along the active fiber.

To achieve gain in a slice of doped fiber, the relationship $\sigma_{e,s} N_2 > \sigma_{a,s} N_1$ must be satisfied within the slice. The cross-sections can be determined experimentally from measurements of fluorescence and absorption of a short section of fiber. Experimental data, from which the cross-sections are derived is shown in Figure 13.7 for erbium in an Al and Ge co-doped silica glass.[22] Note the peak in the absorption and emission cross-sections near 1530 nm and the shift in the absorption spectrum toward shorter wavelengths. The shift towards shorter wavelengths is typical of the rare-earths. This is due to the thermal distribution of energy within each group of Stark-split energy levels favoring the lower energy levels. The addition of the Al codopant tends to broaden the gain peak near 1533 nm and reduce the amplifier gain difference between the 1533 nm and 1550 nm bands. The glass host may also be changed to improve the amplification characteristics of EDFAs. Changing from a silica host to fluorozirconate or fluorophosphate glass has been shown to substantially flatten the overall amplifier gain spectrum.[23,24] Broadband amplification from 1530 nm to 1610 nm using a tellurite glass host has also been demonstrated.[25]

The populations N_1 and N_2 are derived from the solution to the rate equation. The rate equation for the metastable state contains the contributions of pump light absorption, stimulated emission and SE.

$$\frac{dN_2}{dt} = \frac{P_p \sigma_{a,p} N_1}{A h v_p} - \frac{P_s \sigma_{e,s} N_2}{A h v_s} - \frac{N_2}{\tau_{sp}} \qquad (13.5)$$

(N_2 change) = (pump absorption) − (stimulated emission) − (spontaneous emission)

where P_p/A and P_s/A are the pump and signal intensities, $h v_p$, $h v_s$ are the pump and signal photon energies and τ_{sp} is the spontaneous decay time. From Equation 13.5, any change in the upper level, N_2 is due to a change in the relative values of pump absorption, stimulated emission or SE.

In the energy two-level approximation, conservation of the erbium ion population requires that

$$N_t = N_1 + N_2 \qquad (13.6)$$

where N_t is the total ion population. The incremental gain, $g(z)$ is related to the power change across a differential slice of fiber. The simplest case occurs when the pump light

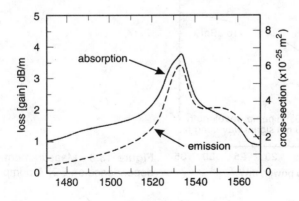

Figure 13.7 Measured absorption and gain characteristics along with calculated absorption and emission cross-sections for Al and Ge codoped EDF [With permission after ref. [22] ©1991 IEEE.]

and signal light propagate in the same direction along the fiber and amplified spontaneous emission is ignored (low gain approximation). The changes in pump and signal powers after passage through a slice of doped fiber are

$$\frac{dP_s}{dz} = g(z)\, P_s(z) \tag{13.7}$$

$$\frac{dP_p}{dz} = \alpha_p(z)\, P_p(z) \tag{13.8}$$

As the magnitude of the signal increases along the fiber, the upper-state population is reduced according to Equation 13.5. This results in increased pump absorption in the increment of fiber as indicated in Equation 13.4. These equations can be integrated numerically to solve for the signal and pump power as a function of length along the doped fiber.

The net amplifier gain is found from Equations 13.2 and 13.3. It depends on the average inversion level of the erbium ion population:

$$G = \exp\{\Gamma_s[\sigma_e[N_2] - \sigma_a[N_1]]L\} \tag{13.9}$$

The average inversion level is set by the pump and signal power levels. The gain dependency on pump power is a figure of merit for different EDFs. Figure 13.8 plots the gain versus pump power for an EDFA for two input powers. The amplifier gain coefficient in units of dB/mW is the maximum slope of the tangent to the curve that passes through the origin. Given the large emission and absorption cross-sections near 1530 nm, the highest gain coefficient is expected (from Equation 13.9) at this wavelength, provided the amplifier is highly inverted. The importance of the overlap factor is also expressed by

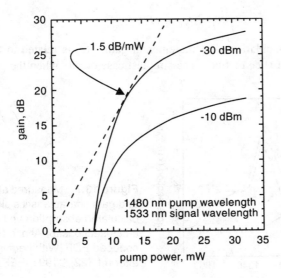

Figure 13.8 Measurement of gain dependence on EDFA pump power.

Equation 13.9. Increasing the fiber numerical aperture increases the overlap factor Γ_s resulting in improvement to the small-signal gain. For this reason, amplifiers designed to obtain the maximum gain per milliwatt of pump power tend to have fibers with high numerical apertures. Values for EDF numerical aperture typically range between 0.2 and 0.4.

13.2.1 Small-Signal Gain

The small-signal region corresponds to input power levels where the signal amplification does not reduce, appreciably, the gain of the amplifier. For the purposes of defining the small-signal gain region, it is useful to estimate the effective input noise $P_{\text{noise, eff}}$ of the amplifier.

$$P_{\text{noise, eff}} \approx 2hvB_o \sim 30\ \text{nW/nm for } \lambda = 1.5\ \mu\text{m} \tag{13.10}$$

where hv is the photon energy (J) and B_o is the optical bandwidth (Hz) of the amplifier. The effective input noise multiplied by the amplifier gain yields the output noise power of the amplifier. As the input signal power becomes significant relative to the input noise power, it plays a larger role in determining the inversion level, N_2. Changes in N_2 result in changes in gain. As long as the input power is small compared to $P_{\text{noise,eff}}$, its affect on the amplifier will be insignificant and the amplifier will be in small signal operation. The amplifier gain can be plotted as a function of input power as shown in Figure 13.9. This type of curve can help identify the small signal input power region. Even at very low input power levels, a reduction in signal gain can occur.

As an example, an amplifier with 10 nm optical bandwidth about 1.55 μm has an effective input noise of ~ 0.3 μW. Therefore, the input signal probe should be less than 30 nW, or −45 dBm, to avoid affecting the amplifier gain. The small-signal gain is sometimes defined as the gain corresponding to a small, but practical input level, (for example, −30 dBm) with the understanding that compression effects may have already occurred to some degree.

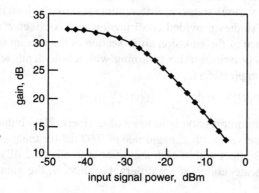

Figure 13.9 EDFA gain dependence on input signal power.

13.2.2 Saturated Gain

The EDFA is usually operated at input signal levels large enough to cause saturation of the amplifier gain. Gain saturation is observed in Figure 13.9 as a reduction in gain with an increase in signal power. The amplifier gain in an ASE-free model can be written implicitly as a function of the ratio of the output power, P_{out} to the saturation power, P_{sat}.[17]

$$G = G_0 \exp\left[-\frac{G-1}{G} \frac{P_{out}}{P_{sat}} \right] \tag{13.11}$$

Where G_0 is the small-signal gain as discussed earlier. The saturation power, P_{sat} at a specific wavelength is the power required to invert a slice of erbium-doped fiber sufficiently to obtain optical transparency in other words zero gain.

P_{sat} is written as:

$$P_{sat} = \frac{Ah\nu}{\sigma_a \tau_{sp}} \tag{13.12}$$

where A is the mode-field area, σ_a is the absorption coefficient as discussed previously, and $\tau_{sp} \sim 10$ ms, is the spontaneous lifetime of the ion in the metastable state.

The saturation power can be modified by increasing the fiber mode-field area A. The amplifier 3 dB compression point is a figure of merit describing the output power capabilities of the amplifier. This is the output power at which the amplifier gain is reduced to 50% of its small signal value. From Equation 13.11, the 3 dB compression power is proportional to the saturation power: $P_{out}^{-3dB} = \ln(2) P_{sat}$.[26] High-power amplifiers tend to have active fibers with larger mode-field diameters to increase the saturation power and hence the 3 dB compression point.

13.2.3 Polarization Hole-Burning

In an experiment where a single wavelength channel was passed through a link employing a large number of optical amplifiers, it was discovered that the small-signal optical gain in the polarization orthogonal to the large-signal polarization was greater than the large signal gain.[5] This polarization-dependent gain, (PDG), occurred even when the large-signal polarization was changed to various states, to differentiate it from the usual polarization-dependent loss. Subsequent studies provided confirmation to this effect whose origin is due to polarization dependence of the emission cross-section of the erbium ions in the silica host. This effect leads to polarization hole-burning with a hole depth which depends on the degree of amplifier compression C_p of :[27]

$$PHB \approx 0.027 C_p - 0.001 C_p^2 \, [\text{dB}] \tag{13.13}$$

for $C_p < 8$ dB. The amplifier compression is in units of decibels. PDG induced by polarized pump light was also observed, with a magnitude of 0.07 dB for the particular amplifier studied.[27] Since the PHB effect within each amplifier is small (~0.2 dB), its impact is more important in large concatenations of amplifiers. Fortunately, the gain recovery for the PHB is slow, and rapid polarization modulation of the input signal has been shown to effectively suppress the effect of PDG.[28,29]

13.2.4 Spectral Hole-Burning

A localized signal power-dependent spectral gain depression is referred to as spectral hole-burning, (SHB).[30] SHB occurs in EDFAs when a strong signal reduces the average ion population contributing to gain at a particular wavelength in excess of the global reduction.

SHB is relatively small in EDFAs since these amplifiers are predominantly homogeneously broadened. A homogeneously broadened amplifier has the property that an input signal at any wavelength in the amplification band can equally access the total energy stored within the amplifier. Homogeneous broadening in EDFAs is caused by the rapid transport of energy across the different Stark-broadened lines within a specific manifold (in other words, $^4I_{15/2}$ or $^4I_{13/2}$ in Figure 13.4).[31] This tends to reduce the extent of the SHB. The presence of phonons (heat exchange) is responsible for the EDFA homogeneous broadening.

Research has shown that at room temperature, SHB is relatively small for EDFAs with a dependency of ~ 0.3 dB per dB increase in gain compression.[32] The effect of SHB tends to be more significant in the 1530 nm wavelength region than the 1550 nm region. A plot of the inhomogeneous gain saturation caused by SHB is shown in Figure 13.10. This measurement was performed using an edge-emitting LED (EELED) probe in combination with a time-domain extinction technique to accurately measure gain as discussed in Section 13.5. The full-width half-maximum of SHB hole-widths are typically in the range of ~ 3 to ~ 10 nm in the EDFA gain spectrum, the narrowest hole-widths occurring near the 1530 nm region.

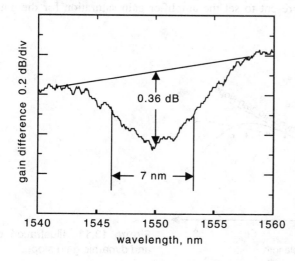

Figure 13.10 EDFA spectral hole-burning at a wavelength of 1550 nm.

13.2.5 Gain Tilt, Gain Slope

The amplifier gain tilt has important implications in systems sensitive to distortion brought on by the combination of laser chirp and amplifier gain slope.[33] In long-haul WDM systems, the amplifier gain spectrum must remain flat to avoid dominance of the power of one channel over the rest. The change or tilt in the amplifier gain spectrum that occurs when wavelength channels are added or dropped to the WDM data stream degrades performance of long-haul telecommunications systems. Gain tilt is defined here as the ratio of the gain change at a test wavelength to the change in gain at a reference wavelength where the gain changes are caused by a change in input conditions. For a homogeneously broadened amplifier, gain tilt is invariant with input power. Once the gain tilt is characterized for one set of input conditions, it can be applied to predict the amplifier gain tilt for other input conditions.

A related concept is the amplifier gain slope. It is important to distinguish between the static gain slope (see Figure 13.11) and the dynamic gain slope. The differences between the static and dynamic gain slopes are due to the change in the amplifier inversion level that results from a change in the wavelength of the strong saturating input signal.

The static gain slope, m_s is defined by

$$m_s(\lambda_o) = \frac{G_s(\lambda_o + \Delta\lambda) - G_s(\lambda_o - \Delta\lambda)}{2\Delta\lambda} \tag{13.14}$$

where $G_s(\lambda_o \pm \Delta\lambda)$ is the gain at the saturating signal wavelength as the saturation signal wavelength is tuned to $\lambda_o \pm \Delta\lambda$.

The dynamic gain slope, m_d is defined as

$$m_d(\lambda_o) = \frac{G_p(\lambda_o + \Delta\lambda) - G_p(\lambda_o - \Delta\lambda)}{2\Delta\lambda} \tag{13.15}$$

where $G_p(\lambda_o \pm \Delta\lambda)$ is the gain of a small signal probe at the wavelength of $\lambda_o \pm \Delta\lambda$. A large input signal maybe present to set the amplifier gain saturation for the gain slope

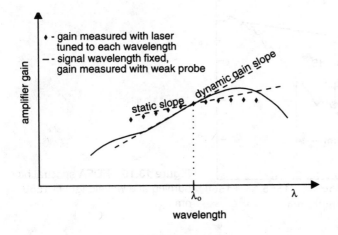

Figure 13.11 Illustration of static and dynamic gain slope.

measurement. The probe for characterizing the dynamic gain slope may be a continuously tunable laser set to low output power, or a broadband light source such as an EELED.

13.3   NOISE

In this section, the noise associated with an amplified optical signal is discussed. Noise in two domains will be considered: optical field noise, and intensity/photocurrent noise. Optical field noise refers to the optical noise spectrum measured with the typical tuning-filter-based (for example, grating) OSA. This type of noise is usually characterized over the EDFA spectral window. ASE from an optical amplifier is the main contributor to this noise. Intensity/photocurrent noise refers to the power or current fluctuations associated with the optical beam. This noise is typically characterized up to tens of GHz in bandwidth. The intensity noise spectrum refers to the power spectrum of the optical intensity prior to detection. The intensity noise spectrum is different than the photocurrent spectrum in subtle ways when shot noise is considered. For this reason, special attention has been applied in the discussion on shot noise in view of its representation in both the intensity noise and photocurrent noise domains. The concepts of power spectral densities, relative intensity noise, and SNR will be used in this section.

13.3.1   Optical Noise

Within the amplifying section of optical fiber, the excited erbium ion can decay to its ground state through stimulated emission caused by a signal photon, or, spontaneously. The spontaneously emitted photon has random direction and phase. Some of the spontaneously emitted photons are captured by the propagating mode of the optical fiber. These captured photons will be amplified as they travel inside the doped fiber. This results in ASE. The total ASE power is summed over all the spatial modes that the optical fiber supports in an optical bandwidth, B_o. In the typical erbium-doped fiber there are two propagating modes of polarization with a total ASE power equal to:

$$P_{\mathrm{ASE}} = 2n_{sp} \, h\nu \, (G - 1)B_o \tag{13.16}$$

where $h\nu$ is the photon energy and G is the amplifier gain.

The spontaneous emission factor, n_{sp}, is given by

$$n_{sp} = \frac{\sigma_e N_2}{\sigma_e N_2 - \sigma_a N_1} \tag{13.17}$$

with σ_e, σ_a, N_1, and N_2 as defined in Section 13.2. The SE factor, n_{sp} is a measure of the quality of the inversion of the optical amplifier. An n_{sp} value near unity is possible with strong pumping in the 980 nm band. This is the lowest value of n_{sp} that can be attained. It corresponds to nearly complete inversion ($N_1 \sim 0$) of the amplifier. Complete inversion, where $N_2 = N_t$ results in the lowest optical noise figure (discussed in Section 13.4).

Depletion of pump power along the erbium-doped fiber causes the N_2 population to vary as well. According to Equation 13.17, the SE factor depends on N_2 and it will vary with length along the active fiber. Usually n_{sp} is defined as the effective or integrated value for the amplifier. When the 1480 nm pump wavelength is used, complete inversion is not possible since the pump and signal share the same ground and excited states. Pump photons are not only absorbed but also contribute to stimulated emission since the emission cross-section is nonzero at this wavelength. The result is an amplifier with incomplete inversion and a higher SE factor, n_{sp}. This translates to a direct increase in the noise figure of the amplifier. The effective values for n_{sp} typically range from 1 (980 pumping) to 4.

In noise figure calculations it is sometimes useful to work with the ASE spectral density (W/Hz) in a single polarization:

$$\rho_{ASE} \overset{\Delta}{=} n_{sp} \, hv(G - 1) \tag{13.18}$$

Given a large enough amplifier gain, the ASE can become significant, resulting in saturation of the amplifier gain by the generated ASE. For this reason EDFAs can be applied as ASE sources for a variety of applications ranging from gyroscopes to "white light" interferometry (Chapter 10).

Example

Find the ASE power generated by an amplifier supporting two polarizations with 20 dB gain, 30 nm bandwidth, a 1.55 μm center wavelength and an n_{sp} factor of 1.5.

Solution

Referring to Equation 13.16, the bandwidth in Hertz is calculated:

$$B_o \approx \frac{c\Delta\lambda}{(\lambda_o)^2} \tag{13.19}$$

This yields a bandwidth B_o of 3.75 THz. The photon energy was computed to be 1.28×10^{-19} J. Using $G = 100$ and Equation 13.16 the total ASE power is 143 μW. The ASE power produced in a 1 nm bandwidth is plotted as a function of optical gain in Figure 13.12 for various effective values of n_{sp}.

13.3.2 Intensity/Photocurrent Noise

Before embarking into the discussion on noise, the basic conversion of light intensity into electrical current by an optical receiver is discussed in terms of the receiver responsivity. The average photocurrent, i_{dc} generated in a photodetector by an optical source of average power $<P>$ is:

$$i_{dc} = \mathcal{R} <P> \tag{13.20}$$

where the detector responsivity is defined as:

$$\mathcal{R} = \frac{\eta q}{hv} \quad [A/W] \tag{13.21}$$

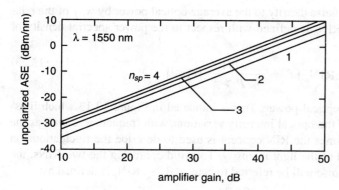

Figure 13.12 ASE power produced in 1 nm bandwidth as a function of length for various spontaneous emission factors. Add 2.3 dB to compute power at a wavelength of 1300 nm.

The light collection quantum efficiency of the receiver is denoted by η. The electronic charge, q is equal to 1.602×10^{-19} coul. At a wavelength of 1.55 μm, the photon energy $h\nu$, is 1.283×10^{-19} J. The light collection quantum efficiency η includes all optical losses that are part of the optical receiver. This can include optical coupling loss as well as the quantum efficiency of the receiver photodetector.

In addition to the average optical power, intensity noise is also present. Intensity noise is a significant limiting factor in optical communications systems. Photodetectors convert intensity noise directly into electrical noise. While the optical field noise can have both amplitude or phase noise, photodetectors do not directly respond to the phase noise. However, phase noise can be converted to intensity noise by interference effects. The following intensity noise types are commonly encountered in optical systems:

- Shot noise,
- Signal-spontaneous beat noise,
- Spontaneous-spontaneous beat noise, and
- Interference noise.

The two beat noises and the interference noise fall in the category of excess noise. It is important to differentiate between shot noise and the excess noise since the resulting photocurrent noise they generate depends differently on the responsivity of the receiver detector. Intensity noise is defined in terms of a power spectral density, $S_p(f)$ of the light intensity fluctuations. The power spectral density of the optical intensity variations is related to the power spectral density of the electrical current $S_i(f)$ variations according to:

$$S_P(f) = \frac{1}{\mathcal{R}^2} S_i(f) \qquad \text{for excess noise}$$

$$S_P(f) = \frac{1}{\mathcal{R}} \frac{h\nu}{q} S_i(f) \qquad \text{for shot noise}$$

(13.22)

$S_i(f)$ is measured with an electrical spectrum analyzer. When $S_p(f)$ is integrated over a bandwidth, it yields the mean-square optical noise power in the integration bandwidth.

It is useful to refer the noise density to the average optical power by way of the relative intensity noise (RIN_o). RIN_o is defined with respect to the power spectral density of the optical intensity as:

$$RIN_o (f) = \frac{S_P (f)}{<P>^2} \quad [Hz^{-1}] \tag{13.23}$$

where $<P>$ is the average optical power. This is indicated in Figure 13.13 which illustrates the spectral density of the optical intensity variations with frequency as well as the average dc intensity. Sometimes the RIN concept is used to describe the fluctuations on the electrical current instead of the light intensity. To avoid confusing the two RINs, the electrical relative intensity noise will be referred to here as RIN_e. RIN_e is defined as:

$$RIN_e (f) = \frac{S_i(f)}{[i_{dc}]^2} \quad [Hz^{-1}] \tag{13.24}$$

Shot Noise.   Shot noise has its origins in the uncertainty of the time of arrival of electrons or photons at a detector. When the dominant noise is due to shot noise it is referred to as shot-noise-limited or quantum-limited. Both the laser signal and ASE contribute shot noise. In this discussion, the shot noise will be examined in both the intensity and electrical domains. This will avoid confusion with respect to the impact of the quantum efficiency η of the optical detection. The shot-noise spectral densities are given by:

$$S_i(f)|_{shot} = 2q \, i_{dc} \quad [A^2/Hz]$$
$$S_P(f)|_{shot} = \frac{2 \, h v i_{dc}}{\mathfrak{R}} = 2hv <P> \quad [W^2/Hz] \tag{13.25}$$

where i_{dc} and $<P>$ are the average values for the electrical current and optical power respectively.

The RIN for the intensity and electrical domains are derived using Equations 13.23, 13.24, and 13.25:

$$RIN_e|_{shot} = \frac{2q}{i_{dc}} \quad [Hz^{-1}]$$
$$RIN_o|_{shot} = \frac{2\eta \, q}{i_{dc}} = \frac{2hv}{<P>} \tag{13.26}$$

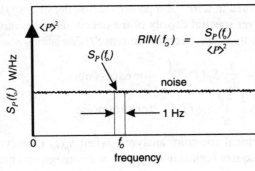

Figure 13.13   Quantification of noise in terms of spectral density and RIN.

$\mathrm{RIN}_e|_{shot}$ is larger than $\mathrm{RIN}_o|_{shot}$ when the quantum efficiency of the optical detection is less than unity. $\mathrm{RIN}_o|_{shot}$ improves (gets smaller) with an increase in optical power. Any optical signal can be made to be shot-noise limited. By attenuating the optical signal (decreasing η), the excess noise sources (discussed later) reduce in magnitude faster than the shot noise, and the shot noise will eventually dominate. The $\mathrm{RIN}_e|_{shot}$ is plotted in Figure 13.14 as a function of photodetector current. From Figure 13.14, the RIN produced by a 1 mW shot-noise-limited laser source is approximately −155 dB/Hz.

A special class of light, known as squeezed-light has an associated intensity noise content below the conventional shot noise level. The intensity noise is compressed or squeezed at the expense of an increase in phase noise. The degree of squeezing is rapidly lost when the squeezed light passes through an optical amplifier or a lossy medium.

Signal-Spontaneous Beat Noise. Interference between signal light and ASE causes intensity fluctuations known as signal-spontaneous beat noise.[34] This noise is unavoidable in EDFA-based systems and is one of the primary noise contributions in optically amplified communications systems. This beat noise is analogous to the case of two frequencies beating in a heterodyne mixer to generate a difference frequency. Recall that the mixing product is polarization dependent, so the signal will beat only with those ASE components in the same polarization as the signal. Since the ASE is typically unpolarized, only one-half will contribute to the sig-sp beat noise density. This mixing process is illustrated in Figure 13.15. The bandwidth of optical receivers is typically less than 50 GHz, (~ 0.4 nm @ λ = 1.55 μm) so only those ASE spectral components within 0.4 nm of the signal wavelength contribute to the detected signal-spontaneous beat noise.

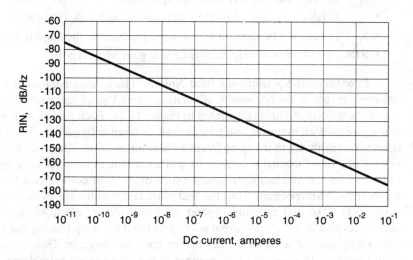

Figure 13.14 Shot-noise RIN dependency on average current.

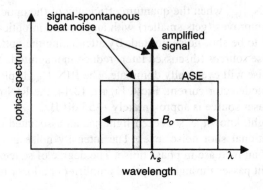

Figure 13.15 Signal-spontaneous beat noise between amplified signal and the spectral components of the ASE.

The detected current noise density measured with an electrical spectrum analyzer has a low-frequency ($f < B_o/2$) value of:

$$S_{i_{sig-sp}}(f) = 4\mathcal{R}^2 G P_s \, \rho_{ASE} \quad [A^2/Hz] \tag{13.27}$$

According to Equations 13.18 and 13.27, the signal-spontaneous beat noise varies as the square of the optical gain and linearly with input signal power. The use of the RIN concept with signal-spontaneous beat noise has the complication due to the average power contribution of the ASE. Taking the average unpolarized ASE power as P_{ASE}, the signal-spontaneous beat noise RIN is given by:

$$RIN_{sig-sp} = \frac{4 G P_s \rho_{ASE}}{(G P_s + P_{ASE})^2} \quad [Hz^{-1}] \tag{13.28}$$

The following observations can be made about sig-sp beat noise:

- $S_{i_{sig-sp}}(f)$ increases linearly with the input signal.
- $S_{i_{sig-sp}}(f)$ does not depend on the ASE spectral width, B_o ($f_o \ll B_o$).
- $S_{i_{sig-sp}}(f)$ can not be reduced by placing a polarizer at the amplifier output.
- RIN_{sig-sp} is approximately independent of gain when $G > 10$.

Spontaneous-Spontaneous Beat Noise. The beating between the different spectral components of the SE results in intensity noise known as spontaneous-spontaneous, (sp-sp) beat noise.[34] This is illustrated in Figure 13.16. Each pair of ASE spectral components generates an intensity beat tone at their difference frequency. Thus, the entire ASE spectrum contributes to the sp-sp intensity beat noise. If the ASE is unpolarized, the ASE in each of two orthogonal polarizations will contribute to the total sp-sp beat noise. From Figure 13.16, the maximum frequency extent of the beat noise is equal to the maximum width of the ASE spectrum. Thus the beat noise could well have an intensity spectrum beyond 1000 GHz, and certainly beyond the bandwidth of electronic receivers. The frequency content of the photocurrent noise generated by sp-sp beating can be significantly reduced by placing an optical filter before the photodetector. This is easily understood from Figure 13.16 where the total number of possible beating pairs decreases as the opti-

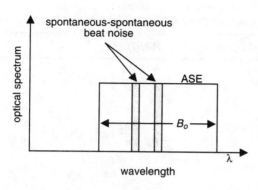

Figure 13.16   Spontaneous-spontaneous beat noise between ASE spectral components.

cal bandwidth decreases. The photocurrent spectrum for the case of an unpolarized ASE spectrum with a rectangular shape is:

$$S_i(f) = 4\mathcal{R}^2\rho_{\text{ASE}}^2\, B_o\Lambda(f/B_o) \quad [A^2/Hz] \tag{13.29}$$

where $\Lambda(f/B_o)$ is the triangle function which has a value of unity at 0 Hz and linearly decreases to 0 at $f = B_o$.

In general, the peak magnitude of the RIN is dependent on the actual shape of ASE spectrum. This is discussed also in Section 13.5.2. When the signal power is small or absent, the RIN caused by sp-sp beat noise varies inversely with the spectral width of the ASE source. A rectangle-shaped ASE spectrum delivers the most RIN of any optical shape[35] with a value equal to:

$$\text{RIN}_{\text{rect}} = \frac{1}{B_o} \quad (\text{for unpolarized light}) \tag{13.30}$$

Reducing the optical bandwidth increases the RIN. This simple relation can also be remembered as follows: RIN is related to the inverse of the number of degrees of freedom the system supports. In the case of ordinary telecom-grade optical fiber, and intensity detection, there are two spatial orientations for polarization, and $B_o/(1\text{ Hz})$ possibilities for bandwidth. For polarized light, the RIN increases by a factor of two since one degree of freedom is absent.

Table 13.2 gives the analytical relations for the frequency dependent sp-sp RIN for several different ASE spectral shapes. The optical field shape refers to the ASE spectral shape as measured with an OSA.

To provide a comparison of the way the different noise types discussed vary with signal power, the shot, sig-sp and sp-sp beat noises are plotted in Figure 13.17. The noise levels are plotted versus amplified signal power for an amplifier having a 5 nm passband and 37 dB gain. Thermal noise is also shown for an electrical system noise figure of 8 dB and a 50 ohm impedance. At low signal levels, the noise is dominated by sp-sp beat noise. Eventually, sig-sp beat noise dominates, increasing linearly with signal power. At low

Table 13.2 Relationship Between Unpolarized Optical Field Spectrum and RIN.

Optical field shape	Normalized $S_E(v)$	RIN(f)	RIN($f = 0$)
Rectangle	$\Pi\left(\dfrac{v - v_o}{B_o}\right)$	$\dfrac{1}{B_o}\Lambda\left(\dfrac{f}{B_o}\right)$	$\dfrac{1}{B_o}$
Gaussian	$\exp\left\{-(4\ln2)\left(\dfrac{v - v_o}{B_o}\right)^2\right\}$	$\dfrac{1}{B_o}\dfrac{\sqrt{2\ln2}}{\sqrt{\pi}}\exp\left\{-(2\ln2)\left(\dfrac{f}{B_o}\right)^2\right\}$	$\dfrac{0.66}{B_o}$
Lorentzian	$\dfrac{B_o^2}{B_o^2 + (2(v - v_o))^2}$	$\dfrac{B_o/\pi}{B_o^2 + f^2}$	$\dfrac{0.32}{B_o}$

signal powers the shot noise is actually dominated by the ASE average power contribution.

Reflection Noise/Multipath Interference.   The presence of optical reflections within the optical amplifier, such as those shown in Figure 13.18 will cause an interferometric conversion of laser phase noise into intensity noise. This intensity noise degrades the SNR at the optical receiver.[6,7] The converted noise is known as multipath interference noise or MPI. Important parameters determining the magnitude of MPI are the reflection levels, the optical gain, the signal linewidth and the time delay between the reflectors. The presence of optical gain can greatly increase the impact of small reflections. The worst-case RIN occurs when the polarizations of the delayed and non-delayed beams are aligned and the average interference phase is near quadrature. The interferometer is said to be in quadrature when the output power is halfway between its minimum and maximum values.

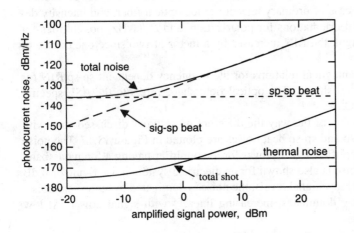

Figure 13.17 Total and individual noise contributions from an EDFA as a function of the amplified signal power.

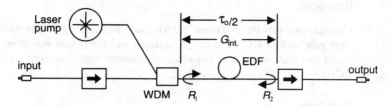

Figure 13.18 Fiber optic amplifier showing optical reflections contributing to multi-path interference noise.

The following relations can be used to estimate the RIN generated by a pair of small optical reflections when illuminated by a laser with a Lorentzian-shaped optical spectrum. The two reflections, denoted by R_1 and R_2 are assumed to satisfy: R_1, $R_2 \ll 1$. The expected RIN for three regimes of coherence (see Chapter 5) and with the interferometer in quadrature is:

any coherence

$$\text{RIN}_{\Delta\phi_{max}}(f) = \frac{4G_{int}^2 R_1 R_2}{\pi} \frac{\Delta\nu}{\Delta\nu^2 + f^2}[1 + e^{-4\pi\Delta\nu\tau_o} - 2\cos(2\pi f\tau_o)e^{-2\pi\Delta\nu\tau o}] \quad (13.31)$$

coherent case

$$\Delta\nu\tau_o < 0.1, \quad f < 1/2\pi\tau_o, \text{ maximum RIN}$$

$$\text{RIN}_{\Delta\phi_{max}}(f = 0) = 16\pi\Delta\nu\,\tau_o^2 G_{int}R_1 R_2 \quad (13.32)$$

incoherent case

$$\Delta\nu\,\tau_o > 1$$

$$\text{RIN}_{\Delta\phi_{max}} = \frac{4G_{int}^2 R_1 R_2}{\pi} \frac{\Delta\nu}{\Delta\nu^2 + f^2} \quad (13.33)$$

where $\Delta\nu$ is the laser FWHM linewidth, f is the baseband frequency, τ_o is the delay time of the reflected light and G_{int} is the optical gain of the medium separating the reflections. The gain-reflection product is assumed to be small (in other words, $G_{int}R_1 R_2 \ll 1$) for the above equations to hold.

In the coherent case, the conversion of phase noise into intensity noise increases as the square of the distance separation between the reflections. Therefore, when using highly coherent lasers in test systems, the lead lengths should be kept as short as possible to reduce the phase noise to intensity noise conversion. As the product of the laser linewidth and the delay τ_o increases, the noise spectrum tends toward a Lorentzian function as defined by Equation 13.33. The gain, G_{int} causes a significant increase in the MPI-induced RIN. Thus reflections must be kept small when optical gain is present. An understanding of the parameters affecting the MPI process can be applied to improve the optical amplifier design or the amplifier test system to limit the effects of this unwanted noise.

Example

Calculate and plot the maximum MPI-induced RIN for a 1.55 μm optical amplifier with a fiber gain section 1.71 m in length. Assume −45 dB optical reflections at each end of the 30 dB gain section and laser linewidths (Lorentzian approximation) varying from 10 MHz to 1 GHz.

Solution

The linewidth-delay time product is calculated first. A fiber refractive index of $n = 1.46$ is assumed. The delay time is calculated ($\tau_o = 2nL/c$) to be 16.6 ns. Thus the smallest and largest $\Delta\nu\tau_o$ products are 0.166 and 166 for linewidths of 1 MHz and 1 GHz respectively. Equation 13.31 is valid for this wide range of the $\Delta\nu\tau$ product. The estimated MPI RIN is shown in Figure 13.19.

From the above discussion, it is interesting to note that the RIN generated by the reflections internal to the amplifier depends on the linewidth of the source. Therefore, the noise generated by the amplifier is a function not only of the laser wavelength and power, but the signal linewidth as well.

13.4 NOISE FIGURE

The amplifier noise figure is a figure of merit quantifying the SNR (related also to the carrier-to-noise ratio, CNR) degradation after passage through the amplifier. Large-noise figures are detrimental to system performance, it causes poor received SNRs, increased jitter in soliton-based systems and ASE accumulation in long-haul amplified links. The main contributors to the noise figure are the effects of amplified spontaneous emission generated within the amplifier and importantly for analog communications, the phase-noise to intensity noise conversion due to internal optical reflections. The ASE manifests

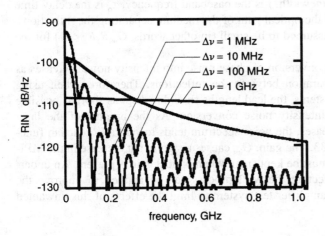

Figure 13.19 Optical amplifier MPI-induced relative intensity noise versus frequency for various signal laser linewidths.

itself through the generation of beat noise. In this section, the noise figure is defined, and later cast in a form that can be readily applied from a measurement standpoint.

13.4.1 Noise-Figure Definition

The degradation of the SNR after passage through an optical amplifier is quantified in terms of the noise figure, F, defined as

$$F = \frac{SNR_{in}}{SNR_{out}} \tag{13.34}$$

In the discussions that follow, the noise figure in decibels is determined according to: $F = 10 \log(F)$. The SNRs are referred to the output of an ideal photodetector which is capable of converting each photon of incident light into electrical current (in other words, 100% quantum efficiency). The input SNR is defined to be that from a shot-noise-limited source. The shot-noise-limited input reference is critical to the definition. If an optical source with a large amount of intensity noise were used to measure the noise figure of an amplifier, the amplified source noise would dominate over the amplifiers own noise contribution and lead to an erroneous noise figure of 0 dB, in other words, no observed SNR degradation caused by the amplifier.

The noise figure concept is illustrated in Figure 13.20. The input SNR is determined with the amplifier bypassed using an idealized source and receiver. The amplifier is inserted and the output SNR is determined. Equation 13.34 is next used to calculate the amplifier noise figure. The idealized source is shot-noise-limited and set to the appropriate

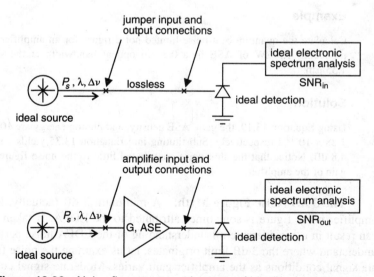

Figure 13.20 Noise figure concept in terms of idealized source and receiver.

power, wavelength, and linewidth. The idealized receiver has a calibrated frequency response and contributes no excess noise of its own. Obviously the real world is not yet ideal and much of the work involving noise figure measurements is in dealing with the source and the receiver non-idealities.

Signal-Spontaneous Beat-Noise-Limited Noise Figure. A commonly used definition of noise figure is the quantum-beat-noise-limited noise figure, sometimes referred to as the sig-sp beat-noise-limited noise figure. This noise figure is more restricted than the general definition defined by Equation 13.34. It doesn't include, for example, the output SNR degradation due to sp-sp beat noise, or MPI. The quantum-beat-noise-limited noise figure is advantageous in simplifying measurement procedures. It is measured with both optical and electrical techniques which are discussed later.

The quantum-beat-noise-limited noise figure is derived using Equations 13.18, 13.25, and 13.27 in Equation 13.34:

$$\mathrm{NF} = \underbrace{\frac{2\rho_{ASE}}{G\,hV}}_{\mathrm{sig-sp}} + \underbrace{\frac{1}{G}}_{\mathrm{shot}} \qquad (13.35)$$

where ρ_{ASE} is the amplifier ASE output density in the same polarization and wavelength as the signal as defined in Equation 13.18. This noise-figure definition is useful because of the ease with which it can be implemented. The ability to correlate noise figure measurement results between different laboratories is improved when this definition is used. The shot noise effect on the noise figure is sometimes excluded for noise calculations involving concatenated amplifiers.

Example

Calculate the quantum beat-noise-limited noise figure for an amplifier with 30 dB of gain, producing 12 μW of ASE in a 0.5 nm optical bandwidth at the signal wavelength of 1.55 μm.

Solution

Using Equation 13.19, the gain, ASE density, and photon energy are 1000, 0.19 f W/Hz, and 1.28×10^{-19} J respectively. Substituting into Equation 13.35 yields a noise figure of 3.0 or 4.8 dB. Notice that the shot noise contributed little to the noise figure because of the high gain of the amplifier.

The 3 dB Noise-Figure Myth. A minimum 3 dB (actually, $\log_{10}(2) = 3.01$ dB) amplifier noise figure is sometimes attributed to the EDFA. If taken out of context this can result in a considerable misunderstanding of the EDFA noise performance. To better understand where the 3 dB limit originates, let us examine the noise figure under moderate signal conditions as the amplifier gain varies. Moderate signal conditions imply that the signal power is much greater than the ASE power in the optical bandwidth of interest. This ensures that the sig-sp beat noise dominates over that of the sp-sp beat noise as indicated by Figure 13.17. Consider the case of a fiber amplifier where initially there are no erbium ions in the "active" optical fiber. Discounting any loss in the optical fiber, the noise figure is unity, in other words, no SNR degradation since the signal passes from amplifier input to output unchanged. As the erbium-ion-doping increases, so does the optical gain, the ASE level, and the signal level. The noise figure increases from 0 dB to 3 dB, or beyond, if other noise sources or optical losses are present. This can be seen from the equation for sig-sp beat noise and shot-noise-limited noise figure derived by substituting Equation 13.18 into Equation 13.35:

$$F = 2n_{sp} \frac{(G-1)}{G} + \frac{1}{G} \tag{13.36}$$

which for large gains yields: $F \approx 2n_{sp}$ where the SE factor, $n_{sp} \geq 1$. A fully inverted amplifier can be achieved with 980 nm pumping resulting in an effective SE factor of unity which leads to a noise figure of 3 dB. Equation 13.36 is plotted versus gain in Figure 13.21. From the figure, a fully inverted amplifier (in other words, $n_{sp} = 1$) with 4 dB of gain and zero input coupling loss has a noise figure near 2 dB. The 3 dB value is the limit for a high-gain amplifier with zero input coupling loss and a fully inverted amplifying fiber. Any loss near the amplifier input, or departure from complete inversion will cause the noise figure to exceed 3 dB.

A special class of amplifiers, referred to as phase-sensitive amplifiers, can achieve a noise figure less than 3 dB. Most optical amplifiers in use, such as EDFAs are phase insensitive, which means that the amplifier gain does not depend on the optical phase of the input signal. Thus the noise generated by the amplifier is amplified in both the in-phase and quadrature phase components. This is the physical origin of the 3 dB limit in high

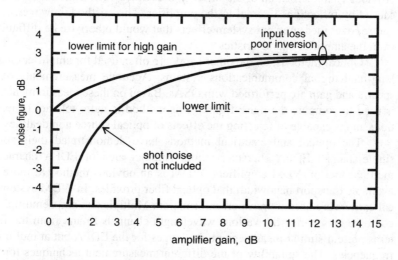

Figure 13.21 Noise figure dependence on optical amplifier gain with and without the shot-noise contribution.

gain amplifiers such as the EDFA. In calculations with concatenations of amplifiers, it is convenient to suppress the shot noise until the signals are analyzed at the detector. In the calculation of noise figure for a single amplifier, failure to include the shot noise will result in significant error in the gain regime below 15 dB.

13.5 CHARACTERIZATION OF GAIN AND NOISE FIGURE

The amplifier gain and noise figure are the fundamental parameters concerning their application to an optical communications link. By measuring these parameters over wavelength, power, input signal polarization, and temperature, the characterization of the amplifier is nearly complete. The measurement of noise figure requires the measurement of gain according to Equations 13.34 and 13.35. The noise-figure measurement techniques are classified into two groups:

- Optical method: optical spectrum analyzer-based,
- Electrical method: electrical spectrum analyzer-based.

Both methods have their merits. The selection of which to use depends on the application of the measurement result and the available instrumentation. Sometimes both electrical and optical methods are used.

The electrical method is often used in optical-amplifier characterization for analog-optical communications. It can be argued that this method provides a more complete noise figure since it directly measures the complete photocurrent noise at the receiver. Hence it provides characterization of amplifier nonidealities such as multipath interference (MPI) caused by reflections internal to the amplifier. This method, however, requires stringent control over measurement system effects that would otherwise be difficult to separate out from the amplifier characteristics.

Optical methods, on the other hand, are often used for amplifier characterization for long-haul digital communications systems. Accurate measurements of amplifier ASE spectra and gain are performed with OSAs. Based on these measurements, Equation 13.35 is used to compute the noise figure. This method is more tolerant of test system reflections and is capable of rejecting the effects of optical source nonidealities.

The optical and electrical methods have demonstrated their ability to perform single-channel EDFA characterization. Another area of EDFA characterization is for multichannel or WDM amplifiers. WDM is an obvious method to more fully utilize the available transport bandwidth that optical fiber provides. In WDM systems, multiple optical carriers are used to transport information. Multiplexers and demultiplexers are used to combine or separate the various wavelength channels to and from the fiber. WDM systems present similar measurement challenges for the EDFA but at multiple optical carrier frequencies. The suitability of the different measurement techniques for the WDM environment are considered separately in this chapter. At the risk of overgeneralization, Table 13.3 is provided to help compare the level of measurement difficulty, as well as the ap-

Table 13.3 Test Method Comparisons

Method	Measure MPI	Applicability for WDM	Difficulty: scale 1 to 5
optical: source subtraction	no	moderate	2
optical: polarization extinction	no	poor	3
optical: time-domain extinction	no	excellent	3
electrical	yes	moderate	4

plicability of the various techniques to WDM measurements. The relative merits of the different methods may change as the measurement art advances.

13.5.1 Amplifier Gain

Several methods of measuring the gain of an optical amplifier are discussed here. While not necessarily a complete list of all possible methods, it covers the most commonly used methods. In the discussions on noise figure measurement that follow, the issue of gain measurement with respect to the specific noise figure measurement techniques is also addressed.

Optical Power Meter: Optical Gain.   Measurement of optical gain can be performed using the simple approach shown in Figure 13.22. In this approach, the incident source power is measured along with the filtered amplifier output using an optical power meter. The system is calibrated by replacing the amplifier with a lossless connection. An important source of measurement error is the presence of ASE incident on the optical power meter. This is reduced significantly by filtering. Filtering the source reduces the effect of source spontaneous emission (SSE) on the amplifier saturation. The combined effects of ASE and amplified SSE contribute to a net gain measurement error. The gain measurement error, defined here as the ratio of the measured gain to the actual gain is given by:

$$\frac{G_m}{G} = 1 + B_o \frac{F\,h\nu}{P_s} \tag{13.37}$$

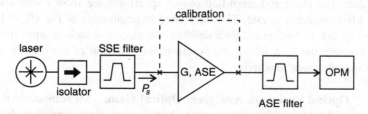

Figure 13.22 Optical amplifier gain measurement with an optical power meter and bandpass filters. OPM: optical power meter.

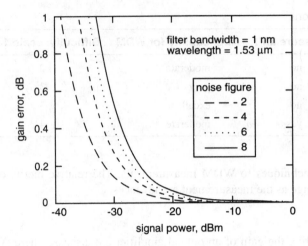

Figure 13.23 Gain measurement error due to additive effect of ASE.

where F is the noise figure, B_o the filter bandwidth, and P_s is the input power. This is plotted in Figure 13.23 for different noise figures assuming a 1 nm filter bandwidth. Input powers below approximately −20 dBm result in appreciable errors for the 1 nm filter bandwidth.

Electrical Spectrum Analyzer: Optical Gain.   One method to measure gain at lower input powers is to use source modulation in conjunction with a frequency selective receiver. An electrical spectrum analyzer can be used to measure a small intensity modulation index imparted onto the optical source. By performing measurements of the photocurrent spectrum at the modulation frequency with and without the optical amplifier, the gain can be determined. The advantage of this approach is that the signal is separated from the ASE by the modulation. The modulation frequency is set to be significantly faster than the inverse of the EDFA gain recovery time (~ 300 μs) so as not to modulate the ASE. The measurement set-up, using an electrical spectrum analyzer, is shown in Figure 13.24a. A measurement using this setup is shown in Figure 13.24b. The laser source (a DFB with an electroabsorption modulator) was sinusoidally modulated at a 10 MHz frequency and passed on to the receiver directly for gain calibration. Next the amplifier was inserted for the gain measurement. The measurement results are shown Figure 13.24b. The input and amplified output spectrums are shown with spectral peaks at the 10 MHz modulation rate. The difference in amplitudes of the 10 MHz modulation tone corresponds to the optical gain squared. An electrical lock-in amplifier may also be used to measure the gain where the synchronous detection of the lock-in amplifier improves the measurement sensitivity.[36]

Optical Spectrum Analyzer: Optical Gain.   Measurement of the amplifier gain using an OSA provides a more general evaluation of the amplifier. Information concerning the ASE spectral shape, source characteristics, and the presence of spurious signals such as pump laser feedthrough is obtained as well. The basic measurement setup is

(a)

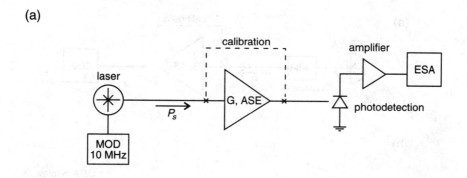

(b)

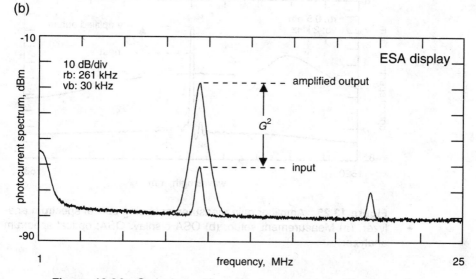

Figure 13.24 Optical amplifier gain measurement. (a) Measurement setup using modulated source. (b) Displayed ESA data. ESA: electrical spectrum analyzer.

shown in Figure 13.25a. A gain measurement of an EDFA under the same input signal conditions as in Figure 13.24b is shown in Figure 13.25b.

The amplifier dynamic gain spectrum can be readily measured by combining a small-probe signal with the laser that sets the EDFA saturation level. This is illustrated in Figure 13.26. The probe signal could be a broadband noise source such as an EELED or a tunable laser.[30,36,37] The EELED approach is more rapid if the tunable laser is not capable of sweeping synchronously in wavelength with the OSA. The tunable laser probe power has less of an impact on the amplifier saturation level for a given measurement SNR as compared with the EELED approach. The effect of the probe on the saturation level of the amplifier should be monitored closely. Preferably the probe power is set to a value less than the effective input noise of the amplifier as given by Equation 13.10, but this is not

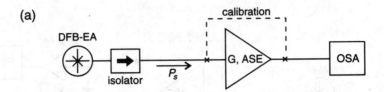

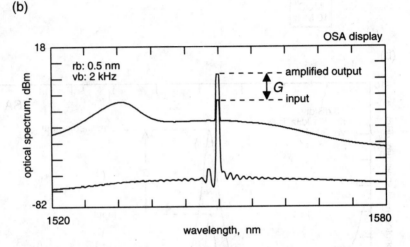

Figure 13.25 Amplifier gain measurement using optical spectrum analyzer. (a) Measurement setup. (b) OSA display. OSA: optical spectrum analyzer.

always practical in view of measurement speed and sensitivity considerations. Actual measurements using the EELED probe technique are presented in the discussion on the time-domain extinction method for noise figure measurement.

13.5.2 Measurement of Noise Figure

Generally, noise figure measurements involve two activities: (1) making noise and gain measurements, (2) removing the test-system noise contribution. Before discussing some of the various techniques available to measure amplifier noise figure, it is worth discussing one of the villains that manifests its presence in electrical and optical methods: laser noise.

    Source Spontaneous Emission. One of the challenges that the various noise figure characterization techniques have had to address in their evolution was how to deal with the excess noise present with optical sources. The excellent noise performance of the EDFA allows laser noise to mask the observed amplifier noise. Recall that the noise fig-

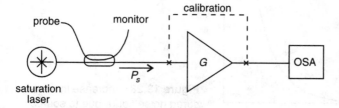

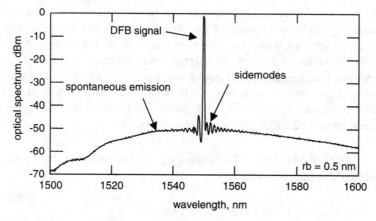

Figure 13.26 Optical amplifier gain measurement with small-signal probe combined with strong signal laser.

ure was defined in terms of a shot-noise limited source. A typical laser source, such as a DFB-LD, consists of an optical amplifier made from semiconductor material with distributed mirrors providing feedback for the lasing process. The DFB-LD internal optical gain generates ASE in the same way as the EDFA. This source noise is referred to as SSE. The SSE is broad in bandwidth, as shown in the measured DFB-LD optical spectrum shown in Figure 13.27. In addition to SSE, laser side-modes are also observed adjacent to the coherent signal. A telltale sign of imminent SSE problems is the observation of source spectral structure in the amplifier output.

   Impact on Optical Methods.   The optical methods of noise figure measurement rely on a measurement of the ASE density at the EDFA output. Therefore, steps must be taken to insure that SSE doesn't cause an overestimate of the amplifier noise figure. The effect of SSE on the noise figure is shown in Figure 13.28. The measured noise figure (in linear units) will be the numerical error shown in Figure 13.28 added to the actual noise

Figure 13.27 Optical spectrum of a DFB laser showing sidemodes and spontaneous emission.

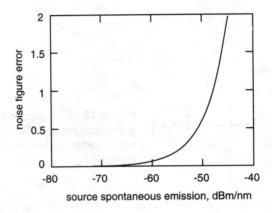

Figure 13.28 Increase in measured noise figure due to source spontaneous emission at amplifier input.

figure. An SSE density of −50 dBm/nm will add approximately 0.6 to the actual noise figure in linear units. It is apparent that to make accurate measurements, the absolute SSE level at the input to a low-noise amplifier should be below approximately −65 dBm/nm. The −50 dBm/nm SSE level of the DFB laser shown in Figure 13.27 would create a large measurement error, if unaccounted for. The simplest method to reduce the SSE is to attenuate the source to achieve the specified measurement accuracy. The short-coming of this approach is the need to measure the amplifier response at relatively high input levels. Reducing the bias current through a semiconductor laser to lower its output power does not alter significantly the SSE generation. An optical attenuator is preferred for setting power levels because it reduces both signal and noise power.

Impact on Electrical Methods. The electrical methods calculate noise figure based on measurements of photocurrent noise. If the optical source illuminating the amplifier is not shot-noise limited, there may be an increase in the measured noise. The error caused by this noise depends on the absolute level of the noise generated by the source at the amplifier input terminals. This error can be plotted in terms of relative intensity noise, (RIN) for various values of input signal power as shown in Figure 13.29. The figure indicates that lowering the input signal power relaxes the requirement for the laser RIN. As an example, an input signal power of −10 dBm from a laser with a RIN of −144 dB/Hz will cause a measurement error of 0.3 if unaccounted for. An amplifier with a noise figure of 2 would measure 2.3 (3.6 dB) with this laser source.

Optical Methods. To measure gain and noise figure, the optical methods must be able to determine the following:

- Gain,
- ASE spectral density,
- Wavelength.

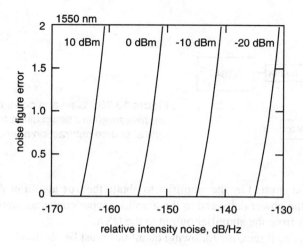

Figure 13.29   Noise figure error due to excess source noise in terms of RIN for different input signal power levels.

Three basic optical methods are used for noise figure measurement:[38] (1) source subtraction technique; (2) polarization nulling; (3) time-domain extinction or pulse method. The differences in these methods are in the ways the source SSE is accounted for and how the noise and gain are measured. The source subtraction method has the simplest setup and allows for rapid measurement of amplifier noise figure. It is best suited for single-channel environments but can be applied in WDM environments. Limitations on the spectral selectivity of OSAs make it difficult to measure close to the channel wavelength with this technique. The polarization extinction method has a more complicated setup with polarization synthesis requirements and longer measurement time. It is useful for single-channel amplifier characterization and allows for noise figure measurements close-in to the actual signal wavelength. The time-domain extinction method is useful for both single or multichannel testing of EDFAs. This method allows for noise figure measurement close-in to the actual signal wavelength and offers very rapid measurement through-put. In all the optical techniques, the amplifier ASE and gain are measured. From these measurements the noise figure is calculated according to Equation 13.35.

In the following discussions on noise figure measurement, the OSA bandwidth, B_o refers to the effective noise bandwidth of the OSA in units of Hertz. It may vary, by up to ~ 20%, from the nominal displayed resolution bandwidth. The reader is referred to Chapter 3 for more detail on this. In the next three sections, the three optical techniques are described in detail.

Optical Source-Subtraction Method.   The optical source subtraction method provides for straightforward characterization of the gain and noise figure performance of optical amplifiers. In its simplest form, the measurement setup consists of a laser to provide the input signal, and an amplitude calibrated OSA as shown in Figure 13.30. As discussed above, the SSE causes an error in the noise figure measurement. With the optical source subtraction approach, the SSE is carefully measured during calibration and later

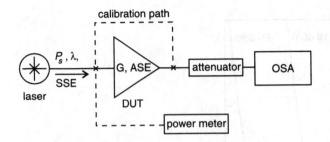

Figure 13.30 Gain and noise figure measurement setup using the optical source subtraction method.

subtracted from the total noise emitted by the amplifier to obtain the true amplifier ASE. In the calibration sequence, the power delivered to the amplifier input connector must be determined as well as the loss from the amplifier output to the OSA.

To measure amplifier noise figure, the following quantities must be obtained:

- ν — optical signal frequency
- P_s — signal power incident at the amplifier input
- P_{out} — total amplifier output power within the OSA resolution bandwidth measured at the signal wavelength including ASE and amplified SSE.
- P_{ASE}—total noise spectral density from the EDFA, including SSE, at the signal wavelength due to both polarizations
- P_{SSE}— SSE spectral density at the signal wavelength caused by both polarizations

From these quantities, the gain and quantum-limited noise figure are calculated according to Equation 13.1 and:

$$NF = \frac{P_{ASE}}{G\,h\nu\,B_o} + \frac{1}{G} - \frac{P_{SSE}}{h\nu\,B_o} \qquad (13.38)$$

noise figure – SSE correction

The last term performs the subtraction of the amplified SSE. The presence of the amplified signal prohibits the measurement of P_{ASE} and P_{SSE} at the signal wavelength. Interpolation is required to estimate these noise powers at the signal wavelength.

Experiment: Interpolation-Source Subtraction. The noise figure and gain of an EDFA was measured at a wavelength of 1.55 μm using an interpolation technique with source subtraction to remove the effects of SSE. The setup is shown in Figure 13.30. The input signal, P_{in}, was provided by a tunable external cavity laser (HP 8168). An optical attenuator (HP 8157) reduced the high powers from the EDFA to an acceptable level for the OSA (HP 71450). The input signal, was first measured with an optical power meter (HP 8153/HP 81532A). Next the OSA and attenuator were calibrated as an ensemble by comparison with the power meter reading. The correction was +1.4 dB. The SSE was measured at the ±1 nm offset interpolation wavelengths using the built-in noise marker func-

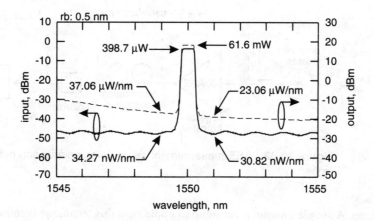

Figure 13.31 Noise figure and gain measurement of an EDFA using the optical source subtraction method combined with ASE interpolation.

tion which refers the measured noise to a 1 nm equivalent noise bandwidth. The corrected signal input and EDFA output spectrums are shown in Figure 13.31. The attenuator was set to 20 dB when the amplifier was inserted into the measurement setup. The EDFA output ASE measurements were made at the interpolation wavelengths and the amplified signal power was measured. The resulting data is tabulated in Table 13.4 along with the noise figure computed according to Equation 13.38.

Polarization Extinction. The polarization extinction method offers an alternative way to reduce the error in the noise figure measurement due to SSE. An additional benefit offered by this method is that for a given OSA wavelength resolution, measurement of ASE can be performed closer to the optical carrier than with the source subtraction tech-

Table 13.4 Noise Figure Measurement Data

Parameter	Value	Units
P_s: power meter	398.7	μW
P_s: OSA	306.2	μW
P_{SSE}*	32.5	nW/nm
P_{out}*	61.6	mW
Gain	21.9	dB
P_{ASE}	30.1	μW/nm
ASE density	25.1	μW/nm
Noise Figure	10.1	dB

*With power meter correction.

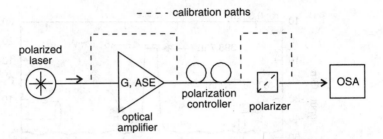

Figure 13.32   ASE measurement setup using polarization nulling technique.

nique. A simple measurement setup to implement this technique is shown in Figure 13.32. For this technique to work, the EDFA ASE should be unpolarized and the SSE must be polarized in the same polarization state as the laser signal. The polarizer in Figure 13.32 allows separation of signal and SE according to polarization states. When the polarization controller is set to suppress the laser signal at the OSA, the SSE will also be suppressed, or extinguished over a certain optical bandwidth. Under these conditions, the OSA measures half the EDFA-produced ASE with reduced measurement corruption by SSE. To obtain the total ASE, the amplitude of the measured spectrum is multiplied by two.

The measurement is calibrated by measuring the loss from the laser to the OSA display as well as characterizing the loss of the polarizer to a signal aligned to the transmitted polarization state. With the system calibrated, the following quantities are determined:

- v — optical signal frequency
- P_s — signal power incident at the amplifier input
- P_{out} — output power measured at the signal wavelength (polarizer bypassed)
- P_{ASE} —total noise at the signal wavelength (polarizer bypassed)
- P_N — ASE power at the signal wavelength (signal nulled)

The OSA amplitude response is calibrated with an average optical power meter. OSAs with noise marker capability simplify the measurement of noise densities by automatically referring the measured noise to a 1 nm effective noise bandwidth. Often it is not practical to completely suppress the amplified signal and so some residual signal may be observed with the OSA. Thus interpolation may be required to estimate P_N, at the signal wavelength.

Once the signal and ASE powers at the signal wavelength are determined, the optical gain and noise figure is calculated according to Equations 13.1 and 13.35 where $\rho_{ASE} = P_N/B_o$.

Impact of Polarization Hole-Burning.   Polarization hole-burning as discussed in Section 13.2.3 results in a lower amplifier gain in the same polarization as the saturating signal.[5] The gain and ASE will be larger in the polarization orthogonal to the saturating signal polarization. Therefore a measurement of the ASE when the signal is nulled will actually be larger than the ASE in the same polarization as the signal laser. Recall that the noise figure depends strictly on the ASE in the same polarization as the signal. If polarization hole-burning is significant, noise figure measurement by polarization extinction will yield an overestimate of the actual noise figure of the amplifier.

Impact of Polarization Mode Dispersion.   Polarization mode dispersion (see Chapter 12) will affect the ability of the measurement systems to reject the SSE over a broad spectral bandwidth. This is caused by the wavelength-dependent birefringence in the measurement system and test amplifier. While the polarization controller in Figure 13.32 can be set to null the signal at one wavelength, the SSE, with its large spectral extent will not be nulled across all wavelengths. A measurement of the null width of the measurement setup of Figure 13.32 is shown in Figure 13.33. In Figure 13.33a the amplifier of Figure 13.32 was bypassed and the laser was set to 1555 nm while the polarization controller was adjusted to obtain a null of approximately 60 dB at the laser wavelength. Next the laser wavelength was tuned from a wavelength of 1520 nm to 1570 nm while the OSA was set to record the maximum observed signal level (bottom trace of Figure 13.33a. Next the polarization controller was adjusted for maximum signal transmission and the laser wavelength was tuned across the band (top trace of Figure 13.33a). It is apparent that PMD in the test system limited the null-width resulting in a null depth in excess of 35 dB across most of the band.

Next the EDFA was inserted and the measurement procedure was repeated. The measured null width was considerably reduced as shown by the bottom trace in Figure 13.33b. The null width for 30 dB extinction was reduced to about 1 nm for this EDFA.

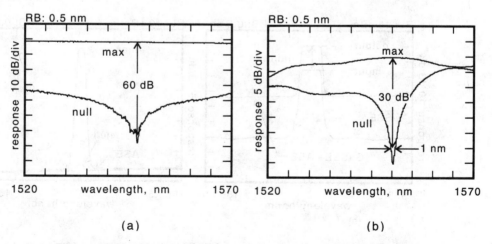

(a)   (b)

Figure 13.33 Measurement of signal rejection using polarization extinction at a wavelength of 1550 nm (a) EDFA bypassed; (b) EDFA inserted.

The top trace corresponds to the case where the polarization controller was set for maximum signal transmission at a wavelength of 1555 nm. From the figure, it is apparent that the degree of suppression of the SSE in the noise figure measurement depends heavily on the PMD in the test system and importantly, on the EDFA. Therefore the null width and depth should be characterized to insure adequate extinction of SSE is obtained prior to the measurement of ASE.

Example: Noise Figure and Gain Measurement

The noise figure and gain of an EDFA was measured using the setup shown in Figure 13.32. The loss through the polarizer was calibrated using the unpolarized ASE from the EDFA as a signal source. The polarizer loss was measured to be 1.1 dB. Next the input signal to the amplifier was characterized by connecting it directly to the OSA, the measurement result is shown in the lower trace of Figure 13.34a. Comparison with a calibrated power meter indicated a correction of 0.47 dB to the OSA readings was required.

The SSE was found to be approximately −53 dBm in a 1 nm noise bandwidth. From Figure 13.28, at an SSE density of −50 dBm/nm (add 3 dB to the SSE since the measured ASE is multiplied by 2) the SSE will add approximately 0.6 to the measured noise figure if not rejected. The gain was measured by connecting the amplifier output directly to the OSA (polarizer was bypassed) as shown by the top trace in Figure 13.34a.

With the polarizer and EDFA in place, the polarization controller was set to maximize the signal displayed on the OSA (Figure 13.34b). This measurement provides a reference to determine the degree of polarization extinguishing achieved by the measurement setup. Next the polarization controller was set to null the signal as indicated by the lower trace in Figure 13.34b. Here the signal could not be completely nulled, thereby requiring the ASE to be estimated by interpolation. With the signal nulled, the ASE was measured at a 0.3 nm spacing from the signal wavelength as shown. While the signal null obtained was approximately

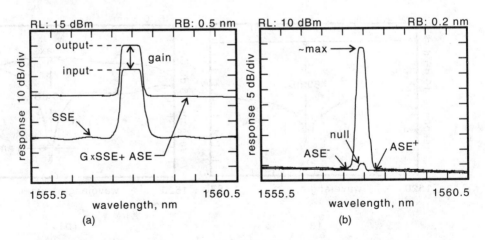

Figure 13.34 Measurement of EDFA gain and noise performance at a wavelength of 1558 nm using polarization extinction. (a) gain (b) ASE generation.

38 dB, the finite null width limited the extinction of SSE to approximately 26 dB at the ASE measurement frequencies. According to Figure 13.28, the 26 dB rejection will make the error due to the presence of SSE negligible. Some of the measurement data and results are shown in Table 13.5.

The noise figure is calculated using the data in Table 13.6 after conversion to MKS units. The ASE density is converted to watts/Hz using Equation 13.19. The noise figure is calculated according to Equation 13.35 yielding 6.1 dB for this amplifier.

Time-Domain Extinction/Pulse Techniques.   The slow gain dynamics of the EDFA may be used advantageously for the measurement of amplifier gain and noise figure using the time-domain extinction (TDE) technique.[39] In this section, the basic TDE concept will be discussed. A discussion of the transient response of the EDFA and its impact on the measurement is also given. Next the TDE method is extended to cover the measurement of dynamic gain (see Section 13.2.5) using an incoherent probe. This extended method has also been referred to as noise gain profiling.[37] TDE with dynamic gain-spectrum measurement enables the user to rapidly characterize the wavelength-dependent gain profile for a particular amplifier saturation condition.

Time-Domain Extinction.   In the TDE method, the input signals, that would otherwise interfere with the measurement, are momentarily gated off.[39,40] The gated light is usually the signal or signals that affect the saturation level of the amplifier. By temporarily blocking the signals, several measurements can be performed: (1) measure the amplifier-generated ASE, (2) probe the amplifier with a small signal to measure the dynamic gain (noise-gain profiling).[37]

The basic measurement setup is shown in Figure 13.35. The cw optical saturating source leading to the amplifier is gated at frequency, f_{TDE}. When the input switch (SW1 in the figure) is opened, the output switch (SW2) is closed allowing the OSA to sample the amplifier ASE output. The switching may be performed either electronically (within the laser source and the OSA) or optically (for example, using acousto-optic modulators). Since the signal is extinguished during the measurement, it does not corrupt the measure-

Table 13.5   Noise Figure Measurement Data

Symbol	Parameter	Value	Units
λ_s	wavelength	1558	nm
P_s	input power	70.4	μW
G	gain	38.2	linear
$\Delta\lambda$	interpolation offset	0.3	nm
ρ^+_{ASE}	ASE density at $\lambda_s + \Delta\lambda$	1.18	μW/nm
ρ^+_{ASE}	ASE density at $\lambda_s - \Delta\lambda$	1.25	μW/nm
$\rho^+_{\text{ASE}} = P_N/B_o$	interpolated ASE density	1.22	μW/nm

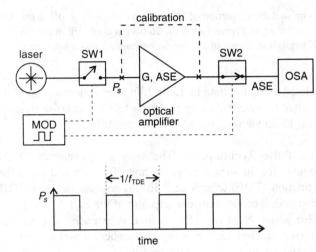

Figure 13.35 Optical amplifier ASE measurement setup for the time-domain extinction technique. OSA measures when signal is gated off. Gain measurement is performed with switches operating in unison.

ment of ASE required for noise figure calculations. This is illustrated for slow gating in Figure 13.36. Spiking in the amplifier output is observed because of the increased energy storage in the amplifier when the signal was gated off. Stimulated emission of the input signal quickly reduces the energy stored (causing spiking) to achieve a steady-state value towards the end of the gated-on period. At low repetition rates (below ~ 1 kHz) and high input powers, spiking may actually cause self-destruction of the EDFA.

When the source is gated at higher frequencies, the spiking is no longer present and the ASE follows a triangle-shaped waveform as shown in Figure 13.37. Increasing the gate frequency reduces the ASE peak-to-peak variation. In the limit of an infinite gate frequency, the triangle waveform converges to its average value. For gate frequencies above

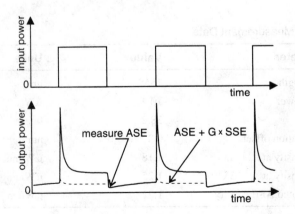

Figure 13.36 Input and output waveforms for low frequency time-domain extinction measurement.

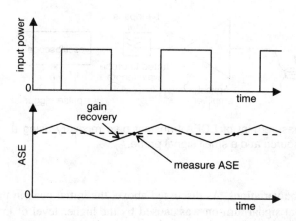

Figure 13.37 Input and output waveforms for high frequency time-domain extinction technique.

approximately ~ 20 kHz *the ASE sampled at the midpoint through the gain recovery time corresponds to the average ASE.* The actual low frequency limit depends on factors such as the applied pump power to the EDFA and the presence of active control circuits for output power leveling. The effective input signal power to the amplifier is the average power of the gated signal as measured with an optical power meter. Once the ASE is determined, the amplifier noise figure is calculated according to Equation 13.35. The amplifier gain can be characterized by a variety of techniques. A broadband EELED probe technique (discussed further on) may be included in the measurement setup to provide a rapid gain measurement across many wavelengths.

EDFA Transient Response.   The mechanisms of pump absorption, energy storage, stimulated emission, and fluorescent decay all contribute to determine the response of the EDFA to changes in signal level.[4] When the signal level is abruptly changed, the EDFA will slowly stabilize to a new level of gain. The recovery time after the signal is gated-off is of the order of 100 μs but depends strongly on the EDFA pump power. In general, the recovery characteristic, measured in terms of the transient response of a small probe signal, varies approximately exponentially with time. Using the approximation: $e^{at} \approx 1 + at$, for small time t, the recovery will follow a linear recovery immediately after the signal is gated off. In the method discussed here, the data sampling must be made in the regime where the gain recovery varies linearly with time to obtain accurate measurements.

Characterizing the EDFA gain recovery is straightforward. The measurement setup used in the following experiment is shown in Figure 13.38. In the measurement, the counter-propagating-pumped EDFA delivered 32 mW of 1480 nm pump light to the erbium-doped amplifying fiber. Isolators at the amplifier ports shielded the unit from reflections. The saturating signal wavelength was set to 1533 nm and the input signal power was square-wave modulated. This caused gain modulation at all wavelengths in the erbium band. The gain modulation was measured with a small signal laser probe tuned to a wavelength of 1552 nm. The magnitude of the gain variation depended strongly on modulation rate as shown in Figure 13.39 where the amplified saturating signal, and the small signal probe are plotted. Where the gain perturbation is small, an expanded scale with off-

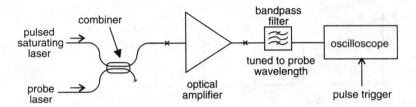

Figure 13.38   Measurement of EDFA transient gain recovery using a pulsed saturating source and a small signal probe.

set is used to show the probe variations. As discussed above, the initial peak in the saturating signal output power at signal turn-on was caused by the higher level of inversion realized after gain recovery. The signal transient can be quite rapid, with amplifier dependent decay-time constants on the order of ~ 10 µs. The reduction of the probe signal gain is evident as the saturating signal depletes the erbium inversion level. As the rate of modulation is increased, the amplifier recovery is truncated, the maximum instantaneous amplifier inversion level decreases, and the gain modulation at the probe wavelength of 1552 nm decreases. With further increases in modulation rate, the gain modulation takes on a triangle waveform profile with a peak-to-peak variation decreasing with modulation rate.

Gain Recovery Errors.   Proper use of the TDE method requires setting the signal gate frequency to the appropriate value. The gate frequency should be set such that the gain recovery has a linear response. If the gate frequency is set too low, the recovery is not linear as indicated by the probe gain recovery in Figure 13.39a. Deviation from a linear response leads to measurement error since the measurement at the midpoint of the gain recovery characteristic is no longer representative of the average high-frequency value.

The error caused by nonlinear gain recovery was measured for an EDFA. The results are shown in Figure 13.40 The error is defined as the ratio of the measured probe gain at the test frequency divided by the probe gain measured at 300 kHz. Below a gate-rate of 10 kHz, the error significantly increases. This was performed for several different values of 1480 nm pump power incident on the erbium-doped fiber. From the data, it can be shown that the error due to nonlinear gain recovery increases linearly with pump power.

Transient Gain Saturation.   Transient gain saturation is responsible for the spiking observed in Figures 13.36 and 13.39. This spiking does not affect the probe gain measurement, but it can impact the measured output signal power, which in turn affects the accuracy of the gain calculation. The error in the determination of output power occurs when the power is measured midway through the non-linear transient gain saturation recovery. This problem is solved by increasing the gate frequency and making an average power measurement. In practice, the gate frequency may be set appropriately by increasing it from ~ 20 kHz, until the transient gain saturation recovery becomes approximately

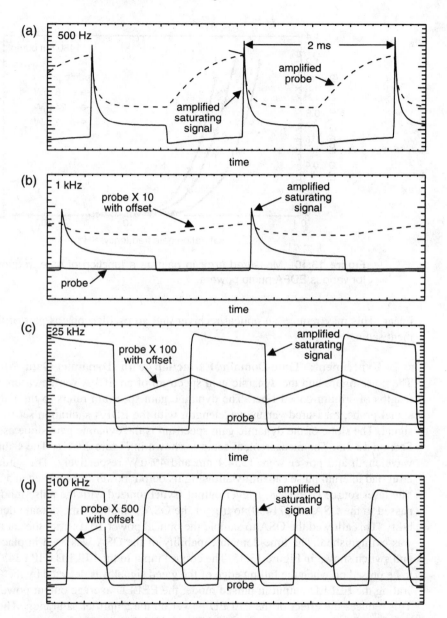

Figure 13.39 Measurement of EDFA transient response for different gate rates. (a) 500 Hz. (b) 1 kHz. (c) 25 kHz. (d) 100 kHz.

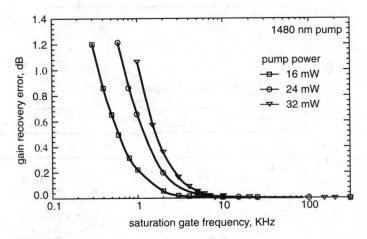

Figure 13.40 Measured error in gain as a function of gate frequency for various EDFA pump powers.

linear. This may require an optical receiver and an oscilloscope as part of the measurement setup.

Experiment: Time-Domain Extinction with Dynamic Gain Measurement.
The noise figure and the dynamic gain spectrum of an EDFA were measured for various lengths of erbium-doped fiber. The dynamic gain spectrum refers to the gain of a small signal probe, measured versus wavelength, with the EDFA saturation set by a saturating signal. The slope of the dynamic gain spectrum is the dynamic gain slope as discussed in Section 13.2.5. The measurement setup is shown in Figure 13.41a. The saturating signal wavelength and power were 1554.4 nm and 40 μW respectively. The saturating signal laser had internal pulse capability allowing its output power to be pulsed at a 25 KHz rate. The laser source (HP 8168) trigger output (synchronized with the pulse modulation) was passed to the OSA (HP 71450) to trigger the OSA's internal time-domain detection capability. This allowed the OSA to sample the optical power 10 μs after the saturating signal was extinguished. The time-domain capability of the OSA was used in place of the second switch shown in Figure 13.35. The control input to the EELED (HP 83437) allowed it to be pulsed on when the falling edge of the gated signal was detected by the OSA. By operating the EELED output in pulsed mode, the EELED average output power is reduced. This reduces the effect of the EELED power on the amplifier saturation. The timing diagram for the measurement is shown in Figure 13.41b. Initially, the measurement was calibrated by bypassing the EDFA as shown in the figure. The EELED was pulsed on by a command from the OSA and its power was measured by the OSA 10 μs after the falling edge of the saturating signal laser. This yielded the wavelength-dependent calibration $P_{cal}(\lambda)$:

$$P_{cal}(\lambda) = P_{LED}(\lambda) \qquad (13.39)$$

(a)

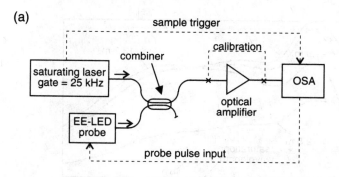

(b)

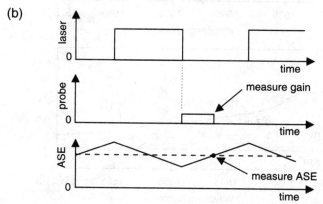

Figure 13.41 Time-domain extinction measurement with dynamic gain spectrum measurement. (a) Measurement setup; (b) optical waveforms for saturating laser, EE-LED probe and EDFA ASE.

Next the EDFA was inserted into the test setup and the OSA measured the output spectrum ($P_1(\lambda)$) 10 μs after the saturating laser was gated off, with the EELED pulsed on. This measurement yielded:

$$P_1(\lambda) = G(\lambda) \times P_{LED}(\lambda) + P_{ASE}(\lambda) \tag{13.40}$$

To solve for gain $G(\lambda)$, the ASE spectrum ($P_2(\lambda)$) is measured with the EELED output off, 10 μs after the saturating signal was gated off, yielding:

$$P_2(\lambda) = P_{ASE}(\lambda) \tag{13.41}$$

The dynamic gain spectrum is calculated according to:

$$G(\lambda) = \frac{P_1(\lambda) - P_2(\lambda)}{P_{cal}(\lambda)} \tag{13.42}$$

The noise figure is computed according to Equation 13.35 using Equation 13.19 and $\rho_{ASE}(\lambda) = P_2(\lambda)/2B_o$.

After each gain and noise figure measurement, the erbium-doped fiber in the amplifier was reduced in length. The results of the noise figure and dynamic gain spectrum for

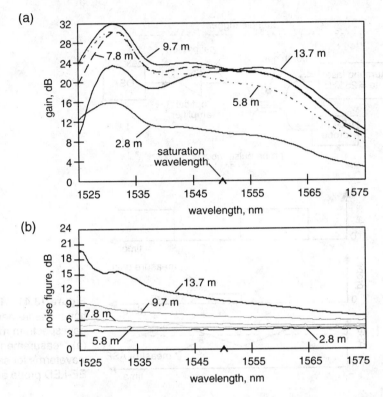

Figure 13.42 EDFA dynamic gain and noise figure measurement for various lengths of erbium-doped fiber. (a) Gain. (b) Noise figure.

the various lengths are shown in Figure 13.42. From the curves, it is apparent that the amplifier can be optimized separately in terms of maximum gain, lowest noise figure, and gain flatness. The gain and noise figure results are plotted, in Figure 13.43, versus length for the two wavelengths of 1530 nm and 1550 nm. From the figure, it is clear that the optimum gain length does not correspond to the lowest noise figure. The erbium-doped fiber length is particularly critical for low noise operation in counter-pumped amplifiers, such as the one measured here, since the noise figure depends heavily on the inversion level near the input. The sensitivity of the amplifier performance at 1530 nm to fiber length (and inversion level near the input) is a reflection of the fact that the absorption cross-section of the erbium-doped fiber peaks near this wavelength (see Figure 13.7).

WDM Characterization. Wavelength division multiplexed (WDM) transmission is a versatile method to increase the transmission capacity of singlemode optical fiber. Laboratory experiments using WDM have demonstrated over 2.6 Tb/s transmission capacity along a single fiber. In Figure 13.44, the input and output spectrums are shown for a very dense 1.1 Tb/s WDM transmission experiment using a transmitter composed of

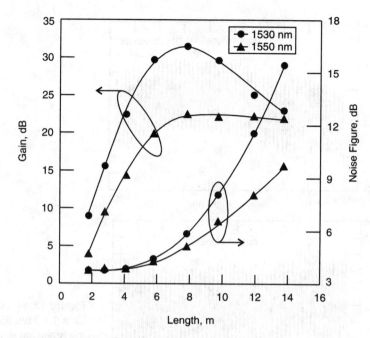

Figure 13.43 Gain and noise figure versus erbium-doped fiber length at the wavelengths of 1530 nm and 1550 nm.

55 lasers, each modulated at a 20 Gb/s rate.[41] The wavelength range of the lasers spanned most of the EDFA gain range from 1529 nm to 1565 nm as shown in the figure. Qualification of EDFAs for WDM applications will require measurement of gain and noise figure over a broad range of wavelengths. In principle, characterization of EDFAs for the WDM environment is an extension of the techniques used for single wavelength test. In WDM gain characterization, the optical gain is measured at each channel wavelength for a set of input conditions. The input conditions include the channel powers and wavelengths or "events" such as an added or dropped channel.

   One of the principle challenges for making WDM noise figure and gain measurements is the required assembly of the large numbers of lasers. This process can be complex and costly to maintain. Measurement of the ASE generation at a specific wavelength also becomes more difficult as channel spacings become narrow.

   In this section, the TDE method combined with dynamic gain-spectrum measurement is discussed for characterization of WDM gain and noise figure. The polarization extinction method is not covered since polarization nulling of a large number of wavelength channels is time-consuming and not practical due to the polarization-mode dispersion (PMD) effects discussed in Section 13.5.2.

   The source-subtraction technique may be used for WDM characterization of EDFAs. Careful attention to the stability of the WDM lasers, and the effects of overlap-

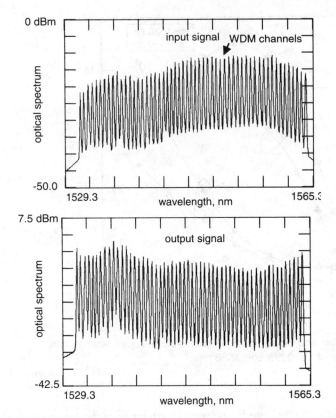

Figure 13.44 Optical spectrums for a 1.1 Tb/s 55 channel WDM transmission experiment: (a) input signal (b) output signal after passage through 150 km optical fiber containing two EDFAs. (With permission, after ref [41] ©1996, Optical Society of America.)

ping SSE from the WDM lasers at the measurement wavelength are required. Because of the similarities with the single-channel method, the reader is referred to the discussion on source-subtraction in Section 13.5.2.

A method allowing the same EDFA saturation effects due to multiple WDM channels while using fewer saturating signal lasers will be presented. This allows a significant simplification of the test system while maintaining amplifier conditions similar to those encountered with the actual number of channels.

Multichannel Method. The EDFA gain is measured by comparing the input and output powers at each channel wavelength. This is similar to the methods used for the single channel case discussed earlier. Instead of gating a single laser, for multichannel TDE test, all the lasers in the WDM transmitter are synchronously gated. The addition of the small signal probe, as discussed in Section 13.5.2 permits measurement of the dynamic gain spectrum allowing gain shape measurement between the WDM channels. The multichannel TDE method with the small-signal gain probe will be compared to the multichannel spot gain measurement method in an experiment with a four channel WDM source.

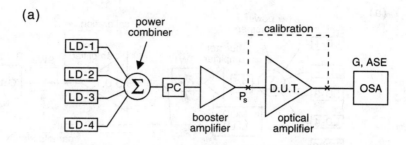

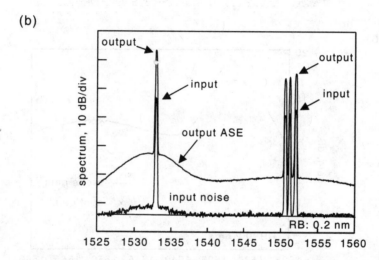

Figure 13.45 Multichannel gain measurement of an EDFA: (a) measurement setup, (b) optical spectrums of input and output signals.

Experiment: WDM Multichannel Gain.   The gain of an EDFA was measured two ways. The setup shown in Figure 13.45a was used in the first experiment to measure the EDFA spot gains at each channel wavelength. Four lasers were combined using fused-fiber directional couplers. The polarization controller (HP 11896A) was set to randomly vary the light-combined lasers. The EDFA booster amplifier compensated for the optical loss (~8 dB) incurred in the power combiner. An OSA measured the optical power spectrums of the input signal to the EDFA and the amplified output spectrum. The input and output spectrums are shown in Figure 13.45b. The EDFA gains at the signal laser wavelengths of 1533.0 nm, 1550.5 nm, 1551.1 nm, and 1552.0 nm were measured to be 16.65 dB, 12.79 dB, 12.70 dB, and 12.65 dB respectively.

In the next experiment, the EDFA gain is measured with a multichannel TDE method using a small-signal probe. The experiment setup is shown in Figure 13.46a. FC/PC connectors were used to connect to the EDFA. The input signal powers and wavelengths were the same as with the previous measurement. An EELED was used as a small

(a)

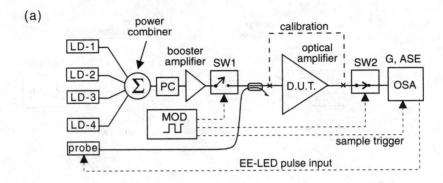

(b)

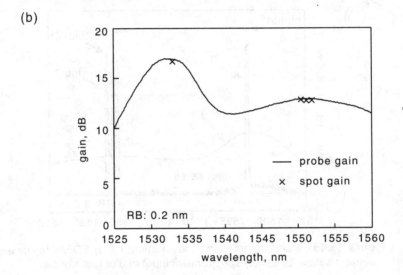

Figure 13.46 Four-channel WDM gain using time-domain extinction with a multichannel stimulus and small signal probe: (a) measurement setup, (b) dynamic gain measurement.

signal probe to measure the wavelength-dependent dynamic gain. The optical switches (acoustooptic modulators) were driven by two pulse generators. One pulse generator provided the trigger for the second pulse generator and the sampling trigger of the OSA (HP 71450). The OSA provided a signal to pulse the EELED output power as discussed in Section (13.5.2). The combination of the optical switch at the EDFA output and the time-domain capability of the OSA resulted in complete extinction of the amplified four-channel WDM source. This permitted a continuous measurement of the EDFA gain across the measurement range as shown in Figure 13.46b. The spot gains measured in the first experiment are plotted in Figure 13.46b for comparison. The two measurement tech-

niques agreed to ~ 0.2 dB. This agreement is within the uncertainty of the connector insertion loss.

Multisource Approximations.   For multichannel measurements, such as those required for WDM applications, the cost and complexity increases with the number of channels. In situations where the test cost/complexity are excessive, an approximate method
may be considered. The reduced-source approximation can be considered for testing
EDFAs in WDM applications.[42] The assumption behind this method is that the amplifier
gain spectrum, or at least some portion of it, is homogeneously broadened. In EDFAs, homogeneous broadening has been shown to be predominant.[30] This is another way of saying that the gain saturation caused by a signal at any wavelength in the EDFA gain band
reduces the amplifier inversion (common energy reservoir) responsible for gain at all
other wavelengths.

It has been shown that spectral hole-burning (SHB) exists in EDFAs to a small degree.[31] To the extent that within a given bandwidth, SHB is insignificant, a single source
can represent the ensemble of signals found in the spectral interval. This concept is shown
in Figure 13.47. Recall from Equation 13.9 that the gain depends on the metastable state
population, N_2. The N_2 population level defines the operating state of the amplifier. To
mimic the effects of several channels, the representative source method must place the
amplifier in the same state that would exist if the channels were actually present. This
yields a requirement on the amplifier inversion level for the two cases

$$N_{2,rs} = N_{2,wdm} \qquad (13.43)$$

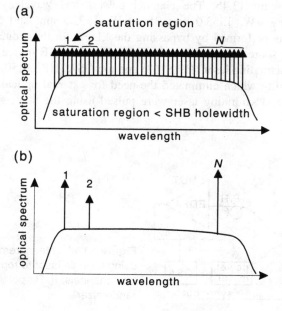

Figure 13.47   Reduced source
concept for multichannel gain characterization. Homogeneous broadening across an interval permits
representation of multiple signals
with a single source.

where $N_{2,rs}$ is the inversion level established with the reduced set of saturating signal sources and $N_{2,wdm}$ is the inversion level established in the presence of all the WDM channels. The steady-state population $N_{2,wdm}$ is derived from a generalization of Equation 13.5 to include multiple signal beams. Equating the excited state populations in the two experiments places conditions on the reduced-source power and wavelength to achieve the best simulation of WDM amplifier performance. This leads to the following requirements for the optical power from a single source to simulate the effects of a cluster of WDM sources about the single source wavelength.[42]

$$P_s = \frac{1}{\lambda_s G_s} \sum_n P_n \lambda_n G_n \tag{13.44}$$

where P_s and P_n are the input single-source and WDM-input channel powers, respectively. The single-source wavelength can be set on the basis of the weighted wavelength

$$\lambda_s = \frac{1}{G_s P_s} \sum_n P_n \lambda_n G_n \tag{13.45}$$

In Equations 13.44 and 13.45, the channel gains are not known initially. Therefore G_n may be set equal to G_s as a starting guess to set the initial values for P_s and λ_s. After the dynamic gain spectrum is measured, better estimates of G_s and G_n are obtained. The measurement procedure converges in about two iterations.

Experiment: Single-Source WDM Gain. The single-source and multichannel WDM gain methods are compared for a four-channel WDM system.[42] The WDM gain measurement was performed with four independent channels combined through a power combiner as shown in Figure 13.46. The channel powers and wavelengths were: 41.1 μW, 1549.6 nm; 39.9 μW, 1553.0 nm; 35.1 μW, 1555.9 nm; and 33.9 μW, 1558.2 nm. Calibration was performed by bypassing the EDFA. In the reduced-source method, a single saturating source was combined with an EELED-ASE source as shown in Figure 13.48. Both the saturating laser (HP 8168) and the OSA (HP 71450) detection circuitry had gating capability which eliminated the need for external optical switches. The EELED (HP 83437) and saturating laser were pulsed using the noise-gain probe

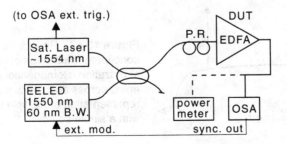

Figure 13.48 Measurement setup using a single laser to represent four channels. P. R.: polarization randomizer.

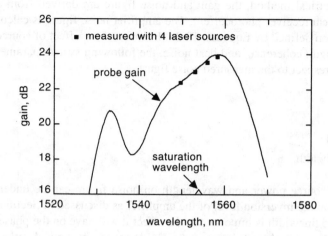

Figure 13.49 Comparison of four-channel WDM gain measurement using four lasers and a single laser with an EE-LED probe.

method described for the TDE technique (Section 13.5.2). The measurement results comparing the four-channel stimulus to the reduced source method with a noise-gain probe are shown in Figure 13.49. The two measurements agree well. There was a maximum of 0.2 dB difference at the third-channel wavelength.

In choosing the number of channels for the reduced source method, the following factors will come to bear:

- Degree of inhomogeneity in the amplifier saturation,
- Total power represented by each reduced-source channel,
- WDM channel spacings.

If the amplifier gain is determined to be completely homogeneously broadened, then a single channel will be sufficient to set the saturation. If inhomogeneous broadening (SHB) is significant and the reduced-source stimulus power becomes large to accommodate the ensemble of WDM channels, a spectral hole will reduce the probe gain in the vicinity of the saturation wavelength. In the wavelength band near 1.55 μm, the FWHM SHB hole-widths in EDFAs are ~ 8 nm which allows the use of fewer saturating signals than at a wavelength of 1.53 μm where the hole-width is ~ 4 nm. Finally, the combination of amplifier inhomogeneity and wide WDM-channel spacings may necessitate multiple sources for the reduced source set to best replicate the amplifier saturation. Ultimately, prior to the application of this technique, a comparison must be made with a complete WDM source to authenticate the measurement. Any significant difference between the measurements could serve as a correction factor for the reduced source method.

Electrical Methods.   Noise figure measured by the electrical method is generally perceived to be a more complete measure of the intensity noise generated by the optical amplifier. It includes effects such as sig-sp beat noise, sp-sp beat noise, and multipath interference noise. In the electrical method, the gain and noise figure are derived from a spectral analysis of the optical receiver photocurrent. The amplifier noise figure is calculated with the general relation defined by Equation 13.34. Because of the effect of source noise, laser power, wavelength, coherence, and beat noise, the following system parameters must be specified with respect to the measured noise figure:

- Source power,
- Source wavelength,
- Source linewidth, and
- Receiver optical bandwidth.

The influence of the source power and wavelength on noise figure can be understood from their influence on the inversion level of the amplifier as discussed in Sections 13.2, and 13.3.1. The source linewidth is important in the effect it will have on the phase-to-intensity noise conversion, or multipath interference (MPI) caused by optical reflections within the EDFA as discussed in Section 13.3.2. Equations 13.31 through 13.33 indicate that the MPI-induced noise varies with the source linewidth, the reflection magnitudes as well as the time delay between the optical reflections. The source linewidth is specified as a stimulus parameter. The receiver optical bandwidth determines the spectral width of the ASE incident on the photodetector. This affects the contribution that the sp-sp beat noise will make to the measured noise. Additionally, the optical filter bandwidth determines the ASE shot noise and the spectral extent of the sig-sp beat noise in the photocurrent power spectrum. Normally this is not observed unless wide bandwidth optoelectronic detection (> 50 GHz) is used along with narrow optical filtering (< 1 nm).

Accurate noise figure measurements require that careful attention is placed on the effect of excess source laser noise, test system MPI noise, and receiver thermal noise. The amplifier noise figure should not depend on these noise contributions. Effects such as the source optical linewidth and the receiver optical bandwidth can result in measurement ambiguities which must be fully specified in the actual measurement. The power levels measured on the electrical spectrum analyzer must be calibrated to an absolute standard to obtain meaningful results. Two methods for this purpose are discussed: the RIN transfer technique and the IM index transfer technique.

To measure the amplifier noise figure the following must be determined:

- Gain,
- Input power,
- Intensity noise density produced by the amplifier: $S_P(f)$, and
- Photon energy.

The gain, input power and photon energy (or wavelength) are assumed to be known according to the methods discussed in Section 13.5.1 and Chapter 3. The principal task is to determine the spectral density of the amplifier intensity noise, $S_P(f)$. Low-noise optical amplifiers are designed to minimize $S_P(f)$.

Measurement of Optical Amplifier Noise.   In this section, a method is discussed to separate the optical amplifier excess noise, $S_P(f)$, from the total noise spectrum given by an electrical spectrum analyzer. The corrupted noise spectrum refers to the optical-amplifier-generated noise, plus the other noise sources or effects that originate from the test system. The basic steps in the noise measurement procedure are[43]:

- Calibrate receiver,
- Measure the corrupted amplifier noise spectrum,
- Remove thermal noise,
- Remove measured shot noise, and
- Calculate $S_P(f)$.

A test system for measuring optical amplifier noise is shown in Figure 13.50. There are a number of factors affecting the measured data as shown in the figure. The receiver thermal noise, detected shot noise, and frequency response of the system will contribute to measurement error. The amplifier noise must be separated from the total measured noise.

The electrical spectrum analyzer display is proportional, by way of $|k(f)|^2$ to the power spectrum of the photocurrent. The bandwidth limitations of the photodetector and electronics are included in $|k(f)|^2$.

With the amplifier under test illuminated by the an optical source with the required wavelength, power and linewidth, the measured photocurrent power spectrum $S_{ESA}(f)|_{mea.}$ is described by:

$$S_{ESA}(f)|_{mea.} = |k(f)|^2 [\mathcal{R}^2 S_P(f) + S_i(f)|_{shot}] + S_{ESA}(f)|_{th} \text{ [W/Hz]} \quad (13.46)$$
(measured noise = excess noise + shot noise + thermal noise)

The terms on the right-hand-side correspond to the excess noise contributions of the optical amplifier, the photocurrent shot noise, and the thermal noise. In an ideal measure-

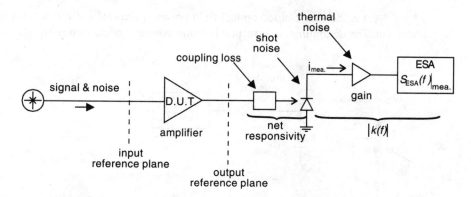

Figure 13.50   Basic measurement setup for measuring optical amplifier noise using the electrical method. ESA: electrical spectrum analyzer.

ment, the thermal noise would be negligeable and $|k(f)|^2$ would be the reference resistance, R_{ESA}, of the electrical spectrum analyzer, which is typically 50 Ω.

Thermal Noise Correction. The thermal noise density, $S_{ESA}(f)|_{th}$ is measured with the input light blocked as shown in Figure 13.51. Under the assumption that the thermal noise density is independent of the magnitude of the optical signal, it is measured by completely blocking the light from the photodetector. The measured thermal noise spectrum is denoted by $S_{ESA}(f)|_{th}$. Subtracting the density $S_{ESA}(f)|_{th}$ from $S_{ESA}(f)|_{mea}$ performs the thermal noise correction.

RIN Transfer: System Transfer Function Correction. The frequency dependent system transfer function, $|\Re k(f)|^2$ is measured next. A calibrated source of excess optical noise replaces our source (for example, amplifier) under test. The setup with the calibration source is shown in Figure 13.52. This step is referred to as a RIN transfer calibration since the optical noise source is characterized by a RIN which is very stable over time. A practical way to implement the noise source is to optically filter the ASE from an EDFA. A typical filter bandwidth is of the order of 1 nm. This will yield a flat intensity noise spectrum up to ~ 2 GHz. The RIN associated with the filtered ASE source is not required to be flat for the calibration to be valid. The only requirements are that the RIN caused by the standard is large compared to the receiver thermal and shot noise, and that the frequency dependence of the RIN is known.

The RIN associated with the filtered source can be derived from an analysis of the optical intensity associated with the filtered optical field spectrum. For an arbitrarily shaped, bandwidth-limited unpolarized ASE source, the frequency-dependent RIN is given by:

$$\text{RIN}_{cal}(f) = \frac{FT\left[\left|FT^{-1}\{S_E(v)\}\right|^2\right]}{\left[\int S_E(v)\,dv\right]^2} \quad [Hz^{-1}] \tag{13.47}$$

where $S_E(v)$ is the single-sided optical field power spectrum, and FT denotes the Fourier transform. The denominator corresponds to the average optical power-squared and the nu-

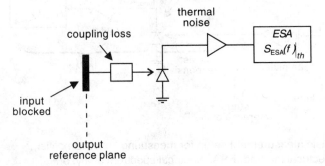

Figure 13.51 Measurement of receiver system thermal noise.

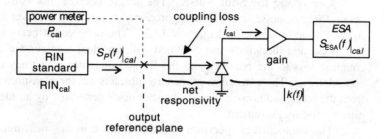

Figure 13.52   RIN transfer technique for receiver calibration.

merator corresponds to the expected spontaneous-spontaneous intensity beat noise. The absolute RIN is determined from a simple measurement of the optical field spectrum using an OSA. The RIN must be calculated numerically, except when the filtered ASE shape follows simple functional forms such as those given in Table 13.2. The RIN depends only on the shape of the ASE spectrum and not on the absolute amplitude. This results in a very stable RIN transfer standard since the bandwidth of the optical filter is fixed.

When the RIN transfer standard is combined with a calibrated optical power meter, the filtered ASE source provides an *absolute spectral noise power reference* according to Equation 13.23. The two fundamental quantities for the reference are the RIN of the filtered ASE source and the average power. The average power measurement is performed accurately with an optical power meter. Thus, according to Equation 13.23, the absolute optical noise reference density is:

$$S_P(f)|_{\text{cal}} = \text{RIN}_{\text{cal}}(f) \times P_{\text{cal}}^2 \quad [W^2/\text{Hz}] \tag{13.48}$$

Where P_{cal} is the RIN standard average output power measured with an optical power meter.

The electrical spectrum analyzer measures the photocurrent noise spectrum $S_{\text{ESA}}(f)|_{\text{cal}}$ due to the RIN transfer standard. Using Equation 13.46, the spectral density is:

$$S_{\text{ESA}}(f)|_{\text{cal}} = |\Re k(f)|^2 \, S_P(f)|_{\text{cal}} \tag{13.49}$$

The thermal and shot noise are not included in the above expression since the photocurrent noise is dominated by the detection of the intensity noise from the RIN transfer standard. The excess noise from the RIN standard dominates over any other noises present.

The unknown calibration constant is found according to:

$$|\Re k(f)|^2 = \frac{S_{\text{ESA}}(f)|_{\text{cal}}}{S_P(f)|_{\text{cal}}} \tag{13.50}$$

With this second calibration step, the system transfer function is now known across the frequency range of interest.

Correcting the Shot Noise. The actual receiver shot noise must be corrected to obtain the shot noise that would be produced by an ideal receiver according to the noise figure definition given in Equation 13.34. The differences between the actual receiver shot noise and the shot noise obtained with an ideal receiver are caused by the optical coupling loss to the receiver and the limited quantum efficiency of the actual receiver photodetector. The correction procedure subtracts out the measurement system shot noise from the measured noise and adds the shot noise generated by an ideal receiver to the amplifier noise measurement.

The photocurrent spectrum of the shot noise in the measurement receiver is given by Equation 13.25. The DC responsivity $\mathcal{R}$ must be found to complete the shot noise subtraction. Two measurements are required to determine $\mathcal{R}$. The setup for the system transfer function measurement shown in Figure 13.52 is used. The average photodetector current, i_{cal}, is measured. The other required measurement, P_{cal}, was already performed (see Equation 13.48). This data permits determination of the net DC responsivity according to:

$$\mathcal{R} = \frac{i_{cal}}{P_{cal}} \tag{13.51}$$

At this point, the calibration process is complete. The next step is to use the calibration data to determine the optical amplifier noise.

Corrected Amplifier Noise. Equation 13.46 can be solved for the unknown intensity noise $S_P(f)$:

$$S_P(f) = \frac{S_{ESA}(f)|_{mea} - S_{ESA}(f)|_{th}}{|\mathcal{R}k(f)|^2} - \frac{2qi_{mea}}{\mathcal{R}^2} \tag{13.52}$$

Equation 13.52 indicates the need for the three operations discussed: thermal noise correction, amplitude frequency response correction, and shot-noise subtraction.

Inserting the calibration coefficients given by Equations 13.50 and 13.51 into 13.52, the excess amplifier noise is found in terms of measured parameters:

$$S_P(f) = \frac{S_{ESA}(f)|_{mea} - S_{ESA}(f)|_{th}}{\dfrac{S_{ESA}(f)|_{cal}}{RIN_{cal}(f)\, P_{cal}^2}} - \frac{2qi_{mea}}{\left(\dfrac{i_{cal}}{P_{cal}}\right)^2} \quad [W^2/Hz] \tag{13.53}$$

This equation is one of the principle results of this section. It shows how the intensity noise $S_P(f)$ (excluding shot noise) from the optical amplifier under test is found by making measurements of: (1) the uncalibrated measurement of the amplifier noise $S_{ESA}(f)|_{mea}$ and photocurrent i_{mea}; (2) the thermal noise; and (3) the average optical power P_{cal} and DC photodetector current i_{cal} when the receiver is connected to the calibrated noise standard. The measurement data and calibration coefficients are summarized in Table 13.6

In the preceding analysis, the effective system quantum efficienty $\mathcal{R}$ was frequency independent. All frequency-dependent system effects have been confined to the electronics which includes transit time effects in the photodetector and bandwidth limitations in

Table 13.6 Variable Definitions for Equation 13.53.

Symbol	Description	Units
$S_P(f)$	Spectral density of optical intensity noise	W²/Hz
$S_{ESA}(f)|_{mea}$	Uncorrected amplifier noise spectrum	W/Hz
$S_{ESA}(f)|_{th}$	Electrical spectrum analyzer noise with input light to photodetector blocked	W/Hz
$S_{ESA}(f)|_{cal}$	Electrical spectrum analyzer noise data with RIN-transfer source applied to measurement system input	W/Hz
$RIN_{cal}(f)$	RIN-transfer standard	Hz⁻¹
P_{cal}	Average power of RIN-transfer standard	W
P_{dc}	Average amplifier output power	W
i_{cal}	Average photocurrent produced by RIN-calibration standard	A
q	Elementary electronic charge: 1.602×10^{-19}	coul.

the electronics. Therefore, the test system must be free from optical reflections which could create significant amplitude ripple, otherwise the measurement technique should be modified to account for the system interference effects. The effect of the reflections between the amplifier and receiver will be to impart a frequency ripple onto the intensity-noise spectrum and to convert source laser phase noise to amplitude noise. Therefore, it is wise to use low-reflection connections and an optical receiver with a low optical back-reflection.

Calculation of Noise Figure. The amplifier noise figure defined in Equation 13.34 requires the determination of the amplifier input and output SNRs. The input SNR is calculated on a shot-noise basis. The output SNR is computed based on the photocurrent signal and noise created in an ideal receiver illuminated by the amplified signal GP_s, and the amplifier excess noise $S_P(f)$ obtained from Equation 13.53. The ideal receiver has a flat frequency response, no thermal noise, and unity quantum detection efficiency ($\mathcal{R}_{ideal} = q/h\nu$).

Input SNR. The shot-noise density given by Equation 13.25 is multiplied by the detection bandwidth B_e to obtain the mean-squared shot noise. The input SNR is calculated by taking the ratio of the signal-to-noise powers at the ideal receiver electrical output:

$$SNR_{in} = \frac{\mathcal{R}_{ideal}^2 P_s^2}{2q i_{dc} B_e} = \frac{P_s}{2h\nu\, B_e} \tag{13.54}$$

Output SNR. The output noise density of the ideal receiver is the sum of the contributions from the amplifier generated noise $S_P(f)$, and the shot noise:

$$S_i(f) = S_i(f)|_{excess} + S_i(f)|_{shot} \tag{13.55}$$

The amplifier contribution to the ideal photocurrent noise is determined by computing the photocurrent spectral density (see Equation 13.22) corresponding to the excess noise in Equation 13.53:

$$S_i(f)|_{\text{excess}} = S_P(f) \times \mathcal{R}_{\text{ideal}}^2 \qquad (13.56)$$

Integrating the excess and shot noises given by Equations 13.56 and 13.25 over the detection bandwidth B_e yields the output SNR:

$$\text{SNR}_{\text{out}} = \frac{<GP_s>^2}{S_P(f)B_e + 2<GP_s> h\upsilon B_e} \qquad (13.57)$$

The noise figure is calculated by taking the ratios of Equations 13.54 and 13.57 with B_e set to 1 Hz:

$$F = \frac{P_s \, S_P(f) \, B_e + 2<GP_s> h\upsilon B_e}{2h\upsilon \quad <GP_s>^2} \qquad (13.58)$$

Equation 13.58 reduces to:

$$F = \frac{S_P(f)}{2h\upsilon \, G^2 P_s} + \frac{1}{G} \qquad (13.59)$$

noise figure = excess noise factor + shot−noise factor

This equation is the main result of the electrical noise figure section. This general relation is composed of two parts. One part contains all the excess-noise contributions to the net-noise figure. The second part is the shot-noise contribution. This result is very general since it does not specify or constrain the type of noises that contributes to the amplifier excess noise. It is valid with sig-sp, sp-sp beat noises as well as MPI-induced phase-to-intensity converted noise. The amplifier excess noise, $S_P(f)$, is measured according to Equation 13.53, P_s is determined using a calibrated optical power meter, and the amplifier gain G is measured using the techniques described in Section 13.5.

As a check for Equation 13.59, the limiting form of this expression is found where sig-sp beat and shot noise are the dominant noise processes. Substituting into Equation 13.59 the sig-sp beat noise from Equation 13.27 and applying the conversion indicated in Equation 13.22, the noise figure obtained is:

$$F = \frac{2\rho_{\text{ASE}}}{G \, h\upsilon} + \frac{1}{G} \qquad (13.60)$$

which is the same relation as defined in Equation 13.35 for the optical method for noise-figure measurement.

Laser Sources with Excess Noise.　Any excess noise (in excess of the usual shot noise) from the source illuminating the amplifier under test must be taken into account in order to make an accurate noise characterization of the amplifier. The effect of the source noise on the accuracy of the electrical method is shown in Figure 13.29. One difficulty with the electrical method is there are not many options available to correct for the excess source noise. If the excess noise from the laser is stable over time, it can be measured and later subtracted from the result obtained when the amplifier is in place.[44] This method will be the focus of the following discussion.

Let the excess source noise be designated by: $S_p(f)|_{source}$. Measurement of the source excess noise is accomplished using the same procedures outlined earlier for characterizing the amplifier excess noise. Any optical attenuation between the source and the receiver should be minimized to obtain the best measurement of the source excess noise. Let T denote the variable optical transmission factor ($0 < T < 1$) between the source and the amplifier under test. T does not include any losses present when the excess source noise $S_P(f)|_{source}$ was measured. T originates from unavoidable optical-coupling losses as well as deliberate optical attenuation provided to control the input signal level P_s to the amplifier. Taking the transmission factor T into account, effective noise power at the amplifier input terminals is:

$$S_P(f)|_s = T^2 S_P(f)|_{source} \qquad (13.61)$$

The noise at the amplifier output is a combination of internally generated noise and the amplified excess noise from the source:

$$S_P(f) = S_P(f)|_{amp} + G^2 T^2 S_P(f)|_{source} \qquad (13.62)$$

The amplifier noise figure is corrected for the excess source noise according to:

$$F = \frac{S_P(f) - G^2 T^2 S_P(f)|_{source}}{2h\upsilon\, G^2 P_s} + \frac{1}{G} \qquad (13.63)$$

Alternate Receiver Calibration Method: IM Index Transfer. An alternative method to the RIN transfer technique for calibrating the optical receiver is the IM index-transfer technique. This technique offers calibration at a single frequency or point by point through multiple calibrations. In this method, the modulation index m_I of a fixed-frequency sinusoidally intensity modulated light is first accurately measured. Let $P(t)$ designate the output intensity versus time of the modulated optical source

$$P(t) = P_{cal}\left(1 + \frac{\Delta P}{P_{cal}} \cos 2\pi f_m t\right) \qquad (13.64)$$

where P_{cal} is the average power detected by the optical receiver. By measuring the DC and modulation strength, the intensity modulation index m_I is obtained at the frequency f_m.

$$\Delta P(f_m) = \; <P_{cal}> \, m_I(f_m) \qquad (13.65)$$

The corresponding power spectral density at the frequency f_m is:

$$S_P(f_m)|_{cal} = m_I^2(f_m) \times P_{cal}^2 \quad [W^2/Hz] \qquad (13.66)$$

When this source of intensity modulation is applied to the optical receiver, the electrical spectrum analyzer will measure:

$$S_{ESA}(f_m)\big|_{cal} = (\mathcal{R}k(f_m))^2 \, S_P(f_m)\big|_{cal} \tag{13.67}$$

The unknown multiplicative constants are next calculated:

$$[\mathcal{R}k(f_m)]^2 = \frac{S_{ESA}(f_m)\big|_{cal}}{S_{\Delta P}(f_m)\big|_{cal}} \tag{13.68}$$

With the IM index-transfer method, the optoelectronic detection chain as well as the electrical spectrum analyzer is calibrated for a deterministic modulation. Note that with this method, the electrical noise bandwidth of the receiver is not calibrated. The electrical spectrum analyzer IF filter noise bandwidth must be calibrated separately.

Noise Figure Experiment.   Using the RIN-transfer calibration technique, the noise figure of an EDFA was measured at a wavelength of 1550 nm with an amplifier input power of −20 dBm. The experiment setup is similar to that shown in Figure 13.50. With the source attenuator set to 0 dB, the source RIN was approximately −149.2 dB/Hz at a power level of −3.6 dBm. The addition of 16.4 dB of attenuation to bring the input signal level to −20 dBm results in an input RIN of −135 dB/Hz. This RIN level according to Figure 13.29 will have no significant impact on the noise figure measurement, the attenuated source is essentially shot-noise-limited.

In the first step, the transfer function from the optical amplifier to the electrical spectrum analyzer display was calibrated. An optically filtered ASE source with an FWHM spectrum of approximately 2.5 nm served as the RIN transfer standard. The optical spectrum of the transfer standard is shown in Figure 13.53a. The RIN for this source was calculated according to Equation 13.47 yielding a low frequency RIN of −116.18 dB/Hz. The calculated RIN spectrum is shown Figure 13.53b. Across the frequency range of interest, the frequency roll-off of the RIN transfer standard is small, and the noise power it generates dominates over all other noise sources.

The gain of the EDFA was measured with an OSA. The optical spectrums of the input and amplified output signals are shown in Figure 13.54a. The EDFA was equipped with an optical filter with an FWHM of approximately 1.6 nm. This filter served to reduce the sp-sp beat noise contribution to the amplifier noise figure. The EDFA produced 15.05 dB of gain as seen from the figure. The three lower traces in Figure 13.54b correspond to the RIN standard, the EDFA, and the thermal noise powers. The thermal noise, $S_{ESA}(f)\big|_{th}$ was subtracted from the EDFA noise $S_{ESA}(f)\big|_{mea.}$ as indicated by Equation 13.53. A measurement of the average power generated by the RIN-transfer standard along with the calibration noise data, $S_{ESA}(f)\big|_{cal}$, shown in Figure 13.54b, permitted absolute amplitude calibration of the display according to Equation 13.53. The shot-noise correction required measurement of the electrical currents produced by the transfer standard and the amplified signal which resulted in an effective responsivity, $\mathcal{R}$, of 0.428. The shot-noise correction for the coupling loss to the detector was small in this experiment, contributing less than 0.1 dB to the noise figure. Using the values for the gain and excess noise, the noise figure was approximately 7.8 dB as shown in Figure 13.54b.

(a)

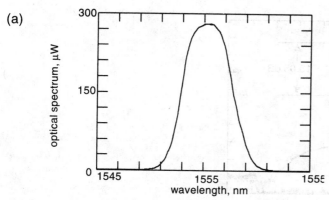

(b)

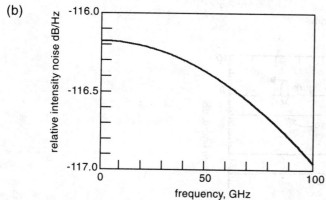

Figure 13.53   RIN transfer standard spectrums. (a) Measured optical power spectrum. (b) Calculated RIN spectrum.

13.6   OTHER TYPES OF OPTICAL AMPLIFIERS

13.6.1   Rare-Earth Doped Fiber Amplifiers

Amplification at a variety of wavelengths has been demonstrated in rare-earth doped optical fibers. The mechanism for gain is similar to that of the EDFA: excitation of the rare-earth ion with pump light causes population inversion between a pair of energy levels. This produces gain at a wavelength corresponding to the energy difference between the inverted levels. A partial energy diagram for the rare-earths in a $LaCl_3$ host is shown in Figure 13.55. This diagram is also useful for other glass hosts since the electronic structure of the trivalent ions provides some shielding of the transitions from the host crystalline fields.

1.3 μm Amplifiers.   When praseodymium ions are doped in a low-phonon energy glass host such as a fluorozirconate fiber, optical gain can be achieved at a number of different wavelengths. Indeed, with this ion, practical demonstrations of amplification or lasing has been achieved from the visible to the infrared.[45-55] The transition responsible for gain near the important 1.3 μm telecom window is designated by 'a' in Figure 13.55. This

(a)

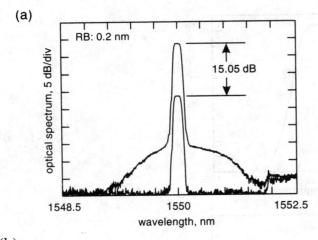

(b)

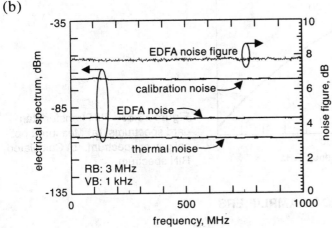

Figure 13.54 Noise figure measurement: (a) optical amplifier gain, (b) electrical spectrum analyzer display and noise figure result.

transition is activated by pumping into the $^1G_4$ band near 1 μm. The praseodymium-doped fluoride fiber amplifier (PDFFA) has demonstrated optical gains in excess of 40 dB, saturated output powers of 20.1 dBm and a sig-sp beat-noise-limited noise figure as low as 3.2 dB.[50–55] The PDFFA has a useful gain spectrum in excess of 30 nm as indicated in Figure 13.56. For communications applications, the problem of making a robust connection between the fluoride fiber and the silica fiber used for transmission has been solved, making commercialization of the PDFFA possible.[55] Alternate decay routes from the $^1G_4$ level reduces the fluorescent lifetime to ~ 110 μs. This is the reason the PFFDA operates most efficiently with stronger input signals. Weak signals must compete with the short fluorescent lifetime for the stored ion energy. For this reason, the PDFFA is usually used as a booster amplifier.

The trivalent neodymium ion $^4F_{3/2} - {}^4I_{13/2}$ transition provides gain about the 1.32 μm wavelength in fluorozirconate fiber. Pumping near the 0.8 μm wavelength provides a

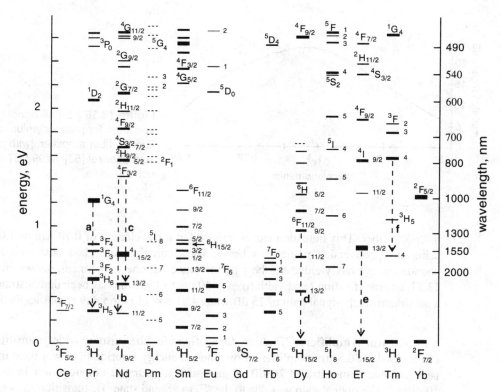

Figure 13.55 Partial energy level diagram for the trivalent rare earth ions. Some notable transitions: a: 1.31 μm, b: 1.06 μm, c: 1.34 μm, d: 1.32 μm, e: 1.53 μm, f: 1.47 μm. [after ref [15]].

ground-state $^4I_{9/2}$ to $^4F_{5/2}$ absorption which then populates the $^4F_{3/2}$-level. This transition is designated by 'c' in Figure 13.55. The efficiency of this amplifier is compromised by the strong 1.05 μm transition, designated by 'b' in the figure, which depopulates the metastable state population. Also upconversion of signal light to the $^4G_{7/2}$ level further competes with the desired transition.[56] Practical use of this transition will require suppression of the ASE generated by the large optical gain of the 1.05 μm transition.

Research is also being directed at the 1.32 μm ($^6H_{9/2}$–$^6F_{11/2}$) doublet to ground state $^6H_{15/2}$ transition of the dysprosium ion in low-phonon energy glasses.[57,58] This transition is designated by 'd' in Figure 13.55. This transition and the glass host is less mature than either the neodymium or praseodymium investigations in fluoride fiber, more work needs to be performed to fully evaluate its potential.

1.47 μm Amplifier. Gain near 1.47 μm from the trivalent thulium $^3F_4$–$^3H_4$ transition (designated by 'f' in Figure 13.55) has a number of applications ranging from transmission line monitoring to the possibility of opening up a new telecommunications band

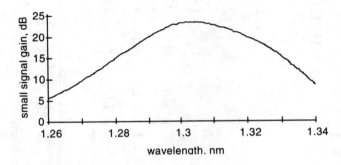

Figure 13.56 Small-signal gain spectrum for praseodymium-doped fluoride fiber amplifier. [with permission after ref [53] ©1993 BT Technol. J.].

in optical fiber. This transition can be pumped at a wavelength of 0.78 μm or 1.064 μm using an upconversion pumping scheme. The fluorozirconate glass host with its low-phonon energy has yielded the best performance. In one experiment, shown in Figure 13.57, an amplification bandwidth from 1440 nm to 1505 nm has been demonstrated with a maximum small-signal gain of 25 dB when 450 mW of pump power was applied.[59]

0.8 μm Amplifier. Thulium-doped fluoride fiber also provides transitions that can be used for 0.8 μm amplifiers. Low phonon energy fluoride fiber was used in an experiment that demonstrated 20 dB of gain with a saturated output power in excess of 10 dBm.[60] The optical gain was due to the $^3F_4$ to ground state $^3H_6$ transition shown in Figure 13.55. The $^3F_4$ energy level was populated by pumping at a wavelength of 790 nm. An amplification wavelength range from 795 nm to 820 nm appears to be feasible using this system.

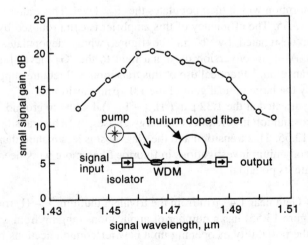

Figure 13.57 Small signal gain spectrum of thulium fluoride fiber amplifier as a function of wavelength [with permission after ref [59] ©1993 IEE].

13.6.2 Gain from Fiber Nonlinearities

Raman Amplifier. When pump-light intensity within an optical fiber becomes large, the glass molecules become excited into different vibrational states. In stimulated Raman scattering (SRS), the difference in energy between two vibrational states can be used to amplify an optical signal. This forms the basis for the Raman fiber amplifier (RFA). Unlike the rare-earth doped fiber amplifiers, the RFA doesn't require the addition of special dopants, the optical fiber in the transmission link acts as the gain medium. The RFA is made by coupling pump light into the transmission fiber. At a signal wavelength of 1.55 μm, pump powers and fiber lengths are of the order of 1 W and tens of kilometers respectively (see Appendix B). The pump laser and signal frequency separation is equal to the Raman shift frequency which depends on the characteristics of the glass fiber. This shift is approximately −12 THz, or +100 nm at a wavelength of 1.55 μm in telecommunications fiber. Therefore to obtain gain in the 1.55 μm wavelength band, the wavelength of the pump laser is set to ~ 1.45 μm. An example of an RFA gain spectrum is shown in Figure 13.58.[61] The useable gain bandwidth is approximately 50 nm (6 THz).

Some simple approximate relations are useful for providing insight into the RFA.[62,63] Assuming that the stimulated Brillouin scattering and/or pump depletion is small, the amplifier gain for a fiber of length L is given by:

$$G \approx e^{\frac{g_r P_o L_{eff}}{A_{eff}} - \alpha_r L} \tag{13.69}$$

where P_o is the input pump power (watts) and L_{eff} is the effective length (meters). The Raman efficiency coefficient C_r is defined by $C_r = g_r / K A_{eff}$ where g_r is the peak Raman gain, $K = 2$ corresponds to nonpolarization maintaining optical fiber and A_{eff} is the effective core area.[62–64] The Raman gain coefficient, C_r depends on wavelength and the type of optical fiber. Increasing the germanium concentration tends to increase C_r. Measure-

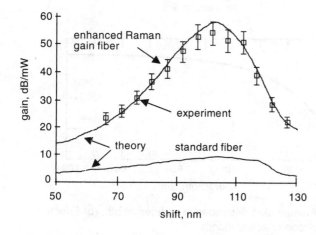

Figure 13.58 Raman gain spectrum for highly doped fiber and standard fiber. Pump wavelength is 1485 nm [with permission after ref [61] ©1990 IEE].

ments of the Raman gain coefficient on different optical fibers in the wavelength range of 1540 nm to 1565 nm yielded a range of values of C_r from 1.8×10^{-4} to 6×10^{-4} W^{-1} m^{-1}.[64] The effective length is reduced compared to the physical length due to fiber losses. L_{eff} is related to the fiber loss at the pump wavelength, α_p (m^{-1}) as:[62]

$$L_{\text{eff}} = \frac{1}{\alpha_p} [1 - e^{-\alpha_p L}] \tag{13.70}$$

Example

Equations 13.69 and 13.70 are illustrated by way of example. The gain and effective length dependence on fiber length are calculated under the following conditions: $C_r = 5 \times 10^{-4}$ W^{-1} m^{-1} (nonpolarization maintaining fiber), pump power is 500 mW, and the losses at the pump and signal wavelengths are assumed to be 0.25 dB/km. The exponential loss factor is first calculated using $\alpha = 0.001 \times \ln(10^{0.25/10})$ m^{-1}. The effective length is next calculated using Equation 13.70 which is used in Equation 13.69. The results are plotted as a function of fiber length in Figure 13.59. Figure 13.59a shows the gain/loss for the case of no pumping and also with 500 mW of pump power applied. Figure 13.59b shows the effective length, as

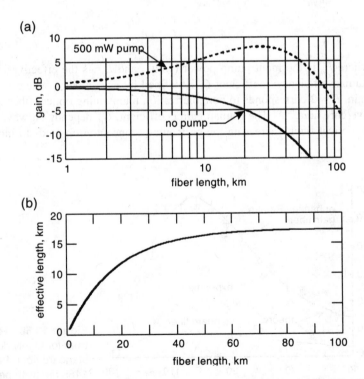

Figure 13.59 (a) Raman gain dependence on fiber length. (b) Effective fiber length as a function of actual length.

limited by propagation loss, as a function of actual fiber length. For a given pump power there is an optimum fiber length. Beyond the optimum length, the RFA gain decreases.

Stimulated Brillouin Scattering: Brillouin Amplifier.   The Brillouin amplifier is based on the acoustic waves (phonons) created when high-intensity light passes through an optical fiber. The optical field causes electrostriction in the fiber which then creates acoustic waves. These acoustic waves scatter light in the backward direction creating optical gain for signals, at the proper frequency, traveling in the opposite direction as the original high intensity light. The gain spectrum is shifted approximately 11 GHz from the pump frequency in silica fiber at 1.55 μm. This is compared to a shift of approximately 6 THz for the Raman effect. The exact shift depends on factors such as the fiber geometry and the residual stress in the glass. The SBS gain spectrum is also quite narrow, on the order of 10 ~ 20 MHz as indicated by the SBS ASE spectrum shown in Figure 13.60. Stimulated Brillouin scattering (SBS) is detrimental in optical links employing narrow linewidth lasers. Even at low power levels, on the order of 1 mW, SBS can create significant backscatter of the forward propagating signal. As a result of the narrow gain bandwidth of the SBS amplifier, it hasn't yet attracted much attention for amplification of telecommunications signals. SBS has received significant attention due to the unwanted optical backscatter it creates.

13.6.3 Semiconductor Amplifiers

The semiconductor optical amplifier (SOA) provides a wavelength flexible solution to optical amplification. It offers efficient economical amplification at every telecommunications wavelength and is a contender for 1.3 and 1.55 μm amplification. There are a number of technical obstacles that have limited their deployment in communications applications. This includes: gain-saturation-induced-crosstalk, gain-ripple, polarization-dependent gain, and poor mode-match to optical fiber resulting in excess optical loss and an increase in noise figure. Device structure improvements to improve the SOA along with the utilization of gain clamping techniques to reduce crosstalk are improving their applicability to communications systems.[65,66,67]

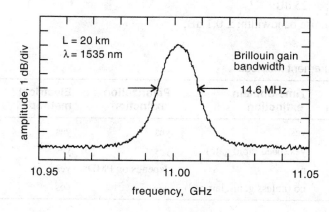

Figure 13.60 Brillouin gain spectrum for 1.3 μm zero dispersion optical fiber.

13.6.4 Measurement of Other Types of Optical Amplifiers

In this chapter, optical and electrical methods for noise figure and gain measurement were discussed with an emphasis on their application to the EDFA. Many of these techniques can be applied to the various non EDFA amplifiers discussed. A brief discussion is given here concerning the applicability of the measurement methods to other types of optical amplifiers.

With the optical method, the optical source subtraction technique can be applied, in principle, to any of the amplifiers discussed above. The polarization nulling and time domain extinction methods require certain assumptions on the amplifier polarization dependence and gain recovery time constants which must be considered. The polarization-nulling method works best for amplifiers with polarization-independent gain and low polarization mode dispersion. The TDE method requires instrumentation capable of gating optical signals rapidly compared to the recovery time of the optical amplifier under test. This may prove very challenging in amplifiers with short gain recovery times (< 1 ns) such as the Raman fiber amplifier (RFA).

The electrical noise figure methods, while more complicated, are quite versatile if performed properly. They can be applied to all optical amplifiers. It is beyond the scope of this text to provide an analysis of each technique for each type of optical amplifier. However, as a suggestive guide, Table 13.7 indicates which technique(s) may be most applicable for the most commonly used amplifiers.

13.7 SOURCES OF MEASUREMENT ERRORS

The characterization of optical amplifiers can involve a variety of instrumentation with many possible configurations and measurement procedures. This tends to make the analysis of measurement error specific to the particular experimental arrangement at hand. There are, however, common sources of measurement error that contribute to the measurement uncertainty.[68] The major sources of measurement uncertainty with some typical values are listed here:

- Connector uncertainty: ± .25 dB;
- OSA effective optical noise bandwidth: ± 0.1 dB;

Table 13.7 Guide to Measurement Methods

Amplifier	Source subtraction	Time-domain extinction	Polarization extinction	Electrical method
EDFA	yes	yes	yes	yes
PDFA	yes	yes (need rapid gate)	yes	yes
Raman	yes	no	depends on PMD	yes
Semiconductor	yes	no (unless gain-clamped)	yes	yes

- OSA scale fidelity: $\pm$ 0.05 dB;
- OSA polarization dependence: $\pm$ 0.1 dB;
- Power meter amplitude accuracy: $\pm$ 0.1 dB; and
- Amplitude instability due to interferometric effects.

From this list, it is clear that eliminating optical connectors at the amplifier and using fusion splices instead will significantly improve the measurement uncertainty. Most of the other uncertainties are addressed by the instrument specifications. The use of polarization randomizers and measurement averaging can reduce the uncertainty due to polarization dependencies in the test equipment.

13.8 USEFUL CONSTANTS FOR EDFA MEASUREMENTS

Table 13.8 includes a list of physical constants that may be useful for optical amplifier noise and gain calculations.

13.9 SUMMARY

In this chapter a variety of measurement techniques were discussed related to amplifier gain and noise figure characterization. The various measurement techniques covered represent a subset, though significant, of the diverse set of methods available for characterization of optical amplifiers.

Some discussion was given to the inner workings of the EDFA to familiarize the reader who is new to this kind of optical amplifier. The discussions on gain and noise in Sections 13.3 and 13.4 provide background information helpful for the discussion on noise figure and measurement techniques. In Section 13.5, the measurement methods for noise figure and gain were grouped into two broad categories: OSA-based, and electrical spectrum analyzer-based. The optical methods included the source subtraction method, the polarization-extinction technique, and the TDE technique. The TDE was extended with the addition of a broadband EELED to permit measurement of the amplifier dynamic gain.

Table 13.8 Useful Physical Constants

Description	Symbol	Value	Units
Electron charge	q	1.602×10^{-19}	coul.
Velocity of light	c	2.99793×10^{8}	ms^{-1}
Planck's constant	h	6.625×10^{-34}	$J\text{-}s^{-1}$
Boltzmann constant	k	1.3806×10^{-23}	$J\text{-}K^{-1}$
1 electron volt	1 eV	1.602×10^{-19}	J

WDM characterization EDFAs was discussed in terms of relevancy of the single-channel techniques previously discussed. An approach was presented to measure the WDM-gain spectrum that uses a reduced set of saturating sources. This technique reduces test costs when large numbers of WDM channels are required.

The electrical technique for noise figure measurement was discussed. This method provides the most complete characterization of the noise performance of an optical amplifier. The electrical method tends to be less tolerant to errors in the measurement technique.

A brief survey of other types of optical amplifiers was given in Section 13.6. This provides a quick appreciation of other existing amplifier technologies. A discussion was also given concerning the relevancy of the measurement methods to these amplifier types. The dominant causes of measurement uncertainty in terms of instrument limitations was outlined in Section 13.7. Finally, the most often used physical constants for noise figure calculations is tabulated in Section 13.8 to save the reader the time required to search this information in texts on physics.

REFERENCES

1. Nilsson, J., Lee, Y.W., Kim, S.J., Lee, S.H., and Choe, W.H., 1996. Analysis of AC gain tilt in erbium-doped fiber amplifiers. *IEEE Photon. Technol. Lett.* 8 (4):515–517.

2. Desurvire, E., Giles, C.R., and Simpson, J.R. 1989. Gain saturation effects in high-speed, multichannel erbium-doped fiber amplifiers at λ=1.53 μm. *J. Lightwave Technol.* 7 (12):2095–2104.

3. Freeman, J. and Conradi, J. 1993. Gain modulation response of erbium-doped fiber amplifiers. *IEEE Photon. Technol. Lett.* 5(2): 224–226.

4. Giles, C.R. and Desurvire, E. 1989. Transient gain and crosstalk in erbium-doped fiber amplifiers. *Optics Lett.* 14:880–882.

5. Taylor, M.G. 1993. Observation of new polarization dependence effect in long haul optically amplified system. *IEEE Photon. Technol. Lett.* 5 (10):1244–1246.

6. Gimlett, J.L., Iqbal, M.Z., Curtis, L., Cheung, N.K., Righetti, A., Fontana, F., Grasso, G. 1989. Impact of multiple reflection noise in Gbit/s lightwave systems with optical fiber amplifiers. *Electron. Lett.* 25 (20):1393–1394.

7. Kobayashi, M., Ishihara, T., and Gotoh, M. 1993. Power penalty due to optical reflections in erbium-doped fiber preamplifier. *IEEE Photon. Technol. Lett.* 5 (8):925–928.

8. Snitzer, E. 1961. Optical maser action of Nd3+ in a barium crown glass. *Phys Rev. Lett.* 7: 444–449.

9. Koester, C.J., Snitzer, E.A. 1964. Amplification in a fiber laser. *Applied Optics* 3 (10):1182.

10. Mears, R.J. et al. 1986. Low threshold, tunable cw and Q-switched fibre laser operating at 1.55 μm. *Electron. Lett.* 22 (3):159.

11. Mears, R. J., Reekie, L. Jauncey, I.M., and Payne, D.N. 1987. High gain rare-earth doped fiber amplifier at 1.54 μm. in *Conference on Optical Fiber Communication/International Conference on Integrated Optics and Optical Fiber Communication Technical Digest Series 1987, Vol. 3,* (Optical Society of America, Washington, DC 1987):167.

12. Delavaux, J-M, P., and Nagel, J.A. 1995. Multi-stage erbium-doped fiber amplifier designs. *J. Lightwave Technol.* 13 (5):703–720.

13. Hill, K.O., Malo, B., Bilodeau, F., Johnson, D.C., and Albert, J. 1993. Bragg gratings fabricated in monomode photosensitive optical fiber by UV exposure through a phase mask. *Appl. Phys. Lett.* 62 (10):1035–1037.

14. Ainslie, B.J. 1991. A review of the fabrication and properties of erbium-doped fibers for optical amplifiers. *J. Lightwave Technol.* 9 (2):220–227.

15. Dieke, G.H., and Crosswhite, H.M. 1963. The spectra of the doubly and triply ionized rare-earths. *Applied Optics,* 2 (7):675–686.

16. Optical fibre lasers and amplifiers, 1996. Edited by P.W. France. Blackie & CRC Press Inc.

17. Desurvire, E. Erbium Doped Fiber Amplifiers, Principles and Applications, John Wiley & Sons, Inc. 1994.

18. Ainslie, B.J., Craig, S.P., and Davey, S.T. 1988. The absorption and fluorescence spectra of rare-earth ions in silica-based monomode fiber. *J. Lightwave Technol.* 6 (2):287–293.

19. Miniscalco, W.J. 1991. Erbium-doped glasses for fiber amplifiers at 1500 nm. *J. Lightwave Technol.* 9 (2):234–250.

20. Horiguichi, M., Shimizu, M., Yamada, M., Yoshino, K., Hanafusa, H. 1990. Highly efficient optical fibre amplifier pumped by a 0.8 μm band laser diode. *Electron. Lett.* 26 (21):1758–1759.

21. Horiguichi, M., Yoshino, M., Shimizu, M., and Yamada, M. 1993. 670 nm semiconductor laser diode pumped erbium-doped fiber amplifers. *Electron. Lett.* 29 (7):593–595.

22. Giles, C.R. and Desurvire, E. 1991. Modeling erbium-doped fiber amplifers. *J. Lightwave Technol.* 9 (2):271–283.

23. Clesca, B., Bayart, D., and Beylat, J.L. 1995. 1.5 μm fluoride-based fiber amplifiers for wide-band multichannel transport networks. *Optical Fiber Techn.* 1:135–157.

24. Ono, H., Nakagawa, K., Yamada, M., and Sudo, S. 1996. Er3+-doped fluorophosphate glass fibre amplifier for WDM systems. *Electron. Lett.* 32 (17):1586–1587.

25. Mori, A., Ohishi, Y., Yamada, M., Ono, H., Nishida, Y., Oikawa, K., and Sudo, S. 1997. *1.5 μm broadband amplification by tellurite-based EDFAs. In Optical Fiber Communication Conference,* Vol. 6, 1997 OSA Technical Digest Series vol. 6 (Optical Society of America, Washington, D.C.), PD1.

26. Agrawal, G.P. 1995. *Non linear fiber optics.* 2nd ed. San Diego, CA: Academic Press.

27. Mazurczyk, V.J., and Zyskind, J.L. 1994. Polarization dependent gain in erbium doped-fiber amplifiers. *IEEE Photon. Technol. Lett.* 6 (5):616–618.

28. Taylor, M.G. and Penticost, S.J., 1994. Improvement in performance of long haul EDFA link using high frequency polarization scrambling modulation. *Electron. Lett.* 30 (10):805–806.

29. Bergano, N.S., Davidson, C.R., and Heismann, F. 1996. Bit-synchronous polarization and phase modulation scheme for improving the performance of optical amplifier transmission systems. *Electron. Lett.* 32 (1):52–54.

30. Desurvire, E., Zyskind, J.L., and Simpson, J.R. 1990. Spectral gain hole-burning at 1.53 μm in erbium-doped fiber amplifiers. *IEEE Photon. Technol. Lett.* 2 (4):246–248.

31. Walker, G.R. 1991. Gain and noise characterisation of erbium-doped fiber amplifiers. *Electron. Lett.* 27 (9):744–745.

32. Srivastava, A.K., Zyskind, J.L., Sulhoff, J.W., Evankow, J.D. Jr., and Mills, M.A. 1996. Room temperature spectral hole-burning in erbium-doped fiber amplifiers. In *OFC'96 Technical Di-*

gest, Optical Fiber Communication Conference, Technical Digest Series Vol. 2 (Optical Society of America, Washington, DC, 1996):33–34.

33. Hansen, S.L., Andreasen, S.B., Thorsen, P., and Dybdal, K. 1993. Experimental verification of new EDFA gain-tilt distortion theory. *IEEE Photon. Technol. Lett.* 5 (12):1433–1435.

34. Olsson, N.A. 1989. Lightwave systems with optical amplifers. *J. Lightwave Technol.* 7 (7):1071–1082.

35. Baney, D.M. and Sorin, W.V. 1995. Broadband frequency characterization of optical receivers using intensity noise. *Hewlett-Packard J.* 46(1):6–12.

36. Bonnedal, D. 1993. Single-setup characterization of optical fiber amplifiers. *IEEE Photon. Technol. Lett.* 5 (10):1193–1195.

37. Chou, H. and Stimple, J. 1995. Inhomogeneous gain saturation of erbium-doped fiber amplifiers. In *Optical Amplifiers and their Applications,* Technical Digest Series Vol. 8 (Optical Society of America, Washington, DC, 1995):92–95.

38. Poole, S. 1994. Noise figure measurement in optical fibre amplifiers. In *NIST Technical Digest—Symposium on optical fiber measurements,* Boulder, CO. NIST special publication 864: 1–6.

39. Baney, D.M., and Dupre, J. 1992. Pulsed-source technique for optical amplifier noise figure measurement. *European Conference on Communications.* Paper WeP2.11. Berlin.

40. Bertilsson, K., Andrekson, P.A., and Olsson, B.E. 1994. Noise figure of erbium-doped fiber amplifiers in the saturated regime. *IEEE Photon. Technol. Lett.* 6(10):199–201.

41. Onaka, H., Miyata, H., Ishikawa, G., Otsuka, K., Ooi, H., Kai, Y., Kinoshita, S., Seino, M., Nishimoto, H., and Chikama, T. 1996. 1.1 Tb/s WDM transmission over a 150 km 1.3 μm zero-dispersion singlemode fiber. In *Optical Fiber Communication Conference,* Technical Digest Series Vol. 2 (Washington, DC: *Optical Society of America,* 1996) PD 19.

42. Baney, D.M. and Stimple, J. 1996. WDM EDFA gain characterization with a reduced set of saturating channels. *IEEE Photon. Technol. Lett.* 8(12):1615–1617.

43. Baney, D.M. and Jungerman, R.L. 1997. Optical Noise Standard for the Electrical Method of Optical Amplifier Noise Figure Measurement" in *Optical Amplifiers and Their Applications, Technical Digest Series* Vol. 2 (*Optical Society of America:* Washington, DC) paper MB3.

44. Willems, F.W., van der Plaats, Hentschel, D., and Leckel, E. 1994. Optical amplifier noise figure determination by signal RIN subtraction. In *NIST Technical Digest—Symposium on optical fiber measurements,* Boulder, CO: NIST Special publication 864:7–9.

45. Smart, R.G., Hanna, D.C., Tropper, A.C., Davey, S.T., Carter, S.F., and Szebesta, D. 1991. CW room temperature upconversion lasing at blue, green and red wavelengths in infrared-pumped Pr^{3+}-doped fluoride fibre. *Electron. Lett.* 27:1307–1309.

46. Piehler, D., Craven, D., Kwong, N.K., and Zarem, H. 1993. Laser-diode-pumped red and green upconversion fibre lasers. *Electron. Lett.* 29:1857–1858.

47. Baney, D.M., Rankin, G., and Chang, K.W. 1996. Simultaneous blue and green upconversion lasing in a laser-diode-pumped Pr^{3+}/Yb3+doped fluoride fiber laser. *Appl. Phys. Lett.* 69(12):1662–1664.

48. Baney, D.M., Rankin, G., and Chang, K.W. 1996. Blue Pr^{3+}–doped ZBLAN fiber upconversion laser. *Optics Lett.* 21(17):1372–1374.

49. Petreski, B.P., Murphy, M.M., Collins, S.F., and Booth, D.J. 1993. Amplification in Pr^{3+}–doped fluorozirconate optical fibre at 632.8 μm. *Electron. Lett.* 29:1421–1423.

50. Durteste, Y., Monerie, M., Allain, J.Y., Poignant, H. 1991. Amplification and lasing at 1.3 μm in praseodymium-doped fluorozirconate fibres. *Electron. Lett.* 27(8):626–628.

51. Miyajima, Y., Sugawa, T., and Fukasaku, Y. 1991. 38.2 dB amplification at 1.31 μm and possibility of 0.98 μm pumping in Pr^{3+}-doped fluoride fibre. *Electron. Lett.* 27(19):626–628.

52. Miyajima, Y., Sugawa, T., Fukasaku, Y. 1992. Noise characteristics of Pr^{3+}-doped fluoride fibre amplifier. *Electron. Lett.* 28(3):246–247.

53. Whitley, T.J., Wyatt, R., Szebesta, D., and Davey, S.T. 1993.Towards a practical 1.3 μm optical fibre amplifier. *BT Technol. J.* 11(2):115–127.

54. Yamada, M. et al. 1995. Low-noise and high-power PR^{3+}-doped fluoride fiber amplifier. *IEEE Photon. Technol. Lett.* 7(8):868–871.

55. Dye, S., Fake, M., and Simmons, T.J. 1994. Practical praseodymium power amplifier with a saturated output power of +18 dBm, in Conference on *Optical Fiber Communications, Technical Digest Series,* Vol. 4 (Optical Society of America, Washington DC, 1994):200.

56. Payne, A.S., Wilke, G.D., Smith, L.K., and Krupke, F. 1994. Auger upconversion losses in Nd-doped laser glasses. *Optics. Comm.* 111:263–268.

57. Page, R., Schaffers, K.I., Wilke, G.D., Waide, P., A., Tassano, J.B., Beach, R.J., Payne, S.A., and Krupke, W.F. 1996. Observation of 1300 nm gain in dysprosium-doped chloride crystals in *Optical Fiber Communication Conference, Technical Digest Series,* Vol. 2 (Optical Society of America, Washington DC, 1996).

58. Samson, B.N., Medeiros, J.A., Neto, R.I., Laming, R.I., and Hewak, D.W. 1994. Dyprosium doped Ga:La:S glass for a efficient optical fibre amplifier operating at 1.3 μm. *Electron. Lett.* 30(19):1617–1619.

59. Komukai, T., Yamamoto, T., Sugawa, T., and Miyajima, Y. 1993. 1.47 μm band Tm^{3+} doped fluoride fibre amplifier using a 1.064 μm upconversion pumping scheme. *Electron. Lett.* 29(1):110–112.

60. Dye, S.P., Fake, M., and Simmons, T.J. 1995. Fully engineered 800 nm thulium-doped fluoride-fiber amplifier. In *Optical Fiber Communications, Technical Digest Series,* Vol. 8 (Optical Society of America, Washington DC, 1995):110.

61. Spirit, D.M., Blank, L.C., Davey, S.T., and Williams, D.L. 1990. System apects of Raman fibre amplifiers. *IEE Proc.* 137, Pt. J. (4):221–224.

62. Stolen, R.H. and Ippen, E.P. 1973. Raman gain in glass optical waveguides. *Appl. Phys. Lett.* 22(6):276–278.

63. Aoki, Y. 1988. Properties of fiber Raman amplifiers and their applicability to digital optical communications systems. *J. Lightwave Technol.* 6(7):1225–1239.

64. da Silva, V.L. 1994. Comparison of Raman efficiencies in optical fibers. In *Conference on Optical Fiber Communications, Technical Digest Series,* Vol. 4 (Optical Society of America, Washington, DC, 1994):136–137.

65. Tiemeijer, L.F., Thijs, P.J.A., and Binsma, J.J.M. 1994. Progress in 1.3 μm polarization insensitive multiple quantum well laser amplifiers. In *Optical Fiber Communications, Technical Digest Series,* Vol. 17 (Optical Society of America, Washington, DC, 1994):234–236.

66. Doussiere, P. 1996. Recent advances in conventional and gain clamped semiconductor optical amplifiers. In *Optical Fiber Communications, Technical Digest Series,* Vol. 19 (Optical Society of America, Washington DC, 1994):220–223.

67. Koren, U., Miller, B.I., Young, M.G., Chien, M., Raybon, G., Brenner, T., Ben-Michael, R., Dreyer, K., and Capik, R.J. 1996. Polarization insensitive semiconductor optical amplifier with integrated electroabsorption modulators. *Electron. Lett.* 32(2):111–112.

68. Leckel, E., Sang, J., Muller, R., Ruck, C., and Hentschel, C. 1995. Erbium-doped fiber amplifier test system. *Hewlett-Packard J.* 46(1):13–19.

APPENDIX

Noise Sources in Optical Measurements

Wayne V. Sorin

This appendix discusses some of the dominant noise sources that limit the sensitivity in both coherent and direct-detection optical receiver configurations. Each noise source will be dealt with independently with the understanding that the total noise is found by summing the squares of the individual noise terms. For comparison purposes all noise sources will be referenced to the photodiode output current (see Figure A.1). This reference position is a convenient location for the comparison of both optically and electrically generated noises. The photocurrent noise can be easily related to optical power sensitivity by use of the photodiode responsivity, which is approximately 1 A/W at wavelengths around 1.55 μm. Except for a change of units, the numerical values for photocurrent and optical power are almost identical. To provide a relatively easy way for comparing the magnitude of different noise sources, the concept of relative intensity noise (RIN) will be introduced. This describes noise as a fractional value, where the noise power in a 1 Hz bandwidth is normalized by the average power.

Each section will first give a general expression for describing the noise source. After this, examples will be given to illustrate how the expressions are used and to give a feeling for their magnitudes. For those who are interested, a simple derivation will be given near the end of each section. This derivation is not intended to be rigorous, but is hoped to provide a physical understanding for the process which generates the noise. The noise associated with optical amplification will be covered in Chapter 13.

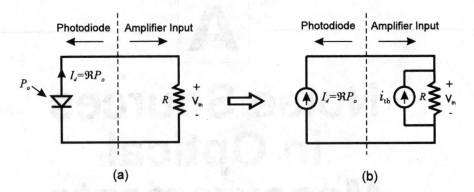

Figure A.1 (a) Simplified illustration of a photodiode connected to an electrical amplifier. (b) The equivalent circuit modeling the photodiode output and thermal resistive noise using ideal current sources.

A.1 ELECTRICAL THERMAL NOISE

One common noise source, which needs to be considered in almost every detection process, is the thermal noise generated in the receiver electronics. If the receiver amplifying process is considered ideal, so that no excess noise is generated, the resulting receiver noise will be determined by the thermal noise (also known as Johnson noise) generated by the resistance first experienced by the photocurrent. As this resistance is made larger the optical power sensitivity is improved. This result will become more evident in the following discussions.

Thermal noise from a resistor can be modeled as being generated by either a voltage or current noise source. Since the signal from a photodiode looks as if it were generated by a current source, it is more convenient to use the current noise-source model for describing thermal noise. This allows the current noise to be directly compared to the generated photocurrent.

Figure A.1a shows the basic configuration for generating a signal voltage using a photodiode and external resistor. Figure A.1b is a simplified equivalent circuit which uses current sources to model the photodiode and thermal noise generated by the resistor. For simplicity the circuit capacitances have been omitted, but they would need to be included for determining the effective noise bandwidth of the circuit. As modeled in Figure A.1b, the thermally generated rms current noise $\hat{i}_{th}$ in a 1 Hz bandwidth is given by

$$\hat{i}_{th} = \sqrt{\frac{4\,kT}{R}} \quad [A/\sqrt{\text{Hz}}] \tag{A.1}$$

where R is the resistance which the photocurrent first experiences, $k = 1.38 \times 10^{-23}$ J/K is Boltzman's constant and T is the temperature of the resistor in Kelvin. The caret above the rms current symbol is used to indicate that the current noise is normalized to a 1 Hz bandwidth. This normalized expression is useful when comparing the magnitude of the thermal noise with the other noise sources in the system. The total rms current (i_{th}) noise is obtained by multiplying Equation A.1 by the square root of the receiver bandwidth ($i_{th} = \hat{i}_{th}\sqrt{\Delta f}$).

As seen from Equation A.1, the thermal current noise (or optical power sensitivity) is reduced by making the resistance larger. This is the opposite result when considering standard, voltage-based electronic circuits. Although a larger resistor reduces receiver noise, the actual value used is usually a compromise between sensitivity and receiver bandwidth. It should be pointed out that for a transimpedance receiver, the resistance in Equation A.1 is the feedback resistance and not the effective input impedance seen by the photodiode.

In practice, the actual noise at the output of the amplifier will be larger due to the excess noise added in the amplifying process. But Equation A.1 is still useful since it predicts the best possible performance given a specific receiver impedance. The room temperature ($T \sim 300$ K) current noise for some representative values of receiver resistance are given in Table A.1

Simple Derivation.   This derivation is not meant to be rigorous or complete but will try to provide a physical understanding for the magnitude of the thermal noise using a few basic concepts. The first of these concepts is that the magnitude of the noise is proportional to the degrees of freedom (or modes) that can exist within the system. Another of the concepts comes from thermodynamics which states that the thermal energy for each degree of freedom is equal to $\frac{1}{2}kT$.

First we will determine the noise power (energy per unit time) associated with the thermal energy of the resistor. This can be obtained by multiplying the thermal energy per mode by the number of modes per second that the system can respond to. The number of modes (degrees of freedom) is equal to the number of orthogonal responses possible by the electrical circuit. In the context of this problem, an electrical circuit with a bandwidth of Δf can generate $2\Delta f$ different orthogonal responses in a 1 s time interval. The factor of two comes about since at any one frequency there are two degrees of freedom corresponding to the two orthogonal phase states ($\cos(\omega t)$ and $\sin(\omega t)$). For example, an electrical

Table A.1   Thermally generated current noise for various resistance values.

R	$\hat{i}_{tn}$ (pA/$\sqrt{Hz}$)
50 Ω	18
100 Ω	13
1 KΩ	3.9
10 KΩ	1.3
100 KΩ	0.39
1 MΩ	0.13
10 MΩ	0.04

circuit with a bandwidth of 1 kHz can generate 2000 independent orthogonal responses in a 1 s time interval. The thermal noise power generated by a resistor is given by

$$P_{\text{thermal}} = \frac{\text{energy}}{\text{mode}} \times \frac{\text{modes}}{\text{second}} = \tfrac{1}{2} kT \times 2\Delta f = kT\Delta f \qquad (A.2)$$

where k is Boltzman's constant and Δf is the effective noise bandwidth of the electrical circuit. At thermal equilibrium, this is the noise power that is delivered from one resistor to an equivalent matched load. It should be noted that this result is independent of the value of the resistance used in the circuit. The resistance comes in later when an equivalent current noise-source is introduced to generate this thermal noise power.

Now we will introduce a fictitious current source in parallel with the resistor which is assumed responsible for generating the thermal noise power. Figure A.2 shows two resistors connected together, each with their equivalent noise current sources. At thermal equilibrium, the noise power delivered from one resistor to the other is equal to the value $kT\Delta f$ obtained from Equation A.2. Since the two resistors are in parallel, the current source for one of the resistors delivers only half of its current to the other resistor. This value of $\tfrac{1}{2}\, i_{th}$ must then be responsible for the thermal noise power $kT\Delta f$ transferred from one resistor to the other. Using this reasoning we get the relationship

$$(\tfrac{1}{2}\, i_{th})^2\, R = kT\Delta f \quad [W] \qquad (A.3)$$

where i_{th} is the rms current generated by the equivalent noise current source in parallel with the resistor. Rearranging Equation A.3 gives the familiar expression of

$$i_{th} = \sqrt{\frac{4kT\Delta f}{R}} \quad [A] \qquad (A.4)$$

This result is valid for frequencies encountered in typical electrical circuits. At very high frequencies, where the photon energy becomes greater than the thermal energy ($hv > kT$) this expression is no longer valid. At room temperature this occurs for frequencies greater than about 6000 GHz ($\lambda < 48\ \mu m$).

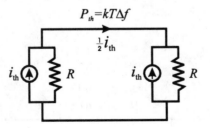

Figure A.2 Thermal noise power transfer between matched resistors at thermal equilibrium.

A.2 OPTICAL INTENSITY NOISE

Another form of noise often encountered in optical measurements is the intensity noise that exists on the optical signal even before the detection process. Intensity noise can originate from several sources. From a fundamental origin, intensity noise occurs from the optical interference between the stimulated laser signal and the spontaneous emission generated within the laser cavity. Laser sources such as distributed feedback (DFB) lasers and Fabry-Perot (FP) laser diodes typically exhibit intensity noise whose value depends on pump levels and feedback conditions. Environmentally varying external feedback can effect the stability of a laser resulting in large variations in its intensity noise.

Intensity noise also exists in nonlaser sources such as edge-emitting light-emitting diodes (EELEDs) and erbium doped-fiber amplifiers (EDFAs). These sources generate amplified spontaneous emission (ASE) whose intensity noise statistics differ from that of lasers. For ASE sources, intensity noise is generated by the interferometric beating between the various frequencies within the spectrum of the ASE. This effect is described in detail at the end of this intensity noise section.

A useful way of describing and comparing intensity noise is to express it as a ratio of noise power in a 1 Hz-bandwidth normalized by the DC signal power. This description is useful since this quantity becomes independent of any attenuation or the absolute power reaching the photodetector. This fractional noise power per bandwidth is often referred to as relative intensity noise (RIN) and is defined as

$$\text{RIN} = \frac{<\Delta \hat{i}^2>}{I_{dc}^2} \quad [\text{Hz}^{-1}] \tag{A.5}$$

where $<\Delta \hat{i}^2>$ is the time-averaged intensity noise power in a 1 Hz bandwidth and I_{dc} is the average DC intensity. Since RIN is a normalized parameter, Equation A.5 is equally valid if the parameters $\Delta \hat{i}$ and I_{dc} refer to optical intensity, detected photocurrent or even receiver output voltage. In practice, RIN can be easily calculated using an electrical spectrum analyzer to measure the time-averaged photocurrent noise power per unit bandwidth $<\Delta \hat{i}^2>$, and a DC ammeter to determine the average DC photocurrent, I_{dc}. The contributions caused by thermal and shot noise should be subtracted from the measured noise power to obtain a more accurate value for the actual intensity noise on the incoming optical signal.

Example

This example illustrates how a typical RIN measurement is made. The output of a DFB laser is detected by a transimpedance receiver which is connected to an electrical spectrum analyzer using a bias-tee to block the DC voltage. Using a voltmeter, the DC voltage from the transimpedance detector is measured to be 5 V. On the spectrum analyzer, the electrical noise power in a 1 Hz noise bandwidth is determined to be −118 dBm. This noise level is typically different than the displayed electrical power since it requires taking the effective noise bandwidth of the analyzer into account. Since the spectrum analyzer calculates electrical power based on a 50 ohm load, −118 dBm (1.6×10^{-15} W) corresponds to a rms noise voltage of 2.8×10^{-7} V in the 1 Hz bandwidth. Dividing this noise voltage by the above 5 V and squaring the ratio gives RIN = 3.1×10^{-15} Hz^{-1} or expressed in decibels, −145 dB/Hz. In this example, we have assumed that the optical intensity noise is the dominant noise source. If this is not the case we would need to subtract the other noise sources before calculating the RIN.

In general, RIN is a function of frequency but for cases where the noise spectrum is flat over the frequency range of interest, it is expressed as a single number. For the case of a flat noise spectrum, the total rms current noise caused by RIN is given by

$$i_{\text{rin}} = I_{dc}\sqrt{\text{RIN } \Delta f} \quad [A] \tag{A.6}$$

where Δf is the effective noise bandwidth of the receiver.

Special Case for ASE Sources. ASE can be generated from sources such as EELEDs, superluminescent diodes (SLD) and EDFAs. These broadband sources typically have very short coherence lengths and are important in performing wavelength dependent insertion-loss measurements (see Chapter 9) and for making optical low-coherence reflectometry measurements (see Chapter 10). The intensity noise from these broadband optical noise sources can also be used for calibrating the frequency response of wide-bandwidth photodetectors.[1] They also show potential as an optical-to-electrical noise standard for calibrating the frequency response of a photodiode and electrical spectrum analyzer combination. This calibration is important when using an electrical spectrum analyzer to measure the noise figure of an EDFA.

An interesting property of these broadband optical noise sources is that their RIN depends only on their optical spectral width Δv_{ase} and is approximately given by the simple expression

$$\text{RIN} \cong \frac{1}{\Delta v_{\text{ase}}} \quad [\text{Hz}^{-1}] \tag{A.7}$$

where it is assumed that the light is unpolarized and exists in a single spatial mode. For polarized light, the RIN is increased by a factor of two. Equation A.7 is valid for frequencies that are small compared to the spectral width of the ASE source. At higher frequencies, the intensity noise decreases in magnitude.[2] Since spectral widths can easily be in the THz range, this roll-off is often not observed on the detected photocurrent. A more complete discussion of this result is given in the simple derivation at the end of this section.

Example

Consider the ASE output from an EDFA without an input signal. Assuming the ASE has a frequency extent of 10 nm centered at 1.55 μm, the spectral width for this noise source is equal to about 1250 GHz. The RIN on this output will be approximately RIN $\cong 8 \times 10^{-13}$ Hz^{-1} or -121 dB/Hz (using Equation A.7) and can be considered spectrally flat over any practical electrical bandwidth. If this signal is sent through a 1 nm interference filter the RIN would increase 10 dB to -111 dB/Hz. This result illustrates that filtering an ASE signal causes it to become more noisy when measuring noise as a fractional quantity. For this case, the absolute noise power actually decreased but at a slower rate than the DC power. Understanding these differences can be important in certain applications.

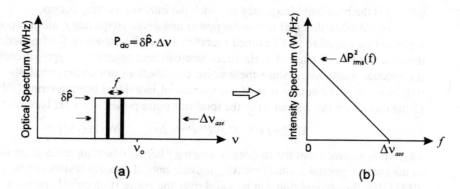

Figure A.3 (a) The optical power spectral density for a thermal light source with a rectangular shaped spectrum. (b) The resulting power spectral density for the optical intensity fluctuations.

Simple Derivation. The purpose of this derivation is to provide a physical understanding for the intensity noise that accompanies a broadband thermal-like optical noise source. These results are also valid for sources which generate ASE such as fiber amplifiers and EELEDs. This example does not describe sources such as lasers, whose statistical properties are different because of gain saturation effects caused within the laser cavity.

To make the analysis simpler, we will assume a thermal light source which has a rectangular shaped optical spectrum as shown in Figure A.3a. This spectral shape could be obtained by filtering an ASE source with a flat-topped bandpass filter such as a grating-based monochrometer. Let the optical power in a 1 Hz slice of bandwidth be given by $\delta\hat{P}$. The total cw optical power P_{dc} can be found by summing the power in all the individual 1 Hz slices to get

$$P_{dc} = \delta\hat{P}\Delta\nu_{\text{ase}} \quad [W] \tag{A.8}$$

where $\Delta\nu_{\text{ase}}$ is the spectral width, in Hertz, of the optical source.

Next we will determine the rms optical noise power (or noise intensity) in a 1 Hz bandwidth centered at a beat frequency f that is small compared to the spectral width, $\Delta\nu_{\text{ase}}$. The origin of this intensity noise comes from optical interference or beating between the various 1 Hz spectral slices what make up the optical spectrum. This is sometimes referred to as spontaneous-spontaneous beat noise.[2] The first step in calculating the total rms noise power is to determine the rms power from the beating of just two of the many 1 Hz spectral slices. After this, all these contributions will be summed to get a total rms noise value. Assuming a polarized optical signal from a singlemode fiber, the rms noise from just two of the spectral slices (separated by a frequency f) is given by

$$\Delta\hat{P}_2(f) = \,<[2\sqrt{\delta\hat{P}\delta\hat{P}}\cos(2\pi f t)]^2>^{\frac{1}{2}} = \sqrt{2}\delta\hat{P}\delta\hat{P} \quad [W/Hz] \tag{A.9}$$

where f is the base-band frequency at which the intensity beating occurs.

To calculate the total rms noise power at a given frequency f, all the spectral slices separated by f need to be combined together. For small values of the frequency spacing, the total number of pairs of 1 Hz slices that can beat together is approximately equal to the spectral width Δv_{ase}. Since these noise contributions are uncorrelated, the total value will increase as the square root of the number of individual beating terms ($\sqrt{N} \approx \sqrt{\Delta v_{ase}}$). Using this reasoning we can write the total rms noise power in a 1 Hz bandwidth as

$$\Delta \hat{P}_{rms}(f) \cong \sqrt{2\delta\hat{P}\delta\hat{P}} \cdot \sqrt{\Delta v_{ase}} \quad [\text{W/Hz}^{-1/2}] \tag{A.10}$$

where it is assumed that the frequency spacing f between beating terms is small compared to the optical spectral width ($f \ll \Delta v_{ase}$). Since optical spectral widths can be greater than 1000 GHz, this assumption can be valid over the entire rf electrical spectrum. For larger frequencies, the noise drops off since there are less terms that can be mixed together to generate the larger beat frequencies. Figure A.3b shows a plot of $\Delta_{Prms}^2(f)$ as a function of frequency. This quantity would be displayed on an electrical spectrum analyzer when measuring the detected photocurrent. For frequencies above Δv_{ase}, there is no intensity noise since the frequency slices would need to be spaced further than the spectral width of the source.

The peak RIN is found by taking the ratio of Equations A.10 and A.8 and squaring the result.

$$\text{RIN} = \frac{\Delta P_{rms}^2(f)}{P_{dc}^2} \cong \frac{2}{\Delta v_{ase}} \quad [\text{Hz}^{-1}] \tag{A.11}$$

This result is valid for the case of polarized light and for frequencies much less than the spectral width of the source ($f \ll \Delta v_{ase}$). If the light source was unpolarized, the RIN would be decreased by a factor of two and we would get the result given in Equation A.7. The decrease in RIN for unpolarized light can be understood by considering the effect of adding an additional uncorrelated (orthogonally polarized) signal of equal power. The total power would double but the noise power would increase by only $\sqrt{2}$, therefore resulting in a smaller RIN. If the spectral shape of the broadband noise source is not rectangular, the RIN will be modified slightly. For the cases of a Gaussian and Lorentzian shaped spectrum, Equation A.11 should be multiplied by 0.66 and 0.32 respectively.[2] For this result, Δv_{ase}, is measured as the full-width-half-maximum (FWHM) spectral width of the source.

One interesting way to think of the RIN from an ASE source is in terms of degrees of freedom. That is, the fractional intensity noise in a 1 Hz bandwidth is inversely proportional to the degrees of freedom the optical signal possesses in a one second time interval. This concept can be useful in predicting how the RIN will change under various conditions. For example, it tells us that if we remove half the degrees of freedom by polarizing an unpolarized signal, the RIN will increase by a factor of two. It also tells us that if a spatially incoherent broadband source (for example, a surface-emitting LED or a Tungsten lamp) excites a multimode fiber the result given by Equation A.11 should be reduced by the number of spatial modes in the fiber.

A.3 PHOTOCURRENT SHOT NOISE

Electrical shot noise occurs because of the random arrival time of the electrons that make up an electrical current. It usually becomes an important noise source when trying to measure a small signal in the presence of a large DC background. This case normally occurs in coherent detection schemes where a small AC current is being measured in the presence of the large background due to the DC local oscillator current. The rms shot-noise current in a 1 Hz bandwidth is given by

$$\hat{i}_{sn} = \sqrt{2qI_{dc}} \quad [A/\sqrt{Hz}] \tag{A.12}$$

where $q = 1.6 \times 10^{-19}\,C$ is the charge of an electron and I_{dc} is the DC photocurrent. Without frequency filtering, shot noise is spectrally flat and therefore has the above value at each measurement frequency. To calculate the total rms shot noise current (i_{sn}) for an electrical circuit with an effective noise bandwidth (Δf), Equation A.12 should be multiplied by the square root of the bandwidth $(i_{sn} = \hat{i}_{sn}\sqrt{\Delta f})$.

An interesting observation can be made when comparing shot noise with thermal noise. Since the shot-noise level depends on signal current, there will be a point for increasing DC current when the shot-noise value exceeds the fixed thermal noise. It turns out that for a photodiode feeding into a resistor, the shot noise starts to exceed the thermal resistor noise when the voltage across the resistor becomes larger than 52 mV. This voltage level is independent of the value of the resistor. This result is useful in practice since it provides an easy method for determining which of the two noise sources is dominant. If the amplifying process generates excess noise, the value of 52 mV needs to be increased accordingly. Another point to mention is the special meaning that the shot-noise limit has in a coherent detection process. In this regime, the receiver has optimum sensitivity with a noise equivalent power equal to a single photon per integration time of the receiver.

Although RIN is defined as the fractional intensity noise on an optical signal (see Equation A.5), it can also be used in a nonconventional way to describe the level of shot-noise on a dc photocurrent. By dividing the shot-noise current by the dc current and squaring the result, we get an expression equivalent to RIN. Expressing the shot noise this way, allows easy comparisons with other noise sources expressed in a similar manner. Using Equations A.5 and A.12, shot noise produces an effective RIN given by

$$RIN_{sn} = \frac{2q}{I_{dc}} \quad [Hz^{-1}] \tag{A.13}$$

This result is useful for determining the required dc photocurrent needed to make an accurate RIN (see Section A.2) measurement on an optical signal. The RIN_{sn} decreases with dc photocurrent while the true optical RIN is independent of the dc signal. To make an accurate RIN measurement, one must ensure that a large enough dc photocurrent is detected to prevent shot noise from being the dominant noise source. For example, to measure a RIN of -155 dB/Hz on a DFB laser requires a photocurrent on the order of $I_{dc} = 1$ mA or greater. Representative values of shot noise for different dc photocurrents are shown in Table A.2.

Table A.2   Representative shot-noise values.

I_{dc}	i_{sn} (pA/$\sqrt{Hz}$)	RIN$_{sn}$ (dB/Hz)
100 nA	0.18	−115
1 μA	0.57	−125
10 μA	1.8	−135
100 μA	5.7	−145
1 mA	18	−155

Simple derivation   The following discussion shows a simple derivation for the shot-noise expression given in Equation A.12. Shot noise can be thought of as being generated by the random arrival time of electrons that make up a dc photocurrent. Figure A.4 illustrates this random arrival-time process. Each vertical arrow represents the detection of a single particle at a specific time. In a nonrigorous manner these particles could also be photons, but for the purpose of this derivation they will be assumed to be electrons.

The above random arrival time can be described by a Poisson probability process. This type of process has the characteristic that in any given time interval, the variance (or rms uncertainty) in the number of electrons is equal to the square root of the average number.

$$\Delta N_{rms} = \sqrt{N} \qquad (A.14)$$

This variation in the average number of arriving electrons during any specified time interval leads to the generation of shot-noise. The rms shot noise current can be written as the rms variation in detected charge per unit time as

$$i_{sn} = \frac{q\Delta N_{rms}}{\Delta t} \quad [A] \qquad (A.15)$$

where q is the charge of the electron and Δt is the measurement time interval. The dc photocurrent can be expressed in a similar manner using the average number of electrons per time interval as

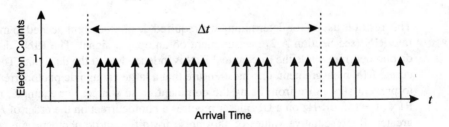

Figure A.4   Random arrival time of photogenerated electrons.

$$I_{dc} = \frac{q\overline{N}}{\Delta t} \quad [A] \tag{A.16}$$

This result allows us to express the rms shot noise current in terms of the dc current using the above three equations.

$$i_{sn} = \sqrt{\frac{qI_{dc}}{\Delta t}} \quad [A] \tag{A.17}$$

To put this expression into a more commonly used form, we must relate the measurement time interval to an equivalent noise bandwidth. Using Fourier analysis it can be shown that the effective noise bandwidth Δf of a flat-topped rectangular gate function of width Δt is given by

$$\Delta f = \frac{1}{2\Delta t} \quad [Hz] \tag{A.18}$$

With this result we can now convert Equation A.17 into the familiar expression

$$i_{sn} = \sqrt{2qI_{dc}\Delta f} \quad [A] \tag{A.19}$$

which is equivalent to the result given in Equation A.12.

An equivalent argument can be made for the shot-noise intensity on an optical signal by considering photons instead of electrons. Now the random arrival time of the photons leads to a fluctuation in the optical power. For this situation, the above equations can be rewritten by replacing the electrical charge with the photon energy ($q \rightarrow h\nu$) and the current with the optical power ($I_{dc} \rightarrow P_{cw}$). When using this classical concept to describe photons, care must be used since it is not rigorous in a quantum mechanical sense and can lead to incorrect results.

Squeezed States   As a final comment on shot-noise, an attempt will be made to describe the concept of optical squeezed states.[3] Squeezed states is a quantum mechanical concept describing the reduction in the "optical shot-noise" on a cw optical signal. Or in other words, removing the randomness in the arrival time (see Figure A.4) of photons and thereby decreasing the associated optical intensity noise. Squeezed states makes use of the uncertainty principle between the position and momentum (in other words, frequency) of a photon. The uncertainty in the photon position can be reduced at the expense of increasing the uncertainty in the photon frequency.

This concept can be understood by replacing the electrons in Figure A.4 with photons. This can be rationalized since for the case of a 100% quantum efficiency detector, there is a one-to-one correspondence between the input photons and the generated electrons. By producing a more equal spacing between photons, the intensity noise is reduced since the uncertainty in the number of photons in a given measurement time becomes less than $\sqrt{N}$. The process of "squeezing" the intensity fluctuations out of an optical signal requires that the positions of the individual photons become well defined. Due to the uncertainty principle, this increases the uncertainty of the photon frequency (or momentum) which leads to a broadening of the optical bandwidth. The more the intensity noise is reduced the larger the optical bandwidth becomes. Although squeezed states have been demonstrated experimentally, reducing the shot noise by more than a several dB becomes extremely difficult.[3]

A.4 OPTICAL-PHASE-NOISE TO INTENSITY-NOISE CONVERSION

The conversion of optical phase noise (frequency fluctuations in the optical carrier) into intensity fluctuations occur when multiple reflectors cause time-delayed portions of the optical signal to interfere with each other. This situation is troublesome whenever small signals must be measured in the presence of large background signals. Phase noise can cause degradation in certain types of communication schemes and reflectometry measurement techniques. In reflectometry applications, phase noise can be important in coherent optical frequency domain techniques, where small sinusoidal signals must be measured in the presence of larger background signals. In this section, we discuss the conversion of phase noise into intensity noise for both the incoherent and coherent cases. These cases are distinguished by comparing the coherence time of the source to the differential delays between the interfering optical signals.

Figure A.5a shows the relationship between optical frequency variations and the detected photocurrent after an optical signal is split and recombined with a relative delay τ_o. This interferometric conversion of frequency variations to intensity variations, is a char-

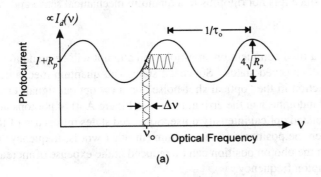

(a)

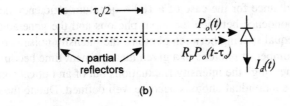

(b)

Figure A.5 (a) Conversion of optical frequency fluctuations into intensity variations due to the coherent interference between time-delayed signals. (b) Interfering signals generated by a pair of partial reflectors.

acteristic of optical circuits such as Michelson and Mach-Zehnder interferometers. These effects can also occur after a laser signal passes through weak etalons caused by unwanted reflections in a transmission link (see Figure A.5b). The detected photocurrent for a cw laser signal that passes through one of the above interferometers can be written as:

$$I_d(t) = \mathcal{R} \, P_o[1 + R_p + 2\sqrt{R_p} \, \cos(2\pi\tau_o\nu(t)] \quad [A] \tag{A.20}$$

where $\mathcal{R}$ is the responsivity of the photodetector, P_o is an optical power, τ_o is the differential time delay, R_p is the ratio of the two interfering optical powers and $\nu(t)$ is the instantaneous optical frequency of the laser source. See Chapter 5.2 for a derivation of this equation. Since Equation A.20 describes a coherent interference effect, it is assumed that the coherence time of the laser is longer than the differential time delay, τ_o. This equation also assumes identical polarization states for the two mixing signals.

As shown in Figure A.5, if the optical carrier frequency is centered at the location of maximum slope (quadrature position), the fractional intensity change due to a small frequency change $\Delta\nu(t)$ is given by

$$\frac{\Delta I(t)}{I_{avg}} \cong K_{fm} \, \Delta\nu(t) \tag{A.21}$$

where $K_{fm} = 4\pi\tau_o\sqrt{R_p}/(1 + R_p)$ is the maximum slope from Equation A.20 and can be thought of as the FM discriminator constant. Since the above expression is a linear approximation obtained from a nonlinear expression, it is only valid if constraints are put on $\Delta\nu(t)$. Restrictions on both the magnitude and modulation frequency of $\Delta\nu(t)$ are required. To keep the accuracy of Equation A.21 to better than 10%, the constraint of $\Delta\nu < 0.1/\tau_o$ should be met. See Chapter 5.3.4 for more details on the constraints assumed for Equation A.21.

Example

Consider a DFB laser in a FSK communication link which uses 1 GHz frequency shifts in the optical carrier frequency. Assume the frequency modulated signal has to pass through a poorly constructed optical component with two Fresnel reflections (each with a reflectivity of 4%) spaced by 1 cm in air. From this information we get $\tau_o \cong 67$ psec, $R_p \cong (.04)^2$, and $\Delta\nu = 1$ GHz. Putting these values into Equation A.21, the resulting modulation or noise on the transmitted power is about 3.4%. Since the magnitude of this modulation signal is not constant but changes with polarization and bias position, this type of noise can cause difficulties in predicting system performance.

Coherent Interference.   The above FM discriminator process can also convert a laser linewidth into a spectral noise density on the detected photocurrent. The term "coherent interference" assumes that the coherence length of the laser is much longer than the differential distance experienced by the interfering signals. An expression for the maximum RIN due to the conversion of phase noise into intensity noise is given by

$$\text{RIN}_{\Delta\phi} \cong \frac{2R_p}{(1 + R_p^2)} \, 8\pi\tau_o^2 \, \Delta\nu_{lw} \, \text{sinc}^2\,(\tau_o f) \quad [\text{Hz}^{-1}] \tag{A.22}$$

where the laser lineshape is assumed to be Lorentzian with a linewidth equal to $\Delta\nu_{lw}$ and the function sinc $(x) = \sin(\pi x)/\pi x$ is used. R_p and τ_o are the same as described in Equation A.20 and A.21. To obtain the maximum value described by Equation A.22, both a quadrature phase relation and matched polarization states are assumed for the two interfering signals. The baseband frequency modulation of the photocurrent is described using the rf frequency variable f. For the assumption of coherent interference to be reasonably valid the constraint of $\Delta\nu_{lw} < 0.1/\tau_o$ should be met. One practical difficulty with this type of noise is that it can fluctuate between its maximum value given by Equation A.22 and zero depending on the environmental variations in the bias condition shown in Figure A.5. This can result in considerable difficulties when trying to trouble shoot noise problems in optical instruments and communication systems.

Incoherent Interference. Another situation that has a relatively simple analytical description is the incoherent case. Incoherent interference occurs when the coherence length of the laser is much shorter than the differential distance experienced by the two interfering signals. For this situation, the relative intensity noise due to the conversion of laser phase noise to intensity fluctuations is given by

$$\text{RIN}_{\Delta\phi} = \frac{2R_p}{(1 + R_p^2)} \frac{2}{\pi\Delta\nu_{lw}} \frac{1}{1 + (f/\Delta\nu_{lw})^2} \quad [\text{Hz}^{-1}] \tag{A.23}$$

where the laser lineshape is again assumed Lorentzian and R_p is equal to the optical power ratio of the two interfering signals. For the assumption of incoherent interference to be valid, the condition $\Delta\nu_{lw} \gg 1/\tau_o$ should be met. The expression in Equation A.23 assumes matched polarization states for maximum interference. If the polarizations states are orthogonal this noise effect goes to zero. Unlike the results given in Equations A.21 and A.22, the assumption of incoherent interference makes the above result independent of environmental changes in the bias of the interferometer. This means that the noise spectrum is much more stable, being independent of path length changes of the interferometer. Using Equation A.23, the laser linewidth can be determined by measuring the 3 dB bandwidth of the current noise spectrum. This technique is often referred to as a delayed self-homodyne linewidth measurement (see Chapter 5.3.3).

Figure A.6 shows the shapes for the power spectral density of the optical intensity noise for both the coherent and incoherent interference cases. The two curves are log plots of Equations A.22 and A.23. For a fixed laser linewidth, the largest low-frequency intensity noise occurs for the incoherent case. Although the noise level for the coherent case is smaller, it can extend to much higher frequencies (much higher than the linewidth of the laser). For the case of partially coherent interference, the noise spectrum will fall somewhere between the coherent and incoherent spectrums plotted in Figure A.6. A more complete result, which is also valid for the case of partial coherence, can be found in Chapter 5.3.4.

An interesting result can be noted for the case of incoherent interference described by Equation A.23. The intensity noise spectrum, assuming equal interfering powers ($R_p = 1$), is identical to that obtained from a thermal-like optical noise source as described near the end of Section A.2. This makes sense intuitively since incoherent interference

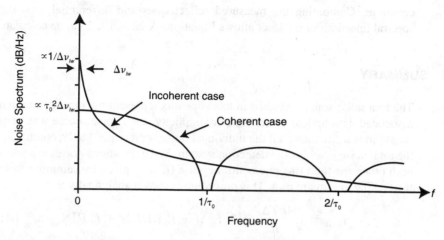

Figure A.6 Power spectral density resulting from the conversion of laser phase-noise into optical intensity noise.

with two equal powers gives a maximally randomized noise signal which is the same as for the case of a thermal noise source. Assuming Lorentzian-shaped optical spectrums, the low-frequency RIN for either case is equal to $2/\pi\Delta\nu_{lw}$ (Equation A.11 modified for a Lorentzian spectrum).

Example

Consider the case of a cw DFB laser with a Lorentzian linewidth of 50 MHz. The laser signal passes through two 4% reflections spaced by 10 cm before reaching a photodetector. For this situation, we have the case of coherent interference and Equation A.22 can be used. Using $\tau_o = 667$ psec, $R_p = (.04)^2$ and $\Delta\nu_{lw} = 50$ MHz, Equation A.22 gives a maximum low frequency $RIN_{\Delta\phi}$ of -117 dB/Hz. This phase-induced intensity noise can dominate over the actual DFB intensity noise which is often on the order of -145 dB/Hz. Also, as mentioned above, this noise can fade in and out depending on environmental variations or shifts in the central laser frequency. The 3 dB bandwidth for this phase noise is determined by the "sinc" function and would be about 750 MHz. In contrast, consider the same two reflectivities (perhaps from two fiber connectors) but now separated by 50 m. For this situation, the interfering signals will be incoherent and the low-frequency phase noise calculated from Equation A.23 becomes $RIN_{\Delta\phi} = -104$ dB/Hz. This noise spectrum has a 3 dB bandwidth of 50 MHz and its magnitude and shape remains relatively stable, independent of environmental drifts.

As the above example illustrates, phase noise can be a dominant noise source. Due to its environmental dependence it may go unnoticed in an initial system test only to become a problem at a later time. The above equations can be used to estimate the maximum effects of phase noise. In practice, optical reflectometry can be used to estimate the optical reflectivities and their associated time delays which are needed to solve these equations. Combining the measured reflectivities and differential time delay with the spectral linewidth of the laser allows Equations A.22 and A.23 to be calculated.

A.5 SUMMARY

The four noise sources studied in this appendix represent the most common noise sources associated with optical detection. For simplicity, each noise source was considered separately. In real situations, all the individual noise sources need to be combined to determine the total noise level for the detection system. Since the above noises are uncorrelated with each other, the total rms photocurrent noise (i_{total}) is given by summing their squares and then taking the square root. This procedure is shown analytically as

$$i_{\text{total}} = \sqrt{\dfrac{4kT\Delta f}{R} + 2qI_{\text{dc}}\Delta f + I_{\text{dc}}^2\text{RIN}\,\Delta f + I_{\text{dc}}^2\text{RIN}_{\Delta\phi}\,\Delta f}\quad [A] \tag{A.24}$$

$$\qquad\qquad\text{(thermal)}\quad\text{(shot)}\quad\text{(intensity)}\qquad\text{(phase)}$$

where the definitions of the parameters can be found in the previous four sections. To keep the expression simple, each noise term is assumed spectrally flat over the bandwidth Δf. If this is not the case, a separate integration over frequency would be required for each term under the square root sign.

Figure A.7 shows the graphical result of combining several of the above noise sources for a high-speed communications receiver. The receiver consists of a room-temperature reversed-biased *pin* photodetector connected to a load impedance of 1 KΩ.

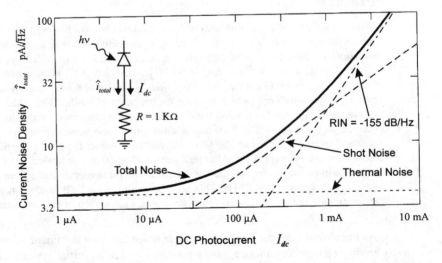

Figure A.7 Total rms photocurrent noise normalized to a 1 Hz bandwidth, caused by the combined effects of thermal, shot, and intensity noise.

To reduce the complexity of the example, the amplification of the photocurrent is assumed ideal so all of the post-detection noise is generated by the thermal noise of the 1 KΩ load resistance. Only the first three noise terms (thermal, shot, and intensity) in Equation A.24 are considered. The optical source is assumed to be a DFB laser with a RIN of −155 dB/Hz. Figure A.7 shows the rms current noise in a 1 Hz bandwidth, as a function of the dc photocurrent. Each of the three noise terms are plotted separately along with the total noise as computed using Equation A.24. For low power levels ($I_{dc} < 52$ μA) the noise is dominated by the thermal noise of the load impedance. For dc currents between 52 μA and 1 mA, the shot noise dominates. And for currents in excess of 1 mA the intensity noise from the DFB source is dominant.

The sensitivity of the receiver to optical power changes can be determined using the total noise current given by Equation A.24. This is simply calculated using

$$\Delta P_{min} = \frac{i_{total}}{\mathcal{R}} \quad [W] \tag{A.25}$$

where $\mathcal{R}$ represents the responsivity (in units amps/watts) of the photodiode. This minimum power sensitivity depends on the square root of the detection bandwidth as shown in Equation A.24.

Example

This example determines the sensitivity of a receiver in the presence of a large dc optical power. Consider the case of a reverse-biased photodiode connected to an electrical circuit with an effective noise bandwidth of 500 MHz. Suppose the incident optical power is $P_o = 1.25$ mW and the resulting photocurrent is $I_{dc} = 1.0$ mA. From these values the photodiode responsivity is calculated to be $\mathcal{R} = I_{dc}/P_o = 0.8$ A/W. Assuming that the dominant noise source is the shot noise from the 1 mA dc current, Table A.2 gives a rms noise current density of 18 $pA/\sqrt{Hz}$. Multiplying this value by the square root of the noise bandwidth gives a total rms current noise of $i_{total} = 0.4$ μA. Equation A.25 can now be used to calculate a receiver sensitivity of 0.5 μW. This sensitivity corresponds to a modulation depth in the input power of 4×10^{-4}.

REFERENCES

1. Baney, D.M., W.V. Sorin, and S.A. Newton. 1994. High-frequency photodiode characterization using a filtered intensity noise technique. *Photon. Tech. Lett.,* 6:1258–1260.

2. Baney, D.M. and W.V. Sorin. 1995. Broadband frequency characterization of optical receivers using intensity noise. *Hewlett-Packard Journal,* 46:6–12.

3. Yamamoto, Y. and W.H. Richardson. 1995. Squeezed States: a closer look at the amplitude and phase of light. *Optics & Photonics News:* 24–29.

APPENDIX

Nonlinear Limits for Optical Measurements

Wayne V. Sorin

Nonlinear effects limit the maximum power that can be transmitted along an optical fiber. This maximum power sets limits on transmission distances and measurement ranges. For example, the ultimate dynamic range for an optical reflectometry measurement is set by the maximum probe power and the receiver sensitivity. Being able to estimate the maximum power that can be transmitted along an optical fiber allows instrument designers to predict the ultimate performance of various measurement techniques. This appendix gives a brief physical description of some of the basic nonlinear effects commonly experienced in fiber optic transmission links. Simplified analytical expressions are presented, allowing maximum power levels to be estimated. Approximate numerical values will be given for standard step-index telecommunications fiber (numerical aperture ~ 0.1) at 1.55 μm. To obtain accurate power limits for specific measurement cases, a more in-depth analysis will probably be required. The provided references[1,2] may help develop a fuller understanding of these nonlinear effects.

B.1 RAMAN LIMIT

Raman nonlinearities occur when a strong optical signal excites molecular resonances within the optical fiber. These molecular vibrations modulate the incident light to generate new optical frequencies. Besides generating new frequencies, these same molecular vibrations also provide optical amplification for the newly generated light. At room temperature, the majority of these new frequencies are generated on the lower frequency side (longer-wavelength side) of the optical carrier. For silica glass, the peak of these new frequencies occurs at about 13 THz below the optical-carrier frequency. This corresponds to

614

a wavelength shift of about 100 nm for signals at 1.55 μm resulting in newly generated wavelengths around 1.65 μm.

The critical power for the Raman nonlinearity is defined as the point where one half of the input signal is lost to newly generated frequencies. Since this critical power depends on the interaction length along the fiber, it is convenient to express this Raman limit as a power-length product. This power-length product can be written as[1]

$$P_{cr} L_{eff} \cong 16 \frac{A_{eff}}{g_R} \sim 17 \text{ W} \cdot \text{km} \qquad \text{(B.1)}$$

where A_{eff} is the effective area of the guided mode, g_R is the Raman gain coefficient, P_{cr} is the critical optical power and L_{eff} is the effective fiber length. The value of 17 W · km is obtained using the values of $g_R = 6.5 \times 10^{-14}$ m/W and $A_{eff} = 70$ μm^2 (mode diameter ~ 9.4 μm) and is given as a representative value for standard telecommunications fiber at 1.55 μm. This result means that if a 17 W signal is coupled into a 1 km length of fiber, the output will consist of about 8.5 W of the original signal and 8.5 W of newly generated frequencies.

A few comments should be made about the effective fiber length used in Equation B.1. For short fiber lengths, this will usually be equal to the physical fiber length. If the fiber is very long, fiber attenuation will limit the effective length. For this situation, the effective length can be calculated using the expression

$$L_{eff} = \frac{1}{\alpha} (1 - e^{-\alpha L}) \qquad \text{(B.2)}$$

where L is the physical fiber length and α is the fiber attenuation constant. For a fiber loss of 0.2 dB/km, the attenuation coefficient is $\alpha = 4.6 \times 10^{-5}$ m^{-1}. This means that a very long fiber, with 0.2 dB/km attenuation, has a maximum effective length of about 21 km. Attenuation is not the only parameter that can limit the effective length. When the optical pump consists of short pulses, fiber dispersion can also set a limit for the effective length. For this case, the effective length will be equal to the walk-off distance between the input pulse and the newly generated Raman frequencies. After the Raman signal no longer spatially overlaps the input pulse, it stops experiencing Raman amplification and, therefore, stops extracting energy from the input pulse.

Besides generating new frequencies, a strong optical signal can provide gain for nearby wavelengths. In applications where high-powered 1480 nm lasers are used to remotely pump erbium-doped fiber amplifiers (EDFAs), significant optical gain can occur for communication signals at 1.55 μm. An experimental demonstration of this effect is shown in the upper part of Figure B.1 A pump power of 150 mW at 1480 nm is launched into a long communications-grade optical fiber. The Raman gain is probed using the amplified spontaneous emission (ASE) from an EDFA. The lower portion of Figure B.1 shows the internal Raman gain as a function of wavelength. The internal gain was measured by recording the change in signal strength when the pump power was turned off and on. For this particular experiment, the peak Raman gain was 8 dB centered at about 108 nm (13.8 THz) from the pump wavelength. The 3 dB gain bandwidth was about

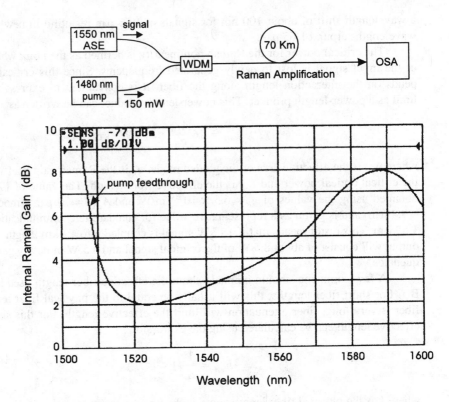

Figure B.1 Raman amplification of a broadband ASE probe signal. High-power lasers at 1480 nm are typically used for remote pumping of erbium-doped fiber amplifiers.

40 nm (5.3 THz). For more information on Raman gain, see Chapter 13 which discusses optical amplification.

B.2 SELF-PHASE MODULATION

High optical intensities temporarily change the refractive index of the optical fiber. The relationship between the optical intensity, I, and fiber index change, δn, can be written as

$$\delta n = n_2 I \qquad (B.3)$$

where n_2 is the nonlinear Kerr coefficient. For standard singlemode silica fiber, the Kerr coefficient has been measured to be approximately 2.3×10^{-20} m^2/W.[3] The optical intensity is equal to the guided power divided by the effective mode area (in other words, $I = P/A_{\text{eff}}$). For a high-intensity input pulse, the Kerr effect results in a time-changing index which causes a self-induced phase modulation of the input pulse. This phase modu-

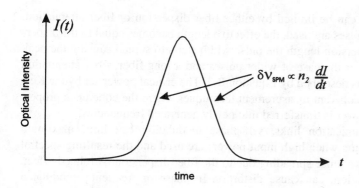

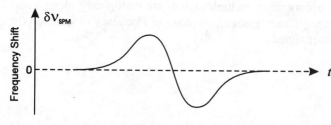

Figure B.2   Self-phase modulation (SPM) generates new frequencies where the intensity changes as a function of time.

lation causes the generation of new optical frequencies. These new frequencies are generated on the rising and falling edges of the pulse where there is a rate-of-change in the intensity. This concept is illustrated in Figure B.2. The bandwidth of the newly generated frequencies is proportional to the product of the nonlinear coefficient and the time derivative of the optical intensity.

The nonlinear Kerr effect plays an important role in the stable propagation of optical soliton pulses. Because of fiber dispersion, newly generated frequencies at the rising and falling edges of the pulse travel at different velocities such that they move towards the center position of the optical pulse. This effect leads to a stable packet of light which does not disperse as it travels along the dispersive fiber.

Just as for the Raman case, we can obtain a power-length product for self-phase modulation. The critical power is the power that causes the optical spectrum of the pulse to double in width due to nonlinear frequency generation.[2] To obtain this relationship, we assume a transform-limited input pulse (in other words, the condition where the product between temporal width and spectral width is a minimum). This power-length product can be approximated as

$$P_{cr}L_{eff} \cong \frac{\lambda A_{eff}}{\pi n_2} \sim 1.5 \text{ W} \cdot \text{km} \qquad \text{(B.4)}$$

where P_{cr} is the critical peak pulse power, L_{eff} is the effective fiber length, λ is the optical wavelength, A_{eff} is the effective mode area, and n_2 is the nonlinear Kerr coefficient. The value of 1.5 W · km is representative of standard telecommunications fiber and is obtained using the values $\lambda = 1.55$ μm, $A_{eff} = 70$ μm^2, and $n_2 = 2.3 \times 10^{-20}$ m^2/W.

The effective length can be limited by either fiber dispersion or fiber attenuation. For example, when short pulses are used, the effective length becomes equal to the dispersion length.[1] After the dispersion length the pulse width starts to spread causing the peak pulse power to decrease. For the case of wider pulses and a long fiber, fiber attenuation limits the effective length as described by Equation B.2. The critical power set by the Kerr effect can be important in coherent measurement techniques where the coherence properties are lost as the probe power is transferred into newly generated frequencies.

In long-haul communication links, self-phase modulation can limit maximum pulsed data rates. This occurs when high input powers are used and the resulting spectral broadening causes excessive pulse spreading due to the fiber dispersion. A related effect, called cross-phase modulation, can cause distortion in phase or frequency modulated communication schemes. This occurs when multiple signals are multiplexed along a single fiber and the intensity changes in one signal cause phase or frequency modulation (in other words, cross-talk) in another signal.

B.3   BRILLOUIN LIMIT

When a narrow-linewidth, high-power signal is guided along an optical fiber it can start to generate an acoustic wave which travels in the same direction as the input optical signal. This acoustic wave has a wavelength equal to approximately one-half the optical wavelength and travels at the acoustic velocity within the fiber. A simplified way to understand the nonlinear Brillouin effect is to think of the acoustic wave as a moving Bragg grating which reflects the input light into the backwards direction. Since the grating is moving in the forward direction, the reflected light is Doppler-shifted to a slightly lower optical frequency. This effect is illustrated in Figure B.3. For a silica fiber at 1.55 μm, the Brillouin frequency shift is about 11 GHz and is determined by the acoustic velocity in the fiber. The linewidth of the reflected signal depends on the losses for the acoustic wave and can range from tens to hundreds of megahertz in standard silica fiber.

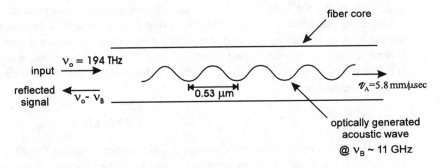

Figure B.3 Brillouin scattering occurs due to an optically generated acoustic wave that acts as a traveling Bragg grating.

As in the cases for Raman scattering and self-phase modulation, we can think of a maximum allowable power set by a power-length product. This power-length product can be written as[1]

$$P_{cr}L_{eff} \cong 21\left(1 + \frac{\Delta\nu}{\Delta\nu_B}\right)\frac{A_{eff}}{g_B} \sim 0.029 \text{ W} \cdot \text{km} \tag{B.5}$$

where $\Delta\nu_B$ is the Brillouin linewidth (typically in the range of 10 to 100 MHz), $\Delta\nu$ is the spectral width of the optical input signal and g_B is the peak Brillouin gain coefficient. The value of 29 mW · km is representative of standard telecommunications fiber at 1.55 μm and was obtained using the values $g_B = 5 \times 10^{-11}$ m/W and $A_{eff} = 70$ μm.[2] This low threshold value assumes a cw input signal with a narrow spectral width ($\Delta\nu < 1$ MHz). When using narrow linewidth laser sources, the critical power caused by Brillouin scattering can be orders of magnitude smaller than Raman scattering or self-phase modulation. For a fiber with a maximum effective length of about 20 km (attenuation of 0.2 dB/km), the critical input power for Brillouin scattering is only 1 or 2 mW.

For pulsed signals, the effective length must be handled differently than for the cases of Raman or self-phase modulation. For Brillouin scattering, the effective length is equal to one-half the pulse-length. This difference occurs since the Brillouin generated optical signal travels in the backwards direction and only interacts with the forward going signal for a limited distance. This effect is very important in long-haul coherent optical time-domain reflectometry measurements. In these measurements, a narrow-linewidth signal ($\Delta\nu < 1$ MHz) is used to probe a long fiber. Based on Equation B.5, the maximum input power would only be 1 or 2 mW. But since the probe signal is pulsed, input powers can be increased by over an order of magnitude before Brillouin scattering starts to deplete the input signal.

B.4 SUMMARY

The following table summarizes the nonlinear effects discussed in the above three sections. Raman nonlinearities usually only need to be considered when very large powers are used. The effects of self-phase modulation can often go unnoticed when using direct

Table B.1   Nonlinearities for standard telecommunications fiber at 1.55 μm.

Nonlinear effect	Power-length product	Frequency shift	Spectral width	Signal type
Raman	~ 17 W · km	~ − 13 THz (~ + 100 nm)	~ 6 THz (~ 50 nm)	cw/pulsed
Self-phase modulation	~ 1.5 W · km	centered on pump	dependent on pulsewidth	pulsed only
Brillouin	~ 0.03 W · km	− 11 GHz (~ + 0.1 nm)	~ 50 MHz	cw/pulsed

detection since the input signal only experiences an increase in its optical spectral width. For narrow linewidth optical signals, Brillouin scattering will most likely be the limiting fiber nonlinearity.

REFERENCES

1. Agrawal, G.P. 1995. *Nonlinear fiber optics.* 2nd ed. San Diego, CA: Academic Press, Inc.

2. Miller, S.E. and A.G. Chynoweth. 1979. *Optical fiber telecommunications.* New York, NY: Academic Press, Inc.

3. da Silva, V.L., Y. Liu, A.J. Antos, G.E. Berkey, and M.A. Newhouse. 1996. *Nonlinear coefficient of optical fibers at 1550 nm.* Boulder, CO: Symposium on Optical Fiber Measurements, NIST pub. 905:61–65.

Fiber Optic Connectors and Their Care

Val McOmber

C.1 BACKGROUND

In the early days of fiber optic development, companies were designing and attempting to field "the best" connector. While there were a few large diameter and plastic fibers in use, the important step of the standardization to a uniform 125 μm diameter glass fiber, for most fiber applications was made. With this standard diameter fiber, one singlemode and twomultimode core sizes (Figure C.1) settled into place. The singlemode core settled on 9 μm (which can sustain singlemode propagation for wavelengths longer than about 1200 nm). The multimode cable needs could not find one core size but settled on both 62.5 and 50 μm cores, with either step or graded-index core-to-cladding transitions. For the purposes of this discussions, the important point is the single, standard outside diameter of the glass. Now connector manufacturers could focus on the mechanical alignment task of holding the fiber and aligning it with another identical size fiber. To achieve this difficult task, there were many different connector types and styles of mechanical assemblies designed. The purpose of each connector is to bring together and hold the two cores of the glass fiber ends. The merits of each are in the ability to hold the close tolerances necessary for a good connection, and in the repeatability of multiple connections. Designers from Japan, Europe, and the U.S. were all rushing to find a way to hold the submicron tolerances necessary, at an acceptable cost.

At this point, another standardization appeared. In the connector, the element that holds the fiber, and provides the alignment positioning, is the ferrule (see Figure C.2). The agreement was to make the ferrule's diameter 2.5 mm. After this, a number of connectors appeared on the market at about the same time, each with a major company or region of the world backing it. The first wave of "common" connectors would include the

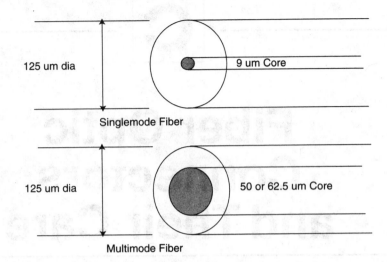

Figure C.1 Common fiber-optic cable has an outside diameter of 125 μm. The active part of the fiber is the core. Most fiber comes in singlemode (with a core of 9 or 10 μm) and multimode (with a core of either 50 or 62.5 μm).

FC, HMS-10, PC, D4, SMA, SC, DIN, ST, and Biconic. Some of these are shown in Figure C.3. Some connectors were designed for performance, while other connectors were designed for cost and volume as the key objective. Often, with these lower cost connectors, very few connections were expected, and after about ten connections the tight tolerances would degrade causing connector performance to degrade.

C.2   CONNECTOR STYLES

While it is acceptable for a particular link or system designer to choose one or another of these connectors (where the system designer could determine the tradeoffs of each connector), it is quite another problem for instrument designers to be able to choose a connector at the exclusion of all others. The instrument designer must either guess the most likely connector (and perhaps lose customers using other connectors), choose the highest-quality connector for the most repeatable measurements (and miss most of the customer base), or design some way to accommodate more than one connector style. The obvious choice required designing another, universal connector adapter, for all connectors. By nature, this connector must be designed for performance and ruggedness. The result of this effort from Hewlett-Packard, with some of the various adapters, is shown in Figure C.4.

After the connector designers learned how to hold connections in close alignment, their attention and efforts shifted to reducing reflections. An example of why reflections became the focus is shown in Figure C.5. This display shows how reflections effect the relative intensity noise (RIN) of a Fabry-Perot (FP) laser. This measure of a laser's noise

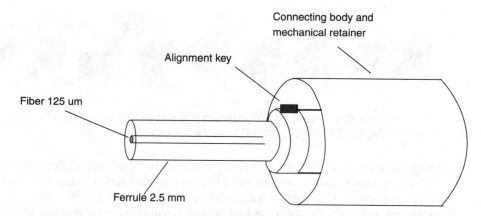

Figure C.2   At the end of a cable is the connector, which contains a mechanical assembly to hold the cable rigid and aligned, and the ferrule which is keyed. Inside, and concentric to the ferrule, is the glass fiber.

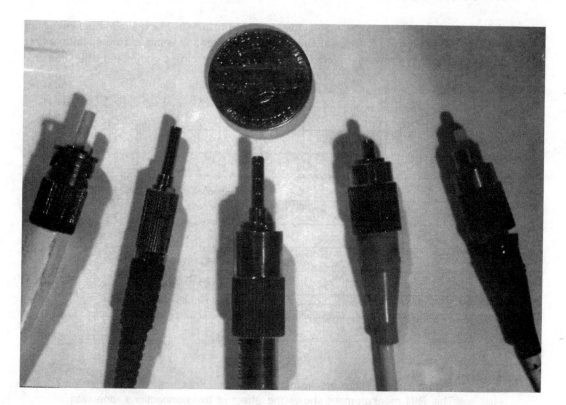

Figure C.3   Photos of five common fiber connectors: ST, DIN, HMS-10, PC, FC.

Figure C.4 Photo of HP connector adapters: HMS-10, FC/PC, D4, SMA, SC/SPC/APC, DIN, PC to angle adapter, ST, Biconic.

performance is one of the critical measurements for the laser source. Reflections getting back into the laser cavity can cause most lasers to be adversely affected. In this example, the FP laser's noise increased as much as 15 dB or more, due to an non-optimized connector. Distributed feedback (DFB) and other lasers can also become unstable with such reflections.

C.3 CONNECTOR DESIGN

A look at what makes up most connectors may be illustrative. The main component of the connector is the ferrule. The ferrule holds the glass fiber and is made from metallic or ceramic material. As shown in Figure C.2, most connectors have a common ferrule diameter

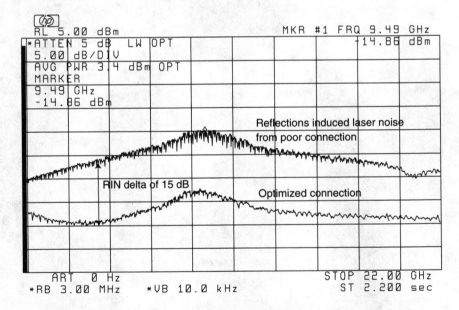

Figure C.5 Reflections from connections affect active components. This RIN measurement shows the effect of two connections, one with more reflections than the other, on a laser source.

of 2.5 mm. The 125 μm glass fiber is centered in the ferrule and exits at its endface where the holding material and fiber are together polished to achieve a smooth endface. Differences in connector types are mainly differences in the mechanical assembly that hold the ferrule in position against another, identical ferrule. The ferrule and the mating sleeve hold the fiber in alignment. The alignment key keeps the fiber from rotating, aids in concentric core matching, and keeps odd shaped cores (such as polarization maintaining fiber) and angled connectors in proper alignment.

Figure C.6 diagrams the major evolution in connector design. In the first (Figure C.6a) is a "fiber connection" or FC connector. These connectors were made with a flat endface. It is polished in such a way that the glass ends up concave and slightly recessed from an otherwise flat face. With this recessed glass there is no glass-to-glass connection. The resulting small gap causes an air-to-glass interface and relatively high reflections (14 dB return loss is the result of a glass-to-air boundary).

The PC, or physical contact connector (Figure C.6b) is polished such that the whole endface, including the glass-fiber tip, ends up with a slight convex shape. This slight rounding (using a radius of about 25 mm) brings the fiber up as the highest point on the endface ensuring a glass-to-glass connection. As long as the fiber and ferrule are kept at the same surface (and clean) the glass-to-glass connection removes the air-gap induced reflections and the return loss is improved from 14 dB to 30 or 35 dB. The FC/PC, ST, SC, APC, and DIN are all physically contacting connectors using the same 2.5 mm diameter ferrule and ferrule construction style. The differences between these connectors are in the mechanical holding assembly that holds the ferrule, the "connection" of the connector. For example the ST is similar to an electrical BNC with a bayonet connection. The FC/PC and HMS-10 uses a threaded sleeve like an electrical SMA connector. The SC is a newer snap-in style.

As applications became more demanding of lower reflections, this PC contact was improved by better polishing. The shape did not change, but a smoother surface allows better glass contact and a return loss of from 45 to as much as 60 dB. Typical construction of this popular cable, called the "super-PC" is shown in Figure C.7. Generally the fiber is held inside a ceramic (Al_2O_3 or ZrO_2) ferrule which is (optionally) contained inside a stainless steel sleeve. Positional concentric accuracy is held to about a micron or less.

It becomes obvious that there is a real danger with these "super-PC" connectors. Any little lint or dust particle can interfere with the glass interface. It doesn't take much to invalidate the critical polish of the fiber end. Finger oil and lint can change a connector with great return loss to a poor one. Worse, the glass face is at the highest point of contact and is very susceptible to damage. Grinding in contaminants and grit can destroy a connector, breaking out pieces of the glass. A cable that is damaged can in turn destroy other connectors and ruin an instrument!

In the quest for improved return-loss performance, the design moved away from very low insertion loss to focus on reflection control. To achieve better return loss, angled fiber ends were introduced (Figure C.6c). With an angled endface, the light could have a large reflection at the surface, but the angle causes the reflected light to go into the cladding and dissipate before the reflected light gets back to the source. This newer connector has a strong future, but the problem of a lack of standardization remains. They come in both an air-gaped

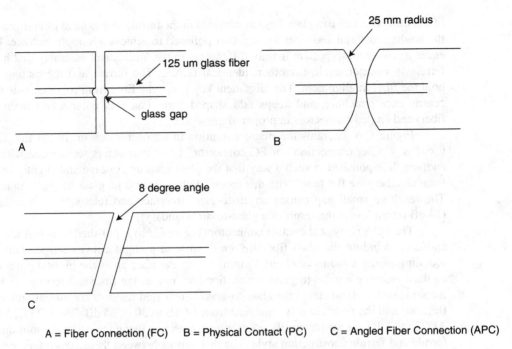

A = Fiber Connection (FC)   B = Physical Contact (PC)   C = Angled Fiber Connection (APC)

Figure C.6   Major developments of fiber-optic endfaces.

and physical contact form (lower loss), and each have different angled orientations possible. Also, while most connectors use an 8 degree angle, there are both "flat" angled and "non-flat" angled versions available (Figure C.8). Thus, trying to mate different angled connectors might not only be ineffective, but could damage the connector.

Given all these choices, the FC/PC style emerged as the most popular connector, capturing the vast majority of instrumentation applications. While not the highest performing connector, it seems to be a good compromise of performance, reliability, and cost. If cleaned well, the PC connector can withstand many insertions and meet most instrumentation needs.

Even with the popularity of the FC/PC, new design efforts are continuing. Just as this book is going to press, a new E2000 connector was announced. Like the SC, it is a push-in, pull-out assembly with the added feature of automatic covering caps for both the fiber and the receptacle. This cover will block dust, fingers, and laser light (eye protection). As Figure C.9 shows, in this new connector style, the design of the endface has also changed. Called a Pilzferrule, this connector has a raised core in the ferrule producing a mesa or button-top (*pilz* means mushroom in German) on the endface. This core uses an active centering operation (like the HMS10) for better fiber position control, better return loss, and lower insertion loss, all at a lower cost.

An example of the risk in choosing a new cable with a "better" connector comes from a service alert issued by Hewlett-Packard for their popular power meter. The

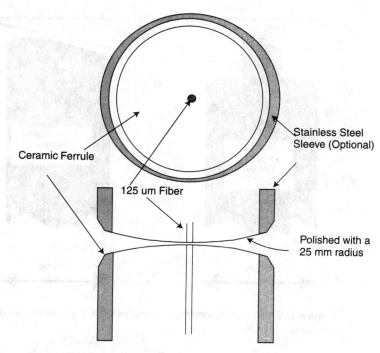

Ceramic Ferrule

Stainless Steel
Sleeve (Optional)

125 um Fiber

Polished with a
25 mm radius

Figure C.7   Construction details of a standard PC connector.

HP 81532A power sensor module uses a special connector with a lens interface to position the fiber end preciously to achieve very accurate power measurements (Figure C.10) from either a PC or a flat-angled connector (the only kind available at its introduction). In this connector, there is a positioning sleeve to hold the end face at a known distance from the lens. However, as Figure C.11 shows, the "nonflat" angled fiber end hits the lens, before the stop ring, causing damage. Also, as in Figure C.12, another "nonstandard" angled fiber connector can cause a missed placement and loss of accuracy in the measurement. The alert also calls out the Pilzferrule-style cable. Similar to the "nonflat" angled case, this Pilzferrule's raised mesa, (Figure C.9), is too high. The mesa top hits the glass before the positioning sleeve stops it. While the Pilzferrule is a very good connector and can be used with many PC style connectors, the design change of this (and other) connectors is unsuited for the power sensor and is damage causing. Thus, while unfortunate, the user must be aware of what kind of connector and fiber to use with each instrument and system for best performance.

So, if reflections are a major concern, and an angled-fiber end cannot be used, what other options and insights are there? One of the causes of reflections is off-center fibers. Singlemode fiber, core is 9 μm. For typical predrilled ceramic-ferrule connectors, the centering, gluing, and other alignment processes used to place the fiber in the center of the end face can leave two fibers as much as a micron off center with each other. To offset

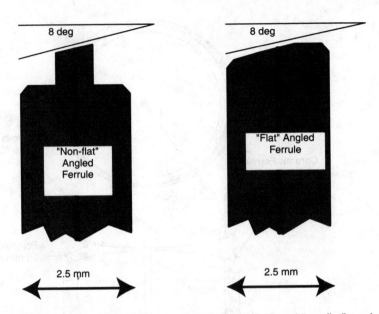

Figure C.8 Angled connectors come in both "flat" and "nonflat" angled ferrules.

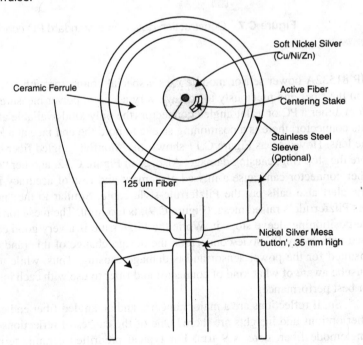

Figure C.9 Construction details of a PC Pilzferrule (mesa or button) connector.

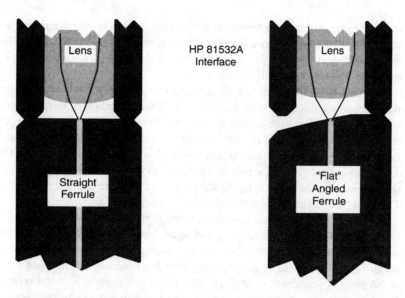

Figure C.10 The HP 81532A works with both straight and flat-angled connectors. The lens is protected and the correct distances are maintained.

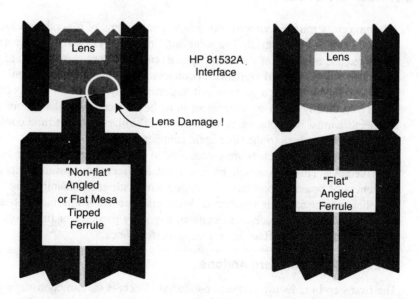

Figure C.11 There is a risk of damaging the lens with some connector styles.

this nonconcentricity, the ferrule is positioned, relative to the connector's alignment key, so that the off-axis fiber is always in the same sector; see Figure C.13. Thus, mating with another connector which is also off center, if that fiber is located in the same sector, the two cores will more closely align.

Instruments that rely on tight tolerances can not afford the "cheaper" connectors with large nonconcentricities. Looking for tighter matches, high-quality measurement instruments typically use a connector like the Diamond (HMS-10) that has concentric tolerances of a few tenths of a micron. Figure C.14 details the HMS-10 connector design. The Diamond process uses a soft nickel silver (Cu/Ni/Zn) inner ferrule that holds the fiber. The process first stakes the soft nickel silver to fix the fiber in a near center location, and then uses a post active staking to shift the fiber into the desired position, within 0.2 μm. This, plus the keyed axis location makes for very good alignments.

The drawback to this connector is the soft endface. Care must be given to keep the endface from excessive scratching and wear. Small wear to the soft material is not a problem if the glass face is not impacted, but, damage can easily occur. The soft material can be pushed with a piece of grit, moving the fiber position, or changing the mating surface. If the glass is scratched, then all of the careful polishing is lost and reflections and insertion loss will soar. Yet, with proper care and cleaning they can last. Some of these HMS-10 soft-center connectors have been in service for years, with many connections per day, maintaining a very good return loss (high glass-surface quality).

C.4   CONNECTOR CARE

All connectors must be viewed like a high quality lens of a good camera, a stray fingerprint can cause a lot of grief. The weak link in many systems, and in instrument reliability, is the use and care of the connector. Because current connectors are so easy to use, there tends to be reduced vigilance in connector care and cleaning. It only takes *one* time of a missed cleaning for a piece of grit to permanently damage the glass and ruin the connector. Too many times an instrument must be serviced to replace damaged connectors. Connectors must be replaced, and thousands of dollars and lost time could be avoided if better care were given to the fiber optic connector.

Fiber optic connectors are susceptible to damage that is not immediately obvious to the naked eye. This damage can have significant effects on measurements being made. An awareness of potential problems, in conjunction with good cleaning practices, can ensure that their performance is optimized. With glass-on-glass interfaces, it is clear that any degradation of a ferrule or fiber endface, any stray particles or finger oil on the endface, can have a significant effect on connector performance.[1]

C.4.1   Connector-Care Actions

The first step in reducing connector-induced effects is care in choosing a connector style. Some questions to ask are: Will the connector need to make multiple connections? What is the reflection tolerance? For example a DFB laser needs high isolation to remain stable

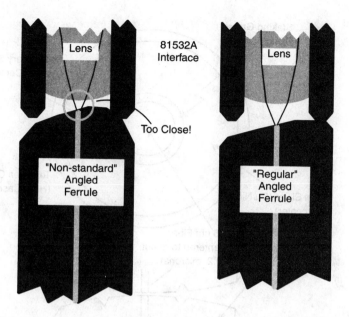

Figure C.12   There is a risk of measurement errors with some non-standard angled ferrules.

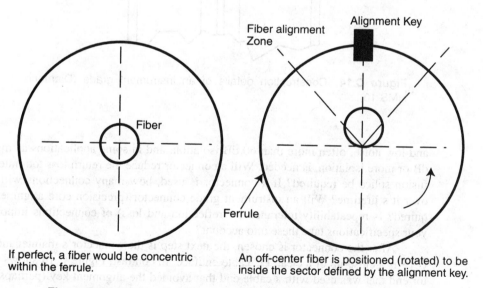

If perfect, a fiber would be concentric within the ferrule.

An off-center fiber is positioned (rotated) to be inside the sector defined by the alignment key.

Figure C.13   Close alignment and placement of the fiber for concentricity is important for low return loss. After cementing the fiber, the ferrule is rotated and keyed so that any nonconcentricity is maintained within a narrow sector for best mating.

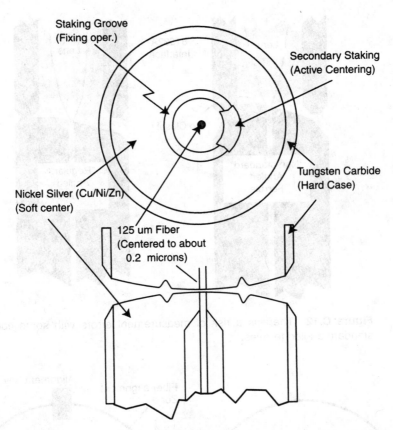

Figure C.14 Construction details of an instrument grade (Diamond HMS-10).

and low noise, often more than 60 dB isolation, and in some applications as much as 90 dB or more isolation, is needed. Will a connector reduce the return loss too much? Will a fusion splice be required? If a connector is used, how many connections will be made over it's lifetime? Will an instrument grade connector (precision core alignment) be required? Is repeatability tolerance (of reflection and loss) of connections important? Do your specifications take these into account?

After the connector is chosen, the next step is the connector's maintenance. Figure C.15 is a close-up of a clean PC cable endface. In contrast, Figure C.16 shows a connector end that was used with a cable end that avoided the alignment key (perhaps a temporary ferrule without a key, or possibly a mismatched long ferrule put into a short connector assemble). Material is smeared and ground into the endface causing light scattering and poor reflection. And, as Figure C.17 shows, if continued, this action will grind off the glass face and destroy the connector.

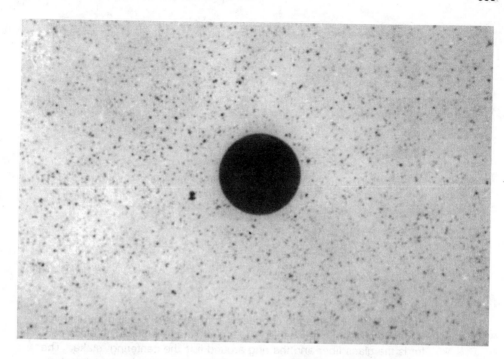

Figure C.15 A photo of a clean and problem-free connector.

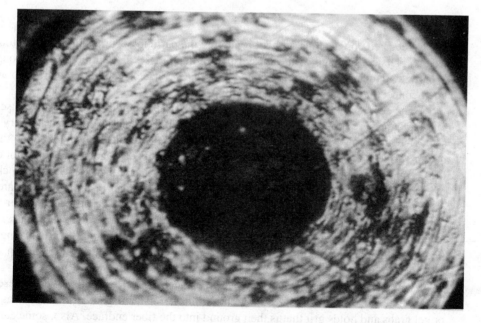

Figure C.16 A photo of a dirty endface from poor cleaning.

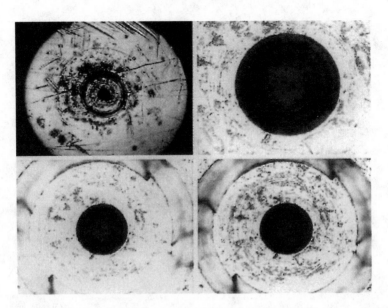

Figure C.17   A damaged connector from abuse and using an unkeyed ferrule to mate with this cable. The top left image is 50×. The black center is the glass fiber and the ring around it is the centering "stake." The bottom two are at 200× and the top right is the glass fiber end at 400× showing the ringed damage on the glass.

Repeated connections without removing the participles, or using improper tools can lead to physical damage of the glass fiber endface as seen in Figures C.18 through 21. Where the damage is severe, the damage on one connector can be transferred to another good connector that comes in contact with it.

The cure is disciplined connector care. Soft swabs should be used—never use metal or sharp objects. Some connectors have a soft endface material that can be very easily scratched and damaged. In Figure C.18, for example, the results of using a pin on the connector surface are seen. In Figure C.19, a close-up of this damaged connector, not only is there gouging in the endface around the fiber, but the glass fiber itself has been chipped off and broken. Broken or damaged glass can lead to a number of problems that are often ascribed to other causes. This connector will probably damage any other fiber that is mated to it. In Figures C.20 and C.21, another connector is seen with severe abuse. In this example, the soft ferrule core was pushed away from the glass and there is pitting in the fiber end. In addition, the glass fiber is sticking up almost 3 μm above the uneven surface (from severe wear of the soft metal that was once flush with the glass fiber).

Another potential problem comes with matching gel and oils. While these often work well on first insertion, they are great dirt magnets. Damage is often caused as the oil or gel grabs and holds grit that is then ground into the fiber endface. Also, some early gels

Figure C.18   A damaged fiber end from using a pin on it.

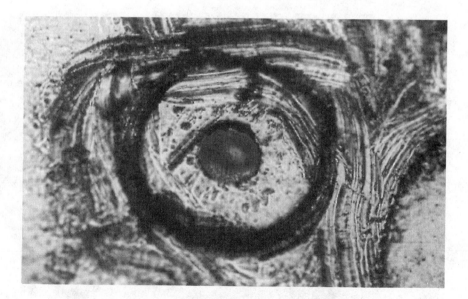

Figure C.19   A close-up (about 100×) of C.18, showing the pin broke off a piece of the glass fiber and moved it out of position.

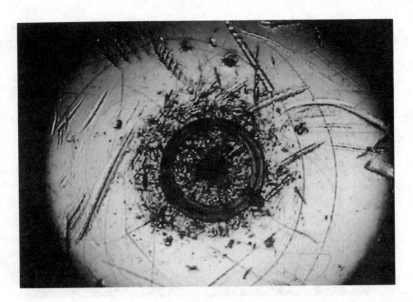

Figure C.20　At 50× the damage to another connector endface, from grit and repeated noncleaning, is seen.

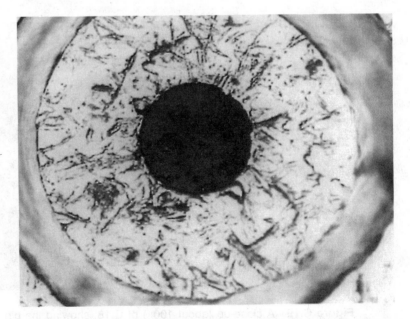

Figure C.21　At 200×, the damage and pitting to the connector in figure C.20 fiber is seen.

were designed for the FC, noncontacting connectors, using small glass spheres. When used in contacting connectors, these glass balls can scratch and pit the fiber. Index matching oil is also discouraged. It runs and migrates into unwanted areas, attracting and holding dust. If an index matching gel or oil must be used, apply it to a freshly cleaned connector, make the measurement, and immediately clean the gel or oil off. Never use a gel for long-term connections and never use it to improve a damaged connector. The gel can mask the damaged end and continued use of a damaged fiber can transfer damage to another connector.

Another tip for good fiber connector mating is: *not too tight.* Unlike electrical connections, tighter is not better. The connector's job is to bring the endfaces of two fibers together. Once they touch, tighter only causes a greater force to be applied to the delicate endface. In some connectors, the end can cock off-axis with a tight connection (due to the curved face) resulting in a worse return loss. Many measurements (such as the RIN measurement described earlier in Figure C.5) are actually improved by backing off the connector pressure. Also, if a piece of grit does happen to get by the cleaning procedure, a tighter connection is more likely to damage the glass. Tighten just until the two fibers touch.

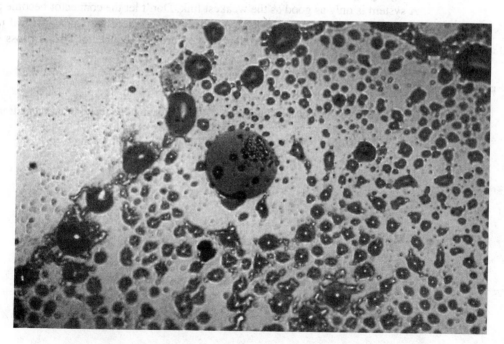

Figure C.22 A good connector was cleaned with a solvent and left to air dry. After evaporation the contaminants are left behind.

C.5 CLEANING PROCEDURES

The basics of cleaning are simple. Use a pure grade of isopropyl alcohol on a clean cotton swab to wipe off the endface and ferrule. (While other devices and methods also work, this one is easy, problem free, and inexpensive.) After the wet scrub, wipe off the wet endface with a dry swab, or blow it off with filtered, nonresidue compressed air. As Figure C.22 shows, leaving the fluid to air dry will leave behind the dissolved oils in little puddles which will interfere with the correct physical contact. Then, when reinserting the cable into the connector, insert it gently, in as straight a line as possible. Tipping, and inserting with an angle, can scrape off material from inside the connector or even break the inside sleeve of some connectors made with ceramic material.

It is not uncommon for a cable or connector to require more than one cleaning. If the performance seems poor, clean again. Often the second cleaning will restore the proper performance where the first did not. The idea is that the first step is to gently remove any grit and oil using the alcohol as a solvent. Remove the solvent by blowing it off or using a gentle dry wipe. Then if there is a caked-on layer of material still present (such as can happen if the beryllium-copper sides of the ferrule retainer gets scraped and deposited on the endface during insertion of the cable), the second cleaning should be harder, with a scrubbing action. But if scrubbed first, grit can be caught in the swab and become a gouging element.

A system is only as good as the weakest link. Don't let the connector become a failure because of poor attention. Use fusion splices on the more permanent critical nodes. Choose the best connector possible. Check (measure) the connector's return loss often, for degradation. Clean every connector, every time.

REFERENCE

1. Luis M. Fernadez and Donald R. Cropper. 1995. Evaluating Fiberoptic Connector Cleanliness for Accurate Measurements. *Fiberoptic Product News.* (6):45, 46.

Index

Absorption cells, 105–108
Acetylene (*see* Absorption cells)
Amplified spontaneous emission (ASE), 374–377, 520
Antireflection coating, 66–67
Autocorrelator, 278
Avalanche photodetector (*see* photodetectors)

Bare fiber adapter, 71, 341
Birefringence, 234
Bit error ratio (BER), 13, 288–289
Brillouin amplifier, 589
Brillouin scattering, 448, 618

Chirp, 5, 173, 208–213
Chromatic dispersion, 23–24, 294, 479–486
Clock recovery, 293
Coherence length, 136, 172, 351–353
Coherence time, 136, 173, 351–353
Coherent detection, 399–401
Coherent FMCW reflectometry, 426–429
Coherent frequency-domain reflectometry (C-OFDR), 425
Coherent speckle, 396
Composite second order distortion (CSO), 268

Composite triple beat distortion (CTB), 269
Connectors, 621–638
Cutback measurement, 342

Dead zone, 442–444
Degree of polarization, 40, 222, 228
Delayed self-heterodyne, 185–188
Delayed self-homodyne, 188–189
Dense wavelength division multiplexing (DWDM), 9–11
Depolarization, 228
Differential group delay, 492
Diffraction grating, 90, 95–98
Diffraction limited spot size, 94
Direct detection, 399–401
Dispersion
    air, 145–149
    chromatic, 23–24, 294, 479–486
    diffraction grating, 98
    intermodal, 17, 476–478
    polarization mode, 8, 487–516, 557
Distortion, 266–269
Distributed feedback laser (DFB)
    (*see* lasers)

Dynamic range
   insertion loss measurements, 363, 367,
   378–379, 381
   optical spectrum analyzer, 110–112
   optical time domain reflectometer, 441–444
   wavelength meter, 157

Edge-emitting LED (EELED) (*see* LEDs)
Erbium doped fiber amplifiers (EDFA), 20,
   125–126, 520–529
External cavity laser (*see* lasers)
Extinction ratio, 297, 316–323
Eye diagram, 298–300
Eyeline diagrams, 311
Eye masks, 324–326

Fabry-perot laser (*see* lasers)
Fabry-perot interferometer (*see* interferometry)
Fiber
   dark fiber, 468
   erbium-doped, 20, 522–525
   single model, 18–20
   multimode, 16–18
   nonlinearities, 614–620
   polarization maintaining, 20, 238
   praseodymium-doped, 583–584
Fizeau interferometer (*see* interferometry)

Gain slope, 532
Gain tilt, 532

Helium-neon laser (*see* lasers)
Heterodyne, 43, 175–177, 279–280
Homodyne, 43–44, 177
Hydrogen cyanide (*see* absorption cells)

Impulse response, 276–279
Index of refraction, 145–149
Insertion loss, 21–22, 339–382, 454–457
Integrating sphere, 70
Intensity modulation, 250, 265
Intensity noise (*see* noise)
Interference, 175, 347–350, 609–610
Interferogram, 135–136
Interferometry
   michelson, 90, 134, 139, 351

fizeau, 163–164
fabry-perot, 88–89, 159–164, 190
Interpolation source subtraction, 554–555
Jitter, 14–15, 326–337
Jones matrix, 224–226

Lasers
   distributed feedback (DFB), 28–29,
   123–125,
     248
   external cavity, 360–361
   fabry-perot (FP), 27–28, 122–123
   helium-neon, 150–151
   mode-locked, 277
   pump, 521
   vertical cavity surface emitting (VCSEL),
   29–30
Light emitting diodes (LEDs)
   edge-emitting, 33–34, 373–374
   surface-emitting, 33, 120–121
Lightwave component analyzer, 46, 253–263
Lightwave signal analyzer, 44, 264–265
Linewidth, 173
Littman OSA, 102–103
Littrow condition, 97

Measurement range
   insertion loss, 363, 367, 378, 381
   optical low coherence reflectometry, 408
   optical time domain reflectometry, 441–442
   power meter, 63
Michelson interferometer (*see* interferometry)
Mode coupling, 490–492
Modulation
   frequency response, 46–47, 252–253
   depth, 45, 265–266
   distortion, 45, 266–269
   non return-to-zero (NRZ), 286
   time domain, 47–48
Modulators
   acousto-optic, 559
   electroabsorption, 32
   mach-zehnder, 31–32, 248–249, 258–261
Monochromator, 90
Mueller matrix, 232–234, 356–358

Noise
intensity noise, 45, 269–275, 534, 601–604
phase noise, 608
shot, 604–607
thermal, 598–600
Noise figure, 23, 542–546
Noise bandwidth, 110
Numerical aperture, 68

Optical low coherence reflectometry (OLCR), 401–417
Optical spectrum analyzer (OSA), 41–42, 87–130
Optical time domain reflectometry (OTDR), 26–27, 420–422

Photodetectors
avalanche photodetectors, 36–37
dark current, 64
p-i-n photodetectors, 34–36, 59–61, 249–250
polarization dependence, 66
responsivity, 59, 61–62
thermal, 56–58
Photon counting OTDR, 422–423
p-i-n photodetector (see photodetectors)
Poincare' sphere, 229–231
Polarimeter, 231
Polarization, 25–26, 39–41, 220–245
Polarization controller, 234–235
Polarization dependent loss (PDL), 346–347, 354–358
Polarization ellipse, 223–224
Polarization diversity receiver, 114, 412–413
Polarization extinction ratio, 240
Polarization hole burning, 530–531
Polarization maintaining fiber (see fiber)
Polarization mode dispersion, 8, 487–516
Power measurement, 37–38, 55–86
Principle states of polarization, 490–492
Pseudo-random binary sequence (PRBS), 291–292

Raman amplifier, 587–589
Raman nonlinearity, 614–616
Raman scattering, 448

Rate equations, 257–258
Rayleigh backscatter, 396–398, 448–453
Reflectance, 461
Relative intensity noise (RIN) (see noise)
Relaxation oscillation frequency, 258
Remote fiber test, 467
Recirculating loops, 126–128
Retardance, 235–236
Reflectometry, 48, 383–474
coherent optical frequency domain, 426–429
direct detection optical time domain reflectometry, 420–422
optical continuous wave, 387–389
optical low coherence, 401–417
photon counting optical time domain, 422–423
Resolution bandwidth, 98, 109, 154
Responsivity, 59, 61–62
Return loss, 387–389, 461

Sampling oscilloscopes, 300
Sampling techniques
equivalent time, 302
microwave transition analysis, 304
random repetitive, 304
real time, 301–302
sequential, 303–304
Self -homodyne, 177–179
Self-phase modulation, 616–618
Semiconductor optical amplifier, 589
Sensitivity
optical low-coherence reflectometer, 408–412
optical time domain reflectometer, 439
power meter, 64–65
wavelength meter, 156
Shot noise, 536–537
Side-mode suppression ratio, 41, 124
Signal-spontaneous beat noise, 537–538
Source-spontaneous emission, 363
Speckle pattern, 17
Spectral hole burning, 531
Spectrometer, 90

Spontaneous emission, 33
Spontaneous-spontaneous beat noise, 538–540
Squeezed states, 607
Stokes parameters, 226–227
Synchronous digital hierarchy (SDH), 5, 285–287
Synchronous optical network (SONET), 5, 285–287

Time division multiplexing (TDM), 5
Time domain extinction, 559–566
Tunable laser, 360–361

Vertical cavity surface emitting laser (VCSEL) (*see* lasers)

Wavelength division multiplexing (WDM), 9–11
Wavelength discriminator, 164–165, 194
Wavelength meters, 42–43, 131–168
Waveplate, 235
White light interferometry (*see* Optical low-coherence reflectometry)

Zero span, 119

plug into
Prentice Hall PTR Online!

Thank you for purchasing this Prentice Hall PTR book. As a professional, we know that having information about the latest technology at your fingertips is essential. Keep up-to-date about Prentice Hall PTR on the World Wide Web.

Visit the Prentice Hall PTR Web page at
http://www.prenhall.com/divisions/ptr/
and get the latest information about:

- New Books, Software & Features of the Month
- New Book and Series Home Pages
- Stores that Sell Our Books
- Author Events and Trade Shows

join prentice hall ptr's new internet mailing lists!

Each month, subscribers to our mailing lists receive two e-mail messages highlighting recent releases, author events, new content on the Prentice Hall PTR web site, and where to meet us at professional meetings. Join one, a few, *or all* of our mailing lists in targeted subject areas in Computers and Engineering.

Visit the Mailroom at http://www.prenhall.com/mail_lists/
to subscribe to our mailing lists in...

COMPUTER SCIENCE:

Programming and Methodologies
Communications
Operating Systems
Database Technologies

ENGINEERING:

Electrical Engineering
Chemical and Environmental Engineering
Mechanical and Civil Engineering
Industrial Engineering and Quality

PTR

PH

get connected with prentice hall ptr online!